Bioprocess Engineering Principles

Pauline M. Doran

Bioprocess Engineering Principles

ELSEVIER
ACADEMIC
PRESS

AMSTERDAM • BOSTON • HEIDELBERG • LONDON • NEW YORK • OXFORD
PARIS • SAN DIEGO • SAN FRANCISCO • SINGAPORE • SYDNEY • TOKYO

This book is printed on acid-free paper

Copyright © 1995, Elsevier Ltd. All rights reserved
Seventh printing 2002
Reprinted 2003, 2004

No part of this publication may be reproduced, stored in a retrieval system,
or transmitted in any form or by any means electronic, mechanical, photocopying,
recording or otherwise, without the prior written permission of the publisher

Permissions may be sought directly from Elsevier's Science & Technology Rights
Department in Oxford, UK: phone: (+44) 1865 843830, fax: (+44) 1865 853333,
e-mail: permissions@elsevier.co.uk. You may also complete your request on-line via
the Elsevier homepage (http://www.elsevier.com), by selecting 'Customer Support'
and then 'Obtaining Permissions'

Elsevier Academic Press
84 Theobald's Road, London WC1X 8RR, UK
http://www.elsevier.com

Elsevier Academic Press
525 B Street, Suite 1900, San Diego, California 92101-4495, USA
http://www.elsevier.com

British Library Cataloguing in Publication Data
A catalogue record for this book is available from the British Library

ISBN 0-12-220855-2
ISBN 0-12-220856-0 (pbk)

Typeset by Columns Design & Production Services Ltd, Reading
Printed and bound in Great Britain by Martins the Printers, Berwick upon Tweed

Contents

Chapter 9

Chapter 10

APPENDICES   393

INDEX   417

Preface

Recent developments in genetic and molecular biology have excited world-wide interest in biotechnology. The ability to manipulate DNA has already changed our perceptions of medicine, agriculture and environmental management. Scientific breakthroughs in gene expression, protein engineering and cell fusion are being translated by a strengthening biotechnology industry into revolutionary new products and services.

Many a student has been enticed by the promise of biotechnology and the excitement of being near the cutting edge of scientific advancement. However, the value of biotechnology is more likely to be assessed by business, government and consumers alike in terms of commercial applications, impact on the marketplace and financial success. Graduates trained in molecular biology and cell manipulation soon realise that these techniques are only part of the complete picture; bringing about the full benefits of biotechnology requires substantial manufacturing capability involving large-scale processing of biological material. For the most part, chemical engineers have assumed the responsibility for bioprocess development. However, increasingly, biotechnologists are being employed by companies to work in co-operation with biochemical engineers to achieve pragmatic commercial goals. Yet, while aspects of biochemistry, microbiology and molecular genetics have for many years been included in chemical-engineering curricula, there has been relatively little attempt to teach biotechnologists even those qualitative aspects of engineering applicable to process design.

The primary aim of this book is to present the principles of bioprocess engineering in a way that is accessible to biological scientists. It does not seek to make biologists into bioprocess engineers, but to expose them to engineering concepts and ways of thinking. The material included in the book has been used to teach graduate students with diverse backgrounds in biology, chemistry and medical science. While several excellent texts on bioprocess engineering are currently available, these generally assume the reader already has engineering training. On the other hand, standard chemical-engineering texts do not often consider examples from bioprocessing and are written almost exclusively with the petroleum and chemical industries in mind. There was a need for a textbook which explains the engineering approach to process analysis while providing worked examples and problems about biological systems. In this book, more than 170 problems and calculations encompass a wide range of bioprocess applications involving recombinant cells, plant- and animal-cell cultures and immobilised biocatalysts as well as traditional fermentation systems. It is assumed that the reader has an adequate background in biology.

One of the biggest challenges in preparing the text was determining the appropriate level of mathematics. In general, biologists do not often encounter detailed mathematical analysis. However, as a great deal of engineering involves formulation and solution of mathematical models, and many important conclusions about process behaviour are best explained using mathematical relationships, it is neither easy nor desirable to eliminate all mathematics from a textbook such as this. Mathematical treatment is necessary to show how design equations depend on crucial assumptions; in other cases the equations are so simple and their application so useful that non-engineering scientists should be familiar with them. Derivation of most mathematical models is fully explained in an attempt to counter the tendency of many students to memorise rather than understand the meaning of equations. Nevertheless, in fitting with its principal aim, much more of this book is descriptive compared with standard chemical-engineering texts.

The chapters are organised around broad engineering sub-disciplines such as mass and energy balances, fluid dynamics, transport phenomena and reaction theory, rather than around particular applications of bioprocessing. That the same fundamental engineering principle can be readily applied to a variety of bioprocess industries is illustrated in the worked examples and problems. Although this textbook is written primarily for senior students and graduates of biotechnology, it should also be useful in food-, environmental- and civil-engineering courses. Because the qualitative treatment of selected topics is at a relatively advanced level, the book is appropriate for chemical-engineering graduates, undergraduates and industrial practitioners.

I would like to acknowledge several colleagues whose advice I sought at various stages of manuscript preparation. Jay Bailey, Russell Cail, David DiBiasio, Noel Dunn and Peter Rogers each reviewed sections of the text. Sections 3.3 and 11.2 on analysis of experimental data owe much to Robert J. Hall who provided lecture notes on this topic. Thanks are also due to Jacqui Quennell whose computer drawing skills are evident in most of the book's illustrations.

Pauline M. Doran

University of New South Wales
Sydney, Australia
January 1994

Part 1
Introduction

I

Bioprocess Development: An Interdisciplinary Challenge

Bioprocessing is an essential part of many food, chemical and pharmaceutical industries. Bioprocess operations make use of microbial, animal and plant cells and components of cells such as enzymes to manufacture new products and destroy harmful wastes.

Use of microorganisms to transform biological materials for production of fermented foods has its origins in antiquity. Since then, bioprocesses have been developed for an enormous range of commercial products, from relatively cheap materials such as industrial alcohol and organic solvents, to expensive specialty chemicals such as antibiotics, therapeutic proteins and vaccines. Industrially-useful enzymes and living cells such as bakers' and brewers' yeast are also commercial products of bioprocessing.

Table 1.1 gives examples of bioprocesses employing whole cells. Typical organisms used and the approximate market size for the products are also listed. The table is by no means exhaustive; not included are processes for wastewater treatment, bioremediation, microbial mineral recovery and manufacture of traditional foods and beverages such as yoghurt, bread, vinegar, soy sauce, beer and wine. Industrial processes employing enzymes are also not listed in Table 1.1; these include brewing, baking, confectionery manufacture, fruit-juice clarification and antibiotic transformation. Large quantities of enzymes are used commercially to convert starch into fermentable sugars which serve as starting materials for other bioprocesses.

Our ability to harness the capabilities of cells and enzymes has been closely related to advancements in microbiology, biochemistry and cell physiology. Knowledge in these areas is expanding rapidly; tools of modern biotechnology such as recombinant DNA, gene probes, cell fusion and tissue culture offer new opportunities to develop novel products or improve bioprocessing methods. Visions of sophisticated medicines, cultured human tissues and organs, biochips for new-age computers, environmentally-compatible pesticides and powerful pollution-degrading microbes herald a revolution in the role of biology in industry.

Although new products and processes can be conceived and partially developed in the laboratory, bringing modern biotechnology to industrial fruition requires engineering skills and know-how. Biological systems can be complex and difficult to control; nevertheless, they obey the laws of chemistry and physics and are therefore amenable to engineering analysis. Substantial engineering input is essential in many aspects of bioprocessing, including design and operation of bioreactors, sterilisers and product-recovery equipment, development of systems for process automation and control, and efficient and safe layout of fermentation factories. The subject of this book, *bioprocess engineering*, is the study of engineering principles applied to processes involving cell or enzyme catalysts.

1.1 Steps in Bioprocess Development: A Typical New Product From Recombinant DNA

The interdisciplinary nature of bioprocessing is evident if we look at the stages of development required for a complete industrial process. As an example, consider manufacture of a new recombinant-DNA-derived product such as insulin, growth hormone or interferon. As shown in Figure 1.1, several steps are required to convert the idea of the product into commercial reality; these stages involve different types of scientific expertise.

The first stages of bioprocess development (Steps 1–11) are concerned with genetic manipulation of the host organism; in this case, a gene from animal DNA is cloned into *Escherichia coli*. Genetic engineering is done in laboratories on a small scale by scientists trained in molecular biology and biochemistry. Tools of the trade include Petri dishes, micropipettes, microcentrifuges, nano- or microgram quantities of restriction enzymes, and electrophoresis gels for DNA and protein fractionation. In terms of bioprocess development, parameters of major importance are stability of the constructed strains and level of expression of the desired product.

After cloning, the growth and production characteristics of

Table 1.1 Major products of biological processing

(Adapted from M.L. Shuler, 1987, Bioprocess engineering. In: Encyclopedia of Physical Science and Technology, vol 2, R.A. Meyers, Ed., Academic Press, Orlando)

Fermentation product	Typical organism used	Approximate world market size (kg yr^{-1})
Bulk organics		
Ethanol (non-beverage)	*Saccharomyces cerevisiae*	2×10^{10}
Acetone/butanol	*Clostridium acetobutylicum*	2×10^{6} (butanol)
Biomass		
Starter cultures and yeasts for food and agriculture	Lactic acid bacteria or bakers' yeast	5×10^{8}
Single-cell protein	*Pseudomonas methylotrophus* or *Candida utilis*	$0.5–1 \times 10^{8}$
Organic acids		
Citric acid	*Aspergillus niger*	$2–3 \times 10^{8}$
Gluconic acid	*Aspergillus niger*	5×10^{7}
Lactic acid	*Lactobacillus delbrueckii*	2×10^{7}
Itaconic acid	*Aspergillus itaconicus*	
Amino acids		
L-glutamic acid	*Corynebacterium glutamicum*	3×10^{8}
L-lysine	*Brevibacterium flavum*	3×10^{7}
L-phenylalanine	*Corynebacterium glutamicum*	2×10^{6}
L-arginine	*Brevibacterium flavum*	2×10^{6}
Others	*Corynebacterium* spp.	1×10^{6}
Microbial transformations		
Steroids	*Rhizopus arrhizus*	
D-sorbitol to L-sorbose (in vitamin C production)	*Acetobacter suboxydans*	4×10^{7}
Antibiotics		
Penicillins	*Penicillium chrysogenum*	$3–4 \times 10^{7}$
Cephalosporins	*Cephalosporium acremonium*	1×10^{7}
Tetracyclines (e.g. 7-chlortetracycline)	*Streptomyces aureofaciens*	1×10^{7}
Macrolide antibiotics (e.g. erythromycin)	*Streptomyces erythreus*	2×10^{6}
Polypeptide antibiotics (e.g. gramicidin)	*Bacillus brevis*	1×10^{6}
Aminoglycoside antibiotics (e.g. streptomycin)	*Streptomyces griseus*	
Aromatic antibiotics (e.g. griseofulvin)	*Penicillium griseofulvum*	
Extracellular polysaccharides		
Xanthan gum	*Xanthomonas campestris*	5×10^{6}
Dextran	*Leuconostoc mesenteroides*	small

Nucleotides

5′-guanosine monophosphate	*Brevibacterium ammoniagenes*	1×10^5

Enzymes

Proteases	*Bacillus* spp.	6×10^5
α-amylase	*Bacillus amyloliquefaciens*	4×10^5
Glucoamylase	*Aspergillus niger*	4×10^5
Glucose isomerase	*Bacillus coagulans*	1×10^5
Pectinase	*Aspergillus niger*	1×10^4
Rennin	*Mucor miehei* or recombinant yeast	1×10^4
All others		5×10^4

Vitamins

B$_{12}$	*Propionibacterium shermanii* or *Pseudomonas denitrificans*	1×10^4
Riboflavin	*Eremothecium ashbyii*	

Ergot alkaloids | *Claviceps paspali* | 5×10^3

Pigments

Shikonin	*Lithospermum erythrorhizon* (plant-cell culture)	60
β-carotene	*Blakeslea trispora*	

Vaccines

Diphtheria	*Corynebacterium diphtheriae*	< 50
Tetanus	*Clostridium tetani*	
Pertussis (whooping cough)	*Bordetella pertussis*	
Poliomyelitis virus	Live attenuated viruses grown in monkey kidney or human diploid cells	
Rubella	Live attenuated viruses grown in baby-hamster kidney cells	
Hepatitis B	Surface antigen expressed in recombinant yeast	

Therapeutic proteins

		< 20
Insulin	Recombinant *Escherichia coli*	
Growth hormone	Recombinant *Escherichia coli* or recombinant mammalian cells	
Erythropoietin	Recombinant mammalian cells	
Factor VIII-C	Recombinant mammalian cells	
Tissue plasminogen activator	Recombinant mammalian cells	
Interferon-α_2	Recombinant *Escherichia coli*	

Monoclonal antibodies | Hybridoma cells | < 20

Insecticides

Bacterial spores	*Bacillus thuringiensis*	
Fungal spores	*Hirsutella thompsonii*	

Figure 1.1   Steps in development of a complete bioprocess for commercial manufacture of a new recombinant-DNA-derived product.

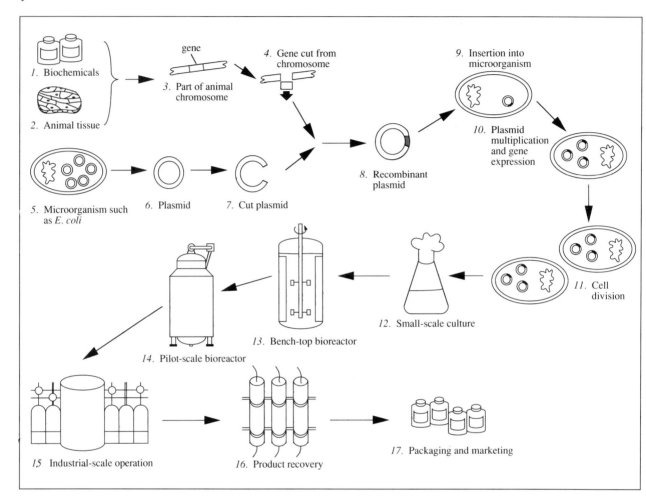

the cells must be measured as a function of culture environment (Step 12). Practical skills in microbiology and kinetic analysis are required; small-scale culture is mostly carried out using shake flasks of 250-ml to 1-litre capacity. Medium composition, pH, temperature and other environmental conditions allowing optimal growth and productivity are determined. Calculated parameters such as cell growth rate, specific productivity and product yield are used to describe performance of the organism.

Once the culture conditions for production are known, scale-up of the process starts. The first stage may be a 1- or 2-litre *bench-top bioreactor* equipped with instruments for measuring and adjusting temperature, pH, dissolved-oxygen concentration, stirrer speed and other process variables (Step

13). Cultures can be more closely monitored in bioreactors than in shake flasks so better control over the process is possible. Information is collected about the oxygen requirements of the cells, their shear sensitivity, foaming characteristics and other parameters. Limitations imposed by the reactor on activity of the organism must be identified. For example, if the bioreactor cannot provide dissolved oxygen to an aerobic culture at a sufficiently high rate, the culture will become oxygen-starved. Similarly, in mixing the broth to expose the cells to nutrients in the medium, the stirrer in the reactor may cause cell damage. Whether or not the reactor can provide conditions for optimal activity of the cells is of prime concern. The situation is assessed using measured and calculated parameters such as mass-transfer coefficients, mixing time, gas hold-up, rate of oxygen uptake, power number, impeller shear-rate, and many others. It must also be decided whether the culture is best operated as a batch, semi-batch or continuous process; experimental results for culture performance under various modes of reactor operation may be examined. The viability of the process as a commercial venture is of great interest; information about activity of the cells is used in further calculations to determine economic feasibility.

Following this stage of process development, the system is scaled up again to a *pilot-scale bioreactor* (Step 14). Engineers trained in bioprocessing are normally involved in pilot-scale operations. A vessel of capacity 100–1000 litres is built according to specifications determined from the bench-scale prototype. The design is usually similar to that which worked best on the smaller scale. The aim of pilot-scale studies is to examine the response of cells to scale-up. Changing the size of the equipment seems relatively trivial; however, loss or variation of performance often occurs. Even though the geometry of the reactor, method of aeration and mixing, impeller design and other features may be similar in small and large fermenters, the effect on activity of cells can be great. Loss of productivity following scale-up may or may not be recovered; economic projections often need to be re-assessed as a result of pilot-scale findings.

If the scale-up step is completed successfully, design of the *industrial-scale operation* commences (Step 15). This part of process development is clearly in the territory of bioprocess engineering. As well as the reactor itself, all of the auxiliary service facilities must be designed and tested. These include air supply and sterilisation equipment, steam generator and supply lines, medium preparation and sterilisation facilities, cooling-water supply and process-control network. Particular attention is required to ensure the fermentation can be carried out aseptically. When recombinant cells or pathogenic organisms are involved, design of the process must also reflect containment and safety requirements.

An important part of the total process is *product recovery* (Step 16), also known as *downstream processing*. After leaving the fermenter, raw broth is treated in a series of steps to produce the final product. Product recovery is often difficult and expensive; for some recombinant-DNA-derived products, purification accounts for 80–90% of the total processing cost. Actual procedures used for downstream processing depend on the nature of the product and the broth; physical, chemical or biological methods may be employed. Many operations which are standard in the laboratory become uneconomic or impractical on an industrial scale. Commercial procedures include filtration, centrifugation and flotation for separation of cells from the liquid, mechanical disruption of the cells if the product is intracellular, solvent extraction, chromatography, membrane filtration, adsorption, crystallisation and drying. Disposal of effluent after removal of the desired product must also be considered. Like bioreactor design, techniques applied industrially for downstream processing are first developed and tested using small-scale apparatus. Scientists trained in chemistry, biochemistry, chemical engineering and industrial chemistry play important roles in designing product recovery and purification systems.

After the product has been isolated in sufficient purity it is packaged and marketed (Step 17). For new pharmaceuticals such as recombinant human growth hormone or insulin, medical and clinical trials are required to test the efficacy of the product. Animals are used first, then humans. Only after these trials are carried out and the safety of the product established can it be released for general health-care application. Other tests are required for food products. Bioprocess engineers with a detailed knowledge of the production process are often involved in documenting manufacturing procedures for submission to regulatory authorities. Manufacturing standards must be met; this is particularly the case for recombinant products where a greater number of safety and precautionary measures is required.

As shown in this example, a broad range of disciplines is involved in bioprocessing. Scientists working in this area are constantly confronted with biological, chemical, physical, engineering and sometimes medical questions.

1.2 A Quantitative Approach

The biological characteristics of cells and enzymes often impose constraints on bioprocessing; knowledge of them is therefore an important prerequisite for rational engineering design. For instance, thermostability properties must be taken into account when choosing the operating temperature of an enzyme reactor, while susceptibility of an organism to substrate inhibition will determine whether substrate is fed to the fermenter all at once or intermittently. It is equally true, however, that biologists working in biotechnology must consider the engineering aspects of bioprocessing; selection or manipulation of organisms should be carried out to achieve the best results in production-scale operations. It would be disappointing, for example, to spend a year or two manipulating an organism to express a foreign gene if the cells in culture produce a highly viscous broth that cannot be adequately mixed or supplied with oxygen in large-scale vessels. Similarly, improving cell permeability to facilitate product excretion has limited utility if the new organism is too fragile to withstand the mechanical forces developed during fermenter operation.

Another area requiring cooperation and understanding between engineers and laboratory scientists is medium formation. For example, addition of serum may be beneficial to growth of animal cells, but can significantly reduce product yields during recovery operations and, in large-scale processes, requires special sterilisation and handling procedures.

All areas of bioprocess development—the cell or enzyme used, the culture conditions provided, the fermentation equipment and product-recovery operations—are interdependent. Because improvement in one area can be disadvantageous to another, ideally, bioprocess development should proceed using an integrated approach. In practice, combining the skills of engineers with those of biologists can be difficult owing to the very different ways in which biologists and engineers are trained. Biological scientists generally have strong experimental technique and are good at testing qualitative models; however, because calculations and equations are not a prominent feature of the life sciences, biologists are usually less familiar with mathematics. On the other hand, as calculations are important in all areas of equipment design and process analysis, quantitative methods, physics and mathematical theories play a central role in engineering. There is also a difference in the way biologists and biochemical engineers think about complex processes such as cell and enzyme function. Fascinating as the minutiae of these biological systems may be, in order to build working reactors and other equipment, engineers must take a simplified and pragmatic approach. It is often disappointing for the biology-trained scientist that engineers seem to ignore the wonder, intricacy and complexity of life to focus only on those aspects which have significant quantitative effect on the final outcome of the process.

Given the importance of interaction between biology and engineering in bioprocessing, these differences in outlook between engineers and biologists must be overcome. Although it is unrealistic to expect all biotechnologists to undertake full engineering training, there are many advantages in understanding the practical principles of bioprocess engineering if not the full theoretical detail. The principal objective of this book is to teach scientists trained in biology those aspects of engineering science which are relevant to bioprocessing. An adequate background in biology is assumed. At the end of this study, you will have gained a heightened appreciation for bioprocess engineering. You will be able to communicate on a professional level with bioprocess engineers and know how to analyse and critically evaluate new processing proposals. You will be able to carry out routine calculations and checks on processes; in many cases these calculations are not difficult and can be of great value. You will also know what type of expertise a bioprocess engineer can offer and when it is necessary to consult an expert in the field. In the laboratory, your awareness of engineering methods will help avoid common mistakes in data analysis and design of experimental apparatus.

As our exploitation of biology continues, there is an increasing demand for scientists trained in bioprocess technology who can translate new discoveries into industrial-scale production. As a biotechnologist, you could be expected to work at the interface of biology and engineering science. This textbook on bioprocess engineering is designed to prepare you for this challenge.

2

Introduction to Engineering Calculations

Calculations used in bioprocess engineering require a systematic approach with well-defined methods and rules. Conventions and definitions which form the backbone of engineering analysis are presented in this chapter. Many of these you will use over and over again as you progress through this text. In laying the foundation for calculations and problem-solving, this chapter will be a useful reference which you may need to review from time to time.

The first step in quantitative analysis of systems is to express the system properties using mathematical language. This chapter begins by considering how physical and chemical processes are translated into mathematics. The nature of physical variables, dimensions and units are discussed, and formalised procedures for unit conversions outlined. You will have already encountered many of the concepts used in measurement, such as concentration, density, pressure, temperature, etc.; rules for quantifying these variables are summarised here in preparation for Chapters 4–6 where they are first applied to solve processing problems. The occurrence of reactions in biological systems is of particular importance; terminology involved in stoichiometric analysis is considered in this chapter. Finally, since equations representing biological processes often involve physical or chemical properties of materials, references for handbooks containing this information are provided.

Worked examples and problems are used to illustrate and reinforce the material described in the text. Although the terminology and engineering concepts used in these examples may be unfamiliar, solutions to each problem can be obtained using techniques fully explained within this chapter. Many of the equations introduced as problems and examples are explained in more detail in later sections of this book; the emphasis in this chapter is on use of basic mathematical principles irrespective of the particular application. At the end of the chapter is a check-list so you can be sure you have assimilated all the important points.

2.1 Physical Variables, Dimensions and Units

Engineering calculations involve manipulation of numbers. Most of these numbers represent the magnitude of measurable *physical variables,* such as mass, length, time, velocity, area, viscosity, temperature, density, and so on. Other observable characteristics of nature, such as taste or aroma, cannot at present be described completely using appropriate numbers; we cannot, therefore, include these in calculations.

From all the physical variables in the world, the seven quantities listed in Table 2.1 have been chosen by international

Table 2.1 Base quantities

Base quantity	Dimensional symbol	Base SI unit	Unit symbol
Length	L	metre	m
Mass	M	kilogram	kg
Time	T	second	s
Electric current	I	ampere	A
Temperature	Θ	kelvin	K
Amount of substance	N	gram-mole	mol or gmol
Luminous intensity	J	candela	cd
Supplementary units			
Plane angle	–	radian	rad
Solid angle	–	steradian	sr

agreement as a basis for measurement [1]. Two further supplementary units are used to express angular quantities. The base quantities are called *dimensions*, and it is from these that the dimensions of other physical variables are derived. For example, the dimensions of velocity, defined as distance travelled per unit time, are LT^{-1}; the dimensions of force, being mass × acceleration, are LMT^{-2}. A list of useful derived dimensional quantities is given in Table 2.2.

Physical variables can be classified into two groups: *substantial variables* and *natural variables*.

2.1.1 Substantial Variables

Examples of substantial variables are mass, length, volume, viscosity and temperature. Expression of the magnitude of substantial variables requires a precise physical standard against which measurement is made. These standards are called *units*. You are already familiar with many units, e.g. metre, foot and mile are units of length; hour and second are units of time. Statements about the magnitude of substantial variables must contain two parts: the number and the unit

Table 2.2 Dimensional quantities (dimensionless quantities have dimension 1)

Quantity	Dimensions	Quantity	Dimensions
Acceleration	LT^{-2}	Osmotic pressure	$L^{-1}MT^{-2}$
Angular velocity	T^{-1}	Partition coefficient	1
Area	L^2	Period	T
Atomic weight	1	Power	L^2MT^{-3}
('relative atomic mass')		Pressure	$L^{-1}MT^{-2}$
Concentration	$L^{-3}N$	Rotational frequency	T^{-1}
Conductivity	$L^{-3}M^{-1}T^3I^2$	Shear rate	T^{-1}
Density	$L^{-3}M$	Shear stress	$L^{-1}MT^{-2}$
Diffusion coefficient	L^2T^{-1}	Specific death constant	T^{-1}
Distribution coefficient	1	Specific gravity	1
Effectiveness factor	1	Specific growth rate	T^{-1}
Efficiency	1	Specific heat capacity	$L^2T^{-2}\Theta^{-1}$
Energy	L^2MT^{-2}	Specific interfacial area	L^{-1}
Enthalpy	L^2MT^{-2}	Specific latent heat	L^2T^{-2}
Entropy	$L^2MT^{-2}\Theta^{-1}$	Specific production rate	T^{-1}
Equilibrium constant	1	Specific volume	L^3M^{-1}
Force	LMT^{-2}	Shear strain	1
Fouling factor	$MT^{-3}\Theta^{-1}$	Stress	$L^{-1}MT^{-2}$
Frequency	T^{-1}	Surface tension	MT^{-2}
Friction coefficient	1	Thermal conductivity	$LMT^{-3}\Theta^{-1}$
Gas hold-up	1	Thermal resistance	$L^{-2}M^{-1}T^3\Theta$
Half life	T	Torque	L^2MT^{-2}
Heat	L^2MT^{-2}	Velocity	LT^{-1}
Heat flux	MT^{-3}	Viscosity (dynamic)	$L^{-1}MT^{-1}$
Heat-transfer coefficient	$MT^{-3}\Theta^{-1}$	Viscosity (kinematic)	L^2T^{-1}
Illuminance	$L^{-2}J$	Void faction	1
Maintenance coefficient	T^{-1}	Volume	L^3
Mass flux	$L^{-2}MT^{-1}$	Weight	LMT^{-2}
Mass-transfer coefficient	LT^{-1}	Work	L^2MT^{-2}
Momentum	LMT^{-1}	Yield coefficient	1
Molar mass	MN^{-1}		
Molecular weight	1		
('relative molecular mass')			

used for measurement. Clearly, reporting the speed of a moving car as 20 has no meaning unless information about the units, say km h^{-1}, is also included.

As numbers representing substantial variables are multiplied, subtracted, divided or added, their units must also be combined. The values of two or more substantial variables may be added or subtracted only if their units are the same, e.g.:

$$5.0 \text{ kg} + 2.2 \text{ kg} = 7.2 \text{ kg}.$$

On the other hand, the values and units of *any* substantial variables can be combined by multiplication or division, e.g.:

$$\frac{1500 \text{ km}}{12.5 \text{ h}} = 120 \text{ km h}^{-1}.$$

The way in which units are carried along during calculations has important consequences. Not only is proper treatment of units essential if the final answer is to have the correct units, units and dimensions can also be used as a guide when deducing how physical variables are related in scientific theories and equations.

2.1.2 Natural Variables

The second group of physical variables are natural variables. Specification of the magnitude of these variables does not require units or any other standard of measurement. Natural variables are also referred to as *dimensionless variables, dimensionless groups* or *dimensionless numbers*. The simplest natural variables are ratios of substantial variables. For example, the aspect ratio of a cylinder is its length divided by its diameter; the result is a dimensionless number.

Other natural variables are not as obvious as this, and involve combinations of substantial variables that do not have the same dimensions. Engineers make frequent use of dimensionless numbers for succinct representation of physical phenomena. For example, a common dimensionless group in fluid mechanics is the Reynolds number, *Re*. For flow in a pipe, the Reynolds number is given by the equation:

$$Re = \frac{Du\rho}{\mu}$$

(2.1)

where ρ is fluid density, u is fluid velocity, D is pipe diameter and μ is fluid viscosity. When the dimensions of these variables are combined according to Eq. (2.1), the dimensions of the

numerator exactly cancel those of the denominator. Other dimensionless variables relevant to bioprocess engineering are the Schmidt number, Prandtl number, Sherwood number, Peclet number, Nusselt number, Grashof number, power number and many others. Definitions and applications of these natural variables are given in later chapters of this book.

In calculations involving rotational phenomena, rotation is described using number of revolutions or radians:

$$\text{number of radians} = \frac{\text{length of arc}}{\text{radius}}$$

(2.2)

$$\text{number of revolutions} = \frac{\text{length of arc}}{\text{circumference}} = \frac{\text{length of arc}}{2\pi r}$$

(2.3)

where r is radius. One revolution is equal to 2π radians. Radians and revolutions are non-dimensional because the dimensions of length for arc, radius and circumference in Eqs (2.2) and (2.3) cancel. Consequently, rotational speed (e.g. number of revolutions per second) and angular velocity (e.g. number of radians per second), as well as frequency (e.g. number of vibrations per second), all have dimensions T^{-1}. Degrees, which are subdivisions of a revolution, are converted into revolutions or radians before application in engineering calculations.

2.1.3 Dimensional Homogeneity in Equations

Rules about dimensions determine how equations are formulated. 'Properly constructed' equations representing general relationships between physical variables must be dimensionally homogeneous. For dimensional homogeneity, the dimensions of terms which are added or subtracted must be the same, and the dimensions of the right-hand side of the equation must be the same as the left-hand side. As a simple example, consider the Margules equation for evaluating fluid viscosity from experimental measurements:

$$\mu = \frac{M}{4\pi b\Omega} \left(\frac{1}{R_o^2} - \frac{1}{R_i^2} \right).$$

(2.4)

The terms and dimensions in this equation are listed in Table 2.3. Numbers such as 4 have no dimensions; the symbol π represents the number 3.1415926536 which is also dimensionless. A quick check shows that Eq. (2.4) is dimensionally homogeneous since both sides of the equation have dimensions

$L^{-1}MT^{-1}$ and all terms added or subtracted have the same dimensions. Note that when a term such as R_o is raised to a power such as 2, the units and dimensions of R_o must also be raised to that power.

For dimensional homogeneity, the argument of any transcendental function, such as a logarithmic, trigonometric or exponential function, must be dimensionless. The following examples illustrate this principle.

(i) An expression for cell growth is:

$$\ln \frac{x}{x_0} = \mu t$$

(2.5)

where x is cell concentration at time t, x_0 is initial cell concentration, and μ is the specific growth rate. The argument of the logarithm, the ratio of cell concentrations, is dimensionless.

(ii) The displacement y due to action of a progressive wave with amplitude A, frequency $\omega/2\pi$ and velocity v is given by the equation:

$$y = A \sin \left[\omega \left(t - \frac{x}{v} \right) \right]$$

(2.6)

where t is time and x is distance from the origin. The argument of the sine function, $\omega \left(t - \frac{x}{v} \right)$, is dimensionless.

(iii) The relationship between α, the mutation rate of *Escherichia coli*, and temperature T, can be described using an Arrhenius-type equation:

$$\alpha = \alpha_0 e^{-E/RT}$$

(2.7)

where α_0 is the mutation reaction constant, E is specific activation energy and R is the ideal gas constant (see Section 2.5). The dimensions of RT are the same as those of E, so the exponent is as it should be: dimensionless.

Dimensional homogeneity of equations can sometimes be masked by mathematical manipulation. As an example, Eq. (2.5) might be written:

$$\ln x = \ln x_0 + \mu t.$$

(2.8)

Inspection of this equation shows that rearrangement of the

Table 2.3 Terms and dimensions of Eq. (2.4)

Term	Dimensions	SI Units
μ (dynamic viscosity)	$L^{-1}MT^{-1}$	pascal second (Pa s)
M (torque)	L^2MT^{-2}	newton metre (N m)
h (cylinder height)	L	metre (m)
Ω (angular velocity)	T^{-1}	radian per second (rad s^{-1})
R_o (outer radius)	L	metre (m)
R_i (inner radius)	L	metre (m)

terms to group $\ln x$ and $\ln x_0$ together recovers dimensional homogeneity by providing a dimensionless argument for the logarithm.

Integration and differentiation of terms affect dimensionality. Integration of a function with respect to x increases the dimensions of that function by the dimensions of x. Conversely, differentiation with respect to x results in the dimensions being reduced by the dimensions of x. For example, if C is the concentration of a particular compound expressed as mass per unit volume and x is distance, dC/dx has dimensions $L^{-4}M$, while d^2C/dx^2 has dimensions $L^{-5}M$. On the other hand, if μ is the specific growth rate of an organism with dimensions T^{-1}, then $\int \mu\, dt$ is dimensionless where t is time.

2.1.4 Equations Without Dimensional Homogeneity

For repetitive calculations or when an equation is derived from observation rather than from theoretical principles, it is sometimes convenient to present the equation in a non-homogeneous form. Such equations are called *equations in numerics* or *empirical equations*. In empirical equations, the units associated with each variable must be stated explicitly. An example is Richards' correlation for the dimensionless gas hold-up ϵ in a stirred fermenter [2]:

$$\left(\frac{P}{V} \right)^{0.4} u^{1/2} = 30\epsilon + 1.33$$

(2.9)

where P is power in units of horsepower, V is ungassed liquid volume in units of ft^3, u is linear gas velocity in units of $ft\ s^{-1}$ and ϵ is fractional gas hold-up, a dimensionless variable. The dimensions of each side of Eq. (2.9) are certainly not the same. For direct application of Eq. (2.9), only those units stated above can be used.

2.2 Units

Several systems of units for expressing the magnitude of physical variables have been devised through the ages. The metric system of units originated from the National Assembly of France in 1790. In 1960 this system was rationalised, and the SI or Système International d'Unités was adopted as the international standard. Unit names and their abbreviations have been standardised; according to SI convention, unit abbreviations are the same for both singular and plural and are not followed by a period. SI prefixes used to indicate multiples and sub-multiples of units are listed in Table 2.4. Despite widespread use of SI units, no single system of units has universal application. In particular, engineers in the USA continue to apply British or imperial units. In addition, many physical property data collected before 1960 are published in lists and tables using non-standard units.

Familiarity with both metric and non-metric units is necessary. Many units used in engineering such as the slug (1 slug = 14.5939 kilograms), dram (1 dram = 1.77185 grams), stoke (a unit of kinematic viscosity), poundal (a unit of force) and erg (a unit of energy), are probably not known to you. Although no longer commonly applied, these are legitimate units which may appear in engineering reports and tables of data.

In calculations it is often necessary to convert units. Units are changed using *conversion factors*. Some conversion factors, such as 1 inch = 2.54 cm and 2.20 lb = 1 kg, you probably already know. Tables of common conversion factors are given in Appendix A at the back of this book. Unit conversions are not only necessary to convert imperial units to metric; some physical variables have several metric units in common use.

For example, viscosity may be reported as centipoise or $kg\ h^{-1}\ m^{-1}$; pressure may be given in standard atmospheres, pascals, or millimetres of mercury. Conversion of units seems simple enough; however difficulties can arise when several variables are being converted in a single equation. Accordingly, an organised mathematical approach is needed.

For each conversion factor, a *unity bracket* can be derived. The value of the unity bracket, as the name suggests, is unity. As an example,

$$1\ \text{lb} = 453.6\ \text{g} \qquad (2.10)$$

can be converted by division of both sides of the equation by 1 lb to give a unity bracket denoted by | |:

$$1 = \left|\ \frac{453.6\ \text{g}}{1\ \text{lb}}\ \right|. \qquad (2.11)$$

Similarly, division of both sides of Eq. (2.10) by 453.6 g gives another unity bracket:

$$1 = \left|\ \frac{1\ \text{lb}}{453.6\ \text{g}}\ \right|. \qquad (2.12)$$

To calculate how many pounds are in 200 g, we can multiply 200 g by the unity bracket in Eq. (2.12) or divide 200 g by the unity bracket in Eq. (2.11). This is permissible since the value

Table 2.4   SI prefixes

(From J.V. Drazil, 1983, Quantities and Units of Measurement, *Mansell, London)*

Factor	Prefix	Symbol	Factor	Prefix	Symbol
10^{-1}	deci*	d	10^{18}	exa	E
10^{-2}	centi*	c	10^{15}	peta	P
10^{-3}	milli	m	10^{12}	tera	T
10^{-6}	micro	μ	10^{9}	giga	G
10^{-9}	nano	n	10^{6}	mega	M
10^{-12}	pico	p	10^{3}	kilo	k
10^{-15}	femto	f	10^{2}	hecto*	h
10^{-18}	atto	a	10^{1}	deka*	da

*   Used for areas and volumes.

of both unity brackets is unity, and multiplication or division by 1 does not change the value of 200 g. Using the option of multiplying by Eq. (2.12):

$$
200 \text{ g} = 200 \cancel{\text{g}} \cdot \left| \frac{1 \text{ lb}}{453.6 \cancel{\text{g}}} \right|
$$

(2.13)

On the right-hand side, cancelling the old units leaves the desired unit, lb. Dividing the numbers gives:

$$
200 \text{ g} = 0.441 \text{ lb}:
$$

(2.14)

A more complicated calculation involving a complete equation is given in Example 2.1.

Example 2.1 Unit conversion

Air is pumped through an orifice immersed in liquid. The size of the bubbles leaving the orifice depends on the diameter of the orifice and the properties of the liquid. The equation representing this situation is:

$$
\frac{g(\rho_L - \rho_G) D_b^3}{\sigma D_o} = 6
$$

where g = gravitational acceleration = 32.174 ft s^{-2}; ρ_L = liquid density = 1 g cm^{-3}; ρ_G = gas density = 0.081 lb ft^{-3}; D_b = bubble diameter; σ = gas–liquid surface tension = 70.8 dyn cm^{-1}; and D_o = orifice diameter = 1 mm.

Calculate the bubble diameter D_b.

Solution:

Convert the data to a consistent set of units, e.g. g, cm, s. From Appendix A, the conversion factors required are:

      1 ft = 0.3048 m;

      1 lb = 453.6 g; and

      1 dyn cm^{-1} = 1 g s^{-2}.

Also: 1 m = 100 cm; and

      10 mm = 1 cm.

Converting units:

$$
g = 32.174 \, \frac{\text{ft}}{\text{s}^2} \cdot \left| \frac{0.3048 \text{ m}}{1 \text{ ft}} \right| \cdot \left| \frac{100 \text{ cm}}{1 \text{ m}} \right| = 980.7 \text{ cm s}^{-2}
$$

$$
\rho_G = 0.081 \frac{\text{lb}}{\text{ft}^3} \cdot \left| \frac{453.6 \text{ g}}{1 \text{ lb}} \right| \cdot \left| \frac{1 \text{ ft}}{0.3048 \text{ m}} \right|^3 \cdot \left| \frac{1 \text{ m}}{100 \text{ cm}} \right|^3 = 1.30 \times 10^{-3} \text{g cm}^{-3}
$$

$$
\sigma = 70.8 \text{ dyn cm}^{-1} \cdot \left| \frac{1 \text{ g s}^{-2}}{1 \text{ dyn cm}^{-1}} \right| = 70.8 \text{ g s}^{-2}
$$

$$
D_o = 1 \text{ mm} \cdot \left| \frac{1 \text{ cm}}{10 \text{ mm}} \right| = 0.1 \text{ cm}.
$$

Rearranging the equation to give an expression for D_b^3:

$$
D_b^3 = \frac{6 \sigma D_o}{g(\rho_L - \rho_G)}.
$$

Substituting values gives:

$$D_b^3 = \frac{6\,(70.8 \text{ g s}^{-2})\,(0.1 \text{ cm})}{980.7 \text{ cm s}^{-2}\,(1 \text{ g cm}^{-3} - 1.30 \times 10^{-3}\text{g cm}^{-3})} = 4.34 \times 10^{-2} \text{ cm}^3.$$

Taking the cube root:

$$D_b = 0.35 \text{ cm}.$$

Note that unity brackets are squared or cubed when appropriate, e.g. when converting ft^3 to cm^3. This is permissible since the value of the unity bracket is 1, and 1^2 or 1^3 is still 1.

2.3 Force and Weight

According to Newton's law, the force exerted on a body in motion is proportional to its mass multiplied by the acceleration. As listed in Table 2.2, the dimensions of force are LMT^{-2}; the *natural units* of force in the SI system are kg m s^{-2}. Analogously, g cm s^{-2} and lb ft s^{-2} are the natural units of force in the metric and British systems, respectively.

Force occurs frequently in engineering calculations, and *derived units* are used more commonly than natural units. In SI, the derived unit is the *newton*, abbreviated as N:

$$1 \text{ N} = 1 \text{ kg m s}^{-2}. \tag{2.15}$$

In the British or imperial system, the derived unit for force is defined as (1 lb mass) $\times$ (gravitational acceleration at sea level and 45° latitude). The derived force-unit in this case is called the *pound-force*, and is denoted lb_f:

$$1 \text{ lb}_f = 32.174 \text{ lb}_m \text{ ft s}^{-2} \tag{2.16}$$

as gravitational acceleration at sea level and 45° latitude is 32.174 ft s^{-2}. Note that pound-mass, represented usually as lb, has been shown here using the abbreviation, lb_m, to distinguish it from lb_f. Use of the pound in the imperial system for reporting both mass and force can be a source of confusion and requires care.

In order to convert force from a defined unit to a natural unit, a special dimensionless unity-bracket called g_c is used. The form of g_c depends on the units being converted. From Eqs (2.15) and (2.16):

$$g_c = 1 = \left| \frac{1 \text{ N}}{1 \text{ kg m s}^{-2}} \right| = \left| \frac{1 \text{ lb}_f}{32.174 \text{ lb}_m \text{ ft s}^{-2}} \right|. \tag{2.17}$$

Application of g_c is illustrated in Example 2.2.

Example 2.2 Use of g_c

Calculate the kinetic energy of 250 lb_m liquid flowing through a pipe at 35 ft s^{-1}. Express your answer in units of ft lb_f.

Solution:
Kinetic energy is given by the equation:

$$\text{kinetic energy} = E_k = \tfrac{1}{2}Mv^2$$

where M is mass and v is velocity. Using the values given:

$$E_k = \tfrac{1}{2}(250 \text{ lb}_m)\left(35\,\frac{\text{ft}}{\text{s}}\right)^2 = 1.531 \times 10^5\,\frac{\text{lb}_m \text{ft}^2}{\text{s}^2}.$$

Multiplying by g_c from Eq. (2.17) gives:

$$E_k = 1.531 \times 10^5\,\frac{\text{lb}_m \text{ft}^2}{\text{s}^2} \cdot \left| \frac{1 \text{ lb}_f}{32.174 \text{ lb}_m \text{ ft s}^{-2}} \right|.$$

Calculating and cancelling units gives the answer:

$$E_k = 4760 \text{ ft lb}_f.$$

Weight is the force with which a body is attracted by gravity to the centre of the earth. It changes according to the value of the gravitational acceleration g, which varies by about 0.5% over the earth's surface. In SI units g is approximately 9.8 m s^{-2}; in imperial units g is about 32.2 ft s^{-2}. Using Newton's law and depending on the exact value of g, the weight of a mass of 1 kg is about 9.8 newtons; the weight of a mass of 1 lb is about 1 lb$_f$. Note that although the value of g changes with position on the earth's surface (or in the universe), the value of g_c within a given system of units does not. g_c is a factor for converting units, not a physical variable.

2.4 Measurement Conventions

Familiarity with common physical variables and methods for expressing their magnitude is necessary for engineering analysis of bioprocesses. This section covers some useful definitions and engineering conventions that will be applied throughout the text.

2.4.1 Density

Density is a substantial variable defined as mass per unit volume. Its dimensions are $L^{-3}M$, and the usual symbol is ρ. Units for density are, for example, g cm^{-3}, kg m^{-3} and lb ft^{-3}. If the density of acetone is 0.792 g cm^{-3}, the mass of 150 cm^3 acetone can be calculated as follows:

$$150 \text{ cm}^3 \left(\frac{0.792 \text{ g}}{\text{cm}^3} \right) = 119 \text{ g}.$$

Densities of solids and liquids vary slightly with temperature. The density of water at 4°C is 1.0000 g cm^{-3}, or 62.4 lb ft^{-3}. The density of solutions is a function of both concentration and temperature. Gas densities are highly dependent on temperature and pressure.

2.4.2 Specific Gravity

Specific gravity, also known as 'relative density', is a dimensionless variable. It is the ratio of two densities, that of the substance in question and that of a specified reference material. For liquids and solids, the reference material is usually water. For gases, air is commonly used as reference, but other reference gases may also be specified.

As mentioned above, liquid densities vary somewhat with temperature. Accordingly, when reporting specific gravity the temperatures of the substance and its reference material are specified. If the specific gravity of ethanol is given as $0.789_{4°C}^{20°C}$, this means that the specific gravity is 0.789 for ethanol at 20°C referenced against water at 4°C. Since the

density of water at 4°C is almost exactly 1.0000 g cm^{-3}, we can say immediately that the density of ethanol at 20°C is 0.789 g cm^{-3}.

2.4.3 Specific Volume

Specific volume is the inverse of density. The dimensions of specific volume are L^3M^{-1}.

2.4.4 Mole

In the SI system, a mole is 'the amount of substance of a system which contains as many elementary entities as there are atoms in 0.012 kg of carbon-12' [3]. This means that a mole in the SI system is about 6.02×10^{23} molecules, and is denoted by the term *gram-mole* or *gmol*. One thousand gmol is called a *kilogram-mole* or *kgmol*. In the American engineering system, the basic mole unit is the *pound-mole* or *lbmol*, which is $6.02 \times 10^{23} \times 453.6$ molecules. The gmol, kgmol and lbmol therefore represent three different quantities. When molar quantities are specified simply as 'moles', gmol is usually meant.

The number of moles in a given mass of material is calculated as follows:

$$\text{gram moles} = \frac{\text{mass in grams}}{\text{molar mass in grams}}$$

(2.18)

$$\text{lb moles} = \frac{\text{mass in lb}}{\text{molar mass in lb}} .$$

(2.19)

Molar mass is the mass of one mole of substance, and has dimensions MN^{-1}. Molar mass is routinely referred to as *molecular weight*, although the molecular weight of a compound is a dimensionless quantity calculated as the sum of the atomic weights of the elements constituting a molecule of that compound. The *atomic weight* of an element is its mass relative to carbon-12 having a mass of exactly 12; atomic weight is also dimensionless. The terms 'molecular weight' and 'atomic weight' are frequently used by engineers and chemists instead of the more correct terms, 'relative molecular mass' and 'relative atomic mass'.

2.4.5 Chemical Composition

Process streams usually consist of mixtures of components or solutions of one or more solutes. The following terms are used to define the composition of mixtures and solutions.

The *mole fraction* of component A in a mixture is defined as:

$$\text{mole fraction A} = \frac{\text{number of moles of A}}{\text{total number of moles}} .$$

$$(2.20)$$

Mole percent is mole fraction × 100. In the absence of chemical reactions and loss of material from the system, the composition of a mixture expressed in mole fraction or mole percent does not vary with temperature.

The *mass fraction* of component A in a mixture is defined as:

$$\text{mass fraction A} = \frac{\text{mass of A}}{\text{total mass}} .$$

$$(2.21)$$

Mass percent is mass fraction × 100; mass fraction and mass percent are also called *weight fraction* and *weight percent*, respectively. Another common expression for composition is weight-for-weight percent (%w/w); although not so well defined, this is usually considered to be the same as weight percent. For example, a solution of sucrose in water with a concentration of 40% w/w contains 40 g sucrose per 100 g solution, 40 tonnes sucrose per 100 tonnes solution, 40 lb sucrose per 100 lb solution, and so on. In the absence of chemical reactions and loss of material from the system, mass and weight percent do not change with temperature.

Because the composition of liquids and solids is usually reported using mass percent, this can be assumed even if not specified. For example, if an aqueous mixture is reported to contain 5% NaOH and 3% $MgSO_4$, it is conventional to assume that there are 5 g NaOH and 3 g $MgSO_4$ in every 100 g solution. Of course, mole or volume percent may be used for liquid and solid mixtures; however this should be stated explicitly, e.g. 10 vol% or 50 mole%.

The *volume fraction* of component A in a mixture is:

$$\text{volume fraction A} = \frac{\text{volume of A}}{\text{total volume}} .$$

$$(2.22)$$

Volume percent is volume fraction × 100. Although not as clearly defined as volume percent, volume-for-volume percent (%v/v) is usually interpreted in the same way as volume percent; for example, an aqueous sulphuric acid mixture containing 30 cm^3 acid in 100 cm^3 solution is referred to as a 30% (v/v) solution. Weight-for-volume percent (%w/v) is also commonly used; a codeine concentration of 0.15% w/v generally means 0.15 g codeine per 100 ml solution.

Compositions of gases are commonly given in volume percent; if percentage figures are given without specification, volume percent is assumed. According to the *International*

Critical Tables [4], the composition of air is 20.99% oxygen, 78.03% nitrogen, 0.94% argon and 0.03% carbon dioxide; small amounts of hydrogen, helium, neon, krypton and xenon make up the remaining 0.01%. For most purposes, all inerts are lumped together with nitrogen; the composition of air is taken as approximately 21% oxygen and 79% nitrogen. This means that any sample of air will contain about 21% oxygen *by volume*. At low pressure, gas volume is directly proportional to number of moles; therefore, the composition of air stated above can be interpreted as 21 *mole%* oxygen. Since temperature changes at low pressure produce the same relative change in partial volumes of constituent gases as in the total volume, volumetric composition of gas mixtures is not altered by variation in temperature. Temperature changes affect the component gases equally, so the overall composition is unchanged.

There are many other choices for expressing the concentration of a component in solutions and mixtures:

(i)   Moles per unit volume, e.g. $gmol\,l^{-1}$, $lbmol\,ft^{-3}$.
(ii) Mass per unit volume, e.g. $kg\,m^{-3}$, $g\,l^{-1}$, $lb\,ft^{-3}$.
(iii) Parts per million, ppm. This is used for very dilute solutions. Usually, ppm is a mass fraction for solids and liquids and a mole fraction for gases. For example, an aqueous solution of 20 ppm manganese contains 20 g manganese per 10^6 g solution. A sulphur dioxide concentration of 80 ppm in air means 80 gmol SO_2 per 10^6 gmol gas mixture. At low pressures this is equivalent to 80 litres SO_2 per 10^6 litres gas mixture.
(iv) Molarity, $gmol\,l^{-1}$.
(v)   Molality, gmol per 1000 g solvent.
(vi) Normality, mole equivalents l^{-1}. A normal solution contains one equivalent gram-weight of solute per litre of solution. For an acid or base, an equivalent gram-weight is the weight of solute in grams that will produce or react with one gmol hydrogen ions. Accordingly, a 1 N solution of HCl is the same as a 1 M solution; on the other hand, a 1 N H_2SO_4 or 1 N $Ca(OH)_2$ solution is 0.5 M.
(vii) Formality, formula gram-weight l^{-1}. If the molecular weight of a solute is not clearly defined, formality may be used to express concentration. A formal solution contains one formula gram-weight of solute per litre of solution. If the formula gram-weight and molecular gram-weight are the same, molarity and formality are the same.

In several industries, concentration is expressed in an indirect way using specific gravity. For a given solute and solvent, the density and specific gravity of solutions are directly dependent on concentration of solute. Specific gravity is conveniently measured using a hydrometer which may be calibrated using special scales. The *Baumé scale*, originally developed in France to measure levels of salt in brine, is in common use. One

Baumé scale is used for liquids lighter than water; another is used for liquids heavier than water. For liquids heavier than water such as sugar solutions:

$$\text{degrees Baumé (°Bé)} = 145 - \frac{145}{G}$$

(2.23)

where G is specific gravity. Unfortunately, the reference temperature for the Baumé and other gravity scales is not standardised world-wide. If the Baumé hydrometer is calibrated at 60°F (15.6°C), G in Eq (2.23) would be the specific gravity at 60°F relative to water at 60°F; however another common reference temperature is 20°C (68°F). The Baumé scale is used widely in the wine and food industries as a measure of sugar concentration. For example, readings of °Bé from grape juice help determine when grapes should be harvested for wine making. The Baumé scale gives only an approximate indication of sugar levels; there is always some contribution to specific gravity from soluble compounds other than sugar.

Degrees Brix (°Brix), or *degrees Balling*, is another hydrometer scale used extensively in the sugar industry. Brix scales calibrated at 15.6°C and 20°C are in common use. With the 20°C scale, each degree Brix indicates 1 gram of sucrose per 100 g liquid.

2.4.6 Temperature

Temperature is a measure of the thermal energy of a body at thermal equilibrium. It is commonly measured in degrees *Celsius* (centigrade) or *Fahrenheit*. In science, the Celsius scale is most common; 0°C is taken as the ice point of water and 100°C the normal boiling point of water. The Fahrenheit scale has everyday use in the USA; 32°F represents the ice point and 212°F the normal boiling point of water. Both Fahrenheit and Celsius scales are *relative temperature scales*, i.e. their zero points have been arbitrarily assigned.

Sometimes it is necessary to use *absolute temperatures*. Absolute-temperature scales have as their zero point the lowest temperature believed possible. Absolute temperature is used in application of the ideal gas law and many other laws of thermodynamics. A scale for absolute temperature with degree units the same as on the Celsius scale is known as the *Kelvin* scale; the absolute-temperature scale using Fahrenheit degree-units is the *Rankine* scale. Units on the Kelvin scale used to be termed 'degrees Kelvin' and abbreviated °K. It is modern practice, however, to name the unit simply 'kelvin'; the SI symbol for kelvin is K. Units on the Rankine scale are denoted °R. 0°R = 0 K = −459.67°F = −273.15°C. Comparison of the four temperature scales is shown in Figure 2.1. One unit on the

Kelvin–Celsius scale corresponds to a temperature difference of 1.8 times a single unit on the Rankine–Fahrenheit scale; the range of 180 Rankine–Fahrenheit degrees between the freezing and boiling points of water corresponds to 100 degrees on the Kelvin–Celsius scale.

Equations for converting temperature units are as follows; T represents the temperature reading:

$$T(\text{K}) = T(°\text{C}) + 273.15$$

(2.24)

$$T(°\text{R}) = T(°\text{F}) + 459.67$$

(2.25)

$$T(°\text{R}) = 1.8\ T(\text{K})$$

(2.26)

$$T(°\text{F}) = 1.8\ T(°\text{C}) + 32.$$

(2.27)

2.4.7 Pressure

Pressure is defined as force per unit area, and has dimensions $L^{-1}MT^{-2}$. Units of pressure are numerous, including pounds per square inch (psi), millimetres of mercury (mmHg), standard atmospheres (atm), bar, newtons per square metre (N m^{-2}), and many others. The SI pressure unit, N m^{-2}, is called a pascal (Pa). Like temperature, pressure may be expressed using absolute or relative scales.

Absolute pressure is pressure relative to a complete vacuum. Because this reference pressure is independent of location, temperature and weather, absolute pressure is a precise and invariant quantity. However, absolute pressure is not commonly measured. Most pressure-measuring devices sense the difference in pressure between the sample and the surrounding atmosphere at the time of measurement. Measurements using these instruments give *relative pressure*, also known as *gauge pressure*. Absolute pressure can be calculated from gauge pressure as follows:

absolute pressure = gauge pressure + atmospheric pressure.

(2.28)

As you know from listening to weather reports, atmospheric pressure varies with time and place and is measured using a *barometer*. Atmospheric pressure or *barometric pressure* should not be confused with the standard unit of pressure called the standard atmosphere (atm), defined as $1.013 \times 10^5\ \text{N m}^{-2}$, 14.70 psi, or 760 mmHg at 0°C. Sometimes the units for pressure include information about whether the pressure is absolute or relative. Pounds per square inch is abbreviated *psia* for absolute pressure or *psig* for gauge pressure. *Atma* denotes standard atmospheres of absolute pressure.

Figure 2.1 Comparison of temperature scales.

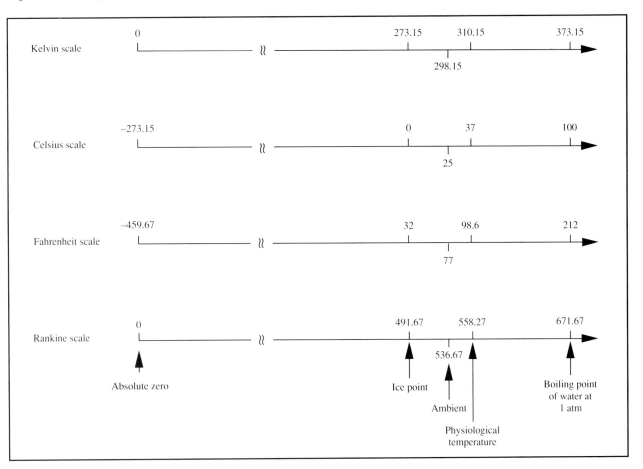

Vacuum pressure is another pressure term, used to indicate pressure below barometric pressure. A gauge pressure of −5 psig, or 5 psi below atmospheric, is the same as a vacuum of 5 psi. A perfect vacuum corresponds to an absolute pressure of zero.

2.5 Standard Conditions and Ideal Gases

A *standard state* of temperature and pressure has been defined and is used when specifying properties of gases, particularly molar volumes. Standard conditions are needed because the volume of a gas depends not only on the quantity present but also on the temperature and pressure. The most widely-adopted standard state is 0°C and 1 atm.

Relationships between gas volume, pressure and temperature were formulated in the 18th and 19th centuries. These correlations were developed under conditions of temperature and pressure so that the average distance between gas molecules

was great enough to counteract the effect of intramolecular forces, and the volume of the molecules themselves could be neglected. Under these conditions, a gas became known as an *ideal gas*. This term now in common use refers to a gas which obeys certain simple physical laws, such as those of Boyle, Charles and Dalton. Molar volumes for an ideal gas at standard conditions are:

$$1 \text{ gmol} = 22.4 \text{ litres} \tag{2.29}$$

$$1 \text{ kgmol} = 22.4 \text{ m}^3 \tag{2.30}$$

$$1 \text{ lbmol} = 359 \text{ ft}^3. \tag{2.31}$$

No real gas is an ideal gas at all temperatures and pressures. However, light gases such as hydrogen, oxygen and air deviate negligibly from ideal behaviour over a wide range of conditions. On the other hand, heavier gases such as sulphur dioxide and hydrocarbons can deviate considerably from ideal, particularly at high pressures. Vapours near the boiling point also deviate markedly from ideal. Nevertheless, for many applications in bioprocess engineering, gases can be considered ideal without much loss of accuracy.

Eqs (2.29)–(2.31) can be verified using the *ideal gas law*:

$$pV = nRT$$

$$(2.32)$$

where p is absolute pressure, V is volume, n is moles, T is absolute temperature and R is the *ideal gas constant*. Eq. (2.32) can be applied using various combinations of units for the physical variables, as long as the correct value and units of R are employed. Table 2.5 gives a list of R values in different systems of units.

Table 2.5   Values of the ideal gas constant, R

(From R.E. Balzhiser, M.R. Samuels and J.D. Eliassen, 1972, Chemical Engineering Thermodynamics, *Prentice-Hall, New Jersey)*

Energy unit	Temperature unit	Mole unit	R
cal	K	gmol	1.9872
J	K	gmol	8.3144
cm^3 atm	K	gmol	82.057
l atm	K	gmol	0.082057
m^3 atm	K	gmol	0.000082057
l mmHg	K	gmol	62.361
l bar	K	gmol	0.083144
$kg_f m^{-2} l$	K	gmol	847.9
$kg_f cm^{-2} l$	K	gmol	0.08479
mmHg ft^3	K	lbmol	998.9
atm ft^3	K	lbmol	1.314
Btu	°R	lbmol	1.9869
psi ft^3	°R	lbmol	10.731
lb_f ft	°R	lbmol	1545
atm ft^3	°R	lbmol	0.7302
in.Hg ft^3	°R	lbmol	21.85
hp h	°R	lbmol	0.0007805
kW h	°R	lbmol	0.0005819
mmHg ft^3	°R	lbmol	555

Example 2.3   Ideal gas law

Gas leaving a fermenter at close to 1 atm pressure and 25°C has the following composition: 78.2% nitrogen, 19.2% oxygen, 2.6% carbon dioxide. Calculate:

(a) the mass composition of the fermenter off-gas; and
(b) the mass of CO_2 in each cubic metre of gas leaving the fermenter.

Solution:

Molecular weights:	nitrogen	=	28
	oxygen	=	32
	carbon dioxide	=	44.

(a) Because the gas is at low pressure, percentages given for composition can be considered mole percentages. Therefore, using the molecular weights, 100 gmol off-gas contains:

$$78.2 \text{ gmol N}_2 \cdot \left| \frac{28 \text{ g N}_2}{1 \text{ gmol N}_2} \right| = 2189.6 \text{ g N}_2$$

$$19.2 \text{ gmol O}_2 \cdot \left| \frac{32 \text{ g O}_2}{1 \text{ gmol O}_2} \right| = 614.4 \text{ g O}_2$$

$$2.6 \text{ gmol CO}_2 \cdot \left| \frac{44 \text{ g CO}_2}{1 \text{ gmol CO}_2} \right| = 114.4 \text{ g CO}_2.$$

Therefore, the total mass is $(2189.6 + 614.4 + 114.4)$ g $= 2918.4$ g. The mass composition can be calculated as follows:

$$\text{Mass percent N}_2 = \frac{2189.6 \text{ g}}{2918.4 \text{ g}} \times 100 = 75.0\%$$

$$\text{Mass percent O}_2 = \frac{614.4 \text{ g}}{2918.4 \text{ g}} \times 100 = 21.1\%$$

$$\text{Mass percent CO}_2 = \frac{114.4 \text{ g}}{2918.4 \text{ g}} \times 100 = 3.9\%.$$

Therefore, the composition of the gas is 75.0 mass% N_2, 21.1 mass% O_2 and 3.9 mass% CO_2.

(b) As the gas composition is given in volume percent, in each cubic metre of gas there must be 0.026 m^3 CO_2. The relationship between moles of gas and volume at 1 atm and 25°C is determined using Eq. (2.32) and Table 2.5:

$$(1 \text{ atm}) (0.026 \text{ m}^3) = n \, (0.000082057 \, \frac{\text{m}^3 \text{ atm}}{\text{gmol K}}) \, (298.15 \text{ K}).$$

Calculating the moles of CO_2 present:

$$n = 1.06 \text{ gmol}.$$

Converting to mass of CO_2:

$$1.06 \text{ gmol} \cdot \left| \frac{44 \text{ g}}{1 \text{ gmol}} \right| = 46.8 \text{ g}.$$

Therefore, each cubic metre of fermenter off-gas contains 46.8 g CO_2.

2.6 Physical and Chemical Property Data

Information about the properties of materials is often required in engineering calculations. Because measurement of physical and chemical properties is time-consuming and expensive, handbooks containing this information are a tremendous resource. You may already be familiar with some handbooks of physical and chemical data, including:

(i) *International Critical Tables* [4]
(ii) *Handbook of Chemistry and Physics* [5]; and
(iii) *Handbook of Chemistry* [6].

To these can be added:

(iv) *Chemical Engineers' Handbook* [7];

and, for information about biological materials,

(v) *Biochemical Engineering and Biotechnology Handbook* [8].

A selection of physical and chemical property data is included in Appendix B.

2.7 Stoichiometry

In chemical or biochemical reactions, atoms and molecules rearrange to form new groups. Mass and molar relationships between the reactants consumed and products formed can be determined using stoichiometric calculations. This information is deduced from correctly-written reaction equations and relevant atomic weights.

As an example, consider the principal reaction in alcohol fermentation: conversion of glucose to ethanol and carbon dioxide:

$$C_6H_{12}O_6 \rightarrow 2\ C_2H_6O + 2\ CO_2. \tag{2.33}$$

This reaction equation states that one molecule of glucose breaks down to give two molecules of ethanol and two molecules of carbon dioxide. Applying molecular weights, the equation shows that reaction of 180 g glucose produces 92 g ethanol and 88 g carbon dioxide. During chemical or biochemical reactions, the following two quantities are conserved:

(i) *total mass*, i.e. total mass of reactants = total mass of products; and

(ii) *number of atoms of each element*, e.g. the number of C, H and O atoms in the reactants = the number of C, H and O atoms, respectively, in the products.

Note that there is no corresponding law for conservation of moles; moles of reactants ≠ moles of products.

Example 2.4 Stoichiometry of amino acid synthesis

The overall reaction for microbial conversion of glucose to L-glutamic acid is:

$$C_6H_{12}O_6 + NH_3 + {}^3/_2O_2 \rightarrow C_5H_9NO_4 + CO_2 + 3\ H_2O.$$
(glucose)   (glutamic acid)

What mass of oxygen is required to produce 15 g glutamic acid?

Solution:
Molecular weights: oxygen   =   32
                                  glutamic acid   =   147

In the following equation, g glutamic acid is converted to gmol using a unity bracket for molecular weight, the stoichiometric equation is applied to convert gmol glutamic acid to gmol oxygen, and finally gmol oxygen are converted to g for the final answer:

$$15\ \text{g glumatic acid} \cdot \left| \frac{1\ \text{gmol glutamic acid}}{147\ \text{g glutamic acid}} \right| \cdot \left| \frac{{}^3/_2\ \text{gmol O}_2}{1\ \text{gmol glutamic acid}} \right| \cdot \left| \frac{32\ \text{g O}_2}{1\ \text{gmol O}_2} \right| = 4.9\ \text{g O}_2.$$

Therefore, 4.9 g oxygen is required. More oxygen will be needed if microbial growth also occurs.

By themselves, equations such as (2.33) suggest that all the reactants are converted into the products specified in the equation, and that the reaction proceeds to completion. This is often not the case for industrial reactions. Because the stoichiometry may not be known precisely, or in order to manipulate the reaction beneficially, reactants are not usually supplied in the exact proportions indicated by the reaction equation. Excess quantities of some reactants may be provided; this excess material is found in the product mixture once the reaction is stopped. In addition, reactants are often consumed in side reactions to make products not described by the principal reaction equation; these side-products also form part of the final reaction mixture. In these circumstances, additional information is needed before the amounts of product formed or reactants consumed can be calculated. Terms used to describe partial and branched reactions are outlined below.

(i) The *limiting reactant* is the reactant present in the smallest *stoichiometric* amount. While other reactants may be present in smaller absolute quantities, at the time when the last molecule of the limiting reactant is consumed, residual amounts of all reactants except the limiting reactant will be present in the reaction mixture. As an illustration, for the glutamic acid reaction of Example 2.4, if 100 g glucose, 17 g NH_3 and 48 g O_2 are provided for conversion, glucose will be the limiting reactant even though a greater mass of it is available compared with the other substrates.

(ii) An *excess reactant* is a reactant present in an amount in excess of that required to combine with all of the limiting reactant. It follows that an excess reactant is one remaining in the reaction mixture once all the limiting reactant is consumed. The *percentage excess* is calculated using the amount of excess material relative to the quantity required for complete consumption of the limiting reactant:

$$\% \text{ excess} = \frac{\left(\begin{array}{c}\text{moles present} - \text{moles required to react}\\\text{completely with the limiting reactant}\end{array}\right)}{\left(\begin{array}{c}\text{moles required to react}\\\text{completely with the limiting reactant}\end{array}\right)} \times 100 \tag{2.34}$$

or

$$\% \text{ excess} = \frac{\left(\begin{array}{c}\text{mass present} - \text{mass required to react}\\\text{completely with the limiting reactant}\end{array}\right)}{\left(\begin{array}{c}\text{mass required to react}\\\text{completely with the limiting reactant}\end{array}\right)} \times 100. \tag{2.35}$$

The *required* amount of a reactant is the stoichiometric quantity needed for complete conversion of the limiting reactant. In the above glutamic acid example, the required amount of NH_3 for complete conversion of 100 g glucose is 9.4 g; therefore if 17 g NH_3 are provided the percent excess NH_3 is 80%. Even if only part of the reaction actually occurs, required and excess quantities are based on the entire amount of the limiting reactant.

Other reaction terms are not as well defined with multiple definitions in common use:

(iii) *Conversion* is the fraction or percentage of a reactant converted into products.

(iv) *Degree of completion* is usually the fraction or percentage of the limiting reactant converted into products.

(v) *Selectivity* is the amount of a particular product formed as a fraction of the amount that would have been formed if all the feed material had been converted to that product.

(vi) *Yield* is the ratio of mass or moles of product formed to the mass or moles of reactant consumed. If more than one product or reactant is involved in the reaction, the particular compounds referred to must be stated, e.g. the yield of glutamic acid from glucose was 0.6 g g^{-1}. Because of the complexity of metabolism and the frequent occurrence of side reactions, yield is an important term in bioprocess analysis. Application of the yield concept for cell and enzyme reactions is described in more detail in Chapter 11.

Example 2.5 Incomplete reaction and yield

Depending on culture conditions, glucose can be catabolised by yeast to produce ethanol and carbon dioxide, or can be diverted into other biosynthetic reactions. An inoculum of yeast is added to a solution containing 10 g l^{-1} glucose. After some time only 1 g l^{-1} glucose remains while the concentration of ethanol is 3.2 g l^{-1}. Determine:

(a) the fractional conversion of glucose to ethanol; and
(b) the yield of ethanol from glucose.

Solution:

(a) To find the fractional conversion of glucose to ethanol, we must first determine exactly how much glucose was directed into ethanol biosynthesis. Using a basis of 1 litre and Eq. (2.33) for ethanol fermentation, we can calculate the mass of glucose required for synthesis of 3.2 g ethanol:

$$3.2 \text{ g ethanol} \cdot \left|\frac{1 \text{ gmol ethanol}}{46 \text{ g ethanol}}\right| \cdot \left|\frac{1 \text{ gmol glucose}}{2 \text{ gmol ethanol}}\right| \cdot \left|\frac{180 \text{ g glucose}}{1 \text{ gmol glucose}}\right| = 6.3 \text{ g glucose}.$$

Therefore, based on the total amount of glucose provided per litre (10 g), the fractional conversion of glucose to ethanol was 0.63. Based on the amount of glucose actually consumed per litre (9 g), the fractional conversion to ethanol was 0.70.

(b) Yield of ethanol from glucose is based on the total mass of glucose consumed. Since 9 g glucose was consumed per litre to provide 3.2 g l^{-1} ethanol, the yield of ethanol from glucose was 0.36 g g^{-1}. We can also conclude that, per litre, 2.7 g glucose was consumed but not used for ethanol synthesis.

2.8 Summary of Chapter 2

Having studied the contents of Chapter 2, you should:

(i) understand dimensionality and be able to convert units with ease;

(ii) understand the terms *mole, molecular weight, density, specific gravity, temperature* and *pressure*, know various ways of expressing *concentration* of solutions and mixtures, and be able to work simple problems involving these concepts;

(iii) be able to apply the ideal gas law;

(iv) know where to find physical and chemical property data in the literature; and

(v) understand reaction terms such as *limiting reactant, excess reactant, conversion, degree of completion, selectivity* and *yield*, and be able to apply stoichiometric principles to reaction problems.

Problems

2.1 Unit conversion

(a) Convert 1.5×10^{-6} centipoise to kg s^{-1} cm^{-1}.

(b) Convert 0.122 horsepower (British) to British thermal units per minute (Btu min^{-1}).

(c) Convert 670 mmHg ft^3 to metric horsepower h.

(d) Convert 345 Btu lb^{-1} to kcal g^{-1}.

2.2 Unit conversion

Using Eq. (2.1) for the Reynolds number, calculate *Re* for the following two sets of data:

Parameter	Case 1	Case 2
D	2 mm	1 in.
u	3 cm s^{-1}	1 m s^{-1}
ρ	25 lb ft^{-3}	12.5 kg m^{-3}
μ	10^{-6} cP	0.14×10^{-4} lb$_m$ s^{-1} ft^{-1}

2.3 Dimensionless groups and property data

The rate at which oxygen is transported from gas phase to liquid phase is a very important parameter in fermenter design. A well-known correlation for transfer of gas is:

$$Sh = 0.31 \ Gr^{1/3} \ Sc^{1/3}$$

where *Sh* is the Sherwood number, *Gr* is the Grashof number and *Sc* is the Schmidt number. These dimensionless numbers are defined as follows:

$$Sh = \frac{k_L D_b}{\mathcal{D}}$$

$$Gr = \frac{D_b^3 \ \rho_G \ (\rho_L - \rho_G)g}{\mu_L^2}$$

$$Sc = \frac{\mu_L}{\rho_L \mathcal{D}}$$

where k_L is mass-transfer coefficient, D_b is bubble diameter, $\mathcal{D}$ is diffusivity of gas in the liquid, ρ_G is density of gas, ρ_L is density of liquid, μ_L is viscosity of liquid, and *g* is gravitational acceleration = 32.17 ft s^{-2}.

A gas sparger in a fermenter operated at 28°C and 1 atm produces bubbles of about 2 mm diameter. Calculate the value of the mass transfer coefficient, k_L. Collect property data from, e.g. *Chemical Engineers' Handbook*, and assume that the culture broth has similar properties to water. (Do you think this is a reasonable assumption?) Report the literature source for any property data used. State explicitly any other assumptions you make.

2.4 Mass and weight

The density of water is 62.4 lb$_m$ ft^{-3}. What is the weight of 10 ft^3 of water:

(a) at sea level and 45° latitude?; and

(b) somewhere above the earth's surface where $g = 9.76$ m s^{-2}?

2.5 Dimensionless numbers

The Colburn equation for heat transfer is:

$$\left(\frac{h}{C_p G}\right)\left(\frac{C_p \mu}{k}\right)^{2/3} = \frac{0.023}{\left(\frac{DG}{\mu}\right)^{0.2}}$$

where C_p is heat capacity, Btu lb^{-1} °F^{-1}; μ is viscosity, lb h^{-1} ft^{-1}; k is thermal conductivity, Btu h^{-1} ft^{-2} (°F ft^{-1})$^{-1}$; D is pipe diameter, ft; and G is mass velocity per unit area, lb h^{-1} ft^{-2}.

The Colburn equation is dimensionally consistent. What are the units and dimensions of the heat-transfer coefficient, h?

2.6 Dimensional homogeneity and g_c

Two students have reported different versions of the dimensionless power number N_P used to relate fluid properties to the power required for stirring:

(i) $N_P = \dfrac{Pg}{\rho N_i^3 D_i^5}$; and

(ii) $N_P = \dfrac{Pg_c}{\rho N_i^3 D_i^5}$

where P is power, g is gravitational acceleration, ρ is fluid density, N_i is stirrer speed, D_i is stirrer diameter and g_c is the force unity bracket. Which equation is correct?

2.7 Molar units

If a bucket holds 20.0 lb NaOH, how many:

(a) lbmol NaOH;
(b) gmol NaOH; and
(c) kgmol NaOH

does it contain?

2.8 Density and specific gravity

(a) The specific gravity of nitric acid is $1.5129^{20°C}_{4°C}$.
   (i) What is its density at 20°C in kg m^{-3}?
   (ii) What is its molar specific volume?
(b) The volumetric flow rate of carbon tetrachloride (CCl$_4$) in a pipe is 50 cm^3 min^{-1}. The density of CCl$_4$ is 1.6 g cm^{-3}.
   (i) What is the mass flow rate of CCl$_4$?
   (ii) What is the molar flow rate of CCl$_4$?

2.9 Molecular weight

Calculate the average molecular weight of air.

2.10 Mole fraction

A solution contains 30 wt% water, 25 wt% ethanol, 15 wt% methanol, 12 wt% glycerol, 10 wt% acetic acid and 8 wt% benzaldehyde. What is the mole fraction of each component?

2.11 Temperature scales

What is -40°F in degrees centigrade? degrees Rankine? kelvin?

2.12 Pressure scales

(a) The pressure gauge on an autoclave reads 15 psi. What is the absolute pressure in the chamber in psi? in atm?
(b) A vacuum gauge reads 3 psi. What is the pressure?

2.13 Stoichiometry and incomplete reaction

For production of penicillin (C$_{16}$H$_{18}$O$_4$N$_2$S) using *Penicillium* mould, glucose (C$_6$H$_{12}$O$_6$) is used as substrate, and phenylacetic acid (C$_8$H$_8$O$_2$) is added as precursor. The stoichiometry for overall synthesis is:

$$1.67\,C_6H_{12}O_6 + 2\,NH_3 + 0.5\,O_2 + H_2SO_4 + C_8H_8O_2$$
$$\rightarrow C_{16}H_{18}O_4N_2S + 2\,CO_2 + 9\,H_2O.$$

(a) What is the maximum theoretical yield of penicillin from glucose?
(b) When results from a particular penicillin fermentation were analysed, it was found that 24% of the glucose had been used for growth, 70% for cell maintenance activities (such as membrane transport and macromolecule turnover), and only 6% for penicillin synthesis. Calculate the yield of penicillin from glucose under these conditions.
(c) Batch fermentation under the conditions described in (b) is carried out in a 100-litre tank. Initially, the tank is filled with nutrient medium containing 50 g l^{-1} glucose and 4 g l^{-1} phenylacetic acid. If the reaction is stopped when the glucose concentration is 5.5 g l^{-1}, determine:
   (i) which is the limiting substrate if NH$_3$, O$_2$ and H$_2$SO$_4$ are provided in excess;
   (ii) the total mass of glucose used for growth;

(iii) the amount of penicillin produced; and

(iv) the final concentration of phenylacetic acid.

2.14 Stoichiometry, yield and the ideal gas law

Stoichiometric equations are used to represent growth of microorganisms provided a 'molecular formula' for the cells is available. The molecular formula for biomass is obtained by measuring the amounts of C, N, H, O and other elements in cells. For a particular bacterial strain, the molecular formula was determined to be $C_{4.4}H_{7.3}O_{1.2}N_{0.86}$.

These bacteria are grown under aerobic conditions with hexadecane ($C_{16}H_{34}$) as substrate. The reaction equation describing growth is:

$$C_{16}H_{34} + 16.28\,O_2 + 1.42\,NH_3$$
$$\rightarrow 1.65\,C_{4.4}H_{7.3}O_{1.2}N_{0.86} + 8.74\,CO_2 + 13.11\,H_2O.$$

(a) Is the stoichiometric equation balanced?

(b) Assuming 100% conversion, what is the yield of cells from hexadecane in g g^{-1}?

(c) Assuming 100% conversion, what is the yield of cells from oxygen in g g^{-1}?

(d) You have been put in charge of a small fermenter for growing the bacteria and aim to produce 2.5 kg of cells for inoculation of a pilot-scale reactor.

  (i) What minimum amount of hexadecane substrate must be contained in your culture medium?

  (ii) What must be the minimum concentration of hexadecane in the medium if the fermenter working volume is 3 cubic metres?

  (iii) What minimum volume of air at 20°C and 1 atm pressure must be pumped into the fermenter during growth to produce the required amount of cells?

References

1. Drazil, J.V. (1983) *Quantities and Units of Measurement*, Mansell, London.

2. Richards, J.W. (1961) Studies in aeration and agitation. *Prog. Ind. Microbiol.* **3**, 141–172.

3. *The International System of Units (SI)* (1977) National Bureau of Standards Special Publication 330, US Government Printing Office, Washington. Adopted by the 14th General Conference on Weights and Measures (1971, Resolution 3).

4. *International Critical Tables* (1926) McGraw-Hill, New York.

5. *Handbook of Chemistry and Physics*, CRC Press, Boca Raton.

6. Dean, J.A. (Ed.) (1985) *Lange's Handbook of Chemistry*, 13th edn, McGraw-Hill, New York.

7. Perry, R.H., D.W. Green and J.O. Maloney (Eds) (1984) *Chemical Engineers' Handbook*, 6th edn, McGraw-Hill, New York.

8. Atkinson, B. and F. Mavituna (1991) *Biochemical Engineering and Biotechnology Handbook*, 2nd edn, Macmillan, Basingstoke.

Suggestions for Further Reading

Units and Dimensions (see also refs 1 and 3)

Ipsen, D.C. (1960) *Units, Dimensions, and Dimensionless Numbers*, McGraw-Hill, New York.

Massey, B.S. (1986) *Measures in Science and Engineering: Their Expression, Relation and Interpretation*, Chapters 1–5, Ellis Horwood, Chichester.

Qasim, S.H. (1977) *SI Units in Engineering and Technology*, Pergamon Press, Oxford.

Ramsay, D.C. and G.W. Taylor (1971) *Engineering in S.I. Units*, Chambers, Edinburgh.

Engineering Variables

Felder, R.M. and R.W. Rousseau (1978) *Elementary Principles of Chemical Processes*, Chapters 2 and 3, John Wiley, New York.

Himmelblau, D.M. (1974) *Basic Principles and Calculations in Chemical Engineering*, 3rd edn, Chapter 1, Prentice-Hall, New Jersey.

Shaheen, E.I. (1975) *Basic Practice of Chemical Engineering*, Chapter 2, Houghton Mifflin, Boston.

Whitwell, J.C. and R.K. Toner (1969) *Conservation of Mass and Energy*, Chapter 2, Blaisdell, Waltham, Massachusetts.

3

Presentation and Analysis of Data

Quantitative information is fundamental to scientific and engineering analysis. Information about bioprocesses, such as the amount of substrate fed into the system, the operating conditions, and properties of the product stream, is obtained by measuring pertinent physical and chemical variables. In industry, data are collected for equipment design, process control, trouble-shooting and economic evaluations. In research, experimental data are used to develop new theories and test theoretical predictions. In either case, quantitative interpretation of data is absolutely essential for making rational decisions about the system under investigation. The ability to extract useful and accurate information from data is an important skill for any scientist. Professional presentation and communication of results is also required.

Techniques for data analysis must take into account the existence of error in measurements. Because there is always an element of uncertainty associated with measured data, interpretation calls for a great deal of judgement. This is especially the case when critical decisions in design or operation of processes depend on data evaluation. Although computers and calculators make data processing less tedious, the data analyst must possess enough perception to use these tools effectively.

This chapter discusses sources of error in data and methods of handling errors in calculations. Presentation and analysis of data using graphs and equations and presentation of process information using flow sheets are described.

3.1 Errors in Data and Calculations

Measurements are never perfect. Experimentally-determined quantities are always somewhat inaccurate due to measurement error; absolutely 'correct' values of physical quantities (time, length, concentration, temperature, etc.) cannot be found. The significance or reliability of conclusions drawn from data must take measurement error into consideration. Estimation of error and principles of error propagation in calculations are important elements of engineering analysis and help prevent misleading representation of data. General principles for estimation and expression of errors are discussed in the following sections.

3.1.1 Significant Figures

Data used in engineering calculations vary considerably in accuracy. Economic projections may estimate market demand

for a new biotechnology product to within ±100%; on the other hand, some properties of materials are known to within ±0.0001% or less. The uncertainty associated with quantities should be reflected in the way they are written. The number of figures used to report a measured or calculated variable is an indirect indication of the precision to which that variable is known. It would be absurd, for example, to quote the estimated income from sales of a new product using ten decimal places. Nevertheless, the mistake of quoting too many figures is not uncommon; display of superfluous figures on calculators is very easy but should not be transferred to scientific reports.

A *significant figure* is any digit, 1–9, used to specify a number. Zero may also be a significant figure when it is not used merely to locate the position of the decimal point. For example, the numbers 6304, 0.004321, 43.55 and 8.063×10^{10} each contain four significant figures. For the number 1200, however, there is no way of knowing whether or not the two zeros are significant figures; a direct statement or an alternative way of expressing the number is needed. For example, 1.2×10^3 has two significant figures, while 1.200×10^3 has four.

A number is rounded to n significant figures using the following rules:

(i)   If the number in the $(n+1)$th position is less than 5, discard all figures to the right of the nth place.

(ii) If the number in the $(n+1)$th position is greater than 5, discard all figures to the right of the nth place, and increase the nth digit by 1.

(iii) If the number in the $(n+1)$th position is exactly 5, discard all figures to the right of the nth place, and increase the nth digit by 1.

For example, when rounding off to four significant figures:

1.426348 becomes 1.426;
1.426748 becomes 1.427; and
1.4265 becomes 1.427.

The last rule is not universal but is engineering convention; most electronic calculators and computers round up halves. Generally, rounding off means that the value may be wrong by up to 5 units in the next number-column not reported. Thus, 10.77 kg means that the mass lies somewhere between 10.765 kg and 10.775 kg, whereas 10.7754 kg represents a mass between 10.77535 kg and 10.77545 kg. These rules apply only to quantities based on measured values; some numbers used in calculations refer to precisely known or counted quantities. For example, there is no error associated with the number $\frac{1}{2}$ in the equation for kinetic energy:

$$\text{kinetic energy} = E_k = \frac{1}{2} M v^2$$

where M is mass and v is velocity.

It is good practice during calculations to carry along one or two extra significant figures for combination during arithmetic operations; final rounding-off should be done only at the end. How many figures should we quote in the final answer? There are several rules-of-thumb for rounding off after calculations, so rigid adherence to all rules is not always possible. However as a guide, after multiplication or division, the number of significant figures in the result should equal the smallest number of significant figures of any of the quantities involved in the calculation. For example:

$$(6.681 \times 10^{-2})(5.4 \times 10^9) = 3.608 \times 10^8 \to 3.6 \times 10^8$$

and

$$\frac{6.16}{0.054677} = 112.6616310 \to 113.$$

For addition and subtraction, look at the position of the last significant figure in each number relative to the decimal point. The position of the last significant figure in the result should be the same as that most to the left, as illustrated below:

$$24.335 + 3.90 + 0.00987 = 28.24487 \to 28.24$$

and

$$121.808 - 112.87634 = 8.93166 \to 8.932.$$

3.1.2 Absolute and Relative Uncertainty

Uncertainty associated with measurements can be stated more explicitly than is possible using the rules of significant figures. For a particular measurement, we should be able to give our best estimate of the parameter value, and the interval representing its range of uncertainty. For example, we might be able to say with confidence that the prevailing temperature lies between 23.7°C and 24.3°C. Another way of expressing this result is 24 ± 0.3°C. The value ± 0.3°C is known as the *uncertainty* or *error* of the measurement, and allows us to judge the quality of the measuring process. Since ± 0.3°C represents the actual temperature range by which the reading is uncertain, it is known as the *absolute error*. An alternative expression for 24 ± 0.3°C is 24°C ± 1.25%; in this case the *relative error* is ± 1.25%.

Because most values of uncertainty must be estimated rather than measured, there is a rule-of-thumb that magnitudes of errors should be given with only one or two significant figures. A flow rate may be expressed as 146 ± 13 gmol h^{-1}, even though this means that two figures, i.e. the '4' and the '6' in the result, are uncertain. The number of digits used to express the result should be compatible with the magnitude of its estimated error. For example, in the statement: 2.1437 ± 0.12 grams, the estimated uncertainty of 0.12 grams shows that the last two digits in the result are superfluous. Use of more than three significant figures in this case gives a false impression of accuracy.

There are rules for combining errors during mathematical operations. The uncertainty associated with calculated results is found from the errors associated with the raw data. For addition and subtraction the rule is: *add absolute errors*. The total of the absolute errors becomes the absolute error associated with the final answer. For example, the sum of 1.25 ± 0.13 and 0.973 ± 0.051 is:

$$(1.25 + 0.973) \pm (0.13 + 0.051) = 2.22 \pm 0.18 = 2.22 \pm 8.1\%.$$

Considerable loss of accuracy can occur after subtraction, especially when two large numbers are subtracted to give an answer of small numerical value. Because the absolute error after subtraction of two numbers always increases, the relative error associated with a small-number answer can be very great. For example, consider the difference between two numbers, each with small relative error: 12 736 ± 0.5% and 12 681 ± 0.5%. For subtraction, the absolute errors are added:

$$(12\ 736 \pm 64) - (12\ 681 \pm 63) = 55 \pm 127 = 55 \pm 230\%.$$

Even though it could be argued that the two errors might almost cancel each other (if one were +64 and the other were −63, for example), we can never be certain that this would occur. 230% represents the worst case or *maximum possible error*. For measured values, any small number obtained by subtraction of two large numbers must be examined carefully and with justifiable suspicion. Unless explicit errors are reported, the large uncertainty associated with results can go unnoticed.

For multiplication and division: *add relative errors*. The total of the relative errors becomes the relative error associated with the answer. For example, 164 ± 1 divided by 790 ± 20 is the same as $164 \pm 0.61\%$ divided by $790 \pm 2.5\%$:

$$\left(^{790}/_{164}\right) \pm (2.5 + 0.61)\% = 4.82 \pm 3.1\% = 4.82 \pm 0.15.$$

Propagation of errors in more complex expressions will not be discussed here; more information and rules for combining errors can be found in other references [1–3].

So far we have considered the error occurring in a single observation. However, as discussed below, better estimates of errors are obtained by taking repeated measurements. Because this approach is useful only for certain types of measurement error, let us consider the various sources of error in experimental data.

3.1.3 Types of Error

There are two broad classes of measurement error: *systematic* and *random*. A systematic error is one which affects all measurements of the same variable in the same way. If the cause of systematic error is identified, it can be accounted for using a correction factor. For example, errors caused by an imperfectly calibrated analytical balance may be identified using standard weights; measurements with the balance can then be corrected to compensate for the error. Systematic errors easily go undetected; performing the same measurement using different instruments, methods and observers is required to detect systematic error [4].

Random or accidental errors are due to unknown causes. Random errors are present in almost all data; they are revealed when repeated measurements of an unchanging quantity give a 'scatter' of different results. As outlined in the next section, scatter from repeated measurements is used in statistical analysis to quantify random error. The term *precision* refers to the *reliability* or *reproducibility* of data, and indicates the extent to which a measurement is free from random error. *Accuracy*, on the other hand, requires both random and systematic errors to be small. Repeated weighings using a poorly-calibrated balance can give results that are very precise (because each reading is similar); however the result would be inaccurate because of the incorrect calibration and systematic error.

During experiments, large, isolated, one-of-a-kind errors can also occur. This type of error is different from the systematic and random errors mentioned above and can be described as a 'blunder'. Accounting for blunders in experimental data requires knowledge of the experimental process and judgement about the likely accuracy of the measurement.

3.1.4 Statistical Analysis

Measurements containing random errors but free of systematic errors and blunders can be analysed using statistical procedures. Details are available in standard texts, e.g. [5]; only the most basic techniques for statistical treatment will be described here. From readings containing random error, we aim to find the best estimate of the variable measured, and to quantify the extent to which random error affects the data.

In the following analysis, errors are assumed to follow a normal or Gaussian distribution. Normally-distributed random errors in a single measurement are just as likely to be positive as negative; thus, if an infinite number of repeated measurements were made of the same variable, random error would completely cancel out from the arithmetic mean of these values. For less than an infinite number of observations, the arithmetic mean of repeated measurements is still regarded as the best estimate of the variable, provided each measurement is made with equal care and under identical conditions. Taking replicate measurements is therefore standard practice in science; whenever possible, several readings of each datum point should be obtained. For variable x measured n times, the *arithmetic mean* is calculated as follows:

$$\bar{x} = \text{mean value of } x = \frac{\sum\limits^{n} x}{n} = \frac{x_1 + x_2 + x_3 + \ldots x_n}{n}.$$

(3.1)

As indicated, the symbol $\sum\limits^{n}$ represents the sum of n values; $\sum\limits^{n} x$ means the sum of n values of parameter x.

In addition to the mean, we need some measure of the precision of the measurements; this is obtained by considering the scatter of individual values about the mean. The deviation of an individual value from the mean is known as the *residual*; an example of a residual is $(x_1 - \bar{x})$ where x_1 is a measurement in a set of replicates. The most useful indicator of the magnitude of the residuals is the *standard deviation*. For a set of experimental data, standard deviation σ is calculated as follows:

$$\sigma = \sqrt{\frac{\sum\limits^{n}(x - \bar{x})^2}{n - 1}}$$

$$= \sqrt{\frac{(x_1 - \bar{x})^2 + (x_2 - \bar{x})^2 + (x_3 - \bar{x})^2 + \ldots (x_n - \bar{x})^2}{n - 1}}.$$

(3.2)

Eq. (3.2) is the definition used by most modern statisticians and manufacturers of electronic calculators; σ as defined in Eq. (3.2) is sometimes called the *sample standard deviation*.

Therefore, to report the results of repeated measurements, we quote the mean as the best estimate of the variable, and the standard deviation as a measure of the confidence we place in the result. The units and dimensions of the mean and standard deviation are the same as those of x. For less than an infinite number of repeated measurements, the mean and standard deviation calculated using one set of observations will produce a different result from that determined using another set. It can be shown mathematically that values of the mean and standard deviation become more reliable as n increases; taking replicate measurements is therefore standard practice in science. A compromise is usually struck between the conflicting demands of precision and the time and expense of experimentation; sometimes it is impossible to make a large number of replicate measurements. Sample size should always be quoted when reporting the outcome of statistical analysis. When substantial improvement in the accuracy of the mean and standard deviation is required, this is generally more effectively achieved by improving the intrinsic accuracy of the measurement rather than by just taking a multitude of repeated readings.

Example 3.1 Mean and standard deviation

The final concentration of L-lysine produced by a regulatory mutant of *Brevibacterium lactofermentum* is measured 10 times. The results in g l^{-1} are: 47.3, 51.9, 52.2, 51.8, 49.2, 51.1, 52.4, 47.1, 49.1 and 46.3. How should the lysine concentration be reported?

Solution:
For this sample, $n = 10$. From Eq. (3.1):

$$\bar{x} = \frac{47.3 + 51.9 + 52.2 + 51.8 + 49.2 + 51.1 + 52.4 + 47.1 + 49.1 + 46.3}{10}$$

$$= 49.84 \text{ g } l^{-1}.$$

Substituting this result into Eq. (3.2) gives:

$$\sigma = \sqrt{\frac{49.24}{9}} = 2.34 \text{ g } l^{-1}.$$

Therefore, from 10 repeated measurements, the lysine concentration was 49.8 g l^{-1} with standard deviation 2.3 g l^{-1}.

Methods for combining standard deviations in calculations are discussed elsewhere [1, 2]. Remember that standard statistical analysis does not account for systematic error; parameters such as the mean and standard deviation are useful only if the error in measurements is random. *The effect of systematic error cannot be minimised using standard statistical analysis or by collecting repeated measurements.*

3.2 Presentation of Experimental Data

Experimental data are often collected to examine relationships between variables. The role of these variables in the experimental process is clearly defined. *Dependent variables* or *response variables* are uncontrolled during the experiment; dependent variables are measured as they respond to changes in one or more *independent variables* which are controlled or fixed. For example, if we wanted to determine how UV radiation affects the frequency of mutation in a culture, radiation dose would be the independent variable and number of mutants the dependent variable.

There are three general methods for presenting data:

(i) tables;
(ii) graphs; and
(iii) equations.

Each has its own strengths and weaknesses. Tables listing data have highest accuracy, but can easily become too long and the overall result or trend of the data cannot be readily visualised. Graphs or plots of data create immediate visual impact since relationships between variables are represented directly. Graphs also allow easy interpolation of data, which can be difficult with tables. By convention, independent variables are plotted along the *abscissa* (the X-axis), while one or more dependent variables are plotted along the *ordinate* (Y-axis). Plots show at a glance the general pattern of data, and can help identify whether there are anomalous points; it is good practice to plot raw experimental data as they are being measured. In addition, graphs can be used directly for quantitative data analysis.

Physical phenomena can be represented using equations or *mathematical models*; for example, balanced growth of microorganisms is described using the model:

$$x = x_0\, e^{\mu t}$$

(3.3)

where x is the cell concentration at time t, x_0 is the initial cell concentration, and μ is the specific growth rate. Mathematical models can be either mechanistic or empirical. *Mechanistic models* are founded on theoretical assessment of the phenomenon being measured. An example is the Michaelis–Menten equation for enzyme reaction:

$$v = \frac{v_{\max}\, s}{K_m + s}$$

(3.4)

where v is rate of reaction, $v_{\max}$ is maximum rate of reaction, K_m is the Michaelis constant, and s is substrate concentration. The Michaelis–Menten equation is based on a loose analysis of reactions supposed to occur during simple enzyme catalysis. On the other hand, *empirical models* are used when no theoretical hypothesis can be postulated. Empirical models may be the only feasible option for correlating data from complicated processes. As an example, the following correlation has been developed to relate the power required to stir aerated liquids to that required in non-aerated systems:

$$\frac{P_g}{P_0} = 0.10 \left(\frac{F_g}{N_i V} \right)^{-0.25} \left(\frac{N_i^2 D_i^4}{g\, W_i\, V^{2/3}} \right)^{-0.20}$$

(3.5)

In Eq. (3.5) P_g is power consumption with sparging, P_0 is power consumption without sparging, F_g is volumetric gas flow rate, N_i is stirrer speed, V is liquid volume, D_i is impeller diameter, g is gravitational acceleration, and W_i is impeller blade width. There is no easy theoretical explanation for this relationship; the equation is based on many observations using different impellers, gas flow rates and rates of stirring. Equations such as Eq. (3.5) are a short, concise means for communicating the results of a large number of experiments. However, they are one step removed from the raw data and can be only an approximate representation of all the information collected.

3.3 Data Analysis

Once experimental data are collected, what we do with them depends on the information being sought. Data are generally collected for one or more of the following reasons:

(i) to visualise the general trend of influence of one variable on another;
(ii) to test the applicability of a particular model to a process;
(iii) to estimate the value of coefficients in process models; and
(iv) to develop new empirical models.

Analysis of data would be enormously simplified if each datum point did not contain error. For example, after an experiment in which the apparent viscosity of a mycelial broth is measured as a function of temperature, if all points on a plot of viscosity versus temperature lay perfectly along a line and there were no scatter, it would be very easy to determine unequivocally the relationship between the variables. In reality, however, procedures for data analysis must be closely linked with statistical mathematics to account for random errors in measurement. Despite their importance, detailed description of statistical analysis is beyond the scope of this book; there are entire texts devoted to the subject. Rather than presenting methods for data analysis as such, the following sections discuss some of the ideas behind interpretation of experiments. Once the general approach is understood, the actual procedures involved can be obtained from the references listed at the end of this chapter.

As we shall see, interpreting experimental data requires a great deal of judgement and sometimes involves difficult decisions. Nowadays most scientists and engineers have access to computers or calculators equipped with software for data processing. These facilities are very convenient and have removed much of the tedium associated with statistical analysis. There is a danger, however, that software packages are applied without appreciation of inherent assumptions in the analysis or its mathematical limitations. Thus, the user cannot know how valuable or otherwise are the generated results.

As already mentioned in Section 3.1.4, standard statistical methods consider only random error, not systematic error. In practical terms, this means that most procedures for data processing are unsuitable when errors are due to poor instrument calibration, repetition of the same mistakes in measurement, or preconceived ideas about the expected result. All effort must be made to eliminate these types of error before treating the data. As also noted in Section 3.1.4, the reliability of results from statistical analysis improves if many readings are taken. No amount of sophisticated mathematical or other type of manipulation can make up for sparse, inaccurate data.

3.3.1 Trends

Consider the data plotted in Figure 3.1 representing consumption of glucose during batch culture of plant cells. If there were serious doubt about the trend of the data we could present the plot as a scatter of individual points without any lines drawn through them. Sometimes data are simply connected using line segments as shown in Figure 3.1(a); the problem with this representation is that it suggests that the ups and downs of glucose concentration are real. If, as with these data, we are assured there is a progressive downward trend in sugar concentration despite the occasional apparent increase, we should *smooth* the data by drawing a curve through the points as shown in Figure 3.1(b). Smoothing moderates the effects of experimental error. By drawing a particular curve we are indicating that, although the scatter of points is considerable, we believe the actual behaviour of the system is smooth

and continuous, and that all of the data without experimental error would lie on that line. Usually there is great flexibility as to where the smoothing curve is placed, and several questions arise. To which points should the curve pass closest? Should all the data points be included, or are some points clearly in error? It soon becomes apparent that many equally-acceptable curves can be drawn through the data.

Various techniques are available for smoothing. A smooth line can be drawn freehand or with French or flexible curves and other drafting equipment; this is called *hand smoothing*. Procedures for minimising bias during hand smoothing can be applied; some examples are discussed further in Chapter 11. The danger involved in smoothing manually: that we tend to smooth the expected response into the data, is well recognised. Another method is to use a computer software package; this is called *machine smoothing*. Computer routines, by smoothing data according to pre-programmed mathematical or statistical principles, eliminate the subjective element but are still capable of introducing bias into the results. For example, abrupt changes in the trend of data are generally not recognised by statistical analysis. The advantage of hand smoothing is that judgements about the significance of individual data points can be taken into account.

Choice of curve is critical if smoothed data are to be applied in subsequent analysis. The data of Figure 3.1 may be used to calculate the *rate* of glucose consumption as a function of time; procedures for this type of analysis are described further in Chapter 11. In rate analysis, different smoothing curves can lead to significantly different results. Because final

Figure 3.1 Glucose concentration during batch culture of plant cells; (a) data connected directly by line segments; (b) data represented by a smooth curve.

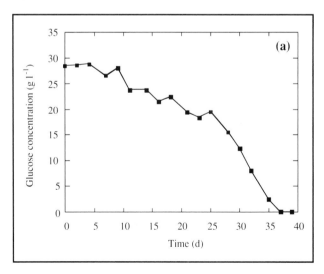
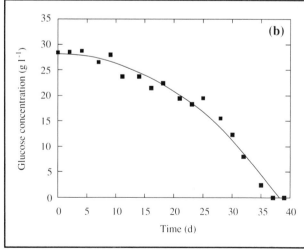

interpretation of the data depends on decisions made during smoothing, it is important to minimise any errors introduced. One obvious way of doing this is to take as many readings as possible; when smooth curves are drawn through too few points it is very difficult to justify the smoothing process.

3.3.2 Testing Mathematical Models

Most applications of data analysis involve correlating measured data with existing mathematical models. The model proposes some functional relationship between two or more variables; our primary objective is to compare the properties of the model with those of the experimental system.

As an example, consider Figure 3.2 which shows the results from experiments in which rates of heat production and oxygen consumption are measured for several microbial cultures [6]. Although there is considerable scatter in these data we could be led to believe that the relationship between rate of heat production and rate of oxygen consumption is linear, as indicated by the straight line in Figure 3.2. However there is an infinite number of ways to represent any set of data; how do

we know that this linear relationship is the best? For instance we might consider whether the data could be fitted by the curve shown in Figure 3.3. This non-linear, oscillating model seems to follow the data reasonably well; should we conclude that there is a more complex non-linear relationship between heat production and oxygen consumption?

Ultimately, we cannot know if a particular relationship holds between variables. This is because we can only test a selection of possible relationships and determine which *of them* fits closest to the data. We can determine which model, linear or oscillating, is the *better* representation of the data in Figure 3.2, but we can never conclude that the relationship between the variables is *actually* linear or oscillating. This fundamental limitation of data analysis has important consequences and must be accommodated in our approach. We must start off with a hypothesis about how the parameters are related and use data to determine whether this hypothesis is supported. A basic tenet in the philosophy of science is that it is only possible to *disprove* hypotheses by showing that experimental data do not conform to the model. The idea that the primary business of science is to falsify theories, not verify

Figure 3.2　Correlation between rate of heat evolution and rate of oxygen consumption for a variety of microbial fermentations. (○) *Escherichia coli*, glucose medium; (◑) *Candida intermedia*, glucose medium; (△) *C. intermedia*, molasses medium; (▽) *Bacillus subtilis*, glucose medium; (■) *B. subtilis*, molasses medium; (◓) *B. subtilis*, soybean-meal medium; (◈) *Aspergillus niger*, glucose medium; (●) *Asp. niger*, molasses medium. (From C.L. Cooney, D.I.C. Wang and R.I. Mateles, Measurement of heat evolution and correlation with oxygen consumption during microbial growth, *Biotechnol. Bioeng.* **11**, 269–281; Copyright © 1968. Reprinted by permission of John Wiley and Sons, Inc.)

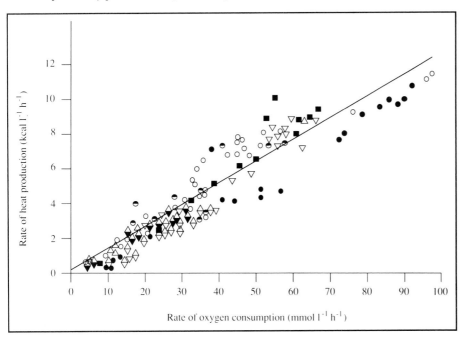

Figure 3.3 Alternative non-linear correlation for the data of Figure 3.2.

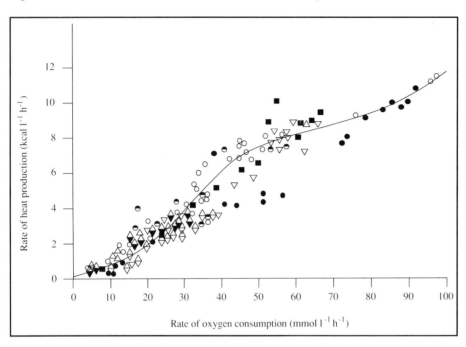

them, was developed this century by Austrian philosopher, Karl Popper. Popper's philosophical excursions into the meaning of scientific truth make extremely interesting reading, e.g. [7, 8]; his theories have direct application in analysis of measured data. We can never deduce with absolute certainty the physical relationships between variables using experiments. Language used to report the results of data analysis must reflect these limitations; particular models used to correlate data cannot be described as 'correct' or 'true' descriptions of the system, only 'satisfactory' or 'adequate' for our purposes and measurement precision.

3.3.3 Goodness of Fit: Least-Squares Analysis

Determining how well data conform to a particular model requires numerical procedures. Generally, these techniques rely on measurement of the deviations or residuals of each datum point from the curve or line representing the model being tested. For example, residuals after correlating cell plasmid content with growth rate using a linear model are shown by the dashed lines in Figure 3.4. A curve or line producing small residuals is considered a good fit of the data.

A popular technique for locating the line or curve which minimises the residuals is *least-squares analysis*. This statistical procedure is based on minimising the *sum of squares of the residuals*. There are several variations of the procedure: Legendre's

method minimises the sum-of-squares of residuals of the dependent variable; Gauss's and Laplace's methods minimise the sum of squares of weighted residuals where the weighting factors depend on the scatter of replicate data points. Each method gives different results; it should be remembered that the curve of 'best' fit is ultimately a matter of opinion. For example, by minimising the sum of squares of the residuals, least-squares analysis could produce a curve which does not pass close to particular data points known beforehand to be more accurate than the rest. Alternatively, we could choose to define the best fit as that which minimises the absolute values of the residuals, or the sum of the residuals raised to the fourth power. The decision to use the sum of squares is an arbitrary one; many alternative approaches are equally valid mathematically.

As well as minimising the residuals, other factors must be taken into account when correlating data. First, the curve used to fit the data should create approximately equal numbers of positive and negative residuals. As shown in Figure 3.4(a), when there are more positive than negative deviations from the points, even though the sum of the residuals is relatively small, the line representing the data cannot be considered a good fit. The fit is also poor when, as shown in Figure 3.4(b), all the positive residuals occur at low values of the independent variable while the negative residuals occur at high values. There should be no significant correlation of the residuals with either the dependent or independent variable. The best

Figure 3.4 Residuals in plasmid content after fitting a straight line to experimental data.

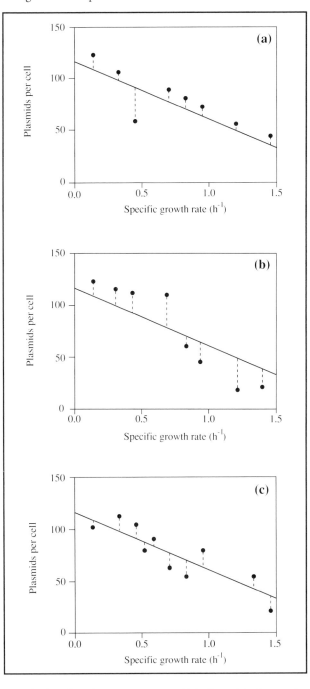

Some data sets contain one or more points which deviate substantially from predicted values, more than is expected from 'normal' random experimental error. These points known as 'outliers' have large residuals and, therefore, strongly influence regression methods using the sum-of-squares approach. It is usually inappropriate to eliminate outliers; they may be legitimate experimental results reflecting the true behaviour of the system and could be explained and fitted using an alternative model not yet considered. The best way to handle outliers is to analyse the data with and without the aberrant values to make sure their elimination does not influence discrimination between models. It must be emphasised that only one point at a time and only very rare data points, if any, should be eliminated from data sets.

Measuring individual residuals and applying least-squares analysis would be very useful in determining which of the two curves in Figures 3.2 and 3.3 fits the data more closely. However, as well as mathematical considerations, other factors can influence choice of model for experimental data. Consider again the data of Figures 3.2 and 3.3. Unless the fit obtained with the oscillatory model were very much improved compared with the linear model, we might prefer the straight-line correlation because it is simple, and because it conforms with what we know about microbial metabolism and the thermodynamics of respiration. It is difficult to find a credible theoretical justification for representing the relationship with an oscillating curve, so we could be persuaded to reject the non-linear model even though it fits the data reasonably well. Choosing between models on the basis of supposed mechanism requires a great deal of judgement. Since we cannot know for sure what the relationship is between the two parameters, choosing between models on the basis of supposed mechanism brings in an element of bias. This type of presumptive judgement is the reason why it is so difficult to overturn established scientific theories; even if data are available to support a new hypothesis there is a tendency to reject it because it does not agree with accepted theory. Nevertheless, if we wanted to fly in the face of convention and argue that an oscillatory relationship between rates of heat evolution and oxygen consumption is more reasonable than a straight-line relationship, we would undoubtedly have to support our claim with more evidence than the data shown in Figure 3.3.

3.3.4 Linear and Non-Linear Models

A straight line is represented by the equation:

$$y = A x + B.$$

$$(3.6)$$

straight-line fit is shown in Figure 3.4(c); the residuals are relatively small, well distributed in both positive and negative directions, and there is no relationship between the residuals and either variable.

B is the *intercept* of the straight line on the ordinate; A is the *slope*. A and B are also called the *coefficients*, *parameters* or *adjustable parameters* of Eq. (3.6). Once a straight line is drawn, A is found by taking any two points (x_1, y_1) and (x_2, y_2) on the line, and calculating:

$$A = \frac{(y_2 - y_1)}{(x_2 - x_1)}.$$

(3.7)

As indicated in Figure 3.5, (x_1, y_1) and (x_2, y_2) are points *on the line* through the data; they are not measured data values. Once A is known, B is calculated as:

$$B = y_1 - A x_1 \quad \text{or} \quad B = y_2 - A x_2.$$

(3.8)

Suppose we measure n pairs of values of two variables x and y, and a plot of the dependent variable y versus the independent variable x suggests a straight-line relationship. In testing correlation of the data with Eq. (3.6), changing the values of A and B will affect how well the model fits the data. Values of A and B giving the best straight line are determined by *linear regression* or *linear least-squares analysis*. This procedure is one of the most frequently used in data analysis; linear-regression routines are part of many computer packages and are available on hand-held calculators. Linear regression methods fit data by finding the straight line which minimises the sum of squares of the residuals. Details of the method can be found in statistics texts [5, 9, 10].

Because linear regression is so accessible, it can be readily applied without proper regard for its appropriateness or the assumptions incorporated in its method. Unless the following points are considered before using regression analysis, biased estimates of parameter values will be obtained.

(i) Least-squares analysis applies only to data containing random error.

(ii) The variables x and y must be independent.

(iii) Simple linear-regression methods are restricted to the special case of all uncertainty being associated with one variable. If the analysis uses a regression of y on x, then y should be the variable involving the largest errors. More complicated techniques are required to deal with errors in x and y simultaneously.

(iv) Simple linear-regression methods assume that each data point has equal significance. Modified procedures must be used if some points are considered more or less important than others, or if the line must pass through some specified point, e.g. the origin. It is also assumed that each point is equally precise, i.e. the standard deviation or random error associated with individual readings is the same. In experiments, the degree of fluctuation in the response variable often changes within the range of interest; for example, measurements may be more or less affected by instrument noise at the high or low end of the scale, so that data collected at the beginning of an experiment will have different errors compared with those measured at the end. Under these conditions, simple least-squares analysis is not appropriate.

(v) As already mentioned with respect to Figures 3.4(a) and 3.4(b), positive and negative residuals should be approximately evenly distributed, and residuals should be independent of both x and y variables.

Correlating data with straight lines is a relatively easy form of data analysis. When experimental data deviate markedly from a straight line, correlation using non-linear models is required. It is usually more difficult to decide which model to test and obtain parameter values when data do not follow linear relationships. As an example, consider growth of *Saccharomyces cerevisiae* yeast, which is expected to follow the non-linear model of Eq. (3.3). We could attempt to check whether measured cell-concentration data are consistent with Eq. (3.3) by plotting the values on linear graph paper as shown in Figure 3.6(a). The data appear to exhibit an exponential response typical of simple growth kinetics but it is not clear that an exponential model is appropriate. It is also difficult to ascertain some of the finer points of culture behaviour; for instance, whether the initial points represent a lag phase or whether exponential growth commenced immediately. Furthermore, the value of μ for this culture is not readily discernible from Figure 3.6(a).

Figure 3.5 Straight-line correlation for calculation of model parameters.

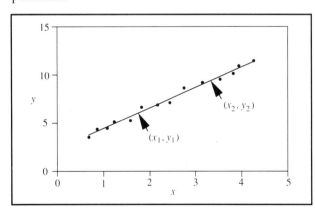

Figure 3.6 Growth curve for *Saccharomyces cerevisiae*
(a) data plotted directly on linear graph-paper;
(b) linearisation of growth data by plotting logarthims of cell concentration versus time.

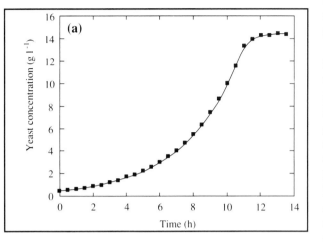

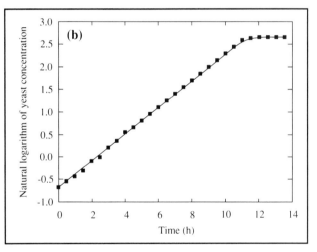

A convenient graphical approach to this problem is to transform the model equation into a linear form. Following the rules for logarithms outlined in Appendix D, taking the natural logarithm of both sides of Eq (3.3) gives:

$$\ln x = \ln x_0 + \mu t.$$

(3.9)

Eq. (3.9) indicates a linear relationship between $\ln x$ and t, with intercept $\ln x_0$ and slope μ. Accordingly, if Eq. (3.3) is a good model of yeast growth, a plot of the natural logarithm of cell concentration versus time should, during the growth phase, yield a straight line. Results of this linear transformation

are shown in Figure 3.6(b). All points before stationary phase appear to lie on a straight line suggesting absence of a lag phase; the value of μ is readily calculated from the slope of the line. Graphical linearisation has the advantage that gross deviations from the model are immediately evident upon visual inspection. Other non-linear relationships and suggested methods for yielding straight-line plots are given in Table 3.1.

Once data have been transformed to produce straight lines, it is tempting to apply linear least-squares analysis to determine the model parameters. We could enter the values of time and logarithms of cell concentration into a computer or calculator programmed for linear regression. This analysis would give us the straight line through the data which minimises the sum-of-squares of the residuals. Most users of linear regression choose this technique because they believe it will automatically give them an objective and unbiased analysis of their data. However, *application of linear least-squares analysis to linearised data can result in biased estimates of model parameters*. The reason is related to the assumption in least-squares analysis that each datum point has equal random error associated with it.

When data are linearised, the error structure is changed so that distribution of errors becomes biased [11, 12]. Although standard deviations for each raw datum point may be approximately constant over the range of measurement, when logarithms are calculated, the error associated with each datum point becomes dependent on its magnitude. This also happens when data are inverted, as in some of the transformations suggested in Table 3.1. Small errors in y lead to enormous errors in $1/y$ when y is small; for large values of y the same errors are barely noticeable in $1/y$. This effect is shown in Figure 3.7; the error bars represent a constant error in y of $\pm$ 0.05 y'. When the magnitude of errors after transformation is dependent on the value of the variable, simple least-squares analysis should not be used.

In such cases, modifications must be made to the analysis. One alternative is to apply *weighted least-squares techniques.* The usual way of doing this is to take replicate measurements of the variable, transform the data, calculate the standard deviations for the transformed variable, and then weight the values by $1/\sigma^2$. Correctly weighted linear regression often gives satisfactory parameter values for non-linear models; details of the procedures can be found elsewhere [2, 9].

Techniques of *non-linear regression* usually give better results than weighted linear regression. In non-linear regression, equations such as those in Table 3.1 are fitted directly to the data. However, determining an optimal set of parameters by non-linear regression can be difficult and reliability of the results more difficult to interpret. The most common non-linear

Table 3.1   Methods for plotting data as straight lines

$y = A x^n$	Plot y vs x on logarithmic coordinates.
$y = A + B x^2$	Plot y vs x^2.
$y = A + B x^n$	First obtain A as the intercept on a plot of y vs x; then plot $(y - A)$ vs x on logarithmic coordinates.
$y = B^x$	Plot y vs x on semi-logarithmic coordinates.
$y = A + (B/x)$	Plot y vs $^1/_x$.
$y = \dfrac{1}{A x + B}$	Plot $^1/_y$ vs x.
$y = \dfrac{x}{(A + B x)}$	Plot $^x/_y$ vs x, or $^1/_y$ vs $^1/_x$.
$y = 1 + (A x^2 + B)^{1/2}$	Plot $(y - 1)^2$ vs x^2.
$y = A + B x + C x^2$	Plot $\dfrac{(y - y_n)}{(x - x_n)}$ vs x, where (y_n, x_n) are the coordinates of any point on a smooth curve through the experimental points.
$y = \dfrac{x}{(A + B x)} + C$	Plot $\dfrac{(x - x_n)}{(y - y_n)}$ vs x, where (y_n, x_n) are the coordinates of any point on a smooth curve through the experimental points.

methods, such as the Gauss–Newton procedure widely available as computer software, are based on gradient, search or linearisation algorithms and use iterative solution techniques. More information about non-linear approaches to data analysis is available in other books [5, 9, 10, 13].

Figure 3.7   Error bars for $^1/_y$ vary in magnitude even though the error in y is the same for each datum point.

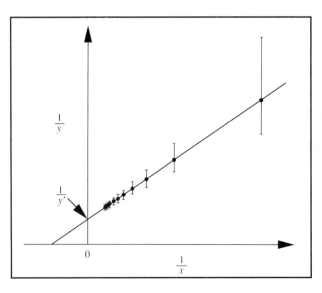

3.4 Graph Paper With Logarithmic Coordinates

Two frequently-occurring non-linear functions are the *power law*, $y = B x^A$, and the *exponential function*, $y = B e^{Ax}$. These relationships are often presented using graph paper with logarithmic coordinates. Before proceeding, it may be necessary for you to review the mathematical rules for logarithms outlined in Appendix D.

3.4.1 Log–Log Plots

When plotted on a linear scale, some data have the form of either 1, 2 or 3 in Figure 3.8(a), all of which are not straight lines. Note also that none of these curves intersects either axis except at the origin. If straight-line representation is required, we must transform the data by calculating logarithms; plots of log or ln x versus log or ln y yield straight lines as shown in Figure 3.8(b). The best straight line through the data can be estimated using suitable non-linear regression analysis as discussed in Section 3.3.4.

When there are many data points, calculating the logarithms of x and y can be very time-consuming. An alternative is to use log–log graph paper. The raw data, not logarithms, are plotted directly on log–log paper; the resulting graph is as if logarithms were calculated to base e. Graph paper with both axes scaled logarithmically is shown in Figure 3.9; each axis in this example covers two logarithmic cycles. On log–log plots

Figure 3.8 Equivalent curves on linear and log–log graph paper. (See text.)

Figure 3.9 Log–log plot.

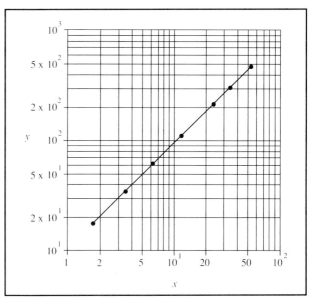

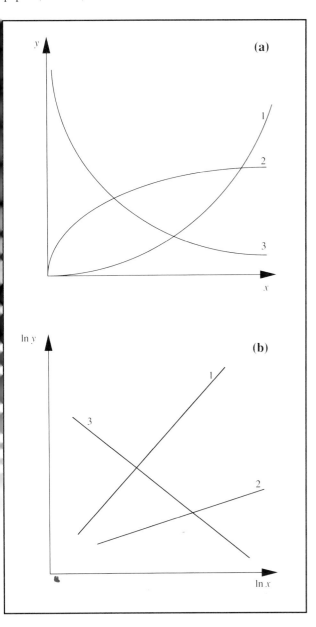

or

$$\ln y = \ln B + A \ln x.$$

(3.11)

Inspection of Eq. (3.10) shows that, if A is positive, $y = 0$ when $x = 0$. Therefore, a positive value of A corresponds to either curve 1 or 2 passing through the origin of Figure 3.8(a). If A is negative, when $x = 0$, y is infinite; therefore, negative A corresponds to curve 3 in Figure 3.8(a) which is asymptotic to both linear axes.

A and B are obtained from a straight line on log–log paper as follows. A can be calculated in two ways:

(i) If the log–log graph paper is drawn so that the ordinate and abscissa scales are the same, i.e. the distance measured with a ruler for a 10-fold change in the y variable is the same as for a 10-fold change in the x variable, A is the actual slope of the line. A is obtained by taking two points (x_1, y_1) and (x_2, y_2) on the line, and measuring the distances between y_2 and y_1 and x_2 and x_1 with a ruler:

$$A = \frac{(\text{distance between } y_2 \text{ and } y_1)}{(\text{distance between } x_2 \text{ and } x_1)}.$$

(3.12)

(ii) Alternatively, A is obtained by reading from the axes the coordinates of two points on the line, (x_1, y_1) and (x_2, y_2), and making the calculation:

the origin (0,0) can never be represented; this is because ln 0 (or $\log_{10} 0$) is not defined.

A straight line on log–log graph paper corresponds to the equation:

$$y = B x^A$$

(3.10)

$$A = \frac{(\ln y_2 - \ln y_1)}{(\ln x_2 - \ln x_1)} = \frac{\ln (y_2/y_1)}{\ln (x_2/x_1)} \; . \tag{3.13}$$

Note that all points x_1, y_1, x_2 and y_2 used in these calculations are points *on the line* through the data; they are not measured data values.

Once A is known, B is calculated from Eq. (3.11) as follows:

$$\ln B = \ln y_1 - A \ln x_1 \quad \text{or} \quad \ln B = \ln y_2 - A \ln x_2 \tag{3.14}$$

where $B = e^{(\ln B)}$. B can also be determined as the value of y when $x = 1$.

3.4.2 Semi-Log Plots

When plotted on linear-scale graph paper, some data show exponential rise or decay as illustrated in Figure 3.10(a). Curves 1 and 2 can be transformed into straight lines if log y or ln y is plotted against x, as shown in Figure 3.10(b).

An alternative to calculating logarithms is a semi-log plot, also known as a *linear-log plot*. As shown in Figure 3.11 raw data, not logarithms, are plotted directly on semi-log paper; the resulting graph is as if all logarithms were calculated to base e. Zero cannot be represented on the log-scale axis of semi-log

plots. In Figure 3.11, values of the dependent variable were fitted within one logarithmic cycle from 10 to 100; semi-log paper with multiple cycles is also available.

A straight line on semi-log paper corresponds to the equation:

$$y = B e^{Ax} \tag{3.15}$$

or

$$\ln y = \ln B + A x. \tag{3.16}$$

Values of A and B are obtained from the straight line as follows. If two points (x_1, y_1) and (x_2, y_2), are located on the line, A is given by:

$$A = \frac{(\ln y_2 - \ln y_1)}{(x_2 - x_1)} = \frac{\ln(y_2/y_1)}{(x_2 - x_1)} \; . \tag{3.17}$$

B is the value of y at $x = 0$, i.e. B is the intercept of the line at the ordinate. Alternatively, once A is known, B is determined as follows:

$$\ln B = \ln y_1 - A x_1 \quad \text{or} \quad \ln B = \ln y_2 - A x_2. \tag{3.18}$$

B is calculated as $e^{(\ln B)}$.

Figure 3.10 Equivalent curves on linear and semi-log graph paper. (See text.)

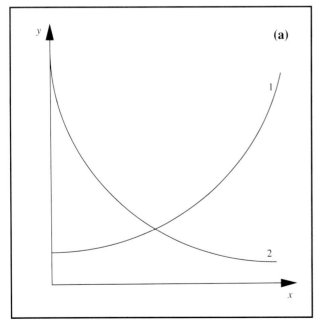
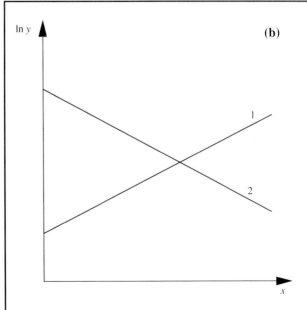

Figure 3.11   Semi-log plot.

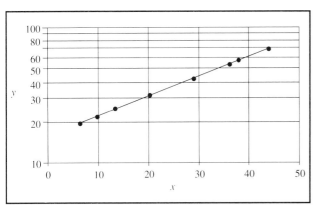

Example 3.2 Cell growth data

Data for cell concentration x versus time t are plotted on semi-log graph paper. Points $(t_1 = 0.5 \text{ h}, x_1 = 3.5 \text{ g l}^{-1})$ and $(t_2 = 15 \text{ h}, x_2 = 10.6 \text{ g l}^{-1})$ fall on a straight line passing through the data.

(a) Determine the equation relating x and t.
(b) What is the value of the specific growth rate for this culture?

Solution:

(a) A straight line on semi-log graph paper means that x and t are correlated with the equation $x = B e^{At}$. A and B are calculated from Eqs (3.17) and (3.18):

$$A = \frac{\ln 10.6 - \ln 3.5}{15 - 0.5} = 0.076$$

and

$$\ln B = \ln 10.6 - (0.076)(15) = 1.215$$

or

$$B = 3.37.$$

Therefore, the equation for cell concentration as a function of time is:

$$x = 3.37 e^{0.076\,t}.$$

This result should be checked, for example, by substituting $t_1 = 0.5$:

$$B e^{At_1} = 3.37 e^{(0.076)(0.5)} = 3.5 = x_1.$$

(b) After comparing the empirical equation with Eq. (3.3), the specific growth rate μ is 0.076 h^{-1}.

3.5 General Procedures For Plotting Data

Axes on plots must be labelled for the graph to carry any meaning. The units associated with all physical variables should be stated explicitly. If more than one curve is plotted on a single graph, each curve must be identified with a label.

It is good practice to indicate the precision of data on graphs using *error bars*. As an example, Table 3.2 lists values of monoclonal-antibody concentration as a function of medium flow rate during continuous culture of hybridoma cells in a stirred fermenter. The flow rate was measured to within ± 0.02 litres per day. Measurement of antibody concentration was more difficult and somewhat imprecise; these values are estimated to involve errors of ± 10 μg ml^{-1}. Errors associated with the data are indicated in Figure 3.12 using error bars to show the possible range of each variable.

Table 3.2 Antibody concentration during continuous culture of hybridoma cells

Flow rate ($l\,d^{-1}$)	Antibody concentration (μg ml^{-1})
0.33	75.9
0.40	58.4
0.52	40.5
0.62	28.9
0.78	22.0
1.05	11.5

3.6 Process Flow Diagrams

This chapter is concerned with ways of presenting and analysing data. Because of the complexity of large-scale manufacturing processes, communicating information about these systems requires special methods. *Flow diagrams* or *flow sheets* are simplified pictorial representations of processes and are used to present relevant process information and data. Flow sheets vary in complexity from simple block diagrams to highly complex schematic drawings showing main and auxiliary process equipment such as pipes, valves, pumps and by-pass loops.

Figure 3.13 is a simplified process flow diagram showing the major operations for production of the antibiotic, bacitracin. This qualitative flow sheet indicates the flow of materials, the sequence of process operations, and the principal equipment in use. When flow diagrams are applied in calculations, the operating conditions, masses and concentrations of material handled by the process are also specified. An example is Figure 3.14, which represents recovery operations for 2,3-butanediol produced commercially by fermentation of whole wheat mash. The quantities and compositions of

Figure 3.12 Error bars for antibody concentration and medium flow rate measured during continuous culture of hybridoma cells.

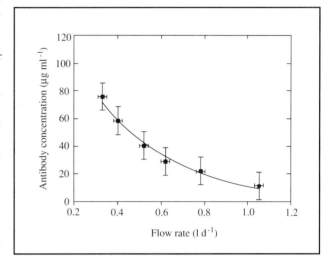

streams undergoing processes such as distillation, evaporation, screening and drying are shown to allow calculation of product yields and energy costs.

Detailed engineering flowsheets such as Figure 3.15 are useful for plant construction work and trouble-shooting because they show all piping, valves, drains, pumps and safety equipment. Standard symbols are adopted to convey the information as concisely as possible. Figure 3.15 represents a pilot-scale fermenter with separate vessels for antifoam, acid and alkali. All air, medium inlet and harvest lines are shown, as are the steam and condensate-drainage lines for *in situ* steam sterilisation of the entire apparatus.

In addition to those illustrated here, other specialised types of flow diagram are used to specify instrumentation for process control networks in large-scale processing plants, and for utilities such as steam, water, fuel and air supplies. We will not be applying complicated or detailed diagrams such as Figure 3.15 in our analysis of bioprocessing; their use is beyond the scope of this book. However, simplified versions of flow diagrams are extremely useful, especially for material- and energy-balance calculations; we will be applying block diagram flow sheets in Chapters 4–6 for this purpose. You should become familiar with flow diagrams for showing data and other process information.

3.7 Summary of Chapter 3

This chapter covers a range of topics related to data presentation and analysis. After studying Chapter 3 you should:

Figure 3.13 Process flowsheet showing the major operations for production of bacitracin. (From G.C. Inskeep, R.E. Bennett, J.F. Dudley and M.W. Shepard, 1951, Bacitracin: product of biochemical engineering, *Ind. Eng. Chem.* **43**, 1488–1498.)

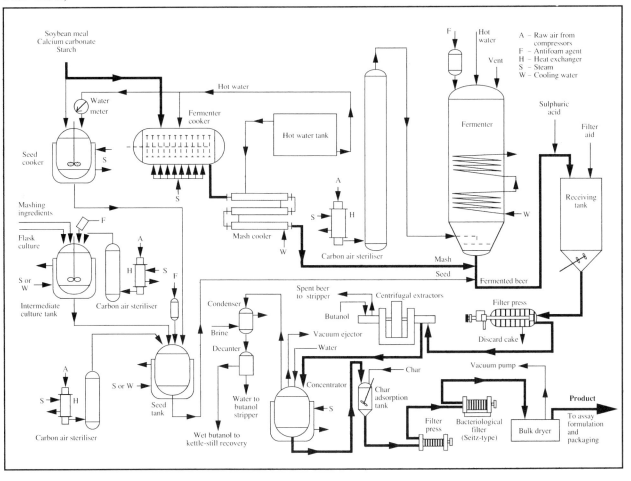

(i)   understand use of *significant figures*;

(ii)   know the types of *error* which affect accuracy of experimental data and which errors can be accounted for with statistical techniques;

(iii)   be able to combine errors in simple calculations;

(iv)   be able to report the results of replicate measurements in terms of the *mean* and *standard deviation*;

(v)   understand the fundamental limitations associated with use of experimental data for testing mathematical models;

(vi)   be familiar with *least-squares analysis* and its assumptions;

(vii)   be able to analyse linear plots and determine model parameters;

(viii)   understand how simple non-linear models can be linearised to obtain straight-line plots;

(ix)   be able to use *log* and *semi-log graph paper* with ease; and

(x)   be familiar with simple *process flow diagrams*.

Problems

3.1 Combination of errors

The oxygen mass-transfer coefficient in fermentation vessels is determined from experimental measurements using the formula:

$$k_L a = \frac{OTR}{C_{AL}^* - C_{AL}}$$

where $k_L a$ is the mass-transfer coefficient, OTR is the oxygen transfer rate, C_{AL}^* is the solubility of oxygen in the fermentation

Figure 3.14 Quantitative flowsheet for the downstream processing of 2,3-butanediol based on fermentation of 1000 bushels wheat per day by *Aerobacillus polymyxa.* (From J.A. Wheat, J.D. Leslie, R.V. Tomkins, H.E. Mitton, D.S. Scott and G.A. Ledingham, 1948, Production and properties of 2,3-butanediol, XXVIII: Pilot plant recovery of *levo*-2,3-butanediol from whole wheat mashes fermented by *Aerobacillus polymyxa, Can. J. Res.* **26F**, 469–496.)

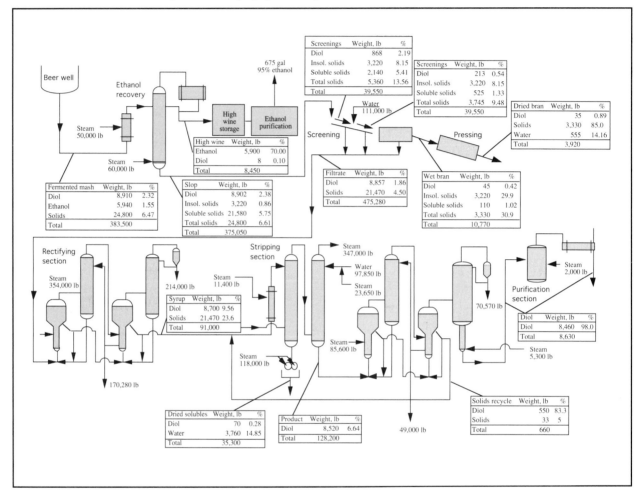

broth, and C_{AL} is the dissolved-oxygen concentration. C_{AL}^* is estimated to be 0.25 mol m^{-3} with an uncertainty of $\pm 4\%$; C_{AL} is measured as 0.183 mol m^{-3} $\pm 4\%$. If the OTR is 0.011 mol m^{-3} s^{-1} $\pm 5\%$, what is the uncertainty associated with $k_L a$?

3.2 Mean and standard deviation

The pH for maximum activity of β-amylase enzyme is measured four times. The results are: 5.15, 5.45, 5.50 and 5.35.

(a) What is the best estimate of optimal pH?
(b) How accurate is this value?

(c) If the experiment were stopped after only the first two measurements were taken, what would be the result and its accuracy?
(d) If an additional four measurements were made with the same results as those above, how would this change the outcome of the experiment?

3.3 Linear and non-linear models

Determine the equation for y as a function of x using the following information. Reference to coordinate point (x, y) means that x is the abscissa value and y is the ordinate value.

Figure 3.15 Detailed equipment diagram for pilot-plant fermentation system. (Reproduced with permission from LH Engineering Ltd, a member of the Inceltech Group of companies. Copyright 1983.)

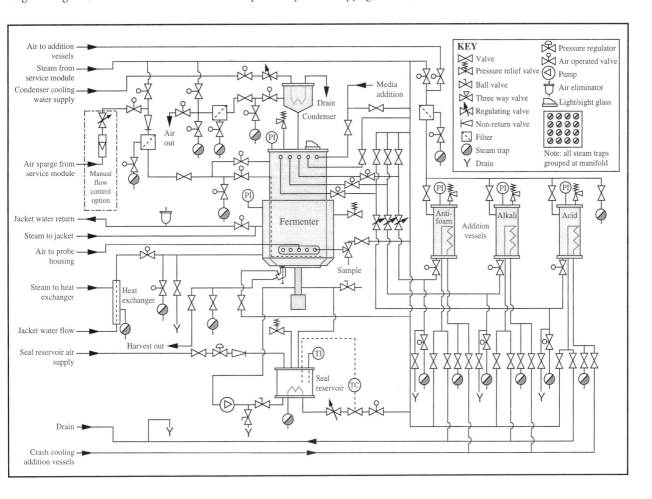

(a) A plot of y versus x on linear graph paper gives a straight line passing through the points $(1, 10)$ and $(8, 0.5)$.
(b) A plot of y versus $x^{1/2}$ on linear graph paper gives a straight line passing through points $(3.2, 14.5)$ and $(8.9, 38.5)$.
(c) A plot of $1/y$ versus x^2 on linear graph paper gives a straight line passing through points $(5, 6)$ and $(1, 3)$.
(d) A plot of y versus x on log-log paper gives a straight line passing through $(0.5, 25)$ and $(550, 2600)$.
(e) A plot of y versus x on semi-log paper gives a straight line passing through $(1.5, 2.5)$ and $(10, 0.036)$.

3.4 Linear curve fitting

Sucrose concentration in a fermentation broth is measured using HPLC. Chromatogram peak areas are measured for five standard sucrose solutions to calibrate the instrument. Measurements are performed in triplicate with results as follows:

Sucrose concentration $(g\,l^{-1})$	Peak area
6.0	55.55, 57.01, 57.95
12.0	110.66, 114.76, 113.05
18.0	168.90, 169.44, 173.55
24.0	233.66, 233.89, 230.67
30.0	300.45, 304.56, 301.11

(a) Determine the mean peak areas for each sucrose concentration, and the standard deviations.
(b) Plot the data. Plot the standard deviations as error bars.
(c) Find an equation for sucrose concentration as a function of peak area.
(d) A sample containing sucrose gives a peak area of 209.86. What is the sucrose concentration?

3.5 Non-linear model: calculation of parameters

The mutation rate of *E. coli* increases with temperature. The following data were obtained by measuring the frequency of mutation of his^- cells to produce his^+ colonies:

Temperature (°C)	Relative mutation frequency, α
15	4.4×10^{-15}
20	2.0×10^{-14}
25	8.6×10^{-14}
30	3.5×10^{-13}
35	1.4×10^{-12}

Mutation frequency is expected to obey an Arrhenius-type equation:

$$\alpha = \alpha_0\, e^{-E/RT}$$

where α_0 is the mutation rate parameter, E is activation energy, R is the universal gas constant and T is absolute temperature.

(a) Test the model using an appropriate plot on log graph paper.
(b) What is the activation energy for the mutation reaction?
(c) What is the value of α_0?

3.6 Linear regression: distribution of residuals

Medium conductivity is sometimes used to monitor cell growth during batch culture. Experiments are carried out to relate decrease in conductivity to increase in plant-cell biomass during culture of *Catharanthus roseus* in an airlift fermenter. The results are tabulated below.

Decrease in medium conductivity $(mS\,cm^{-1})$	Increase in biomass concentration $(g\,l^{-1})$
0	0
0.12	2.4
0.31	2.0
0.41	2.8
0.82	4.5
1.03	5.1
1.40	5.8
1.91	6.0
2.11	6.2
2.42	6.2
2.44	6.2
2.74	6.6
2.91	6.0
3.53	7.0
4.39	9.8
5.21	14.0
5.24	12.6
5.55	14.6

(a) Plot the points on linear coordinates and obtain an equation for the 'best' straight line through the data using linear least-squares analysis.
(b) Plot the residuals in biomass increase versus conductivity change after comparing the model equation with the

actual data. What do you conclude about the goodness of fit for the straight line?

3.7 Discriminating between rival models

In bioreactors where the liquid contents are mixed by sparging air into the vessel, the liquid velocity is dependent directly on the gas velocity. The following results were obtained from experiments with 0.15 M NaCl solution.

Gas superficial velocity, u_G (m s^{-1})	Liquid superficial velocity, u_L (m s^{-1})
0.02	0.060
0.03	0.066
0.04	0.071
0.05	0.084
0.06	0.085
0.07	0.086
0.08	0.091
0.09	0.095
0.095	0.095

(a) How well are these data fitted with a linear model? Determine the equation for the 'best' straight line relating gas and liquid velocities.

(b) It has been reported in the literature that fluid velocities in air-driven reactors are related using the power equation:

$$u_L = \alpha\, u_G^{\,v}$$

where α and v are constants. Is this model an appropriate description of the experimental data?

(c) Which equation, linear or non-linear, is the better model for the reactor system?

3.8 Non-linear model: calculation of parameters

When nutrient medium is autoclaved, the number of viable microorganisms decreases with time spent in the autoclave. An experiment is conducted to measure the number of viable cells N in a bottle of glucose solution after various sterilisation times t. Triplicate measurements are taken of cell number; the mean and standard deviation of each measurement are listed in the following table.

t (min)	Mean N	Standard deviation of N
5	3.6×10^3	0.20×10^3
10	6.3×10^2	0.40×10^2
15	1.07×10^2	0.09×10^2
20	1.8×10^1	0.12×10^1
30	< 1	–

From what is known about thermal death kinetics for microorganisms, it is expected that the relationship between N and t is of the form:

$$N = N_0\, e^{-k_d t}$$

where k_d is the specific death constant and N_0 is the number of viable cells present before autoclaving begins.

(a) Plot the results on suitable graph paper to obtain a straight line through the data.
(b) Plot the standard deviations on the graph as error bars.
(c) What are the values of k_d and N_0?
(d) What are the units and dimensions of k_d and N_0?

References

1. Massey, B.S. (1986) *Measures in Science and Engineering: Their Expression, Relation and Interpretation*, Ellis Horwood, Chichester.
2. Baird, D.C. (1988) *Experimentation: An Introduction to Measurement Theory and Experiment Design*, 2nd edn, Prentice-Hall, New Jersey.
3. Barry, B.A. (1978) *Errors in Practical Measurement in Science, Engineering, and Technology*, John Wiley, New York.
4. Youden, W.J. (1962) Systematic errors in physical constants. *Technometrics* 4, 111–123.
5. Walpole, R.E. and R.H. Myers (1972) *Probability and Statistics For Engineers and Scientists*, Macmillan, New York.
6. Cooney, C.L., D.I.C. Wang and R.I. Mateles (1968) Measurement of heat evolution and correlation with oxygen consumption during microbial growth. *Biotechnol. Bioeng.* 11, 269–281.
7. Popper, K.R. (1972) *The Logic of Scientific Discovery*, Hutchinson, London.
8. Popper, K.R. (1972) *Conjectures and Refutations: The Growth of Scientific Knowledge*, Routledge and Kegan Paul, London.

9. Himmelblau, D.M. (1970) *Process Analysis by Statistical Methods*, John Wiley, New York.
10. Draper, N.R. and H. Smith (1981) *Applied Regression Analysis*, 2nd edn, John Wiley, New York.
11. Sagnella, G.A. (1985) Model fitting, parameter estimation, linear and non-linear regression. *Trends Biochem. Sci.* **10**, 100–103.
12. Dowd, J.E. and D.S. Riggs (1965) A comparison of estimates of Michaelis–Menten kinetic constants from various linear transformations. *J. Biol. Chem.* **240**, 863–869.
13. Magar, M.E. (1972) *Data Analysis in Biochemistry and Biophysics*, Academic Press, New York.

Suggestions for Further Reading

Measurement and Error (see also refs 1–4)

Barford, N.C. (1985) *Experimental Measurements: Precision, Error and Truth*, 2nd edn, John Wiley, Chichester.
Lyon, A.J. (1970) *Dealing With Data*, Pergamon Press, Oxford.
Mickley, H.S., T.K. Sherwood and C.E. Reed (1957) *Applied Mathematics in Chemical Engineering*, 2nd edn, Chapter 2, McGraw-Hill, New York.

Statistical Analysis of Data (see also refs 5, 9–11)

Garfinkel, D. and K.A. Fegley (1984) Fitting physiological models to data. *Am. J. Physiol.* **246**, R641–R650.
Mannervik, B. (1982) Regression analysis, experimental error, and statistical criteria in the design and analysis of experiments for discrimination between rival kinetic models. *Meth. Enzymol.* **87**, 370–390.

Graphing Techniques

Felder, R.M. and R.W. Rousseau (1978) *Elementary Principles of Chemical Processes*, Chapter 4, John Wiley, New York.
Mickley, H.S., T.K. Sherwood and C.E. Reed (1957) *Applied Mathematics in Chemical Engineering*, 2nd edn, Chapter 1, McGraw-Hill, New York.

Process Flow Diagrams

Vilbrandt, F.C and C.E. Dryden (1959) *Chemical Engineering Plant Design*, 4th edn, Chapters 3 and 9, McGraw-Hill, New York.

Part 2 Material and Energy Balances

4

Material Balances

One of the simplest concepts in process engineering is the material or mass balance. Because mass in biological systems is conserved at all times, the law of conservation of mass provides the theoretical framework for material balances.

In steady-state material balances, masses entering a process are summed up and compared with the total mass leaving the system; the term 'balance' implies that masses entering and leaving should be equal. Essentially, material balances are accounting procedures: total mass entering must be accounted for at the end of the process, even if it undergoes heating, mixing, drying, fermentation or any other operation (except nuclear reaction) within the system. Usually it is not feasible to measure the masses and compositions of all streams entering and leaving a system; unknown quantities can be calculated using mass-balance principles. Mass-balance problems have a constant theme: given the masses of some input and output streams, calculate the masses of others.

Mass balances provide a very powerful tool in engineering analysis. Many complex situations are simplified by looking at the movement of mass and equating what comes out to what goes in. Questions such as: what is the concentration of carbon dioxide in the fermenter off-gas? what fraction of the substrate consumed is not converted into products? how much reactant is needed to produce *x* grams of product? how much oxygen must be provided for this fermentation to proceed? can be answered using mass balances. This chapter explains how the law of conservation of mass is applied to atoms, molecular species and total mass, and sets up formal techniques for solving material-balance problems with and without reaction. Aspects of metabolic stoichiometry are also discussed for calculation of nutrient and oxygen requirements during fermentation processes.

4.1 Thermodynamic Preliminaries

Thermodynamics is a fundamental branch of science dealing with the properties of matter. Thermodynamic principles are useful in setting up material balances; some terms borrowed from thermodynamics are defined below.

4.1.1 System and Process

In thermodynamics, a *system* consists of any matter identified for investigation. As indicated in Figure 4.1, the system is set apart from the *surroundings*, which are the remainder of the *universe*, by a *system boundary*. The system boundary may be

real and tangible, such as the walls of a beaker or fermenter, or imaginary. If the boundary does not allow mass to pass from system to surroundings and vice versa, the system is a *closed* system with constant mass. Conversely, a system able to exchange mass with its surroundings is an *open* system.

A *process* causes changes in the system or surroundings. Several terms are commonly used to describe processes.

(i) A *batch process* operates in a closed system. All materials are added to the system at the start of the process; the system is then closed and products removed only when the process is complete.

(ii) A *semi-batch process* allows either input or output of mass, but not both.

Figure 4.1 Thermodynamic system.

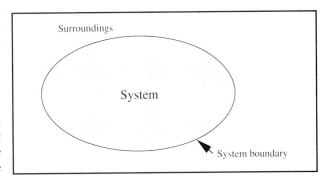

(iii) A *fed-batch process* allows input of material to the system but not output.

(iv) A *continuous process* allows matter to flow in and out of the system. If rates of mass input and output are equal, continuous processes can be operated indefinitely.

4.1.2 Steady State and Equilibrium

If all properties of a system, such as temperature, pressure, concentration, volume, mass, etc. do not vary with time, the process is said to be at *steady state*. Thus, if we monitor any variable of a steady-state system, its value will be unchanging with time.

According to this definition of steady state, batch, fed-batch and semi-batch processes cannot operate under steady-state conditions. Mass of the system is either increasing or decreasing with time during fed-batch and semi-batch processes; in batch processes, even though the total mass is constant, changes occurring inside the system cause the system properties to vary with time. Such processes are called *transient* or *unsteady-state* processes. On the other hand, continuous processes may be either steady state or transient. It is usual to run continuous processes as close to steady state as possible; however, unsteady-state conditions will exist during start-up and for some time after any change in operating conditions.

Steady state is an important and useful concept in engineering analysis. However, it is often confused with another thermodynamic term, *equilibrium*. A system at equilibrium is one in which all opposing forces are exactly counter-balanced so that the properties of the system do not change with time. From experience we know that systems tend to approach an equilibrium condition when they are isolated from their surroundings. At equilibrium there is no net change in either the system or the universe. Equilibrium implies that there is no net driving force for change; the energy of the system is at a minimum and, in rough terms, the system is 'static', 'unmoving' or 'inert'. For example, when liquid and vapour are in equilibrium in a closed vessel, although there may be constant exchange of molecules between the phases, there is no net change in either the system or the surroundings.

To convert raw materials into useful products there must be an overall change in the universe. Because systems at equilibrium produce no net change, equilibrium is of little value in processing operations. The best strategy is to avoid equilibrium by continuously disturbing the system in such a way that raw material will always be undergoing transformation into the desired product. In continuous processes at steady state, mass is constantly exchanged with the surroundings; this disturbance drives the system away from equilibrium so that a net

change in both the system and the universe can occur. Large-scale equilibrium does not often occur in engineering systems; steady states are more common.

4.2 Law of Conservation of Mass

Mass is conserved in ordinary chemical and physical processes. Consider the system of Figure 4.2 operating as a continuous process with input and output streams containing glucose. The mass flow rate of glucose into the system is $\hat{M}_i$ kg h^{-1}; the mass flow rate out is $\hat{M}_o$ kg h^{-1}. If $\hat{M}_i$ and $\hat{M}_o$ are different, there are four possible explanations:

(i) measurements of $\hat{M}_i$ and $\hat{M}_o$ are wrong;

(ii) the system has a leak allowing glucose to enter or escape undetected;

(iii) glucose is consumed or generated by chemical reaction within the system; or

(iv) glucose accumulates within the system.

If we assume that the measurements are correct and there are no leaks, the difference between $\hat{M}_i$ and $\hat{M}_o$ must be due to consumption or generation by reaction, and/or accumulation. A mass balance for the system can be written in a general way to account for these possibilities:

$$\begin{Bmatrix} \text{mass in} \\ \text{through} \\ \text{system} \\ \text{boundaries} \end{Bmatrix} - \begin{Bmatrix} \text{mass out} \\ \text{through} \\ \text{system} \\ \text{boundaries} \end{Bmatrix} + \begin{Bmatrix} \text{mass} \\ \text{generated} \\ \text{within} \\ \text{system} \end{Bmatrix} - \begin{Bmatrix} \text{mass} \\ \text{consumed} \\ \text{within} \\ \text{system} \end{Bmatrix} = \begin{Bmatrix} \text{mass} \\ \text{accumulated} \\ \text{within} \\ \text{system} \end{Bmatrix}.$$

(4.1)

The accumulation term in the above equation can be either positive or negative; negative accumulation represents depletion of pre-existing reserves. Eq. (4.1) is known as the *general mass-balance equation*. The mass referred to in the equation can be total mass, mass of a particular molecular or atomic species, or biomass. Use of Eq. (4.1) is illustrated in Example 4.1.

Figure 4.2 Flow sheet for a mass balance on glucose.

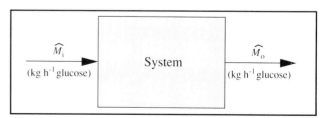

Example 4.1 General mass-balance equation

A continuous process is set up for treatment of wastewater. Each day, 10^5 kg cellulose and 10^3 kg bacteria enter in the feed stream, while 10^4 kg cellulose and 1.5×10^4 kg bacteria leave in the effluent. The rate of cellulose digestion by the bacteria is 7×10^4 kg d^{-1}. The rate of bacterial growth is 2×10^4 kg d^{-1}; the rate of cell death by lysis is 5×10^2 kg d^{-1}. Write balances for cellulose and bacteria in the system.

Solution:
Cellulose is not generated by the process, only consumed. Using a basis of 1 day, the cellulose balance in kg from Eq. (4.1) is:

$$(10^5 - 10^4 + 0 - 7 \times 10^4) = \text{accumulation}.$$

Therefore, 2×10^4 kg cellulose accumulates in the system each day.
  Performing the same balance for bacteria:

$$(10^3 - 1.5 \times 10^4 + 2 \times 10^4 - 5 \times 10^2) = \text{accumulation}.$$

Therefore, 5.5×10^3 kg bacterial cells accumulate in the system each day.

4.2.1 Types of Material Balance

The general mass-balance equation (4.1) can be applied with equal ease to two different types of mass-balance problem, depending on the data provided. For continuous processes it is usual to collect information about the process referring to a particular instant in time. Amounts of mass entering and leaving the system are specified using flow rates, e.g. molasses enters the system at a rate of 50 lb h^{-1}; at the same instant in time, fermentation broth leaves at a rate of 20 lb h^{-1}. These two quantities can be used directly in Eq. (4.1) as the input and output terms. A mass balance based on rates is called a *differential balance*.

An alternative approach is required for batch and semi-batch processes. Information about these systems is usually collected over a period of time rather than at a particular instant. For example: 100 kg substrate is added to the reactor; after 3 days' incubation, 45 kg product is recovered. Each term of the mass-balance equation in this case is a quantity of mass, not a rate. This type of balance is called an *integral balance*.

In this chapter, we will be using differential balances for continuous systems operating at steady state, and integral balances for batch and semi-batch systems between initial and final states. Calculation procedures for the two types of material balance are very similar.

4.2.2 Simplification of the General Mass-Balance Equation

Eq. (4.1) can be simplified in certain situations. If a continuous process is at steady state, the accumulation term on the right-hand side must be zero. This follows from the definition of steady state: because all properties of the system, including its mass, must be unchanging with time, a system at steady state cannot accumulate mass. Under these conditions, Eq. (4.1) becomes:

$$\text{mass in} + \text{mass generated} = \text{mass out} + \text{mass consumed}. \tag{4.2}$$

Eq. (4.2) is called the *general steady-state mass-balance equation*. Eq. (4.2) also applies over the entire duration of batch and fed-batch processes; 'mass out' in this case is the total mass harvested from the system so that at the end of the process there is no accumulation.

If reaction does not occur in the system, or if the mass balance is applied to a substance that is neither a reactant nor product of reaction, the generation and consumption terms in Eqs (4.1) and (4.2) are zero. Because total mass can be neither created nor destroyed except in nuclear reaction, generation and consumption terms must also be zero in balances applied to total mass. Similarly, generation and consumption of atomic species such as C, N, O, etc. cannot occur in normal chemical reaction. Therefore at steady state, for balances on total mass or atomic species or when reaction does not occur, Eq. (4.2) can be further simplified to:

$$\text{mass in} = \text{mass out}. \tag{4.3}$$

Table 4.1 summarises the types of material balance for which direct application of Eq. (4.3) is valid. Because total number of moles does not balance in systems with reaction, we will carry out all material balances using mass.

Table 4.1 Application of the simplified mass balance Eq. (4.3)

Material	At steady state, does mass in = mass out?	
	Without reaction	With reaction
Total mass	yes	yes
Total number of moles	yes	no
Mass of a molecular species	yes	no
Number of moles of a molecular species	yes	no
Mass of an atomic species	yes	yes
Number of moles of an atomic species	yes	yes

4.3 Procedure For Material-Balance Calculations

The first step in material-balance calculations is to understand the problem. Certain information is available about a process; the task is to calculate unknown quantities. Because it is sometimes difficult to sort through all the details provided, it is best to use standard procedures to translate process information into a form that can be used in calculations.

Material balances should be carried out in an organised manner; this makes the solution easy to follow, check, or use by others. In this chapter, a formalised series of steps is followed for each mass-balance problem. For easier problems these procedures may seem long-winded and unnecessary; however a standard method is helpful when you are first learning mass-balance techniques. The same procedures are used in the next chapter as a basis for energy balances.

These points are essential.

(i) *Draw a clear process flow diagram showing all relevant information.* A simple box diagram showing all streams entering or leaving the system allows information about a process to be organised and summarised in a convenient way. All given quantitative information should be shown on the diagram. Note that the variables of interest in material balances are masses, mass flow rates and mass compositions; if information about particular streams is given using volume or molar quantities, mass flow rates and compositions should be calculated before labelling the flow sheet.

(ii) *Select a set of units and state it clearly.* Calculations are easier when all quantities are expressed using consistent units. Units must also be indicated for all variables shown on process diagrams.

(iii) *Select a basis for the calculation and state it clearly.* In approaching mass-balance problems it is helpful to focus on a specific quantity of material entering or leaving the system. For continuous processes at steady state we usually base the calculation on the amount of material entering or leaving the system within a specified period of time. For batch or semi-batch processes, it is convenient to use either the total amount of material fed to the system or the amount withdrawn at the end. Selection of a basis for calculation makes it easier to visualise the problem; the way this works will become apparent in the worked examples of the next section.

(iv) *State all assumptions applied to the problem.* To solve problems in this and the following chapters, you will need to apply some 'engineering' judgement. Real-life situations are complex, and there will be times when one or more assumptions are required before you can proceed with calculations. To give you experience with this, problems posed in this text may not give you all the necessary information. The details omitted can be assumed, provided your assumptions are reasonable. Engineers make assumptions all the time; knowing when an assumption is permissible and what constitutes a reasonable assumption is one of the marks of a skilled engineer. When you make assumptions about a problem it is vitally important that you state them exactly. Other scientists looking through your calculations need to know the conditions under which your results are applicable; they will also want to decide whether your assumptions are acceptable or whether they can be improved.

In this chapter, differential mass balances on continuous processes are performed with the understanding that the system is at steady state; we can assume that mass flow rates and compositions do not change with time and the accumulation term of Eq. (4.1) is zero. If steady state does not prevail in continuous processes, information about the rate of accumulation would be required for solution of mass-balance problems. This is discussed further in Chapter 6.

Another assumption we must make in mass-balance problems is that the system under investigation does not leak. In totalling up all the masses entering and leaving the system, we must be sure that all streams are taken into account. When analysing real systems it is always a good idea to check for leaks before carrying out mass balances.

(v) *Identify which components of the system, if any, are involved in reaction.* This is necessary for determining which mass-balance equation (4.2) or (4.3), is appropriate. The

Example 4.2 Setting up a flow sheet

Humid air enriched with oxygen is prepared for a gluconic acid fermentation. The air is prepared in a special humidifying chamber. 1.5 l h^{-1} liquid water enters the chamber at the same time as dry air and 15 gmol min^{-1} dry oxygen gas. All the water is evaporated. The outflowing gas is found to contain 1% (w/w) water. Draw and label the flow sheet for this process.

Solution:

Let us choose units of g and min for this process; the information provided is first converted to mass flow rates in these units. The density of water is taken to be 10^3 g l^{-1}; therefore:

$$1.5 \, \text{l h}^{-1} = \frac{1.5 \, \text{l}}{\text{h}} \left(\frac{10^3 \, \text{g}}{\text{l}} \right) \left| \frac{1 \, \text{h}}{60 \, \text{min}} \right| = 25 \, \text{g min}^{-1}.$$

As the molecular weight of O$_2$ is 32:

$$15 \, \text{gmol min}^{-1} = \frac{15 \, \text{gmol}}{\text{min}} \cdot \left| \frac{32 \, \text{g}}{1 \, \text{gmol}} \right| = 480 \, \text{g min}^{-1}.$$

Figure 4E2.1   Flowsheet for oxygen enrichment and humidification of air.

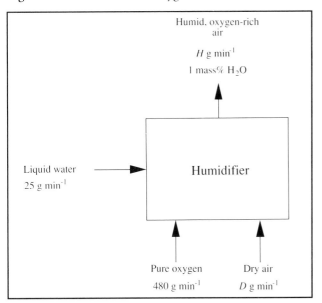

Unknown flow rates are represented with symbols. As shown in Figure 4E2.1, the flow rate of dry air is denoted D g min^{-1} and the flow rate of humid, oxygen-rich air is H g min^{-1}. The water content in the humid air is shown as 1 mass%.

simpler Eq. (4.3) can be applied to molecular species which are neither reactants nor products of reaction.

4.4 Material-Balance Worked Examples

Procedures for setting out mass-balance calculations are out-lined in this section. Although not the only way to attack these problems, the method shown will assist your problem-solving efforts by formalising the mathematical approach. Mass-balance calculations are divided into four steps: *assemble, analyse, calculate* and *finalise*. Differential and integral mass balances with and without reaction are illustrated below.

Example 4.3 Continuous filtration

A fermentation slurry containing *Streptomyces kanamyceticus* cells is filtered using a continuous rotary vacuum filter. 120 kg h^{-1} slurry is fed to the filter; 1 kg slurry contains 60 g cell solids. To improve filtration rates, particles of diatomaceous-earth filter aid are added at a rate of 10 kg h^{-1}. The concentration of kanamycin in the slurry is 0.05% by weight. Liquid filtrate is collected at a rate of 112 kg h^{-1}; the concentration of kanamycin in the filtrate is 0.045% (w/w). Filter cake containing cells and filter aid is continuously removed from the filter cloth.

(a) What percentage liquid is the filter cake?
(b) If the concentration of kanamycin in the filter-cake liquid is the same as in the filtrate, how much kanamycin is absorbed per kg filter aid?

Solution:
1. *Assemble*
    (i) *Draw the flowsheet showing all data with units.*
    This is shown in Figure 4E3.1.

Figure 4E3.1 Flowsheet for continuous filtration.

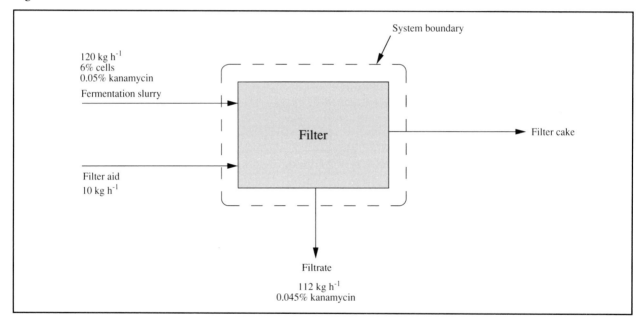

    (ii) *Define the system boundary by drawing on the flowsheet.*
    The system boundary is shown in Figure 4E3.1.
2. *Analyse*
    (i) *State any assumptions.*
        — process is operating at steady state
        — system does not leak
        — filtrate contains no solids
        — cells do not absorb or release kanamycin during filtration
        — filter aid is dry
        — the liquid phase of the slurry, excluding kanamycin, can be considered water
    (ii) *Collect and state any extra data needed.*
    No extra data are required.

(iii) *Select and state a basis.*

The calculation is based on 120 kg slurry entering the filter, or 1 hour.

(iv) *List the compounds, if any, which are involved in reaction.*

No compounds are involved in reaction.

(v) *Write down the appropriate general mass-balance equation.*

The system is at steady state and no reaction occurs; therefore Eq. (4.3) is appropriate:

mass in = mass out.

3. *Calculate*

(i) *Set up a calculation table showing all components of all streams passing across system boundaries. State the units used for the table. Enter all known quantities.*

As shown in Figure 4E3.1, four streams cross the system boundaries: fermentation slurry, filter aid, filtrate and filter cake. The components of these streams: cells, kanamycin, filter aid and water are represented in Table 4E3.1. The table is divided into two major sections: In and Out. Masses entering or leaving the system each hour are shown in the table; the units used are kg. Because filtrate and filter cake flow out of the system, there are no entries for these streams on the left hand-side of the table. Conversely, there are no entries for fermentation slurry or filter aid on the Out side of the table. The total mass of each stream is given in the last column on each side of the table. The total amount of each component flowing in and out of the system is shown in the last row. With all known quantities entered, several masses remain unknown; these quantities are indicated by question marks.

Table 4E3.1 Mass-balance table (kg)

Stream	In					Out				
	Cells	*Kanamycin*	*Filter aid*	*Water*	*Total*	*Cells*	*Kanamycin*	*Filter aid*	*Water*	*Total*
Fermentation slurry	7.2	0.06	0	?	120	–	–	–	–	–
Filter aid	0	0	10	0	10	–	–	–	–	–
Filtrate	–	–	–	–	–	0	0.05	0	?	112
Filter cake	–	–	–	–	–	?	?	?	?	?
Total	?	?	?	?	?	?	?	?	?	?

(ii) *Calculate unknown quantities, apply the mass-balance equation.*

To complete Table 4E3.1, let us consider each row and column separately. In the row representing fermentation slurry, the total mass of the stream is known as 120 kg and the masses of each component except water are known. The entry for water can therefore be determined as the difference between 120 kg and the sum of the known components: $(120 - 7.2 - 0.06 - 0)$ kg = 112.74 kg. This mass for water has been entered in Table 4E3.2. The row for filter aid is already complete in Table 4E3.1; no cells or kanamycin are present in the diatomaceous earth, which we assume is dry. We can fill in the final row of the In side of the table; numbers in this row are obtained by adding the values in each vertical column. The total mass of cells input to the system in all streams is 7.2 kg, the total kanamycin entering is 0.06 kg, etc. The total mass of all components fed into the system is the sum of the last column of the left-hand side: $(120 + 10)$ kg = 130 kg. On the Out side, we can complete the row for filtrate. We have assumed there are no solids such as cells or filter aid in the filtrate; therefore the mass of water in the filtrate is $(112 - 0.05)$ kg = 111.95 kg. As yet, the entire composition and mass of the filter cake remain unknown.

Table 4E3.2 Completed mass-balance table (kg)

Stream	In					Out				
	Cells	*Kanamycin*	*Filter aid*	*Water*	*Total*	*Cells*	*Kanamycin*	*Filter aid*	*Water*	*Total*
Fermentation slurry	7.2	0.06	0	112.74	120	–	–	–	–	–
Filter aid	0	0	10	0	10	–	–	–	–	–
Filtrate	–	–	–	–	–	0	0.05	0	111.95	112
Filter cake	–	–	–	–	–	7.2	0.01	10	0.79	18
Total	7.2	0.06	10	112.74	130	7.2	0.06	10	112.74	130

To complete the table, we must consider the mass-balance equation relevant to this problem, Eq. (4.3). In the absence of reaction, this equation can be applied to total mass and to the masses of each component of the system.

Total mass balance
130 kg total mass in = total mass out.
∴ Total mass out = 130 kg.

Cell balance
7.2 kg cells in = cells out.
∴ Cells out = 7.2 kg.

Kanamycin balance
0.06 kg kanamycin in = kanamycin out.
∴ Kanamycin out = 0.06 kg.

Filter-aid balance
10 kg filter aid in = filter aid out.
∴ Filter aid out = 10 kg.

Water balance
112.74 kg water in = water out.
∴ Water out = 112.74 kg.

These results are entered in the last row of the Out side of Table 4E3.2; in the absence of reaction this row is always identical to the final row of In side. The component masses for filter cake can now be filled in as the difference between numbers in the final row and the masses of each component in the filtrate. Take time to look over Table 4E3.2; you should understand how all the numbers were obtained.

(iii) *Check that your results are reasonable and make sense.*
Mass-balance calculations must be checked. Make sure that all columns and rows of Table 4E3.2 add up to the totals shown.

4. *Finalise*
(i) *Answer the specific questions asked in the problem.*
The percentage liquid in the filter cake can be calculated from the results in Table 4E3.2. Dividing the mass of water in the filter cake by the total mass of this stream, the percentage liquid is:

$$\frac{0.79 \text{ kg}}{18 \text{ kg}} \times 100 = 4.39\%.$$

If the concentration of kanamycin in this liquid is only 0.045%, the mass of kanamycin is very close to:

$$\frac{0.045}{100} \times 0.79 \, \text{kg} = 3.6 \times 10^{-4} \, \text{kg}.$$

However, we know from Table 4E3.2 that a total of 0.01 kg kanamycin is contained in the filter cake; therefore $(0.01 - 3.6 \times 10^{-4}) \, \text{kg} = 0.0096 \, \text{kg}$ kanamycin is unaccounted for. Following our assumption that kanamycin is not adsorbed by the cells, 0.0096 kg kanamycin must be retained by the filter aid. 10 kg filter aid is present; therefore the kanamycin absorbed per kg filter aid is:

$$\frac{0.0096 \, \text{kg}}{10 \, \text{kg}} = 9.6 \times 10^{-4} \, \text{kg kg}^{-1}.$$

(ii) *State the answers clearly and unambiguously, checking significant figures.*
    (a) The liquid content of the filter cake is 4.4%.
    (b) The amount of kanamycin absorbed by the filter aid is $9.6 \times 10^{-4} \, \text{kg kg}^{-1}$.

Note in Example 4.3 that the complete composition of the fermentation slurry was not provided. Cell and kanamycin concentrations were given; however the slurry most probably contained a variety of other components such as residual carbohydrate, minerals, vitamins, amino acids and additional fermentation products. These components were ignored in the mass balance; the liquid phase of the slurry was considered to be water only. This assumption is reasonable as the concentration of dissolved substances in fermentation broths is usually very small; water in spent broth usually accounts for more than 90% of the liquid phase.

Note also in this problem that the masses of some of the components were different by several orders of magnitude, e.g. the mass of kanamycin in the filtrate was of the order 10^{-2} whereas the total mass of this stream was of the order 10^2. Calculation of the mass of water by difference therefore involved subtracting a very small number from a large one and carrying more significant figures than warranted. This is an unavoidable feature of most mass balances for biological processes, which are characterised by dilute solutions, low product concentrations and large amounts of water. However, although excess significant figures were carried in the mass-balance table, the final answers were reported with due regard to data accuracy.

The above example illustrates mass-balance procedures for a simple steady-state process without reaction. An integral mass balance for a batch system without reaction is illustrated in Example 4.4.

Example 4.4 Batch mixing

Corn-steep liquor contains 2.5% invert sugars and 50% water; the rest can be considered solids. Beet molasses containing 50% sucrose, 1% invert sugars, 18% water and the remainder solids, is mixed with corn-steep liquor in a mixing tank. Water is added to produce a diluted sugar mixture containing 2% (w/w) invert sugars. 125 kg corn-steep liquor and 45 kg molasses are fed into the tank.

(a) How much water is required?
(b) What is the concentration of sucrose in the final mixture?

Solution:

1. *Assemble*
    (i) *Flow sheet.*
    The flow sheet for this batch process is shown in Figure 4E4.1. Unlike in Figure 4E3.1 where the streams represented continuously-flowing inputs and outputs, the streams in Figure 4E4.1 represent masses added and removed at the beginning and end of the mixing process, respectively.

Figure 4E4.1   Flowsheet for batch mixing process.

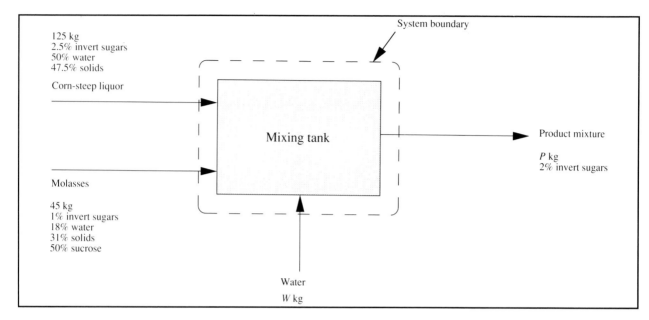

(ii)   *System boundary.*
The system boundary is indicated in Figure 4E4.1.

2.   *Analyse*
   (i)   *Assumptions.*
         — no leaks
         — no inversion of sucrose to reducing sugars, or any other reaction
   (ii)   *Extra data.*
   No extra data are required.
   (iii)   *Basis.*
   125 kg corn-steep liquor.
   (iv)   *Compounds involved in reaction.*
   No compounds are involved in reaction.
   (v)   *Mass-balance equation.*
   The appropriate mass-balance equation is Eq. (4.3):

   mass in = mass out.

3.   *Calculate*
   (i)   *Calculation table.*
   Table 4E4.1 shows all given quantities in kg. Rows and columns on each side of the table have been completed as much as possible from the information provided. Two unknown quantities are given symbols; the mass of water added is denoted W, the total mass of product mixture is denoted P.

Table 4E4.1   Mass-balance table (kg)

Stream	In					Out				
	Invert sugars	*Sucrose*	*Solids*	*H_2O*	*Total*	*Invert sugars*	*Sucrose*	*Solids*	*H_2O*	*Total*
Corn-steep liquor	3.125	0	59.375	62.5	125	–	–	–	–	–
Molasses	0.45	22.5	13.95	8.1	45	–	–	–	–	–
Water	0	0	0	W	W	–	–	–	–	–
Product mixture	–	–	–	–	–	0.02 P	?	?	?	P
Total	3.575	22.5	73.325	70.6 + W	170 + W	0.02 P	?	?	?	P

(ii) *Mass-balance calculations.*
Total mass balance

$(170 + W)$ kg total mass in $= P$ kg total mass out.
$\therefore 170 + W = P.$

$$(1)$$

Invert sugars balance

3.575 kg invert sugars in $= (0.02\ P)$ kg invert sugars out.
$\therefore 3.575 = 0.02\ P$
$P = 178.75$ kg.

Using this result in (1):

$W = 8.75$ kg.

$$(2)$$

Sucrose balance

22.5 kg sucrose in = sucrose out.
$\therefore$ Sucrose out = 22.5 kg.

Solids balance

73.325 kg solids in = solids out.
$\therefore$ Solids out = 73.325 kg.

H_2O balance

$(70.6 + W)$ kg in $= H_2O$ out.

Using the result from (2):

$$79.35 \text{ kg } H_2O \text{ in} = H_2O \text{ out.}$$
$$\therefore H_2O \text{ out} = 79.35 \text{ kg.}$$

These results allow the mass-balance table to be completed, as shown in Table 4E4.2.

Table 4E4.2 Completed mass-balance table (kg)

Stream	In					Out				
	Invert sugars	Sucrose	Solids	H₂O	Total	Invert sugars	Sucrose	Solids	H₂O	Total
Corn-steep liquor	3.125	0	59.375	62.5	125	–	–	–	–	–
Molasses	0.45	22.5	13.95	8.1	45	–	–	–	–	–
Water	0	0	0	8.75	8.75	–	–	–	–	–
Product mixture	–	–	–	–	–	3.575	22.5	73.325	79.35	178.75
Total	3.575	22.5	73.325	79.35	178.75	3.575	22.5	73.325	79.35	178.75

(iii) *Check the results.*
All columns and rows of Table 4E4.2 add up correctly.

4. *Finalise*
   (i) *The specific questions.*
   The water required is 8.75 kg. The sucrose concentration in the product mixture is:

   $$\frac{22.5}{178.75} \times 100 = 12.6\%$$

   (ii) *Answers.*
   (a) 8.75 kg water is required.
   (b) The product mixture contains 13% sucrose.

Material balances on reactive systems are slightly more complicated than Examples 4.3 and 4.4. To solve problems with reaction, stoichiometric relationships must be used in conjunction with mass-balance equations. These procedures are illustrated in Examples 4.5 and 4.6.

Example 4.5 Continuous acetic acid fermentation

Acetobacter aceti bacteria convert ethanol to acetic acid under aerobic conditions. A continuous fermentation process for vinegar production is proposed using non-viable *A. aceti* cells immobilised on the surface of gelatin beads. The production target is 2 kg h^{-1} acetic acid; however the maximum acetic acid concentration tolerated by the cells is 12%. Air is pumped into the fermenter at a rate of 200 gmol h^{-1}.

(a) What minimum amount of ethanol is required?
(b) What minimum amount of water must be used to dilute the ethanol to avoid acid inhibition?
(c) What is the composition of the fermenter off-gas?

Solution:

1. *Assemble*

    (i) *Flow sheet.*
    The flow sheet for this process is shown in Figure 4E5.1.

Figure 4E5.1   Flow sheet for continuous acetic acid fermentation.

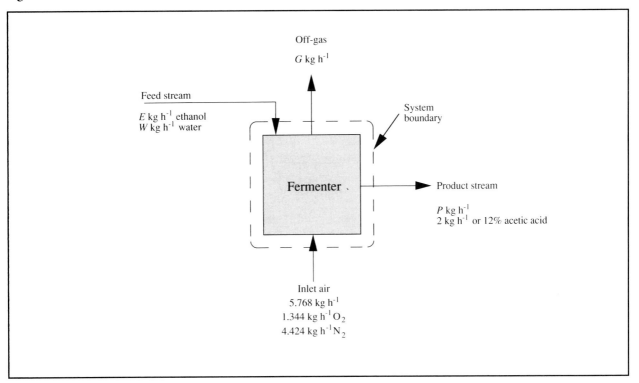

(ii) *System boundary.*
The system boundary is shown in Figure 4E5.1.

(iii) *Write down the reaction equation.*
In the absence of cell growth, maintenance or other metabolism of substrate, the reaction equation is:

$$C_2H_5OH + O_2 \rightarrow CH_3COOH + H_2O.$$
$$\text{(ethanol)} \qquad\qquad \text{(acetic acid)}$$

2. *Analyse*

    (i) *Assumptions.*
    —steady state
    —no leaks
    —inlet air is dry
    —gas volume% = mole%
    —no evaporation of ethanol, H_2O or acetic acid
    —complete conversion of ethanol
    —ethanol is used by the cells for synthesis of acetic acid only; no side-reactions occur

—oxygen transfer is sufficiently rapid to meet the demands of the cells
—concentration of acetic acid in the product stream is 12%

(ii) *Extra data.*

Molecular weights:　ethanol $= 46$

acetic acid $= 60$

$O_2 = 32$

$N_2 = 28$

$H_2O = 18$

Composition of air:　21% O_2, 79% N_2.

(iii) *Basis.*

The calculation is based on 2 kg acetic acid leaving the system, or 1 hour.

(iv) *Compounds involved in reaction.*

The compounds involved in reaction are ethanol, acetic acid, O_2 and H_2O. N_2 is not involved in reaction.

(v) *Mass-balance equations.*

For ethanol, acetic acid, O_2 and H_2O, the appropriate mass-balance equation is Eq. (4.2):

mass in + mass generated = mass out + mass consumed.

For total mass and N_2, the appropriate mass-balance equation is Eq. (4.3):

mass in = mass out.

3.　*Calculate*

(i) *Calculation table.*

The mass-balance table with data provided is shown as Table 4E5.1; the units are kg. EtOH denotes ethanol; HAc is acetic acid. If 2 kg acetic acid represents 12 mass% of the product stream, the total mass of the product stream must be $^2/_{0.12} = 16.67$ kg. If we assume complete conversion of ethanol, the only components of the product stream are acetic acid and water; therefore water must account for 88 mass% of the product stream $= 14.67$ kg. In order to represent what is known about the inlet air, some preliminary calculations are needed.

$$O_2 \text{ content} = (0.21)\,(200 \text{ gmol}) \cdot \left| \frac{32 \text{ g}}{\text{gmol}} \right| = 1344 \text{ g} = 1.344 \text{ kg}$$

$$N_2 \text{ content} = (0.79)\,(200 \text{ gmol}) \cdot \left| \frac{28 \text{ g}}{\text{gmol}} \right| = 4424 \text{ g} = 4.424 \text{ kg}.$$

Therefore, the total mass of air in $= 5.768$ kg. The masses of O_2 and N_2 can now be entered in the table, as shown.

Table 4E5.1　Mass-balance table (kg)

Stream	In						Out					
	EtOH	*HAc*	*H_2O*	*O_2*	*N_2*	*Total*	*EtOH*	*HAc*	*H_2O*	*O_2*	*N_2*	*Total*
Feed stream	E	0	W	0	0	$E + W$	–	–	–	–	–	–
Inlet air	0	0	0	1.344	4.424	5.768	–	–	–	–	–	–
Product stream	–	–	–	–	–	–	0	2	14.67	0	0	16.67
Off-gas	–	–	–	–	–	–	0	0	0	?	?	G
Total	E	0	W	1.344	4.424	5.768 + $E + W$	0	2	14.67	?	?	16.67 + G

E and *W* denote the unknown quantities of ethanol and water in the feed stream, respectively; *G* represents the total mass of off-gas. The question marks in the table show which other quantities must be calculated.

(ii) *Mass-balance and stoichiometry calculations.*
As N_2 is a tie component, its mass balance is straightforward.

N_2 balance

$$4.424 \text{ kg } N_2 \text{ in} = N_2 \text{ out.}$$
$$\therefore N_2 \text{ out} = 4.424 \text{ kg.}$$

To deduce the other unknowns, we must use stoichiometric analysis as well as mass balances.

HAc balance

$$0 \text{ kg HAc in} + \text{HAc generated} = 2 \text{ kg HAc out} + 0 \text{ kg HAc consumed.}$$
$$\therefore \text{ HAc generated} = 2 \text{ kg.}$$

$$2 \text{ kg} = 2 \text{ kg.} \left| \frac{1 \text{ kgmol}}{60 \text{ kg}} \right| = 3.333 \times 10^{-2} \text{ kgmol.}$$

From reaction stoichiometry, we know that generation of 3.333×10^{-2} kgmol HAc requires 3.333×10^{-2} kgmol each of EtOH and O_2, and is accompanied by generation of 3.333×10^{-2} kgmol H_2O:

$$\therefore 3.333 \times 10^{-2} \text{ kgmol.} \left| \frac{46 \text{ kg}}{1 \text{ kgmol}} \right| = 1.533 \text{ kg EtOH is consumed}$$

$$3.333 \times 10^{-2} \text{ kgmol.} \left| \frac{32 \text{ kg}}{1 \text{ kgmol}} \right| = 1.067 \text{ kg } O_2 \text{ is consumed}$$

$$3.333 \times 10^{-2} \text{ kgmol.} \left| \frac{18 \text{ kg}}{1 \text{ kgmol}} \right| = 0.600 \text{ kg } H_2O \text{ is generated.}$$

We can use this information to complete the mass balances for EtOH, O_2 and H_2O.

EtOH balance

$$\text{EtOH in} + 0 \text{ kg EtOH generated} = 0 \text{ kg EtOH out} + 1.533 \text{ kg EtOH consumed.}$$
$$\therefore \text{EtOH in} = 1.533 \text{ kg} = E.$$

O_2 balance

$$1.344 \text{ kg } O_2 \text{ in} + 0 \text{ kg } O_2 \text{ generated} = O_2 \text{ out} + 1.067 \text{ kg } O_2 \text{ consumed.}$$
$$\therefore O_2 \text{ out} = 0.277 \text{ kg.}$$

Therefore, summing the O_2 and N_2 components of the off-gas:

$$G = (0.277 + 4.424) \text{ kg} = 4.701 \text{ kg.}$$

H_2O balance

$$W \text{ kg } H_2O \text{ in} + 0.600 \text{ kg } H_2O \text{ generated} = 14.67 \text{ kg } H_2O \text{ out} + 0 \text{ kg } H_2O \text{ consumed.}$$
$$\therefore W = 14.07 \text{ kg.}$$

These results allow us to complete the mass-balance table, as shown in Table 4E5.2.

Table 4E5.2   Completed mass-balance table (kg)

Stream	In						Out					
	EtOH	HAc	H_2O	O_2	N_2	Total	EtOH	HAc	H_2O	O_2	N_2	Total
Feed stream	1.533	0	14.07	0	0	15.603	–	–	–	–	–	–
Inlet air	0	0	0	1.344	4.424	5.768	–	–	–	–	–	–
Product stream	–	–	–	–	–	–	0	2	14.67	0	0	16.67
Off-gas	–	–	–	–	–	–	0	0	0	0.277	4.424	4.701
Total	1.533	0	14.07	1.344	4.424	21.371	0	2	14.67	0.277	4.424	21.371

(iii) *Check the results.*
All rows and columns of Table 4E5.2 add up correctly.

4.   *Finalise*
(i)   *The specific questions.*
The ethanol required is 1.533 kg. The water required is 14.07 kg. The off-gas contains 0.277 kg O_2 and 4.424 kg N_2. Since gas compositions are normally expressed using volume or mole%, we must convert these values to moles:

$$O_2 \text{ content} = 0.277 \text{ kg} \cdot \left| \frac{1 \text{ kgmol}}{32 \text{ kg}} \right| = 8.656 \times 10^{-3} \text{ kgmol}$$

$$N_2 \text{ content} = 4.424 \text{ kg} \cdot \left| \frac{1 \text{ kgmol}}{28 \text{ kg}} \right| = 0.1580 \text{ kgmol}.$$

Therefore, the total molar quantity of off-gas is 0.1667 kgmol. The off-gas composition is:

$$\frac{8.656 \times 10^{-3} \text{ kgmol}}{0.1667 \text{ kgmol}} \times 100 = 5.19\% \ O_2$$

$$\frac{0.1580 \text{ kgmol}}{0.1667 \text{ kgmol}} \times 100 = 94.8\% \ N_2.$$

(ii)   *Answers.*
Quantities are expressed in kg h^{-1} rather than kg to reflect the continuous nature of the process and the basis used for calculation.

   (a)   1.5 kg h^{-1} ethanol is required.
   (b)   14.1 kg h^{-1} water must be used to dilute the ethanol in the feed stream.
   (c)   The composition of the fermenter off-gas is 5.2% O_2 and 94.8% N_2.

There are several points to note about the problem and calculation of Example 4.5. First, cell growth and its requirement for substrate were not considered because the cells used in this process were non-viable. For fermentation with live cells, growth and other metabolic activity must be taken into account in the mass balance. This requires knowledge of growth stoichiometry, which is considered in Example 4.6 and discussed in more detail in Section 4.6. Use of non-growing immobilised cells in Example 4.5 meant that the cells were not components of any stream flowing in or out of the process, nor were they generated in reaction. Therefore, cell mass did not have to be included in the calculation.

Example 4.5 illustrates the importance of phase separations. Unreacted oxygen and nitrogen were assumed to leave the system as off-gas rather than as components of the liquid product stream. This assumption is reasonable due to the very poor solubility of oxygen and nitrogen in aqueous liquids; although the product stream most likely contains some dissolved gas, the quantities are relatively small. This assumption may need to be reviewed for gases with higher solubility, e.g. ammonia.

In the above problem, nitrogen did not react, nor were there more than one stream in and one stream out carrying nitrogen. A material which goes directly from one stream to another is called a *tie component*; the mass balance for a tie component is relatively simple. Tie components are useful because they can provide partial solutions to mass-balance problems making subsequent calculations easier. More than one tie component may be present in a particular process.

One of the listed assumptions in Example 4.5 is rapid oxygen transfer. Because cells use oxygen in dissolved form, oxygen must be transferred into the liquid phase from gas bubbles supplied to the fermenter. The speed of this process depends on the culture conditions and operation of the fermenter as described in more detail in Chapter 9. In mass-balance problems we assume that all oxygen required by the stoichiometric equation is immediately available to the cells.

Sometimes it is not possible to solve for unknown quantities in mass balances until near the end of the calculation. In such cases, symbols for various components rather than numerical values must be used in the balance equations. This is illustrated in the integral mass-balance of Example 4.6 which analyses batch culture of growing cells for production of xanthan gum.

Example 4.6 Xanthan gum production

Xanthan gum is produced using *Xanthomonas campestris* in batch culture. Laboratory experiments have shown that for each gram of glucose utilised by the bacteria, 0.23 g oxygen and 0.01 g ammonia are consumed, while 0.75 g gum, 0.09 g cells, 0.27 g gaseous CO_2 and 0.13 g H_2O are formed. Other components of the system such as phosphate can be neglected. Medium containing glucose and ammonia dissolved in 20 000 litres water is pumped into a stirred fermenter and inoculated with *X. campestris*. Air is sparged into the fermenter; the total amount of off-gas recovered during the entire batch culture is 1250 kg. Because of the high viscosity and difficulty in handling xanthan-gum solutions, the final gum concentration should not be allowed to exceed 3.5 wt%.

(a) How much glucose and ammonia are required?
(b) What percentage excess air is provided?

Solution:
1.   *Assemble*
     (i)   *Flow sheet.*
     The flow sheet for this process is shown in Figure 4E6.1.
     (ii) *System boundary.*
     The system boundary is shown in Figure 4E6.1.
     (iii) *Reaction equation.*

$$1\text{ g glucose} + 0.23\text{ g O}_2 + 0.01\text{ g NH}_3 \rightarrow 0.75\text{ g gum} + 0.09\text{ g cells} + 0.27\text{ g CO}_2 + 0.13\text{ g H}_2\text{O}.$$

Figure 4E6.1 Flowsheet for xanthan gum fermentation.

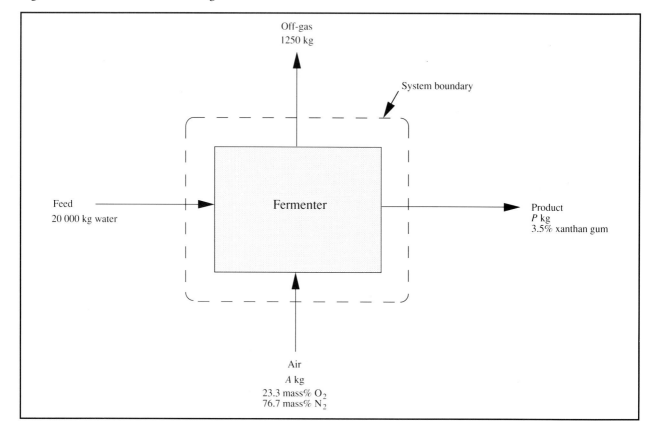

2. *Analyse*
   (i) *Assumptions.*
      —no leaks
      —inlet air and off-gas are dry
      —conversion of glucose and NH_3 is 100% complete
      —CO_2 leaves in the off-gas
   (ii) *Extra data.*
   Molecular weights: $O_2 = 32$
                        $N_2 = 28$
   (iii) *Basis.*
   1250 kg off-gas.
   (iv) *Compounds involved in reaction.*
   The compounds involved in reaction are glucose, O_2, NH_3, gum, cells, CO_2 and H_2O. N_2 is not involved in reaction.
   (v) *Mass-balance equations.*
   For glucose, O_2, NH_3, gum, cells, CO_2 and H_2O, the appropriate mass-balance equation is Eq. (4.2):

   mass in + mass generated = mass out + mass consumed.

   For total mass and N_2, the appropriate mass-balance equation is Eq. (4.3):

   mass in = mass out.

3. *Calculate*

(i) *Calculation table.*

Some preliminary calculations are required to start the mass-balance table. First, using 1 kg l^{-1} as the density of water, 20 000 litres water is equivalent to 20 000 kg. Let A be the unknown mass of air added. At low pressure, air is composed of 21 mol% O_2 and 79 mol% N_2; we need to determine the composition of air as mass fractions. In 100 gmol air:

$$O_2 \text{ content} = 21 \text{ gmol} . \left| \frac{32 \text{ g}}{1 \text{ gmol}} \right| = 672 \text{ g}.$$

$$N_2 \text{ content} = 79 \text{ gmol} . \left| \frac{28 \text{ g}}{1 \text{ gmol}} \right| = 2212 \text{ g}.$$

If the total mass of air in 100 gmol is $(2212 + 672) = 2884$ g, the composition of air is:

$$\frac{672 \text{ g}}{2884 \text{ g}} \times 100 = 23.3 \text{ mass\% } O_2.$$

$$\frac{2212 \text{ g}}{2884 \text{ g}} \times 100 = 76.7 \text{ mass\% } N_2.$$

Therefore, the mass of O_2 in the inlet air is $0.233A$; the mass of N_2 is $0.767A$. Let F denote the total mass of feed medium added; let P denote the total mass of product. We will perform the calculation to produce the maximum-allowable gum concentration; therefore, the mass of gum in the product is $0.035P$. With the assumption of 100% conversion of glucose and NH_3, these compounds are not present in the product. Quantities known at the beginning of the problem are shown in Table 4E6.1.

Table 4E6.1   Mass-balance table (kg)

Stream	In									Out								
	Glucose	O_2	N_2	CO_2	Gum	Cells	NH_3	H_2O	Total	Glucose	O_2	N_2	CO_2	Gum	Cells	NH_3	H_2O	Total
Feed	?	0	0	0	0	0	?	20 000	F	–	–	–	–	–	–	–	–	–
Air	0	0.233A	0.767A	0	0	0	0	0	A	–	–	–	–	–	–	–	–	–
Off-gas	–	–	–	–	–	–	–	–	–	0	?	?	?	0	0	0	0	1250
Product	–	–	–	–	–	–	–	–	–	0	0	0	0	0.035P	?	0	0	P
Total	?	0.233A	0.767A	0	0	0	?	20 000	$F+A$	0	?	?	?	0.035P	?	0	?	1250 + P

(ii) *Mass-balance and stoichiometry calculations.*

Total mass balance

$(F + A)$ kg total mass in $= (1250 + P)$ kg total mass out.
$\therefore F + A = 1250 + P$.

$$(1)$$

Gum balance

0 kg gum in + gum generated $= (0.035P)$ kg gum out + 0 kg gum consumed.
$\therefore$ Gum generated $= (0.035P)$ kg.

From reaction stoichiometry, synthesis of $(0.035P)$ kg gum requires:

$$\frac{0.035P}{0.75}\ (1\ \text{kg}) = (0.0467P)\ \text{kg glucose}$$

$$\frac{0.035P}{0.75}\ (0.23\ \text{kg}) = (0.0107P)\ \text{kg O}_2$$

$$\frac{0.035P}{0.75}\ (0.01\ \text{kg}) = (0.00047P)\ \text{kg NH}_3$$

and produces:

$$\frac{0.035P}{0.75}\ (0.09\ \text{kg}) = (0.0042P)\ \text{kg cells}$$

$$\frac{0.035P}{0.75}\ (0.27\ \text{kg}) = (0.0126P)\ \text{kg CO}_2$$

$$\frac{0.035P}{0.75}\ (0.13\ \text{kg}) = (0.00607P)\ \text{kg H}_2\text{O}.$$

O$_2$ balance

$(0.233A)$ kg O_2 in + 0 kg O_2 generated = O_2 out + $(0.0107P)$ kg O_2 consumed.
∴ O_2 out = $(0.233A - 0.0107P)$ kg.

(2)

N$_2$ balance
N_2 is a tie component.

$(0.767A)$ kg N_2 in = N_2 out.
∴ N_2 out = $(0.767A)$ kg.

(3)

CO$_2$ balance

0 kg CO_2 in + $(0.0126P)$ kg CO_2 generated = CO_2 out + 0 kg CO_2 consumed.
∴ CO_2 out = $(0.0126P)$ kg.

(4)

The total mass of gas out is 1250 kg. Therefore, adding the amounts of O_2, N_2 and CO_2 out from (2), (3) and (4):

$1250 = (0.233A - 0.0107P) + (0.767A) + (0.0126P)$
$\quad\ = A + 0.0019P$
∴ $A\ = 1250 - 0.0019P.$

(5)

Glucose balance

glucose in + 0 kg glucose generated = 0 kg glucose out + $(0.0467P)$ kg glucose consumed.
∴ Glucose in = $(0.0467\ P)$ kg.

(6)

NH₃ balance

> NH_3 in $+ 0$ kg NH_3 generated $= 0$ kg NH_3 out $+ (0.00047P)$ kg NH_3 consumed.
> $\therefore NH_3$ in $= (0.00047P)$ kg.

(7)

We can now calculate the total mass of the feed, F:

> $F =$ glucose in $+ NH_3$ in $+$ water in.

From (6) and (7):

> $F = (0.0467P)$ kg $+ (0.00047P)$ kg $+ 20\ 000$ kg
> $\quad = (20\ 000 + 0.04717P)$ kg.

(8)

We can now use (8) and (5) in (1):

> $(20\ 000 + 0.04717P) + (1250 - 0.0019P) = (1250 + P)$
> $20\ 000 = 0.95473\ P$
> $\therefore P = 20\ 948.3$ kg.
> $\therefore$ Gum out $= 733.2$ kg.

Substituting this result in (5) and (8):

> $A = 1210.2$ kg
> $F = 20\ 988.1$ kg.

From Table 4E6.1:

> O_2 in $= 282.0$ kg
> N_2 in $= 928.2$ kg.

Using the results for P, A and F in (2), (3), (4), (6) and (7):

> O_2 out $= 57.8$ kg
> N_2 out $= 928.2$ kg
> CO_2 out $= 263.9$ kg
> Glucose in $= 978.3$ kg
> NH_3 in $= 9.8$ kg.

Cell balance

> 0 kg cells in $+ (0.0042P)$ kg cells generated $=$ cells out $+ 0$ kg cells consumed.
> $\therefore$ Cells out $= (0.0042P)$ kg
> Cells out $= 88.0$ kg.

H₂O balance

> $20\ 000$ kg H_2O in $+ (0.00607P)$ kg H_2O generated $= H_2O$ out $+ 0$ kg H_2O consumed.
> $\therefore H_2O$ out $= 20\ 000 + (0.00607P)$ kg.
> H_2O out $= 20\ 127.2$ kg.

These entries are included in Table 4E6.2.

Table 4E6.2 Completed mass-balance table (kg)

Stream	In									Out								
	Glucose	*O₂*	*N₂*	*CO₂*	*Gum*	*Cells*	*NH₃*	*H₂O*	*Total*	*Glucose*	*O₂*	*N₂*	*CO₂*	*Gum*	*Cells*	*NH₃*	*H₂O*	*Total*
Feed	978.3	0	0	0	0	0	9.8	20000	20988.1	–	–	–	–	–	–	–	–	–
Air	0	282.0	928.2	0	0	0	0	0	1210.2	–	–	–	–	–	–	–	–	–
Off-gas	–	–	–	–	–	–	–	–	–	0	57.8	928.2	263.9	0	0	0	0	1250
Product	–	–	–	–	–	–	–	–	–	0	0	0	0	733.2	88.0	0	20127.2	20948.3
Total	978.3	282.0	928.2	0	0	0	9.8	20000	22198.3	0	57.8	928.2	263.9	733.2	88.0	0	20127.2	22198.3

(iii) *Check the results.*
All the columns and rows of Table 4E6.2 add up correctly to within round-off error.

4. *Finalise*
(i) *The specific questions.*
From the completed mass-balance table, 978.3 kg glucose and 9.8 kg NH_3 are required. Calculation of percentage excess air is based on oxygen as oxygen is the reacting component of air. Percentage excess can be calculated using Eq. (2.35) in units of kg:

$$\% \text{ excess air} = \frac{\left(\begin{array}{c} \text{kg } O_2 \text{ present} - \text{kg } O_2 \text{ required to react} \\ \text{completely with the limiting substrate} \end{array} \right)}{\left(\begin{array}{c} \text{kg } O_2 \text{ required to react completely} \\ \text{with the limiting substrate} \end{array} \right)} \times 100.$$

In this problem, both glucose and ammonia are limiting substrates. From stoichiometry and the mass-balance table, the mass of oxygen required to react completely with 978.3 kg glucose and 9.8 kg NH_3 is:

$$\frac{978.3 \text{ kg}}{1 \text{ kg}} (0.23 \text{ kg}) = 225.0 \text{ kg } O_2.$$

The mass provided is 282.0 kg; therefore:

$$\% \text{ excess air} = \frac{282.0 - 225.0}{225.0} \times 100 = 25.3\%.$$

(ii) *Answers.*
   (a) 980 kg glucose and 9.8 kg NH_3 are required.
   (b) 25% excess air is provided.

4.5 Material Balances With Recycle, By-Pass and Purge Streams

So far, we have performed mass balances on simple single-unit processes. However, steady-state systems incorporating recycle, by-pass and purge streams are common in bioprocess industries; flow sheets illustrating these modes of operation are shown in Figure 4.3. Material-balance calculations for such systems can be somewhat more involved than those in Examples 4.3 to 4.6; several balances are required before all mass flows can be determined.

As an example, consider the system of Figure 4.4. Because cells are the catalysts in fermentation processes, it is often advantageous to recycle biomass from spent fermentation broth. Cell recycle requires a separation device, such as a centrifuge or gravity settling tank, to provide a concentrated recycle stream under aseptic conditions. The flow sheet for cell recycle is shown in Figure 4.5; as indicated, at least four

Figure 4.3   Flow sheet for processes with (**a**) recycle, (**b**) by-pass and (**c**) purge streams.

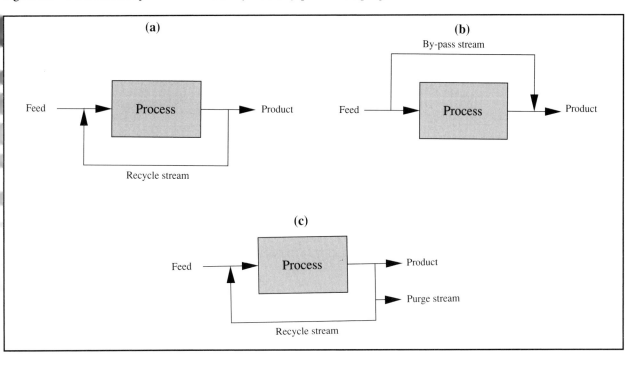

Figure 4.4   Fermenter with cell recycle.

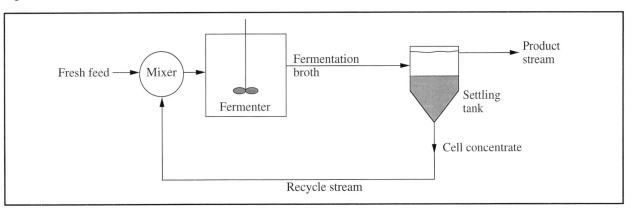

different system boundaries can be defined. System I represents the overall recycle process; only the fresh feed and final product streams cross this system boundary. In addition, separate material balances can be performed over each process unit: the mixer, the fermenter and the settler. Other system boundaries could also be defined; for example, we could group the mixer and fermenter, or settler and fermenter, together. Material balances with recycle involve carrying out individual mass-balance calculations for each designated system.

Depending on which quantities are known and what information is sought, analysis of more than one system may be required before the flow rates and compositions of all streams are known.

Mass balances with recycle, by-pass or purge usually involve longer calculations than for simple processes, but are not more difficult conceptually. Accordingly, we will not treat these types of process further. Examples of mass-balance procedures for multi-unit processes can be found in standard chemical-engineering texts, e.g. [1–3].

Figure 4.5   System boundaries for cell-recycle system.

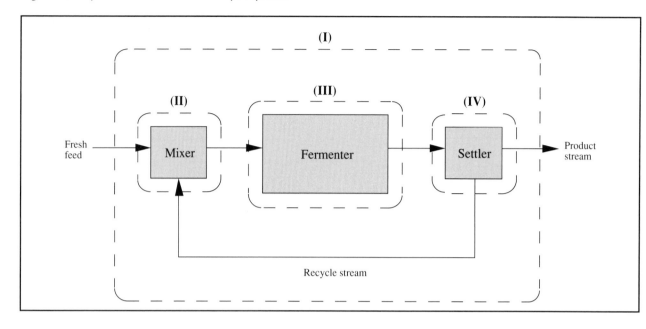

4.6 Stoichiometry of Growth and Product Formation

So far in this chapter, the law of conservation of mass has been used to determine unknown quantities entering or leaving bioprocesses. For mass balances with reaction such as Examples 4.5 and 4.6, the stoichiometry of conversion must be known before the mass balance can be solved. When cell growth occurs, cells are a product of reaction and must be represented in the reaction equation. In this section we will discuss how reaction equations for growth and product synthesis are formulated. Metabolic stoichiometry has many applications in bioprocessing: as well as in mass and energy balances, it can be used to compare theoretical and actual product yields, check the consistency of experimental fermentation data, and formulate nutrient medium.

4.6.1 Growth Stoichiometry and Elemental Balances

Despite its complexity and the thousands of intracellular reactions involved, cell growth obeys the law of conservation of matter. All atoms of carbon, hydrogen, oxygen, nitrogen and other elements consumed during growth are incorporated into new cells or excreted as products. Confining our attention to those compounds taken up or produced in significant quantity, if the only extracellular products formed are CO_2 and

H_2O, we can write the following equation for aerobic cell growth:

$$C_wH_xO_yN_z + a\,O_2 + b\,H_gO_hN_i \rightarrow c\,CH_\alpha O_\beta N_\delta + d\,CO_2 + e\,H_2O. \tag{4.4}$$

In Eq. (4.4), $C_wH_xO_yN_z$ is the chemical formula for the substrate (e.g. for glucose $w = 6$, $x = 12$, $y = 6$ and $z = 0$), $H_gO_hN_i$ is the chemical formula for the nitrogen source, and $CH_\alpha O_\beta N_\delta$ is the chemical 'formula' for dry biomass. a, b, c, d and e are stoichiometric coefficients. Eq. (4.4) is written on the basis of one mole of substrate; therefore a moles O_2 are consumed and d moles CO_2 are formed per mole substrate reacted, etc. As illustrated in Figure 4.6, the equation represents a macroscopic view of metabolism; it ignores the detailed structure of the system and considers only those components which have net interchange with the environment. Despite its simplicity, the macroscopic approach provides a powerful tool for thermodynamic analysis. Eq. (4.4) does not include a multitude of compounds such as ATP and NADH which are integral to metabolism and undergo exchange cycles in cells, but are not subject to net exchange with the environment. Compounds such as vitamins and minerals taken up during metabolism could be included; however, since these growth factors are generally consumed in small quantity we assume here that their contribution to the stoichiometry and energetics of

Figure 4.6   Conversion of substrate, oxygen and nitrogen for cell growth.

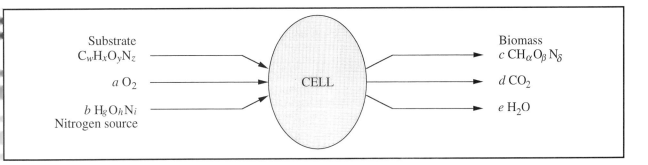

reaction can be neglected. Other substrates and products can easily be added if appropriate.

In Eq. (4.4), biomass is represented by the formula $CH_\alpha O_\beta N_\delta$. There is no fundamental objection to having a molecular formula for cells, even if it is not widely applied in biology. The formula is a reflection of the biomass composition. As shown in Table 4.2, microorganisms such as *Escherichia coli* contain a wide range of elements; however 90–95% of biomass can be accounted for by four major elements: C, H, O and N. Compositions of several species in terms of these four elements are listed in Table 4.3. The formulae in Table 4.3 refer to dry biomass and are based on one C atom; the total amount of biomass formed during growth can be accounted for by the stoichiometric coefficient c in Eq.

(4.4). Bacteria tend to have slightly higher nitrogen contents (11–14%) than fungi (6.3–9.0%) [4]. For a particular species, cell composition depends also on culture conditions and substrate utilised, hence the different entries in Table 4.3 for the same organism. However, the results are remarkably similar for different cells and conditions; $CH_{1.8}O_{0.5}N_{0.2}$ can be used as a general formula when composition analysis is not available. The average 'molecular weight' of biomass based on C, H, O and N content is therefore 24.6, although 5–10% residual ash is often added to account for those elements not included in the formula.

Eq. (4.4) is not complete unless the stoichiometric coefficients a, b, c, d and e are known. Once a formula for biomass is obtained, these coefficients can be evaluated using normal procedures for balancing equations, i.e. elemental balances and solution of simultaneous equations.

C balance:	$w = c + d$	
		(4.5)
H balance:	$x + b\,g = c\,\alpha + 2\,e$	
		(4.6)
O balance:	$y + 2\,a + b\,h = c\,\beta + 2\,d + e$	
		(4.7)
N balance:	$z + b\,i = c\,\delta.$	
		(4.8)

Notice that we have five unknown coefficients (a, b, c, d and e) but only four balance equations. This means that additional information is required before the equations can be solved. Usually this information is obtained from experiments. A useful measurable parameter is the *respiratory quotient* (*RQ*):

$$\text{respiratory quotient, } RQ = \frac{\text{moles } CO_2 \text{ produced}}{\text{moles } O_2 \text{ consumed}} = \frac{d}{a}.$$

(4.9)

Table 4.2   Elemental composition of *Escherichia coli* bacteria

(From R.Y. Stanier, E.A. Adelberg and J. Ingraham, 1976, The Microbial World, 4th edn, Prentice-Hall, New Jersey)

Element	% dry weight
C	50
O	20
N	14
H	8
P	3
S	1
K	1
Na	1
Ca	0.5
Mg	0.5
Cl	0.5
Fe	0.2
All others	0.3

Table 4.3   Elemental composition and degree of reduction for selected organisms

(From J.A. Roels, 1980, Application of macroscopic principles to microbial metabolism, Biotechnol. Bioeng. **22**, *2457–2514)*

Organism	Elemental formula	Degree of reduction γ (relative to NH_3)
Escherichia coli	$CH_{1.77}O_{0.49}N_{0.24}$	4.07
Klebsiella aerogenes	$CH_{1.75}O_{0.43}N_{0.22}$	4.23
Kl. aerogenes	$CH_{1.73}O_{0.43}N_{0.24}$	4.15
Kl. aerogenes	$CH_{1.75}O_{0.47}N_{0.17}$	4.30
Kl. aerogenes	$CH_{1.73}O_{0.43}N_{0.24}$	4.15
Pseudomonas $C_{12}B$	$CH_{2.00}O_{0.52}N_{0.23}$	4.27
Aerobacter aerogenes	$CH_{1.83}O_{0.55}N_{0.25}$	3.98
Paracoccus denitrificans	$CH_{1.81}O_{0.51}N_{0.20}$	4.19
P. denitrificans	$CH_{1.51}O_{0.46}N_{0.19}$	3.96
Saccharomyces cerevisiae	$CH_{1.64}O_{0.52}N_{0.16}$	4.12
S. cerevisiae	$CH_{1.83}O_{0.56}N_{0.17}$	4.20
S. cerevisiae	$CH_{1.81}O_{0.51}N_{0.17}$	4.28
Candida utilis	$CH_{1.83}O_{0.54}N_{0.10}$	4.45
C. utilis	$CH_{1.87}O_{0.56}N_{0.20}$	4.15
C. utilis	$CH_{1.83}O_{0.46}N_{0.19}$	4.34
C. utilis	$CH_{1.87}O_{0.56}N_{0.20}$	4.15
Average	$CH_{1.79}O_{0.50}N_{0.20}$	4.19
		(standard deviation = 3%)

When an experimental value of *RQ* is available, Eqs (4.5) to (4.9) can be solved to determine the stoichiometric coefficients. The results, however, are sensitive to small errors in *RQ*, which must be measured very accurately. When Eq. (4.4) is completed, the quantities of substrate, nitrogen and oxygen required for production of biomass can be determined directly.

Example 4.7 Stoichiometric coefficients for cell growth

Production of single-cell protein from hexadecane is described by the following reaction equation:

$$C_{16}H_{34} + a\,O_2 + b\,NH_3 \;\rightarrow\; c\,CH_{1.66}O_{0.27}N_{0.20} + d\,CO_2 + e\,H_2O$$

where $CH_{1.66}O_{0.27}N_{0.20}$ represents the biomass. If $RQ = 0.43$, determine the stoichiometric coefficients.

Solution:

C balance: $16 = c + d$ (1)

H balance: $34 + 3\,b = 1.66\,c + 2\,e$ (2)

O balance: $2\,a = 0.27\,c + 2\,d + e$ (3)

N balance: $b = 0.20\,c$ (4)

RQ: $0.43 = {}^d/_a.$ (5)

We must solve this set of simultaneous equations. Solution can be achieved in many different ways; usually it is a good idea to express each variable as a function of only one other variable. b is already written simply as a function of c in (4); let us try expressing the other variables solely in terms of c. From (1):

$$d = 16 - c.$$

(6)

From (5):

$$a = \frac{d}{0.43} = 2.326 \, d.$$

(7)

Combining (6) and (7) gives an expression for a in terms of c only:

$$a = 2.326 \, (16 - c)$$
$$a = 37.22 - 2.326 \, c.$$

(8)

Substituting (4) into (2) gives:

$$34 + 3 \, (0.20 \, c) = 1.66 \, c + 2 \, e$$
$$34 = 1.06 \, c + 2 \, e$$
$$e = 17 - 0.53 \, c.$$

(9)

Substituting (8), (6) and (9) into (3) gives:

$$2 \, (37.22 - 2.326 \, c) = 0.27 \, c + 2 \, (16 - c) + (17 - 0.53 \, c)$$
$$25.44 = 2.39 \, c$$
$$c = 10.64.$$

Using this result for c in (8), (4), (6) and (9) gives:

$$a = 12.48$$
$$b = 2.13$$
$$d = 5.37$$
$$e = 11.36.$$

Check that these coefficient values satisfy Eqs (1)–(5).
The complete reaction equation is:

$$C_{16}H_{34} + 12.5 \, O_2 + 2.13 \, NH_3 \rightarrow 10.6 \, CH_{1.66}O_{0.27}N_{0.20} + 5.37 \, CO_2 + 11.4 \, H_2O.$$

Although elemental balances are useful, the presence of water in Eq. (4.4) causes some problems in practical application. Because water is usually present in great excess and changes in water concentration are inconvenient to measure or experimentally verify, H and O balances can present difficulties. Instead, a useful principle is conservation of reducing power or available electrons, which can be applied to determine quantitative relationships between substrates and products. An electron balance shows how available electrons from the substrate are distributed in reaction.

4.6.2 Electron Balances

Available electrons refers to the number of electrons available for transfer to oxygen on combustion of a substance to CO_2, H_2O and nitrogen-containing compounds. The number of available electrons found in organic material is calculated from the valence of the various elements: 4 for C, 1 for H, –2 for O, 5 for P, and 6 for S. The number of available electrons for N depends on the reference state: –3 if ammonia is the reference, 0 for molecular nitrogen N_2, and 5 for nitrate. The reference state for cell growth is usually chosen to be the same as the nitrogen source in the medium. In the following discussion it will be assumed for convenience that ammonia is used as nitrogen source; this can easily be changed if other nitrogen sources are employed [5].

Degree of reduction, γ, is defined as the number of equivalents of available electrons in that quantity of material containing 1 g atom carbon. Therefore, for substrate $C_wH_xO_yN_z$, the number of available electrons is $(4w + x - 2y - 3z)$. The degree of reduction for the substrate, γ_S, is therefore $(4w + x - 2y - 3z)/w$. Degrees of reduction relative to NH_3 and N_2 for several biological compounds are given in Table B.2 in Appendix B. Degree of reduction for CO_2, H_2O and NH_3 is zero.

Electrons available for transfer to oxygen are conserved during metabolism. In a balanced growth equation, number of available electrons is conserved by virtue of the fact that the amounts of each chemical element are conserved. Applying this principle to Eq. (4.4) with ammonia as nitrogen source, the available-electron balance is:

$$w\gamma_S - 4a = c\gamma_B$$

$$(4.10)$$

where γ_S and γ_B are the degrees of reduction of substrate and biomass, respectively. Note that the available-electron balance is not independent of the complete set of elemental balances; if the stoichiometric equation is balanced in terms of each element including H and O, the electron balance is implicitly satisfied.

4.6.3 Biomass Yield

Typically, Eq. (4.10) is used with carbon and nitrogen balances Eqs (4.5) and (4.8) and a measured value of *RQ* for evaluation

of stoichiometric coefficients. However, one electron balance, two elemental balances and one measured quantity are still inadequate information for solution of five unknown coefficients; another experimental quantity is required. As cells grow there is, as a general approximation, a linear relationship between the amount of biomass produced and the amount of substrate consumed. This relationship is expressed quantitatively using the *biomass yield*, Y_{XS}:

$$Y_{XS} = \frac{\text{g cells produced}}{\text{g substrate consumed}} .$$

$$(4.11)$$

A large number of factors influences biomass yield, including medium composition, nature of the carbon and nitrogen sources, pH and temperature. Biomass yield is greater in aerobic than in anaerobic cultures; choice of electron acceptor, e.g. O_2, nitrate or sulphate, can also have a significant effect [5, 6].

When Y_{XS} is constant throughout growth, its experimentally-determined value can be used to determine the stoichiometric coefficient c in Eq. (4.4). Eq. (4.11) expressed in terms of the stoichiometric Eq. (4.4) is:

$$Y_{XS} = \frac{c\,(\text{MW cells})}{(\text{MW substrate})}$$

$$(4.12)$$

where MW is molecular weight and 'MW cells' means the biomass formula-weight plus any residual ash. However, before applying measured values of Y_{XS} and Eq. (4.12) to evaluate c, we must be sure that the experimental culture system is well represented by the stoichiometric equation. For example, we must be sure that substrate is not used to synthesise extracellular products other than CO_2 and H_2O. One complication with real cultures is that some fraction of substrate consumed is always used for *maintenance activities* such as maintenance of membrane potential and internal pH, turnover of cellular components and cell motility. These metabolic functions require substrate but do not necessarily produce cell biomass, CO_2 and H_2O in the way described by Eq. (4.4). It is important to account for maintenance when experimental information is used to complete stoichiometric equations; maintenance requirements and the difference between observed and true yields are discussed further in Chapter 11. For the time being, we will assume that available values for biomass yield reflect substrate consumption for growth only.

4.6.4 Product Stoichiometry

Consider formation of an extracellular product $C_jH_kO_lN_m$ during growth. Eq. (4.4) can be extended to include product synthesis as follows:

$$C_wH_xO_yN_z + aO_2 + bH_gO_hN_i$$
$$\rightarrow c\,CH_\alpha O_\beta N_\delta + d\,CO_2 + e\,H_2O + f\,C_jH_kO_lN_m \tag{4.13}$$

where f is the stoichiometric coefficient for product. Product synthesis introduces one extra unknown stoichiometric coefficient to the equation; thus, an additional relationship between coefficients is required. This is usually provided as another experimentally-determined yield coefficient, the *product yield from substrate*, Y_{PS}:

$$Y_{PS} = \frac{\text{g product formed}}{\text{g substrate consumed}} = \frac{f(\text{MW product})}{(\text{MW substrate})}. \tag{4.14}$$

As mentioned above with regard to biomass yields, we must be sure that the experimental system used to measure Y_{PS} conforms to Eq. (4.13). Eq. (4.13) does not hold if product formation is not directly linked with growth; accordingly it cannot be applied for secondary-metabolite production such as penicillin fermentation, or for biotransformations such as steroid hydroxylation which involve only a small number of enzymes in cells. In these cases, independent reaction equations must be used to describe growth and product synthesis.

4.6.5 Theoretical Oxygen Demand

Oxygen demand is an important parameter in bioprocessing as oxygen is often the limiting substrate in aerobic fermentations. Oxygen demand is represented by the stoichiometric coefficient a in Eqs (4.4) and (4.13). Oxygen requirement is related directly to the electrons available for transfer to oxygen; the oxygen demand can therefore be derived from an appropriate electron balance. When product synthesis occurs as represented by Eq. (4.13), the electron balance is:

$$w\gamma_S - 4a = c\gamma_B + fj\gamma_P \tag{4.15}$$

where γ_P is the degree of reduction of the product. Rearranging gives:

$$a = {}^1/_4\,(w\gamma_S - c\gamma_B - fj\gamma_P). \tag{4.16}$$

Eq. (4.16) is a very useful equation. It means that if we know which organism (γ_B), substrate (w and γ_S) and product (j and γ_P) are involved in cell culture, and the yields of biomass (c) and product (f), we can quickly calculate the oxygen demand. Of course we could also determine a by solving for all the stoichiometric coefficients of Eq. (4.13) as described in Section 4.6.1. Eq. (4.16) allows more rapid evaluation and does not require that the quantities of NH_3, CO_2 and H_2O involved in the reaction be known.

4.6.6 Maximum Possible Yield

From Eq. (4.15) the fractional allocation of available electrons in the substrate can be written as:

$$1 = \frac{4a}{w\gamma_S} + \frac{c\gamma_B}{w\gamma_S} + \frac{fj\gamma_P}{w\gamma_S}. \tag{4.17}$$

In Eq. (4.17), the first term on the right-hand side is the fraction of available electrons transferred from substrate to oxygen, the second term is the fraction of available electrons transferred to biomass, and the third term is the fraction of available electrons transferred to product. This relationship can be used to obtain upper bounds for the yields of biomass and product from substrate.

Let us define ζ_B as the fraction of available electrons in the substrate transferred to biomass:

$$\zeta_B = \frac{c\gamma_B}{w\gamma_S}. \tag{4.18}$$

In the absence of product formation, if all available electrons were used for biomass synthesis, ζ_B would equal unity. Under these conditions, the maximum value of the stoichiometric coefficient c is:

$$c_{max} = \frac{w\gamma_S}{\gamma_B}. \tag{4.19}$$

c_{max} can be converted to a biomass yield with mass units using Eq. (4.12). Therefore, even if we do not know the stoichiometry of growth, we can quickly calculate an upper limit for biomass yield from the molecular formulae for substrate and product. If the composition of the cells is unknown, γ_B can be taken as 4.2 corresponding to the average biomass formula $CH_{1.8}O_{0.5}N_{0.2}$. Maximum biomass yields for several substrates are listed in Table 4.4; maximum biomass yield can be expressed in terms of mass ($Y_{XS,max}$), or as number of C atoms in the biomass per substrate C atom consumed ($^{c}max/_{w}$). These quantities are sometimes known as *thermodynamic maximum biomass yields*. Table 4.4 shows that substrates with high energy content, indicated by high γ_S values, give high maximum biomass yields.

Likewise, the maximum possible product yield in the absence of biomass synthesis can be determined from Eq. (4.17):

$$f_{max} = \frac{w\gamma_S}{j\gamma_P}.$$

(4.20)

Eq. (4.20) allows us to quickly calculate an upper limit for product yield from the molecular formulae for substrate and product.

Table 4.4   Thermodynamic maximum biomass yields

(Adapted from L.E. Erickson, I.G. Minkevich and V.K. Eroshin, 1978, Application of mass and energy balance regularities in fermentation, Biotechnol. Bioeng. **20***, 1595–1621)*

Substrate	Formula	γ_S	Thermodynamic maximum yield corresponding to $\zeta_B = 1$	
			Carbon yield ($^{c}max/_{w}$)	Mass yield $Y_{XS,max}$
Alkanes				
Methane	CH_4	8.0	1.9	2.9
Hexane (*n*)	C_6H_{14}	6.3	1.5	2.6
Hexadecane (*n*)	$C_{16}H_{34}$	6.1	1.5	2.5
Alcohols				
Methanol	CH_4O	6.0	1.4	1.1
Ethanol	C_2H_6O	6.0	1.4	1.5
Ethylene glycol	$C_2H_6O_2$	5.0	1.2	0.9
Glycerol	$C_3H_8O_3$	4.7	1.1	0.9
Carbohydrates				
Formaldehyde	CH_2O	4.0	0.95	0.8
Glucose	$C_6H_{12}O_6$	4.0	0.95	0.8
Sucrose	$C_{12}H_{22}O_{11}$	4.0	0.95	0.8
Starch	$(C_6H_{10}O_5)_x$	4.0	0.95	0.9
Organic acids				
Formic acid	CH_2O_2	2.0	0.5	0.3
Acetic acid	$C_2H_4O_2$	4.0	0.95	0.8
Propionic acid	$C_3H_6O_2$	4.7	1.1	1.1
Lactic acid	$C_3H_6O_3$	4.0	0.95	0.8
Fumaric acid	$C_4H_4O_4$	3.0	0.7	0.6
Oxalic acid	$C_2H_2O_4$	1.0	0.24	0.1

Example 4.8 Product yield and oxygen demand

The chemical reaction equation for respiration of glucose is:

$$C_6H_{12}O_6 + 6\,O_2 \rightarrow 6\,CO_2 + 6\,H_2O.$$

Candida utilis cells convert glucose to CO_2 and H_2O during growth. The cell composition is $CH_{1.84}O_{0.55}N_{0.2}$ plus 5% ash. Yield of biomass from substrate is 0.5 g g^{-1}. Ammonia is used as nitrogen source.

(a) What is the oxygen demand with growth compared to that without?

(b) *C. utilis* is also able to grow with ethanol as substrate, producing cells of the same composition as above. On a mass basis, how does the maximum possible biomass yield from ethanol compare with the maximum possible yield from glucose?

Solution:

Molecular weights: glucose = 180
                            ethanol = 46

MW biomass is (25.44 + ash); since ash accounts for 5% of the total weight, 95% of the total = 25.44. Therefore, MW biomass = 25.44/0.95 = 26.78. From Table B.2, γ_S for glucose is 4.00; γ_S for ethanol is 6.00. $\gamma_B = (4 \times 1 + 1 \times 1.84 - 2 \times 0.55 - 3 \times 0.2)$ = 4.14. For glucose $w = 6$; for ethanol $w = 2$.

(a) $Y_{XS} = 0.5$ g g^{-1}. Converting this mass yield to a molar yield:

$$Y_{XS} = \frac{0.5 \text{ g biomass}}{\text{g glucose}} \cdot \left| \frac{180 \text{ g glucose}}{1 \text{ gmol glucose}} \right| \cdot \left| \frac{1 \text{ gmol biomass}}{26.78 \text{ g biomass}} \right|$$

$$Y_{XS} = 3.36 \, \frac{\text{gmol biomass}}{\text{gmol glucose}} = c.$$

Oxygen demand is given by Eq. (4.16). In the absence of product formation:

$$a = {}^1/_4 \left[6 \, (4.00) - 3.36 \, (4.14) \right] = 2.52.$$

Therefore, the oxygen demand for glucose respiration with growth is 2.5 gmol O_2 per gmol glucose consumed. By comparison with the chemical reaction equation for respiration, this is only about 42% that required in the absence of growth.

(b) Maximum possible biomass yield is given by Eq. (4.19). Using the data above, for glucose:

$$c_{max} = \frac{6(4.00)}{4.14} = 5.80.$$

Converting this to a mass basis:

$$Y_{XS,max} = \frac{5.80 \text{ g biomass}}{\text{gmol glucose}} \cdot \left| \frac{1 \text{ gmol glucose}}{180 \text{ g glucose}} \right| \cdot \left| \frac{26.78 \text{ g biomass}}{1 \text{ gmol biomass}} \right|$$

$$Y_{XS,max} = 0.86 \, \frac{\text{g biomass}}{\text{g glucose}}.$$

For ethanol:

$$c_{max} = \frac{2(6.00)}{4.14} = 2.90$$

and

$$Y_{XS,max} = \left| \frac{2.90 \text{ gmol biomass}}{\text{gmol ethanol}} \right| \cdot \left| \frac{1 \text{ gmol ethanol}}{46 \text{ g ethanol}} \right| \cdot \left| \frac{26.78 \text{ g biomass}}{1 \text{ gmol biomass}} \right|$$

$$Y_{XS,max} = 1.69 \frac{\text{g biomass}}{\text{g ethanol}} .$$

Therefore, on a mass basis, the maximum possible amount of biomass produced per gram ethanol consumed is roughly twice that per gram glucose consumed. This result is consistent with the data listed in Table 4.4.

Example 4.8 illustrates two important points. First, the chemical reaction equation for conversion of substrate without growth is a poor approximation of overall stoichiometry when cell growth occurs. When estimating yields and oxygen requirements for any process involving cell growth, the full stoichiometric equation including biomass should be used. Second, the chemical nature or oxidation state of the substrate has a major influence on product and biomass yield through the number of available electrons.

4.7 Summary of Chapter 4

At the end of Chapter 4 you should:

(i) understand the terms: *system*, *surroundings*, *boundary* and *process* in thermodynamics;
(ii) be able to identify *open* and *closed systems*, and *batch, semi-batch, fed-batch* and *continuous processes*;
(iii) understand the difference between *steady state* and *equilibrium*;
(iv) be able to write appropriate equations for conservation of mass for processes with and without reaction;
(v) be able to solve simple mass-balance problems with and without reaction; and
(vi) be able to apply stoichiometric principles for macroscopic analysis of cell growth and product formation.

Problems

4.1 Cell concentration using membranes

A battery of cylindrical hollow-fibre membranes is operated at steady state to concentrate a bacterial suspension from a fermenter. 350 kg min^{-1} fermenter broth is pumped through a stack of hollow-fibre membranes as shown in Figure 4P1.1. The broth contains 1% bacteria; the rest may be considered water. Buffer solution enters the annular space around the membrane tubes at a flow rate of 80 kg min^{-1}; because broth in the membrane tubes is under pressure, water is forced across the membrane into the buffer. Cells in the broth are too large to pass through the membrane and pass out of the tubes as a concentrate.

Figure 4P1.1 Hollow-fibre membrane for concentration of cells.

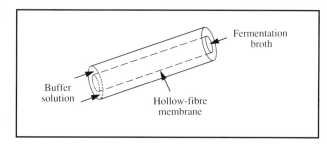

The aim of the membrane system is to produce a cell suspension containing 6% biomass.

(a) What is the flow rate from the annular space?
(b) What is the flow rate of cell suspension from the membrane tubes?

Assume that the cells are not active, i.e. they do not grow. Assume further that the membrane does not allow any molecules other than water to pass from annulus to inner cylinder, or vice versa.

4.2 Membrane reactor

A battery of cylindrical membranes similar to that shown in Figure 4P1.1 is used for an extractive bioconversion. Extractive bioconversion means that fermentation and extraction of product occur at the same time.

Yeast cells are immobilised within the membrane walls. A 10% glucose in water solution is passed through the annular space at a rate of 40 kg h^{-1}. An organic solvent, such as 2-ethyl-1,3-hexanediol, enters the inner tube at a rate of 40 kg h^{-1}.

Because the membrane is constructed of a polymer which repels organic solvents, the hexanediol cannot penetrate the membrane and the yeast is relatively unaffected by its toxicity. On the other hand, because glucose and water are virtually insoluble in 2-ethyl-1,3-hexanediol, these compounds do not enter the inner tube to an appreciable extent. Once immobilised in the membrane, the yeast cannot reproduce but convert glucose to ethanol according to the equation:

$$C_6H_{12}O_6 \rightarrow 2\,C_2H_6O + 2\,CO_2.$$

Ethanol is soluble in 2-ethyl-1,3-hexanediol; it diffuses into the inner tube and is carried out of the system. CO_2 gas exits from the membrane unit through an escape valve. In the aqueous stream leaving the annular space, the concentration of unconverted glucose is 0.2% and the concentration of ethanol is 0.5%. If the system operates at steady state:

(a) What is the concentration of ethanol in the hexanediol stream leaving the reactor?
(b) What is the mass flow rate of CO_2?

4.3 Ethanol distillation

Liquid from a brewery fermenter can be considered to contain 10% ethanol and 90% water. 50 000 kg h^{-1} of this fermentation product are pumped to a distillation column on the factory site. Under current operating conditions a distillate of 45% ethanol and 55% water is produced from the top of the column at a rate one-tenth that of the feed.

(a) What is the composition of the waste 'bottoms' from the still?
(b) What is the rate of alcohol loss in the bottoms?

4.4 Removal of glucose from dried egg

The enzyme, glucose oxidase, is used commercially to remove glucose from dehydrated egg to improve colour, flavour and shelf-life. The reaction is:

$$C_6H_{12}O_6 + O_2 + H_2O \rightarrow C_6H_{12}O_7 + H_2O_2.$$
$$\text{(glucose)} \qquad\qquad\quad \text{(gluconic acid)}$$

A continuous-flow reactor is set up using immobilised-enzyme beads which are retained inside the vessel. Dehydrated egg slurry containing 2% glucose, 20% water and the remainder unreactive egg solids, is available at a rate of 3 000 kg h^{-1}. Air is pumped through the reactor contents so that 18 kg oxygen are delivered per hour. The desired glucose level in the

dehydrated egg product leaving the enzyme reactor is 0.2%. Determine:

(a) which is the limiting substrate;
(b) the percentage excess substrate/s;
(c) the composition of the reactor off-gas; and
(d) the composition of the final egg product.

4.5 Azeotropic distillation

Absolute or 100% ethanol is produced from a mixture of 95% ethanol and 5% water using the Keyes distillation process. A third component, benzene, is added to lower the volatility of the alcohol. Under these conditions, the overhead product is a constant-boiling mixture of 18.5% ethanol, 7.4% H_2O and 74.1% benzene. The process is outlined in Figure 4P5.1.

Figure 4P5.1 Flowsheet for Keyes distillation process.

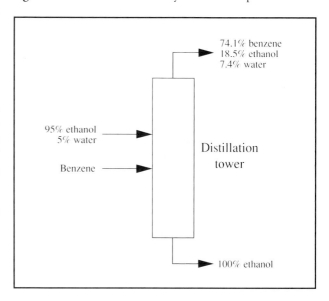

Use the following data to calculate the volume of benzene which should be fed to the still in order to produce 250 litres absolute ethanol: ρ (100% alcohol) = 0.785 g cm^{-3}; ρ (benzene) = 0.872 g cm^{-3}.

4.6 Culture of plant roots

Plant roots produce valuable chemicals *in vitro*. A batch culture of *Atropa belladonna* roots at 25°C is established in an air-driven reactor as shown in Figure 4P6.1. Because roots cannot be removed during operation of the reactor, it is proposed to monitor growth using mass balances.

Figure 4P6.1 Reactor for culture of plant roots.

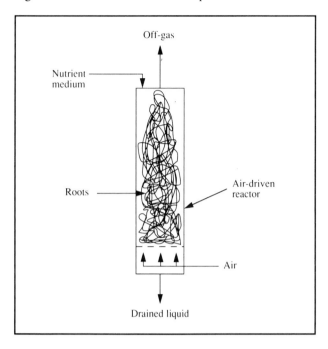

1425 g nutrient medium containing 3% glucose and 1.75% NH_3 is fed into the reactor; the remainder of the medium can be considered water. Air at 25°C and 1 atm pressure is sparged into the fermenter at a rate of 22 $cm^3\,min^{-1}$. During a 10-day culture period, 47 litres O_2 and 15 litres CO_2 are collected in the off-gas. After 10 days, 1110 g liquid containing 0.063% glucose and 1.7% dissolved NH_3 is drained from the vessel. The ratio of fresh weight to dry weight for roots is known to be 14:1.

(a) What dry mass of roots is produced in 10 days?
(b) Write the reaction equation for growth, indicating the approximate chemical formula for the roots, $CH_{\alpha}O_{\beta}N_{\delta}$.
(c) What is the limiting substrate?
(d) What is the yield of roots from glucose?

4.7 Oxygen requirement for growth on glycerol

Klebsiella aerogenes is produced from glycerol in aerobic culture with ammonia as nitrogen source. The biomass contains 8% ash, 0.40 g biomass is produced for each g glycerol consumed, and no major metabolic products are formed. What is the oxygen requirement for this culture in mass terms?

4.8 Product yield in anaerobic digestion

Anaerobic digestion of volatile acids by methane bacteria is represented by the equation:

$$CH_3COOH + NH_3 \rightarrow biomass + CO_2 + H_2O + CH_4.$$
(acetic acid)   (methane)

The composition of methane bacteria is approximated by the empirical formula $CH_{1.4}O_{0.40}N_{0.20}$. For each kg acetic acid consumed, 0.67 kg CO_2 is evolved. How does the yield of methane under these conditions compare with the maximum possible yield?

4.9 Stoichiometry of single-cell protein synthesis

(a) *Cellulomonas* bacteria used as single-cell protein for human or animal food are produced from glucose under anaerobic conditions. All carbon in the substrate is converted into biomass; ammonia is used as nitrogen source. The molecular formula for the biomass is $CH_{1.56}O_{0.54}N_{0.16}$; the cells also contain 5% ash. How does the yield of biomass from substrate in mass and molar terms compare with the maximum possible biomass yield?
(b) Another system for manufacture of single-cell protein is *Methylophilus methylotrophus*. This organism is produced aerobically from methanol with ammonia as nitrogen source. The molecular formula for the biomass is $CH_{1.68}O_{0.36}N_{0.22}$; these cells contain 6% ash.
  (i) How does the maximum yield of biomass compare with (a) above? What is the main reason for the difference?
  (ii) If the actual yield of biomass from methanol is 42% the thermodynamic maximum, what is the oxygen demand?

4.10 Ethanol production by yeast and bacteria

Both *Saccharomyces cerevisiae* yeast and *Zymomonas mobilis* bacteria produce ethanol from glucose under anaerobic conditions without external electron acceptors. The biomass yield from glucose is 0.11 $g\,g^{-1}$ for yeast and 0.05 $g\,g^{-1}$ for *Z. mobilis*. In both cases the nitrogen source is NH_3. Both cell compositions are represented by the formula $CH_{1.8}O_{0.5}N_{0.2}$.

(a) What is the yield of ethanol from glucose in both cases?
(b) How do the yields calculated in (a) compare with the thermodynamic maximum?

4.11 Detecting unknown products

Yeast growing in continuous culture produce 0.37 g biomass per g glucose consumed; about 0.88 g O_2 is consumed per g cells formed. The nitrogen source is ammonia, and the biomass composition is $CH_{1.79}O_{0.56}N_{0.17}$. Are other products also synthesised?

4.12 Medium formulation

Pseudomonas 5401 is to be used for production of single-cell protein for animal feed. The substrate is fuel oil. The composition of *Pseudomonas* 5401 is $CH_{1.83}O_{0.55}N_{0.25}$. If the final cell concentration is 25 g l^{-1}, what minimum concentration of $(NH_4)_2SO_4$ must be provided in the medium if $(NH_4)_2SO_4$ is the sole nitrogen source?

4.13 Oxygen demand for production of recombinant protein

Production of recombinant protein by a genetically-engineered strain of *Escherichia coli* is proportional to cell growth. Ammonia is used as nitrogen source for aerobic respiration of glucose. The recombinant protein has an overall formula $CH_{1.55}O_{0.31}N_{0.25}$. The yield of biomass from glucose is measured at 0.48 g g^{-1}; the yield of recombinant protein from glucose is about 20% that for cells.

(a) How much ammonia is required?

(b) What is the oxygen demand?

(c) If the biomass yield remains at 0.48 g g^{-1}, how much different are the ammonia and oxygen requirements for wild-type *E. coli* unable to synthesise recombinant protein?

4.14 Effect of growth on oxygen demand

The chemical reaction equation for conversion of ethanol (C_2H_6O) to acetic acid ($C_2H_4O_2$) is:

$$C_2H_6O + O_2 \rightarrow C_2H_4O_2 + H_2O.$$

Acetic acid is produced from ethanol during growth of *Acetobacter aceti*, which has the composition $CH_{1.8}O_{0.5}N_{0.2}$. Biomass yield from substrate is 0.14 g g^{-1}; product yield from substrate is 0.92 g g^{-1}. Ammonia is used as nitrogen source. How does growth in this culture affect oxygen demand for acetic acid production?

References

1. Felder, R.M. and R.W. Rousseau (1978) *Elementary Principles of Chemical Processes*, Chapter 5, John Wiley, New York.

2. Himmelblau, D.M. (1974) *Basic Principles and Calculations in Chemical Engineering*, 3rd edn, Chapter 2, Prentice-Hall, New Jersey.

3. Whitwell, J.C. and R.K. Toner (1969) *Conservation of Mass and Energy*, Chapter 4, Blaisdell, Waltham, Massachusetts.

4. Cordier, J.-L., B.M. Butsch, B. Birou and U. von Stockar (1987) The relationship between elemental composition and heat of combustion of microbial biomass. *Appl. Microbiol. Biotechnol.* **25**, 305–312.

5. Roels, J.A. (1983) *Energetics and Kinetics in Biotechnology*, Chapter 3, Elsevier Biomedical Press, Amsterdam.

6. Atkinson, B. and F. Mavituna (1991) *Biochemical Engineering and Biotechnology Handbook*, 2nd edn, Chapter 4, Macmillan, Basingstoke.

Suggestions for Further Reading

Process Mass Balances (see also refs 1–3)

Hougen, O.A., K.M. Watson and R.A. Ragatz (1954) *Chemical Process Principles: Material and Energy Balances*, 2nd edn, Chapter 7, John Wiley, New York.

Shaheen, E.I. (1975) *Basic Practice of Chemical Engineering*, Chapter 4, Houghton Mifflin, Boston.

Metabolic Stoichiometry (see also ref 5)

Erickson, L.E., I.G. Minkevich and V.K. Eroshin (1978) Application of mass and energy balance regularities in fermentation. *Biotechnol. Bioeng.* **20**, 1595–1621.

Heijnen, J.J. and J.A. Roels (1981) A macroscopic model describing yield and maintenance relationships in aerobic fermentation processes. *Biotechnol. Bioeng.* **23**, 739–763.

Nagai, S. (1979) Mass and energy balances for microbial growth kinetics. *Adv. Biochem. Eng.* **11**, 49–83.

Roels, J.A. (1980) Application of macroscopic principles to microbial metabolism. *Biotechnol. Bioeng.* **22**, 2457–2514.

Energy Balances

Unlike many chemical processes, bioprocesses are not particularly energy intensive. Fermenters and enzyme reactors are operated at temperatures and pressures close to ambient; energy input for downstream processing is minimised to avoid damaging heat-labile products. Nevertheless, energy effects are important because biological catalysts are very sensitive to heat and changes in temperature. In large-scale processes, heat released during reaction can cause cell death or denaturation of enzymes if it is not quickly removed. For rational design of temperature-control facilities, energy flows in the system must be determined using energy balances. Energy effects are also important in other areas of bioprocessing such as steam sterilisation.

The law of conservation of energy means that an energy accounting system can be set up to determine the amount of steam or cooling water required to maintain optimum process temperatures. In this chapter, after the necessary thermodynamic concepts are explained, an energy-conservation equation applicable to biological processes is derived. The calculation techniques outlined in Chapter 4 are then extended for solution of simple energy-balance problems.

5.1 Basic Energy Concepts

Energy takes three forms:

(i)   kinetic energy, E_k;
(ii) potential energy, E_p; and
(iii) internal energy, U.

Kinetic energy is the energy possessed by a moving system because of its velocity. *Potential energy* is due to the position of the system in a gravitational or electromagnetic field, or due to the conformation of the system relative to an equilibrium position (e.g. compression of a spring). *Internal energy* is the sum of all molecular, atomic and sub-atomic energies of matter. Internal energy cannot be measured directly or known in absolute terms; we can only quantify change in internal energy.

Energy is transferred as either heat or work. *Heat* is energy which flows across system boundaries because of a temperature difference between the system and surroundings. *Work* is energy transferred as a result of any driving force other than temperature difference. There are two types of work: *shaft work* W_s, which is work done by a moving part within the system, e.g., an impeller mixing a fermentation broth, and *flow work* W_f. Flow work is the energy required to push matter into the system. In a flow-through process, fluid at the inlet has work done on it by fluid just outside of the system, while fluid at the outlet does work on the fluid in front to push the flow along. Flow work is given by the expression:

$$W_f = pV$$

(5.1)

where p is pressure and V is volume. (Convince yourself that pV has the same dimensions as work and energy.)

5.1.1 Units

The SI unit for energy is the *joule* (J): 1 J = 1 newton metre (N m). Another unit is the *calorie* (cal), which is defined as the heat required to raise the temperature of 1 g pure water by 1°C at 1 atm pressure. The quantity of heat according to this definition depends somewhat on the temperature of the water; because there has been no universal agreement on a reference temperature, there are several slightly different calorie-units in use. The *international table calorie* (cal_{IT}) is fixed at 4.1868 J exactly. In imperial units, the British thermal unit (Btu) is common; this is defined as the amount of energy required to raise the temperature of 1 lb water by 1°F at 1 atm pressure. As with the calorie, a reference temperature is required for this definition; 60°F is common although other temperatures are sometimes used.

5.1.2 Intensive and Extensive Properties

Properties of matter fall into two categories: those whose magnitude depends on the quantity of matter present and those whose magnitude does not. Temperature, density, and mole fraction are examples of properties which are independent of the size of the system; these quantities are called *intensive variables*. On the other hand, mass, volume and energy are *extensive variables* which change if mass is added to or removed from the system. Extensive variables can be converted to *specific* quantities by dividing by the mass of the system; for example, specific volume is volume divided by mass. Because specific properties are independent of the mass of the system, they are also intensive variables. In this chapter, for extensive properties denoted by an upper-case symbol, the specific property is given in lower-case notation. Therefore if U is internal energy, u denotes specific internal energy with units, e.g. kJ g^{-1}. Although, strictly speaking, the term 'specific' refers to the quantity per unit mass, we will use the same lower-case symbols for molar quantities, e.g. with units kJ gmol^{-1}.

5.1.3 Enthalpy

Enthalpy is a property used frequently in energy-balance calculations. It is defined as the combination of two energy terms:

$$H = U + pV \tag{5.2}$$

where H is enthalpy, U is internal energy, p is pressure and V is volume. Specific enthalpy h is therefore:

$$h = u + pv \tag{5.3}$$

where u is specific internal-energy and v is specific volume. Since internal energy cannot be measured or known in absolute terms, neither can enthalpy.

5.2 General Energy-Balance Equations

The principle underlying all energy-balance calculations is the law of conservation of energy, which states that energy can be neither created nor destroyed. Although this law does not apply to nuclear reactions, conservation of energy remains a valid principle for bioprocesses because nuclear rearrangements are not involved. In the following sections, we will derive the equations used for solution of energy-balance problems.

The law of conservation of energy can be written as:

$$\begin{Bmatrix} \text{energy in through} \\ \text{system boundaries} \end{Bmatrix} - \begin{Bmatrix} \text{energy out through} \\ \text{system boundaries} \end{Bmatrix} = \begin{Bmatrix} \text{energy accumulated} \\ \text{within the system} \end{Bmatrix}. \tag{5.4}$$

Figure 5.1 Flow system for energy-balance calculations.

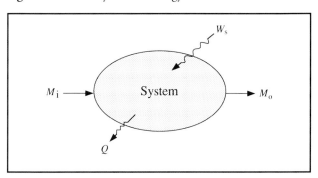

For practical application of this equation, consider the system depicted in Figure 5.1. Mass M_i enters the system while mass M_o leaves. Both these masses have energy associated with them in the form of internal, kinetic and potential energy; flow work is also being done. Energy leaves the system as heat Q; shaft work W_s is done on the system by the surroundings. We will assume that the system is homogeneous without charge or surface-energy effects.

To apply Eq. (5.4), we must identify which forms of energy are involved in each term of the expression. If we group together the extensive properties and express them as specific variables multiplied by mass, Eq. (5.4) can be written:

$$M_i (u + e_k + e_p + pv)_i - M_o (u + e_k + e_p + pv)_o - Q + W_s = \Delta E \tag{5.5}$$

where subscripts i and o refer to inlet and outlet conditions, respectively, and ΔE represents the total change or accumulation of energy in the system. u is specific internal energy, e_k is specific kinetic energy, e_p is specific potential energy, p is pressure, and v is specific volume. All energies associated with masses crossing the system boundary are added together; the energy-transfer terms Q and W_s are considered separately. Shaft work appears explicitly in Eq. (5.5) as W_s; flow work done the by inlet and outlet streams is represented as pv multiplied by mass.

Energy flows represented by Q and W_s can be directed either into or out of the system; appropriate signs must be used to indicate the direction of flow. Because it is usual in bioprocesses that shaft work be done on the system by external sources, in this text we will adopt the convention that work is positive when energy flows *from* the surroundings *to* the system as shown in Figure 5.1. Conversely, work will be considered negative when the system supplies work energy to the surroundings. On the other hand, we will regard heat as positive when the surroundings receives energy from the system, i.e. when the temperature of the system is higher than the surroundings. Therefore, when W_s and Q are positive quantities, W_s makes a positive contribution to the energy content of the system while Q causes a reduction. These effects are accounted for in Eq. (5.5) by the signs preceding Q and W_s. The opposite sign convention is sometimes used in thermodynamics texts. Choice of sign convention is arbitrary if used consistently.

Eq. (5.5) refers to a process with only one input and one output stream. A more general equation is Eq. (5.6), which can be used for any number of separate material flows:

$$\sum_{\substack{\text{input}\\\text{streams}}} M(u + e_k + e_p + pv) - \sum_{\substack{\text{output}\\\text{streams}}} M(u + e_k + e_p + pv)$$
$$- Q + W_s = \Delta E.$$
$$(5.6)$$

The symbol Σ means summation; the internal, kinetic, potential and flow work energies associated with all output streams are added together and subtracted from the sum for all input streams. Eq. (5.6) is a basic form of the *first law of thermodynamics*, a simple mathematical expression of the law of conservation of energy. The equation can be shortened by substituting enthalpy h for $u + pv$ as defined by Eq. (5.3):

$$\sum_{\substack{\text{input}\\\text{streams}}} M(h + e_k + e_p) - \sum_{\substack{\text{output}\\\text{streams}}} M(h + e_k + e_p) - Q + W_s = \Delta E.$$
$$(5.7)$$

5.2.1 Special Cases

Eq. (5.7) can be simplified considerably if the following assumptions are made:

(i)   kinetic energy is negligible; and
(ii) potential energy is negligible.

These assumptions are acceptable for bioprocesses, in which high-velocity motion and large changes in height or electromagnetic field do not generally occur. Thus, the energy-balance equation becomes:

$$\sum_{\substack{\text{input}\\\text{streams}}} (Mh) - \sum_{\substack{\text{output}\\\text{streams}}} (Mh) - Q + W_s = \Delta E.$$
$$(5.8)$$

Eq. (5.8) can be simplified further in the following special cases:

(i)   *Steady-state flow process.* At steady state, all properties of the system are invariant. Therefore, there can be no accumulation or change in the energy of the system: $\Delta E = 0$.

The steady-state energy-balance equation is:

$$\sum_{\substack{\text{input}\\\text{streams}}} (Mh) - \sum_{\substack{\text{output}\\\text{streams}}} (Mh) - Q + W_s = 0.$$

$$(5.9)$$

Eq. (5.9) can also be applied over the entire duration of batch and fed-batch processes if there is no energy accumulation; 'output streams' in this case refers to the harvesting of all mass in the system at the end of the process. Eq. (5.9) is used frequently in bioprocess energy balances.

(ii) *Adiabatic process.* A process in which no heat is transferred to or from the system is termed *adiabatic*; if the system has an *adiabatic wall* it cannot receive or release heat to the surroundings. Under these conditions $Q = 0$ and Eq. (5.8) becomes:

$$\sum_{\substack{\text{input}\\\text{streams}}} (Mh) - \sum_{\substack{\text{output}\\\text{streams}}} (Mh) + W_s = \Delta E.$$

$$(5.10)$$

Eqs (5.8)–(5.10) are energy-balance equations which allow us to predict, for example, how much heat must be removed from a fermenter to maintain optimum conditions, or the effect of evaporation on cooling requirements. To apply the equations we must know the specific enthalpy h of flow streams entering or leaving the system. Methods for calculating enthalpy are outlined in the following sections.

5.3 Enthalpy Calculation Procedures

Irrespective of how enthalpy changes occur, certain conventions are used in enthalpy calculations.

5.3.1 Reference States

Specific enthalpy h appears explicitly in energy-balance equations. What values of h do we use in these equations if enthalpy cannot be measured or known in absolute terms? Because energy balances are actually concerned with the *difference* in enthalpy between incoming and outgoing streams, we can overcome any difficulties by working always in terms of enthalpy change. In many energy-balance problems, changes in enthalpy are evaluated relative to reference states that must be defined at the beginning of the calculation.

Because H cannot be known absolutely, it is convenient to assign $H = 0$ to some reference state. For example, when 1 gmol carbon dioxide is heated at 1 atm pressure from 0°C to 25°C, the change in enthalpy of the gas can be calculated (using methods explained later) as $\Delta H = 0.91$ kJ. If we assign $H = 0$ for CO_2 gas at 0°C, H at 25°C can be considered to be 0.91 kJ. This result does not mean that the absolute value of enthalpy at 25°C is 0.91 kJ; we can say only that the enthalpy at 25°C is 0.91 kJ relative to the enthalpy at 0°C.

We will use various reference states in energy-balance calculations to determine enthalpy change. Suppose for example we want to calculate the change in enthalpy as a system moves from State 1 to State 2. If the enthalpies of States 1 and 2 are known relative to the same reference condition H_{ref}, ΔH is calculated as follows:

$$State\ 1 \xrightarrow{\Delta H} State\ 2$$
$$Enthalpy = H_1 - H_{ref} \qquad Enthalpy = H_2 - H_{ref}$$
$$\Delta H = (H_2 - H_{ref}) - (H_1 - H_{ref}) = H_2 - H_1.$$

ΔH is therefore independent of the reference state because H_{ref} cancels out in the calculation.

5.3.2 State Properties

Values of some variables depend only on the state of the system and not on how that state was reached. These variables are called *state properties* or *functions of state*; examples include temperature, pressure, density and composition. On the other hand, work is a *path function* since the amount of work done depends on the way in which the final state of the system is obtained from previous states.

Enthalpy is a state function. This property of enthalpy is very handy in energy-balance calculations; it means that change in enthalpy for a process can be calculated by taking a series of hypothetical steps or *process path* leading from the initial state and eventually reaching the final state. Change in enthalpy is calculated for each step; the total enthalpy change for the process is then equal to the sum of changes in the hypothetical path. This is true even though the process path used for calculation is not the same as that actually undergone by the system. As an example, consider the enthalpy change for the process shown in Figure 5.2 in which hydrogen peroxide is converted to oxygen and water by catalase enzyme. The enthalpy change for the direct process at 35°C can be calculated using an alternative pathway in which hydrogen peroxide is first cooled to 25°C, oxygen and water are formed by reaction at 25°C, and the products then heated to 35°C. Because the initial and final states for both actual and hypothetical paths are the same, the total enthalpy change is also identical:

$$\Delta H = \Delta H_1 + \Delta H_2 + \Delta H_3$$

$$(5.11)$$

Figure 5.2 Hypothetical process path for calculation of enthalpy change.

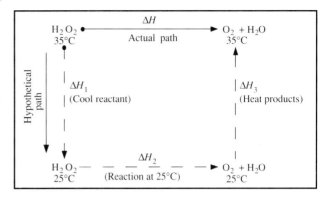

The reason for using hypothetical rather than actual pathways to calculate enthalpy change will be come apparent later in the chapter.

5.4 Enthalpy Change in Non-Reactive Processes

Change in enthalpy can occur as a result of:

(i)   temperature change;
(ii) change of phase;
(iii) mixing or solution; and
(iv) reaction.

In the remainder of this section we will consider enthalpy changes associated with (i), (ii) and (iii). We will then consider how the results are used in energy-balance calculations. Processes involving reaction will be discussed in Sections 5.8–5.11.

5.4.1 Change in Temperature

Heat transferred to raise or lower the temperature of a material is called *sensible heat*; change in the enthalpy of a system due to variation in temperature is called *sensible heat change*. In energy-balance calculations, sensible heat change is determined using a property of matter called the *heat capacity at constant pressure*, or just *heat capacity*. We will use the symbol C_p for heat capacity; units for C_p are, e.g. J gmol^{-1} K^{-1}, cal g^{-1} °C^{-1} and Btu lb^{-1} °F^{-1}. The term *specific heat* refers to heat capacity expressed on a per-unit-mass basis. Heat capacity must be known before enthalpy changes from heating or cooling can be determined. Tables B.3–B.6 in Appendix B list C_p values for several organic and inorganic compounds. Additional C_p data and information about estimating heat capacities can be found in references such as *Chemical*

Engineers' Handbook [1], *Handbook of Chemistry and Physics* [2] and *International Critical Tables* [3].

There are several methods for calculating enthalpy change using C_p values. When C_p is constant, the change in enthalpy of a substance due to change in temperature at constant pressure is:

$$\Delta H = MC_p \Delta T = MC_p (T_2 - T_1)$$

$$(5.12)$$

where M is either mass or moles of the substance depending on the dimensions of C_p, T_1 is the initial temperature and T_2 is the final temperature. The corresponding change in specific enthalpy is:

$$\Delta h = C_p \Delta T = C_p (T_2 - T_1).$$

$$(5.13)$$

Example 5.1 Sensible heat change with constant C_p

What is the enthalpy of 150 g formic acid at 70°C and 1 atm relative to 25°C and 1 atm?

Solution:
From Table B.5, C_p for formic acid in the temperature range of interest is 0.524 cal g^{-1} °C^{-1}. Substituting into Eq (5.12):

$$\Delta H = (150 \text{ g}) (0.524 \text{ cal g}^{-1} \text{ °C}^{-1}) (70 - 25)\text{°C}$$
$$\Delta H = 3537.0 \text{ cal}$$

or

$$\Delta H = 3.54 \text{ kcal.}$$

Relative to $H = 0$ at 25°C, the enthalpy of formic acid at 70°C is 3.54 kcal.

Heat capacities for most substances vary with temperature. This means that when we calculate enthalpy change due to change in temperature, the value of C_p itself varies over the range of ΔT. Heat capacities are often tabulated as polynomial functions of temperature, such as:

$$C_p = a + bT + cT^2 + dT^3.$$

$$(5.14)$$

Coefficients a, b, c and d for a number of substances are given in Table B.3 in Appendix B.

Sometimes we can assume that heat capacity is constant; this will give results for sensible heat change which approximate the true value. Because the temperature range of interest in bioprocessing is relatively small, assuming constant heat capacity for some materials does not introduce large errors. C_p data may not be available at all temperatures; heat capacities like most of those listed in Tables B.5 and B.6 are applicable only at a specified temperature or temperature range. As an example, in Table B.5 the heat capacity for liquid acetone between 24.2°C and 49.4°C is given as 0.538 cal g^{-1} °C^{-1} even though this value will vary within the temperature range. A useful rule of thumb for organic liquids near room temperature is that C_p increases by 0.001–0.002 cal g^{-1} °C^{-1}

for each Celsius-degree temperature increase [4].

One method for calculating sensible heat change when C_p varies with temperature involves use of the *mean heat capacity*, C_{pm}. Table B.4 in Appendix B lists mean heat capacities for several common gases. These values are based on changes in enthalpy relative to a single reference temperature, $T_{ref} = 0$°C. To determine the change in enthalpy for a change in temperature from T_1 to T_2, read the values of C_{pm} at T_1 and T_2 and calculate:

$$\Delta H = M \left[(C_{pm})_{T_2} (T_2 - T_{ref}) - (C_{pm})_{T_1} (T_1 - T_{ref}) \right].$$

$$(5.15)$$

5.4.2 Change of Phase

Phase changes, such as vaporisation and melting, are accompanied by relatively large changes in internal energy and enthalpy as bonds between molecules are broken and reformed. Heat transferred to or from a system causing change of phase at constant temperature and pressure is known as *latent heat*. Types of latent heat are:

(i) *latent heat of vaporisation* (Δh_v): heat required to vaporise a liquid;

(ii) *latent heat of fusion* (Δh_f): heat required to melt a solid; and

(iii) *latent heat of sublimation* (Δh_s): heat required to directly vaporise a solid.

Condensation of gas to liquid requires removal rather than addition of heat; the latent heat evolved in condensation is $-\Delta h_v$. Similarly, the latent heat evolved in freezing or solidification of liquid to solid is $-\Delta h_f$.

Latent heat is a property of substances and, like heat capacity, varies with temperature. Tabulated values of latent heats usually apply to substances at their normal boiling, melting or sublimation point at 1 atm, and are called *standard heats of phase change*. Table B.7 in Appendix B lists latent heats for selected compounds; more values may be found in *Chemical Engineers' Handbook* [1] and *Handbook of Chemistry and Physics* [2].

The change in enthalpy resulting from phase change is calculated directly from the latent heat. For example, increase in enthalpy due to evaporation of liquid mass M at constant temperature is:

$$\Delta H = M \Delta h_v.$$

$$(5.16)$$

Example 5.2 Enthalpy of condensation

50 g benzaldehyde vapour is condensed at 179°C. What is the enthalpy of the liquid relative to the vapour?

Solution:

From Table B.7, the molecular weight of benzaldehyde is 106.12, the normal boiling point is 179.0°C, and the standard heat of vaporisation is 38.40 kJ gmol^{-1}. For condensation the latent heat is -38.40 kJ gmol^{-1}. The enthalpy change is:

$$\Delta H = 50 \text{ g} \left(-38.40 \text{ kJ gmol}^{-1}\right) . \left| \frac{1 \text{gmol}}{106.12 \text{ g}} \right| = -18.09 \text{ kJ}.$$

Therefore, the enthalpy of 50 g benzaldehyde liquid relative to the vapour at 179°C is -18.09 kJ. As heat is released during condensation, the enthalpy of the liquid is lower than the vapour.

Phase changes often occur at temperatures other than the normal boiling, melting or sublimation point; for example, water can evaporate at temperatures higher or lower than 100°C. How can we determine ΔH when the latent heat at the actual temperature of the phase change is not listed in the tables? This problem is overcome by using a hypothetical process path as described in Section 5.3.2. Suppose a substance is vaporised isothermally at 30°C although tabulated values for standard heat of vaporisation refer to 60°C. As shown in Figure 5.3, we can consider a process whereby liquid is heated from 30°C to 60°C, vaporised at 60°C, and the vapour cooled to 30°C. The total enthalpy change for this process is the same as if vaporisation occurred directly at 30°C. ΔH_1 and ΔH_3 are sensible heat changes and can be calculated using heat-capacity values and the methods described in Section 5.4.1. ΔH_2 is the latent heat at standard conditions available from tables. Because enthalpy is a state property, ΔH for the actual path is the same as $\Delta H_1 + \Delta H_2 + \Delta H_3$.

5.4.3 Mixing and Solution

So far we have considered enthalpy changes for pure compounds. For an *ideal solution* or *ideal mixture* of several compounds, the thermodynamic properties of the mixture are a simple sum of contributions from the individual components. However, when compounds are mixed or dissolved, bonds between molecules in the solvent and solute are broken and reformed. In *real solutions* a net absorption or release of energy accompanies these processes resulting in changes in the internal energy and enthalpy of the mixture. Dilution of sulphuric acid with water is a good example; in this case energy is

Figure 5.3 Process path for calculating latent-heat change at a temperature other than the normal boiling point.

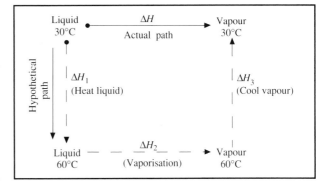

released. For real solutions there is an additional energy term to consider in evaluating enthalpy: the *integral heat of mixing* or *integral heat of solution*, Δh_m. The integral heat of solution is defined as the change in enthalpy which occurs as one mole of solute is dissolved at constant temperature in a given quantity of solvent. The enthalpy of a non-ideal mixture of two compounds A and B is:

$$H_{mixture} = H_A + H_B + \Delta H_m$$

$$(5.17)$$

where ΔH_m is the heat of mixing.

Heat of mixing is a property of the solution components and is dependent on the temperature and concentration of the mixture. As a solution becomes more and more dilute, an asymptotic value of Δh_m is reached. This value is called the *integral heat of solution at infinite dilution*. When water is the primary component of solutions, Δh_m at infinite dilution can be used to calculate the enthalpy of the mixture. Δh_m values for selected aqueous solutions are listed in *Chemical Engineers' Handbook* [1], *Handbook of Chemistry and Physics* [2] and *Biochemical Engineering and Biotechnology Handbook* [5].

Example 5.3 Heat of solution

Malonic acid and water are initially at 25°C. If 15 g malonic acid is dissolved in 5 kg water, how much heat must be added for the solution to remain at 25°C? What is the solution enthalpy relative to the components?

Solution:

The molecular weight of malonic acid is 104. Because the solution is very dilute (< 0.3% w/w), we can use the integral heat of solution at infinite dilution. From handbooks, Δh_m at room temperature is 4.493 kcal gmol^{-1}. This value is positive; therefore the mixture enthalpy is greater than the components and heat is absorbed during solution. The heat required for the solution to remain at 25°C is:

$$\Delta H = 4.493 \text{ kcal gmol}^{-1} (15 \text{ g}) \cdot \left| \frac{1 \text{ gmol}}{104 \text{ g}} \right| = 0.648 \text{ kcal.}$$

Relative to $H = 0$ for water and malonic acid at 25°C, the enthalpy of the solution at 25°C is 0.648 kcal.

In biological systems, significant changes in enthalpy due to heats of mixing do not often occur. Most solutions in fermentation and enzyme processes are dilute aqueous mixtures; in energy-balance calculations these solutions are usually considered ideal without much loss of accuracy.

5.5 Steam Tables

Steam tables have been used for many years by engineers designing industrial processes and power stations. These tables list the thermodynamic properties of water, including specific volume, internal energy and enthalpy. As we are mainly concerned here with enthalpies, a list of enthalpy values for steam and water under various conditions has been extracted from the steam tables and given in Appendix C [6]. All enthalpy values must have a reference point; in the steam tables of Appendix C, $H = 0$ for liquid water at the triple point: 0.01°C and 0.6112 kPa pressure. (The triple point is an invariant condition of pressure and temperature at which ice, liquid water and water vapour are in equilibrium with each other.) Steam tables from other sources may have different reference states. Steam tables eliminate the need for sensible-heat and latent-heat calculations for water and steam, and can be used directly in energy-balance calculations.

Tables C.1 and C.2 in Appendix C list enthalpy values for liquid water and *saturated steam*. When liquid and vapour are in equilibrium with each other, they are saturated; a gas saturated with water contains all the water it can hold at the prevailing temperature and pressure. For a pure substance such as water, once the temperature is specified, saturation occurs at only one pressure. For example, from Table C.2 saturated steam at 188°C has a pressure of 1200 kPa. Also from the table, the enthalpy of this steam relative to the triple point of water is 2782.7 kJ kg^{-1}; liquid water in equilibrium with the steam has an enthalpy of 798.4 kJ kg^{-1}. The latent heat of vaporisation under these conditions is the difference between liquid and vapour enthalpies; as indicated in the middle enthalpy column, Δh_v is 1984.3 kJ kg^{-1}. Table C.1 lists enthalpies of saturated water and steam by temperature; Table C.2 lists these enthalpies by pressure.

It is usual when using steam tables to ignore the effect of pressure on the enthalpy of liquid water. For example, the enthalpy of water at 40°C and 1 atm (101.3 kPa) is found by looking up the enthalpy of saturated water at 40°C in Table C.1, and assuming the value is independent of pressure. This assumption is valid at low pressure, i.e. less than about 50 atm. The enthalpy of liquid water at 40°C and 1 atm is therefore 167.5 kJ kg^{-1}.

Enthalpy values for *superheated steam* are given in Table C.3. If the temperature of saturated vapour is increased (or the pressure decreased at constant temperature), the vapour is said to be superheated. A superheated vapour cannot condense until it is returned to saturation conditions. The difference between the temperature of a superheated gas and its saturation temperature is called the *degrees of superheat* of the gas. In Table C.3, enthalpy is listed as a function of temperature at 15 different pressures from 10 kPa to 50 000 kPa; for example, superheated steam at 1000 kPa pressure and 250°C has an enthalpy of 2943 kJ kg^{-1} relative to the triple point. Table C.3 also lists properties at saturation conditions; at 1000 kPa the saturation temperature is 179.9°C, the enthalpy of liquid water under these conditions is 762.6 kJ kg^{-1}, and the enthalpy of saturated vapour is 2776.2 kJ kg^{-1}. Thus, the degrees of superheat for superheated steam at 1000 kPa and 250°C can be calculated as $(250 - 179.9) = 70.1$ centigrade degrees. Water under pressure remains liquid even at relatively high temperatures. Enthalpy values for liquid water up to 350°C are found in the upper region of Table C.3 above the line extending to the critical pressure.

5.6 Procedure For Energy-Balance Calculations Without Reaction

Methods described in Section 5.4 for evaluating enthalpy can be used to solve energy-balance problems for systems in which reactions do not occur. Many of the points described in Section 4.3 for material balances also apply when setting out an energy balance.

(i) A properly drawn and labelled *flow diagram* is essential to identify all inlet and outlet streams and their compositions. For energy balances, the temperatures, pressures and phases of the material should also be indicated if appropriate.

(ii) The *units* selected for the energy balance should be stated; these units are also used when labelling the flow diagram.

(iii) As in mass balance problems, a *basis* for the calculation must be chosen and stated clearly.

(iv) The *reference state* for $H = 0$ is determined. In the absence of reaction, reference states for each molecular species in the system can be arbitrarily assigned.

(v) State all *assumptions* used in solution of the problem. Assumptions such as absence of leaks and steady-state operation for continuous processes are generally applicable.

Following on from (v), other assumptions commonly made for energy balances include the following:

(a) The system is homogeneous or well mixed. Under these conditions, product streams including gases leave the system at the system temperature.

(b) Heats of mixing are often neglected for mixtures containing compounds of similar molecular structure. Gas mixtures are always considered ideal.

(c) Sometimes shaft work can be neglected even though the system is stirred by mechanical means. This assumption may not apply when vigorous agitation is used or when the liquid being stirred is very viscous. When shaft work is not negligible you will need to know how much mechanical energy is input through the stirrer.

(d) Evaporation in liquid systems may be considered negligible if the components are not particularly volatile or if the operating temperature is relatively low.

(e) Heat losses from the system to the surroundings are often ignored; this assumption is generally valid for large insulated vessels when the operating temperature is close to ambient.

5.7 Energy-Balance Worked Examples Without Reaction

As illustrated in the following examples, the format described in Chapter 4 for material balances can be used as a foundation for energy-balance calculations.

Example 5.4 Continuous water heater

Water at 25°C enters an open heating tank at a rate of 10 kg h^{-1}. Liquid water leaves the tank at 88°C at a rate of 9 kg h^{-1}; 1 kg h^{-1} water vapour is lost from the system through evaporation. At steady state, what is the rate of heat input to the system?

Solution:
1. *Assemble*
   (i) *Select units for the problem.*
   kg, h, kJ, °C.
   (ii) *Draw the flowsheet showing all data and units.*
   The flowsheet is shown in Figure 5E4.1.

Figure 5E4.1   Flowsheet for continuous water heater.

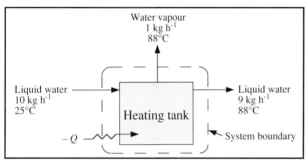

   (iii) *Define the system boundary by drawing on the flowsheet.*
   The system boundary is indicated in Figure 5E4.1.

2. *Analyse*
   (i) *State any assumptions.*
      —process is operating at steady state
      —system does not leak
      —system is homogeneous
      —evaporation occurs at 88°C
      —vapour is saturated
      —shaft work is negligible
      —no heat losses
   (ii) *Select and state a basis.*
   The calculation is based on 10 kg water entering the system, or 1 hour.
   (iii) *Select and state a reference state.*
   The reference state for water is the same as that used in the steam tables: 0.01°C and 0.6112 kPa.
   (iv) *Collect any extra data needed.*
      h (liquid water at 88°C) = 368.5 kJ kg^{-1} (Table C.1)
      h (saturated steam at 88°C) = 2656.9 kJ kg^{-1} (Table C.1)
      h (liquid water at 25°C) = 104.8 kJ kg^{-1} (Table C.1).
   (v) *Determine which compounds are involved in reaction.*
   No reaction occurs.
   (vi) *Write down the appropriate mass-balance equation.*
   The mass balance is already complete.
   (vii) *Write down the appropriate energy-balance equation.*
   At steady state, Eq. (5.9) applies:

$$\underset{\substack{\text{input}\\\text{streams}}}{\sum (Mh)} - \underset{\substack{\text{output}\\\text{streams}}}{\sum (Mh)} - Q + W_s = 0.$$

3. *Calculate*
    (i) *Identify the terms of the energy-balance equation.*
    For this problem $W_s = 0$. The energy-balance equation becomes:

    $$(Mh)_{\text{liq in}} - (Mh)_{\text{liq out}} - (Mh)_{\text{vap out}} - Q = 0.$$

    Substituting the information available:

    $$(10\ \text{kg})\ (104.8\ \text{kJ kg}^{-1}) - (9\ \text{kg})\ (368.5\ \text{kJ kg}^{-1}) - (1\ \text{kg})\ (2656.9\ \text{kJ kg}^{-1}) - Q = 0$$

    $$Q = -4925.4\ \text{kJ}.$$

    Q has a negative value. Thus, according to the sign convention outlined in Section 5.2, heat must be supplied to the system from the surroundings.

4. *Finalise*
    Answer the specific questions asked in the problem; check the number of significant figures; state the answers clearly.
    The rate of heat input is $4.93 \times 10^3\ \text{kJ h}^{-1}$.

Example 5.5 Cooling in downstream processing

In downstream processing of gluconic acid, concentrated fermentation broth containing 20% (w/w) gluconic acid is cooled in a heat exchanger prior to crystallisation. 2000 kg h^{-1} liquid leaving an evaporator at 90°C must be cooled to 6°C. Cooling is achieved by heat exchange with 2700 kg h^{-1} water initially at 2°C. If the final temperature of the cooling water is 50°C, what is the rate of heat loss from the gluconic acid solution to the surroundings? Assume the heat capacity of gluconic acid is 0.35 cal g^{-1} °C^{-1}.

Solution:
1. *Assemble*
    (i) *Units.*
    kg, h, kJ, °C.
    (ii) *Flowsheet.*
    The flowsheet is shown in Figure 5E5.1.

Figure 5E5.1 Flowsheet for cooling gluconic acid product stream.

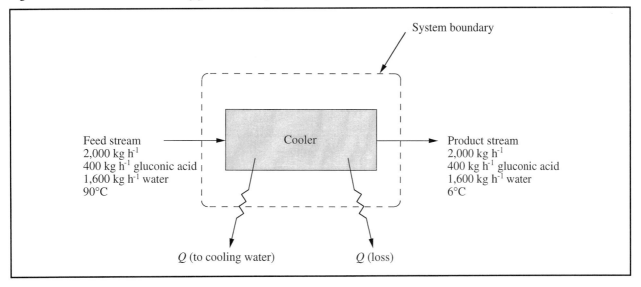

(iii) *System boundary.*

The system boundary indicated in Figure 5E5.1 separates the gluconic acid solution from the cooling water

2.   *Analyse*

(i)   *Assumptions.*

—steady state

—no leaks

—other components of the fermentation broth can be considered water

—gluconic acid and water form an ideal mixture

—no shaft work

(ii)   *Basis.*

2000 kg feed stream, or 1 hour.

(iii)   *Reference state.*

$H = 0$ for gluconic acid at 90°C

$H = 0$ for water at its triple point.

(iv)   *Extra data.*

The heat capacity of gluconic acid is 0.35 cal g^{-1} °C^{-1}; we will assume this C_p remains constant over the temperature range of interest. Converting units:

$$C_p \text{ (gluconic acid)} = \frac{0.35 \text{ cal}}{\text{g} \,°\text{C}} \cdot \left| \frac{4.187 \text{ J}}{1 \text{ cal}} \right| \cdot \left| \frac{1 \text{ kJ}}{1000 \text{ J}} \right| \cdot \left| \frac{1000 \text{ g}}{1 \text{ kg}} \right|$$

C_p (gluconic acid) = 1.47 kJ kg^{-1} °C^{-1}.

h (liquid water at 90°C) = 376.9 kJ kg^{-1} (Table C.1)

h (liquid water at 6°C) = 25.2 kJ kg^{-1} (Table C.1)

h (liquid water at 2°C) = 8.4 kJ kg^{-1} (Table C.1)

h (liquid water at 50°C) = 209.3 kJ kg^{-1} (Table C.1).

(v)   *Compounds involved in reaction.*

No reaction occurs.

(vi)   *Mass-balance equation.*

The mass-balance equation for total mass, gluconic acid and water is:

mass in = mass out.

The mass flow rates are as shown in Figure 5E5.1.

(vii)   *Energy-balance equation.*

$$\underset{\substack{\text{input} \\ \text{streams}}}{\sum (Mh)} - \underset{\substack{\text{output} \\ \text{streams}}}{\sum (Mh)} - Q + W_s = 0.$$

3.   *Calculate*

(i)   *Identify terms in the energy-balance equation.*

$W_s = 0$. There are two heat flows out of the system: one to the cooling water (Q) and one representing loss to the surroundings (Q_{loss}). With symbols W = water and G = gluconic acid, the energy-balance equation is:

$$(Mh)_{\text{W in}} + (Mh)_{\text{G in}} - (Mh)_{\text{W out}} - (Mh)_{\text{G out}} - Q_{loss} - Q = 0.$$

$(Mh)_{\text{W in}} = (1600 \text{ kg}) (376.9 \text{ kJ kg}^{-1}) = 6.03 \times 10^5 \text{ kJ}.$

$(Mh)_{\text{G in}} = 0$ (reference state).

$(Mh)_{\text{W out}} = (1600 \text{ kg}) (1.47 \text{ kJ kg}^{-1}) = 4.03 \times 10^4 \text{ kJ}.$

$(Mh)_{\text{G out}}$ at 6°C is calculated as a sensible heat change from 90°C using Eq. (5.12):

$(Mh)_{\text{G out}} = MC_p (T_2 - T_1) = (400 \text{ kg}) (1.47 \text{ kJ kg}^{-1} \text{°C}^{-1}) (6 - 90)\text{°C}$

$\therefore (Mh)_{\text{G out}} = -4.94 \times 10^4 \text{ kJ}.$

The heat removed to the cooling water, Q, is equal to the enthalpy change of the cooling water between 2°C and 50°C:

$Q = (2700 \text{ kg}) (209.3 - 8.4) \text{ kJ kg}^{-1} = 5.42 \times 10^5 \text{ kJ}.$

These results can now be substituted into the energy-balance equation:

$(6.03 \times 10^5 \text{ kJ}) + (0 \text{ kJ}) - (4.03 \times 10^4 \text{ kJ}) - (-4.94 \times 10^4 \text{ kJ}) - Q_{\text{loss}} - 5.42 \times 10^5 \text{ kJ} = 0$

$\therefore Q_{\text{loss}} = 7.01 \times 10^4 \text{ kJ}.$

4. *Finalise*
The rate of heat loss to the surroundings is $7.0 \times 10^4 \text{ kJ h}^{-1}$.

It is important to recognise that the final answers to energy-balance problems do not depend on the choice of reference states for the components. Although values of h depend on the reference states, as discussed in Section 5.3.1 this dependence disappears when the energy-balance equation is applied and the difference between input and output enthalpies determined. To prove the point, any of the examples in this chapter can be repeated using different reference conditions to obtain the same final answers.

5.8 Enthalpy Change Due to Reaction

Reactions in bioprocesses occur as a result of enzyme activity and cell metabolism. During reaction, relatively large changes in internal energy and enthalpy occur as bonds between atoms are rearranged. *Heat of reaction* ΔH_{rxn} is the energy released or absorbed during reaction, and is equal to the difference in enthalpy of reactants and products:

$$\Delta H_{\text{rxn}} = \sum_{\text{products}} Mh - \sum_{\text{reactants}} Mh$$

(5.18)

or

$$\Delta H_{\text{rxn}} = \sum_{\text{products}} nh - \sum_{\text{reactants}} nh$$

(5.19)

where Σ denotes the sum, M is mass, n is number of moles, and h is specific enthalpy expressed on either a per-mass or per-mole basis. Note that M and n represent the mass and moles

actually involved in the reaction, not the total amount present in the system. In an *exothermic reaction* the energy required to hold the atoms of product together is less than for the reactants; surplus energy is released as heat and ΔH_{rxn} is negative in value. On the other hand, energy is absorbed during *endothermic reactions*; the enthalpy of the products is greater than the reactants and ΔH_{rxn} is positive.

The specific heat of reaction Δh_{rxn} is a property of matter. The value of Δh_{rxn} depends on which reactants and products are involved in the reaction and the temperature and pressure. Because any molecule can participate in a large number of reactions, it is not feasible to tabulate all possible Δh_{rxn} values. Instead, Δh_{rxn} is calculated from the heats of combustion of individual compounds.

5.8.1 Heat of Combustion

Heat of combustion Δh_{c} is defined as the heat evolved during reaction of a substance with oxygen to yield certain oxidation products such as CO_2 gas, H_2O liquid and N_2 gas. The *standard heat of combustion* $\Delta h_{\text{c}}^{\circ}$ is the specific enthalpy change associated with this reaction at standard conditions, usually 25°C and 1 atm pressure. By convention, $\Delta h_{\text{c}}^{\circ}$ is zero for the products of oxidation, i.e. CO_2 gas, H_2O liquid, N_2 gas, etc.; standard heats of combustion for other compounds are always negative. Table B.8 in Appendix B lists selected values; heats of combustion for other materials can be found in *Chemical Engineers' Handbook* [1] and *Handbook of Chemistry and Physics* [2]. As an example, the standard heat of combustion for citric acid is given in Table B.8 as $-1962.0 \text{ kJ gmol}^{-1}$; this refers to the heat evolved at 25°C and 1 atm in the following reaction:

$$C_6H_8O_7 \text{ (s)} + 4^1/_2 O_2 \text{ (g)} \rightarrow 6\,CO_2 \text{ (g)} + 4\,H_2O \text{ (l)}.$$

Standard heats of combustion are used to calculate the *standard heat of reaction* ΔH^o_{rxn} for reactions involving combustible reactants and combustion products:

$$\Delta H^o_{rxn} = \sum_{reactants} n\,\Delta h^o_c - \sum_{products} n\,\Delta h^o_c$$

(5.20)

where n is moles of reactant or product involved in the reaction, and Δh^o_c is the standard heat of combustion per mole. The standard heat of reaction is the difference between the heats of combustion of reactants and products.

Example 5.6 Calculation of heat of reaction from heats of combustion

Fumaric acid is produced from malic acid using the enzyme, fumarase. Calculate the standard heat of reaction for the following enzyme transformation:

$$\underset{\text{(malic acid)}}{C_4H_6O_5} \rightarrow \underset{\text{(fumaric acid)}}{C_4H_4O_4} + H_2O.$$

Solution:
$\Delta h^o_c = 0$ for liquid water. From Eq. (5.20):

$$\Delta H^o_{rxn} = (n\,\Delta h^o_c)_{\text{malic acid}} - (n\,\Delta h^o_c)_{\text{fumaric acid}}$$

Table B.8 lists the standard heats of combustion for these compounds:

$$(\Delta h^o_c)_{\text{malic acid}} = -1328.8 \text{ kJ gmol}^{-1}$$

$$(\Delta h^o_c)_{\text{fumaric acid}} = -1334.0 \text{ kJ gmol}^{-1}.$$

Therefore, using a basis of 1 gmol malic acid converted:

$$\Delta H^o_{rxn} = (1 \text{ gmol}) (-1328.8 \text{ kJ gmol}^{-1}) - (1 \text{ gmol}) (-1334.0 \text{ kJ gmol}^{-1})$$

$$\Delta H^o_{rxn} = 5.2 \text{ kJ}.$$

As ΔH^o_{rxn} is positive, the reaction is endothermic and heat is absorbed.

5.8.2 Heat of Reaction at Non-Standard Conditions

Example 5.6 shows how to calculate the heat of reaction at standard conditions. However, most reactions do not occur at 25°C, and the standard heat of reaction calculated using Eq. (5.20) may not be the same as the actual heat of reaction at the reaction temperature.

Consider the following reaction between compounds A, B, C and D occurring at temperature T:

$$A + B \rightarrow C + D.$$

The standard heat of reaction at 25°C is known from tabulated heat of combustion data. ΔH_{rxn} at temperature T can be calculated using the alternative reaction pathway outlined in Figure 5.4, in which reaction occurs at 25°C and the reactants and products are heated or cooled between 25°C and T before and after the reaction. Because the initial and final states for the actual and hypothetical paths are the same, the total enthalpy change is also identical. Therefore:

$$\Delta H_{rxn} \text{ (at } T) = \Delta H_1 + \Delta H^o_{rxn} + \Delta H_3$$

(5.21)

where ΔH_1 and ΔH_3 are changes in sensible heat and ΔH^o_{rxn} is the standard heat of reaction at 25°C. ΔH_1 and ΔH_3 are evaluated using heat capacities and the methods described in Section 5.4.1.

Depending on how much T deviates from 25°C and the magnitude of ΔH^o_{rxn}, ΔH_{rxn} may not be much different from

Figure 5.4   Hypothetical process path for calculating heat of reaction at non-standard temperature.

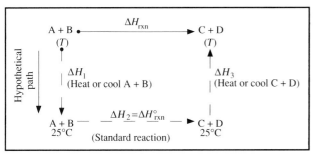

ΔH°_{rxn}. For example, consider the reaction for respiration of glucose:

$$C_6H_{12}O_6 + 6\,O_2 \rightarrow 6\,CO_2 + 6\,H_2O.$$

ΔH°_{rxn} for this conversion is -2805.0 kJ; if the reaction occurs at 37°C, ΔH_{rxn} is -2801.7 kJ. Contributions from sensible heat amount to only 3.3 kJ, which is insignificant compared with the total magnitude of ΔH°_{rxn} and can be ignored without much loss of accuracy. With reference to Figure 5.4, $\Delta H_1 = -4.8$ kJ for cooling 1 gmol glucose and 6 gmol oxygen from 37°C to 25°C; $\Delta H_3 = 8.1$ kJ for heating the products back to 37°C. Having opposite signs, ΔH_1 and ΔH_3 act to cancel each other. This situation is typical of most reactions in bioprocessing where the actual temperature of reaction is not sufficiently different from 25°C to warrant concern about sensible heat changes. When heat of reaction is substantial compared with other types of enthalpy change, ΔH_{rxn} can be assumed equal to ΔH°_{rxn} irrespective of reaction temperature.

A major exception to this general rule are single-enzyme conversions. Because most single-enzyme reactions involve only small molecular rearrangements, heats of reaction are relatively small. For instance, per mole of substrate, the fumarase reaction of Example 5.6 involves a standard enthalpy change of only 5.2 kJ; other examples are 8.7 kJ gmol^{-1} for the glucose isomerase reaction, -26.2 kJ gmol^{-1} for hydrolysis of sucrose, and -29.4 kJ per gmol glucose for hydrolysis of starch. For conversions such as these, sensible energy changes of 5 to 10 kJ are clearly significant and should not be ignored. Calculated enthalpy changes for enzyme reactions are often imprecise; being the difference between two or more large numbers, the small ΔH°_{rxn} for these conversions can carry considerable uncertainty depending on the accuracy of the heats of combustion data. When coupled with usual assumptions such as constant C_p and Δh_m within the temperature

range of interest, this uncertainty means that estimates of heating and cooling requirements for enzyme reactors are sometimes quite rough.

5.9 Heat of Reaction For Processes With Biomass Production

Biochemical reactions in cells do not occur in isolation but are linked in a complex array of metabolic transformations. Catabolic and anabolic reactions take place at the same time, so that energy released in one reaction is used in other energy-requiring processes. Cells use chemical energy quite efficiently; however some is inevitably released as heat. How can we estimate the heat of reaction associated with cell metabolism and growth?

5.9.1 Thermodynamics of Microbial Growth

As described in Section 4.6.1, a macroscopic view of cell growth is represented by the equation:

$$C_wH_xO_yN_z + a\,O_2 + b\,H_gO_hN_i \rightarrow c\,CH_\alpha O_\beta N_\delta + d\,CO_2 + e\,H_2O$$
$$(4.4)$$

where a, b, c, d and e are stoichiometric coefficients, $C_wH_xO_yN_z$ is the substrate, $H_gO_hN_i$ is the nitrogen source, and $CH_\alpha O_\beta N_\delta$ is dry biomass. Once the stoichiometric coefficients or yields are determined, Eq. (4.4) can be used as the reaction equation in energy-balance calculations. We need, however, to determine the heat of reaction for this conversion.

Heats of reaction for cell growth can be estimated using microbial stoichiometry and the concept of available electrons (Section 4.6.2). It has been found empirically that the energy content of organic compounds is related to degree of reduction as follows:

$$\Delta h^{\circ}_c = -q\,\gamma x_C$$
$$(5.22)$$

where Δh°_c is the molar heat of combustion at standard conditions, q is the heat evolved per mole of available electrons transferred to oxygen during combustion, γ is the degree of reduction of the compound defined with respect to N_2, and x_C is the number of carbon atoms in the molecular formula. The coefficient q relating Δh°_c and γ is relatively constant for a large number of compounds. Patel and Erickson [7] assigned a value of 111 kJ gmol^{-1} to q; in another analysis, Roels [8] determined a value of 115 kJ gmol^{-1}. The correlation found by Roels is based on analysis of several chemical and biochemical compounds including biomass; the results are shown in Figure 5.5.

5.9.2 Heat of Reaction With Oxygen as Electron Acceptor

The direct proportionality between heat of combustion and degree of reduction as shown in Figure 5.5 has important implications for determining the heat of reaction in aerobic culture. Degree of reduction is related directly to the amount of oxygen required for complete combustion of a substance; therefore, heat produced in reaction of compounds for which Figure 5.5 applies must be directly proportional to oxygen consumption. In aerobic cultures with oxygen the principal acceptor of electrons, because molecular oxygen O_2 accepts four electrons, if one mole O_2 is consumed during respiration, four moles of electrons must be transferred. Accepting the value of 115 kJ energy released per gmol electrons transferred, the amount of energy released from consumption of one gmol O_2 is therefore (4×115) kJ, or 460 kJ. This result, that the heat of reaction for aerobic metabolism is approximately -460 kJ per gmol O_2 consumed, is verified by the experimental data of Cooney, *et al.* [9] shown in Figure 3.2 (p. 33). The value is quite accurate for a wide range of conditions,

Figure 5.5 Relationship between degree of reduction and heat of combustion for various organic compounds. (From J.A. Roels, 1987, Thermodynamics of growth. In: J. Bu'Lock and B. Kristiansen, Eds, *Basic Biotechnology*, Academic Press, London.)

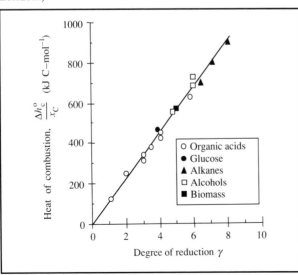

including fermentations involving product formation. Thus, once the amount of oxygen taken up during aerobic cell culture is known, the heat of reaction can be evaluated immediately.

5.9.3 Heat of Reaction With Oxygen Not the Principal Electron Acceptor

If a fermentation uses electron acceptors other than oxygen, for example in anaerobic culture, the simple relationship for heat of reaction derived in Section 5.9.2 does not apply. Heats of combustion must be used to estimate the heat of reaction for anaerobic conversions. Consider the following reaction equation for anaerobic growth with product formation:

$$C_w H_x O_y N_z + b H_g O_h N_i \rightarrow c CH_\alpha O_\beta N_\delta + d CO_2 + e H_2O + f C_j H_k O_l N_m \tag{5.23}$$

where $C_j H_k O_l N_m$ is an extracellular product and f is its stoichiometric coefficient. With ammonia as nitrogen source and heats of combustion of H_2O and CO_2 zero, from Eq. (5.20) the equation for standard heat of reaction is:

$$\Delta H^\circ_{rxn} = (n\Delta h^\circ_c)_{substrate} + (n\Delta h^\circ_c)_{NH_3} - (n\Delta h^\circ_c)_{biomass} - (n\Delta h^\circ_c)_{product} \tag{5.24}$$

where n is number of moles and Δh°_c is the standard molar heat of combustion. Heats of combustion for substrate, NH_3 and product are available from tables; what is the heat of combustion of biomass?

As shown in Table 4.3 (p. 76), the elemental composition of biomass does not vary a great deal. If we assume an average biomass molecular formula of $CH_{1.8}O_{0.5}N_{0.2}$, the reaction equation for combustion of cells to CO_2, H_2O and N_2 is:

$$CH_{1.8}O_{0.5}N_{0.2} + 1.2 O_2 \rightarrow CO_2 + 0.9 H_2O + 0.1 N_2.$$

From Table B.2, the degree of reduction of the biomass relative to N_2 is 4.80. Assuming an average of 5% ash associated with the biomass, the cell molecular weight is 25.9. Heat of combustion is obtained by applying Eq. (5.22):

$$(\Delta h^\circ_c)_{biomass} = (-115 \text{ kJ gmol}^{-1})\, (4.80)\, (1) \cdot \left| \frac{1 \text{ gmol}}{25.9 \text{ g}} \right|$$
$$= -21.3 \text{ kJ g}^{-1}. \tag{5.25}$$

Actual heats of combustion measured by Cordier *et al.* [10] for a range of microorganisms and culture conditions are listed in Table 5.1. The difference in Δh_c values for bacteria and yeast reflects their different elemental compositions. When the composition of a particular organism is unknown, the heat of combustion for bacteria can be taken as -23.2 kJ g^{-1}; for yeast Δh_c is approximately -21.2 kJ g^{-1}. These experimentally determined values compare well with that calculated in Eq. (5.25). Once the heat of combustion of biomass is known, it can be used with the heats of combustion for the other products and substrates to determine the heat of reaction.

5.10 Energy-Balance Equation For Cell Culture

In fermentations, the heat of reaction so dominates the energy balance that small enthalpy effects due to sensible heat change and heats of mixing can generally be ignored. In this section we incorporate these observations into a simplified energy-balance equation for cell processes.

Consider Eq. (5.9) applied to a continuous fermenter. What are the major factors responsible for the enthalpy difference between input and output streams in fermentations? Because cell-culture media are usually dilute aqueous solutions with behaviour close to ideal, even though the composition of the broth may change as substrates are consumed and products formed, changes in heats of mixing of these solutes are generally negligible. Similarly, even though there may be a temperature difference between input and output streams, the overall change in enthalpy due to sensible heat is also small. Usually, heat of reaction, latent heat of phase change and shaft work are the only energy effects worth considering in fermentation energy balances. Evaporation is the most likely phase change in fermenter operation; if evaporation is controlled then latent heat effects can also be ignored. Per cubic metre of fermentation broth, metabolic reactions typically generate $5-20$ kJ heat per second for growth on carbohydrate, and up to 60 kJ s^{-1} for growth on hydrocarbon substrates. By way of comparison, in aerobic cultures sparged with dry air, evaporation of the fermentation broth removes only about

Table 5.1 Heats of combustion for bacteria and yeast

(From J.-L. Cordier, B.M. Butsch, B. Birou and U. von Stockar, 1987, The relationship between elemental composition and heat of combustion of microbial biomass. Appl. Microbiol. Biotechnol. 25, 305–312)

Organism	*Substrate*	Δh_c (kJ g^{-1})
Bacteria		
Escherichia coli	glucose	-23.04 ± 0.06
	glycerol	-22.83 ± 0.07
Enterobacter cloacae	glucose	-23.22 ± 0.14
	glycerol	-23.39 ± 0.12
Methylophilus methylotrophus	methanol	-23.82 ± 0.06
Bacillus thuringiensis	glucose	-22.08 ± 0.03
Yeast		
Candida lipolytica	glucose	-21.34 ± 0.16
Candida boidinii	glucose	-20.14 ± 0.18
	ethanol	-20.40 ± 0.14
	methanol	-21.52 ± 0.09
Kluyveromyces fragilis	lactose	-21.54 ± 0.07
	galactose	-21.78 ± 0.10
	glucose	-21.66 ± 0.19
	glucose*	-21.07 ± 0.07
		-21.30 ± 0.10
		-20.66 ± 0.26
		-21.22 ± 0.14

* Chemostat rather than batch culture: dilution rates were 0.036 h^{-1}, 0.061 h^{-1}, 0.158 h^{-1} and 0.227 h^{-1}, respectively.

0.5 kJ s^{-1} m^{-3} as latent heat. Energy input due to shaft work varies between 0.5 and 5 kJ s^{-1} m^{-3} in large-scale vessels and 10–20 kJ s^{-1} m^{-3} in small vessels. Sensible heats and heats of mixing are generally several orders of magnitude smaller.

For cell processes, therefore, we can simplify energy-balance calculations by substituting expressions for heat of reaction and latent heat for the first two terms of Eq. (5.9). By the definition of Eq. (5.18), ΔH_{rxn} is the difference between product and reactant enthalpies. As the products are contained in the output flow and the reactants in the input, ΔH_{rxn} is approximately equal to the difference in enthalpy between input and output streams. If evaporation is also significant, the enthalpy of vapour leaving the system will be greater than that of liquid entering by $M_v \Delta h_v$, where M_v is the mass of liquid evaporated and Δh_v is the latent heat of vaporisation. The energy-balance equation can be modified as follows:

$$-\Delta H_{rxn} - M_v \Delta h_v - Q + W_s = 0.$$

$$(5.26)$$

ΔH_{rxn} has a negative sign in Eq. (5.26) because ΔH_{rxn} is equal to [enthalpy of products − enthalpy of reactants] whereas the energy-balance equation refers to [enthalpy of inlet streams − enthalpy of outlet streams]. Eq. (5.26) applies even if some proportion of the reactants remains unconverted or if there are components in the system which do not react. At steady state, any material added to the system that does not participate in reaction must leave in the output stream; ignoring enthalpy effects due to change in temperature or solution and unless the material volatilises, the enthalpy of unreacted material in the output stream must be equal to its inlet enthalpy. As sensible heat effects are considered negligible, the difference between ΔH^o_{rxn} and ΔH_{rxn} at the reaction temperature can be ignored.

It must be emphasised that Eq. (5.26) is greatly simplified and, as discussed in Section 5.8.2, may not be applicable to single-enzyme conversions. It is, however, a very useful equation for fermentation processes.

5.11 Fermentation Energy-Balance Worked Examples

For processes involving cell growth and metabolism the enthalpy change accompanying reaction is relatively large. Energy balances for aerobic and anaerobic cultures can therefore be carried out using the modified energy-balance equation (5.26). Because this equation contains no enthalpy terms, it is not necessary to define reference states. Application of Eq. (5.26) to anaerobic fermentation is illustrated in Example 5.7.

Example 5.7 Continuous ethanol fermentation

Saccharomyces cerevisiae is grown anaerobically in continuous culture at 30°C. Glucose is used as carbon source; ammonia is the nitrogen source. A mixture of glycerol and ethanol is produced. At steady state, mass flows to and from the reactor at steady state are as follows:

glucose in	36.0 kg h^{-1}
NH$_3$ in	0.40 kg h^{-1}
cells out	2.81 kg h^{-1}
glycerol out	7.94 kg h^{-1}
ethanol out	11.9 kg h^{-1}
CO$_2$ out	13.6 kg h^{-1}
H$_2$O out	0.15 kg h^{-1}

Estimate the cooling requirements.

Solution:
1. *Assemble*
    (i) *Units.*
    kg, kJ, h, °C.
    (ii) *Flowsheet.*
    The flowsheet for this process is shown in Figure 5E7.1.

Figure 5E7.1 Flowsheet for anaerobic yeast fermentation.

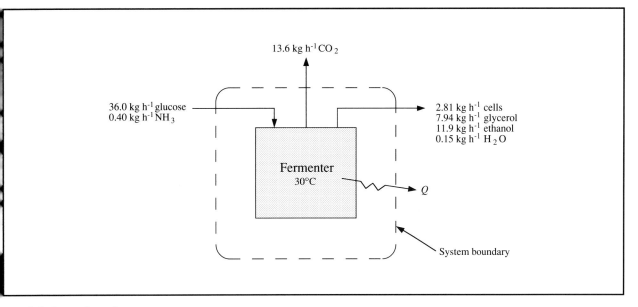

(iii) *System boundary.*

The system boundary is shown in Figure 5E7.1.

2. *Analyse*

   (i) *Assumptions.*

      —steady state

      —no leaks

      —system is homogeneous

      —heat of combustion for yeast is -21.2 kJ g^{-1}

      —ideal solutions

      —negligible sensible heat change

      —no shaft work

      —no evaporation

   (ii) *Basis.*

      36.0 kg glucose, or 1 hour.

   (iii) *Extra data.*

      MW glucose $\qquad = 180$

      MW NH_3 $\qquad = 17$

      MW glycerol $\qquad = 92$

      MW ethanol $\qquad = 46$

      Heats of combustion:

         $(\Delta h_c^o)_{\text{glucose}} = -2805.0 \text{ kJ gmol}^{-1}$ (Table B.8)

         $(\Delta h_c^o)_{\text{NH}_3} = -382.6 \text{ kJ gmol}^{-1}$ (Table B.8)

         $(\Delta h_c^o)_{\text{glycerol}} = -1655.4 \text{ kJ gmol}^{-1}$ (Table B.8)

         $(\Delta h_c^o)_{\text{ethanol}} = -1366.8 \text{ kJ gmol}^{-1}$ (Table B.8).

   (iv) *Reaction.*

      glucose $+ NH_3 \rightarrow$ biomass $+$ glycerol $+$ ethanol $+ CO_2 + H_2O$.

All components are involved in reaction.

(v) *Mass-balance equation.*

The mass balance is already complete.

(vi) *Energy-balance equation.*

For cell metabolism, the modified energy-balance equation is Eq. (5.26):

$$-\Delta H_{rxn} - M_v \Delta h_v - Q + W_s = 0.$$

3. *Calculate*

(i) *Identify terms in the energy-balance equation.*

$W_s = 0$; $M_v = 0$. Therefore:

$$-\Delta H_{rxn} - Q = 0.$$

Evaluate the heat of reaction using Eq. (5.20). As the heat of combustion of H_2O and CO_2 is zero, the heat of reaction is:

$$\Delta H_{rxn} = (n\Delta h_c^o)_G + (n\Delta h_c^o)_A - (n\Delta h_c^o)_B - (n\Delta h_c^o)_{Gly} - (n\Delta h_c^o)_E$$

where G = glucose, A = ammonia, B = cells, Gly = glycerol and E = ethanol. Converting to a mass basis:

$$\Delta H_{rxn} = (M\Delta h_c^o)_G + (M\Delta h_c^o)_A - (M\Delta h_c^o)_B - (M\Delta h_c^o)_{Gly} - (M\Delta h_c^o)_E$$

where Δh_c^o is expressed per unit mass. Converting the Δh_c^o data to kJ kg^{-1}:

$$(\Delta h_c^o)_G = -2805.0 \ \frac{kJ}{gmol} \cdot \left| \frac{1 \ gmol}{180 \ g} \right| \cdot \left| \frac{1000 \ g}{1 \ kg} \right| = -1.558 \times 10^4 \ kJ \ kg^{-1}$$

$$(\Delta h_c^o)_A = -382.6 \ \frac{kJ}{gmol} \cdot \left| \frac{1 \ gmol}{17 \ g} \right| \cdot \left| \frac{1000 \ g}{1 \ kg} \right| = -2.251 \times 10^4 \ kJ \ kg^{-1}$$

$$(\Delta h_c^o)_B = -21.2 \ \frac{kJ}{g} \cdot \left| \frac{1000 \ g}{1 \ kg} \right| = -2.120 \times 10^4 \ kJ \ kg^{-1}$$

$$(\Delta h_c^o)_{Gly} = -1655.4 \ \frac{kJ}{gmol} \cdot \left| \frac{1 \ gmol}{92 \ g} \right| \cdot \left| \frac{1000 \ g}{1 \ kg} \right| = -1.799 \times 10^4 \ kJ \ kg^{-1}$$

$$(\Delta h_c^o)_E = -1366.8 \ \frac{kJ}{gmol} \cdot \left| \frac{1 \ gmol}{46 \ g} \right| \cdot \left| \frac{1000 \ g}{1 \ kg} \right| = -2.971 \times 10^4 \ kJ \ kg^{-1}$$

Therefore:

$$\Delta H_{rxn} = (36.0 \ kg) (-1.558 \times 10^4 \ kJ \ kg^{-1}) + (0.4 \ kg) (-2.251 \times 10^4 \ kJ \ kg^{-1}) - (2.81 \ kg) (-2.120 \times 10^4 \ kJ \ kg^{-1})$$
$$- (7.94 \ kg) (-1.799 \times 10^4 \ kJ \ kg^{-1}) - (11.9 \ kg) (-2.971 \times 10^4 \ kJ \ kg^{-1})$$
$$= -1.392 \times 10^4 \ kJ.$$

Substituting this result into the energy-balance equation:

$$Q = 1.392 \times 10^4 \ kJ.$$

Q is positive indicating that heat must be removed from the system.

4. *Finalise*
1.4×10^4 kJ heat must be removed from the fermenter per hour.

In Example 5.7, the water used as solvent for components of the nutrient medium was ignored. This water was effectively a tie component, moving through the system unchanged and not contributing to the energy balance. Cooling requirements could be determined directly from the heat of reaction.

For aerobic cultures, we can relate the heat of reaction directly to oxygen consumption, providing a short-cut method for determining ΔH_{rxn}. Heats of combustion are not required in these calculations. Also, as long as the amount of oxygen consumed is known, the mass balance for the problem need not be completed. The procedure for energy-balance problems involving aerobic fermentation is illustrated in Example 5.8.

Example 5.8 Citric acid production

Citric acid is manufactured using submerged culture of *Aspergillus niger* in a batch reactor operated at 30°C. Over a period of two days, 2500 kg glucose and 860 kg oxygen are consumed to produce 1500 kg citric acid, 500 kg biomass and other products. Ammonia is used as nitrogen source. Power input to the system by mechanical agitation of the broth is about 15 kW; approximately 100 kg water is evaporated over the culture period. Estimate the cooling requirements.

Solution:
1. *Assemble*
    (i)   *Units.*
    kg, kJ, h, °C.
    (ii) *Flowsheet.*
    The flowsheet is shown in Figure 5E8.1.

Figure 5E8.1   Flowsheet for microbial production of citric acid.

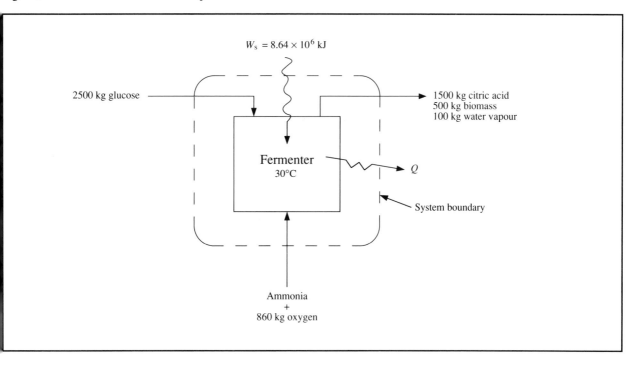

(iii) *System boundary.*
The system boundary is shown in Figure 5E8.1.

2.   *Analyse*
     (i)   *Assumptions.*
           —system is homogeneous
           —no leaks
           —ideal solutions
           —negligible sensible heat
           —heat of reaction at 30°C is -460 kJ gmol^{-1} O_2 consumed
     (ii) *Basis.*
     1500 kg citric acid produced, or 2 days.
     (iii) *Extra data.*
     Δh_v water at 30°C = 2430.7 kJ kg^{-1} (Table C.1)
     (iv) *Reaction.*

$$\text{glucose} + O_2 + NH_3 \rightarrow \text{biomass} + CO_2 + H_2O + \text{citric acid.}$$

All components are involved in reaction.
     (v)   *Mass balance.*
     The mass balance need not be completed as the sensible energy associated with inlet and outlet streams is negligible.
     (vi) *Energy balance.*
     The aim of the integral energy balance for batch culture is to calculate the amount of heat which must be removed to produce zero accumulation of energy in the system. Eq. (5.26) is appropriate:

$$-\Delta H_{rxn} - M_v \Delta h_v - Q + W_s = 0$$

where each term refers to the two-day culture period.

3.   *Calculate*
     (i)   *Identify terms in the energy-balance equation.*
           ΔH_{rxn} is related to the amount of oxygen consumed:

$$\Delta H_{rxn} = (-460 \text{ kJ gmol}^{-1}) (860 \text{ kg}) \cdot \left| \frac{1000 \text{ g}}{1 \text{ kg}} \right| \cdot \left| \frac{1 \text{ gmol}}{32 \text{ g}} \right|$$

$$= -1.24 \times 10^7 \text{ kJ.}$$

Heat lost through evaporation is:

$$M_v \Delta h_v = (100 \text{ kg}) (2430.7 \text{ kJ kg}^{-1}) = 2.43 \times 10^5 \text{ kJ.}$$

Power input by mechanical agitation is 15 kW or 15 kJ s^{-1}. Over a period of 2 days:

$$W_s = (15 \text{ kJ s}^{-1}) (2 \text{ days}) \cdot \left| \frac{3600 \text{ s}}{1 \text{ h}} \right| \cdot \left| \frac{24 \text{ h}}{1 \text{ day}} \right| = 2.59 \times 10^6 \text{ kJ.}$$

These results can now be substituted into the energy-balance equation.

$$-(-1.24 \times 10^7 \text{ kJ}) - (2.43 \times 10^5 \text{ kJ}) - Q + (2.59 \times 10^6 \text{ kJ}) = 0.$$

$$Q = 1.47 \times 10^7 \text{ kJ}.$$

Q is positive, indicating that heat must be removed from the system. Note the relative magnitudes of the energy contributions from heat of reaction, shaft work and evaporation; the effects of evaporation can often be ignored.

4. *Finalise*
    1.5×10^7 kJ heat must be removed from the fermenter per 1500 kg citric acid produced.

5.12 Summary of Chapter 5

At the end of Chapter 5 you should:

(i)   know which forms of energy are common in bioprocesses;

(ii)   know the *general energy balance* in words and as a mathematical equation, and the simplifications that can be made for bioprocesses;

(iii) be familiar with *heat capacity* tables and be able to calculate *sensible heat changes*;

(iv)   be able to calculate *latent heat changes*;

(v)   understand *heats of mixing* for non-ideal solutions;

(vi)   be able to use *steam tables*;

(vii) be able to determine *standard heats of reaction* from *heats of combustion*;

(viii) know how to determine heats of reaction for aerobic and anaerobic cell cultures; and

(ix)   be able to carry out energy-balance calculations for biological systems with and without reaction.

Problems

5.1 Sensible energy change

Calculate the enthalpy change associated with the following processes:

(a) *m*-cresol is heated from 25°C to 100°C;

(b) ethylene glycol is cooled from 20°C to 10°C;

(c) succinic acid is heated from 15°C to 120°C; and

(d) air is cooled from 150°C to 65°C.

5.2 Heat of vaporisation

Nitrogen is sometimes bubbled into fermenters to maintain anaerobic conditions. It does not react, and leaves in the fermenter off-gas. However it can strip water from the fermenter, so that water vapour also leaves in the off-gas. In a continuous fermenter operated at 33°C, 20 g h^{-1} water is evaporated in this way. How much heat must be put into the system to compensate for evaporative cooling?

5.3 Steam tables

Use the steam tables to find:

(a) the heat of vaporisation of water at 85°C;

(b) the enthalpy of liquid water at 35°C relative to $H = 0$ at 10°C;

(c) the enthalpy of saturated water vapour at 40°C relative to $H = 0$ at the triple point; and

(d) the enthalpy of superheated steam at 2.5 atm absolute pressure and 275°C relative to $H = 0$ at the triple point.

5.4 Pre-heating nutrient medium

Steam is used to heat nutrient medium in a continuous-flow process. Saturated steam at 150°C enters a coil on the outside of the heating vessel and is completely condensed. Medium enters the vessel at 15°C and leaves at 44°C. Heat losses from the jacket to the surroundings are estimated as 0.22 kW. If the flow rate of medium is 3250 kg h^{-1} and the heat capacity is $C_p = 0.9$ cal g^{-1} °C^{-1}, how much steam is required?

5.5 Production of glutamic acid

Immobilised cells of a genetically-improved strain of *Brevibacterium lactofermentum* are used to convert glucose to glutamic acid for production of MSG (monosodium glutamate). The immobilised cells are unable to grow, but metabolise glucose according to the equation:

$$C_6H_{12}O_6 + NH_3 + 1^1/_2\, O_2 \rightarrow C_5H_9NO_4 + CO_2 + 3H_2O.$$

A feed stream of 4% glucose in water enters a 25 000-litre reactor at 25°C at a flow rate of 2000 kg h^{-1}. A gaseous mixture of 12% NH$_3$ in air is sparged into the reactor at 1 atm and 15°C at a flow rate of 4 vvm (1 vvm means 1 vessel volume per minute). The product stream from the reactor contains residual sugar at a concentration of 0.5%.

(a) Estimate the cooling requirements.

(b) How important is cooling in this fermentation? For

example, assuming the reaction rate remains constant irrespective of temperature, if cooling were not provided and the reactor operated adiabatically, what would be the temperature? (In fact, the rate of conversion will decline rapidly at high temperatures due to cell death and enzyme deactivation.)

5.6 Bacterial production of alginate

Azotobacter vinelandii is investigated for production of alginate from sucrose. In a continuous fermenter at 28°C with ammonia as nitrogen source, the yield of alginate was found to be 4 g g^{-1} oxygen consumed. It is planned to produce alginate at a rate of 5 kg h^{-1}. Since the viscosity of alginate in aqueous solution is considerable, energy input due to mixing the broth cannot be neglected. The fermenter is equipped with a flat-bladed disc turbine; at a satisfactory mixing speed and air flow-rate, power requirements are estimated at 1.5 kW. Estimate the cooling requirements.

5.7 Acid fermentation

Propionibacterium species are tested for commercial-scale production of propionic acid. Propionic and other acids are synthesised in anaerobic culture using sucrose as substrate and ammonia as nitrogen source. Overall yields from sucrose are as follows:

propionic acid	40% (w/w)
acetic acid	20% (w/w)
butyric acid	5% (w/w)
lactic acid	3.4% (w/w)
biomass	12% (w/w)

Bacteria are inoculated into a vessel containing sucrose and ammonia; a total of 30 kg sucrose is consumed over a period of 10 days. What are the cooling requirements?

5.8 Ethanol fermentation

A crude fermenter is set up in a shed in the backyard of a suburban house. Under anaerobic conditions with ammonia as nitrogen source, about 0.45 g ethanol are formed per g glucose consumed. At steady state, the production rate of ethanol averages 0.4 kg h^{-1}.

The owner of this enterprise decides to reduce her electricity bill by using the heat released during the fermentation to warm water as an adjunct to the household hot-water system. 2.5 litres h^{-1} cold water at 10°C is fed into a jacket surrounding the fermenter. To what temperature is the water heated?

Heat losses from the system are negligible. Use a biomass composition of $CH_{1.75}O_{0.58}N_{0.18}$ plus 8% ash.

5.9 Production of bakers' yeast

Bakers' yeast is produced in a 50 000-litre fermenter under aerobic conditions. The carbon substrate is sucrose; ammonia is provided as nitrogen source. The average biomass composition is $CH_{1.83}O_{0.55}N_{0.17}$ with 5% ash. Under efficient growth conditions, biomass is the only major product; the biomass yield from sucrose is 0.5 g g^{-1}. If the specific growth-rate is 0.45 h^{-1}, estimate the rate of heat removal required to maintain constant temperature in the fermenter when the yeast concentration is 10 g l^{-1}.

References

1. Perry, R.H., D.W. Green and J.O. Maloney (Eds) (1984) *Chemical Engineers' Handbook*, 6th edn, McGraw-Hill, New York.
2. *Handbook of Chemistry and Physics*, CRC Press, Boca Raton.
3. *International Critical Tables* (1926) McGraw-Hill, New York.
4. Perry, R.H. and C.H. Chilton (Eds) (1973) *Chemical Engineers' Handbook*, 5th edn, McGraw-Hill, Tokyo.
5. Atkinson, B. and F. Mavituna (1991) *Biochemical Engineering and Biotechnology Handbook*, 2nd edn, Macmillan, Basingstoke.
6. Haywood, R.W. (1972) *Thermodynamic Tables in SI (Metric) Units*, 2nd edn, Cambridge University Press, Cambridge.
7. Patel, S.A. and L.E. Erickson (1981) Estimation of heats of combustion of biomass from elemental analysis using available electron concepts. *Biotechnol. Bioeng.* **23**, 2051–2067.
8. Roels, J.A. (1983) *Energetics and Kinetics in Biotechnology*, Chapter 3, Elsevier Biomedical Press, Amsterdam.
9. Cooney, C.L., D.I.C. Wang and R.I. Mateles (1968) Measurement of heat evolution and correlation with oxygen consumption during microbial growth. *Biotechnol. Bioeng.* **11**, 269–281.
10. Cordier, J.-L., B.M. Butsch, B. Birou and U. von Stockar (1987) The relationship between elemental composition and heat of combustion of microbial biomass. *Appl. Microbiol. Biotechnol.* **25**, 305–312.

Suggestions for Further Reading

Process Energy Balances

Felder, R.M. and R.W. Rousseau (1978) *Elementary Principles of Chemical Processes*, Chapters 8–10, John Wiley, New York.

Himmelblau, D.M. (1974) *Basic Principles and Calculations in Chemical Engineering*, 3rd edn, Chapter 4, Prentice-Hall, New Jersey.

Shaheen, E.I. (1975) *Basic Practice of Chemical Engineering*, Chapter 5, Houghton Mifflin, Boston.

Whitwell, J.C. and R.K. Toner (1969) *Conservation of Mass and Energy*, Chapters 6–9, Blaisdell, Waltham, Massachusetts.

Metabolic Thermodynamics

Erickson, L.E. (1980) Biomass elemental composition and energy content. *Biotechnol. Bioeng.* **22**, 451–456.

Marison, I. and U. von Stockar (1987) A calorimetric investigation of the aerobic cultivation of *Kluyveromyces fragilis* on various substrates. *Enzyme Microbiol. Technol.* **9**, 33–43.

Minkevich, I.G. and V.K. Eroshin (1973) Productivity and heat generation of fermentation under oxygen limitation. *Folia Microbiol.* **18**, 376–385.

Nagai, S. (1979) Mass and energy balances for microbial growth kinetics. *Adv. Biochem. Eng.* **11**, 49–83.

Roels, J.A. (1980) Application of macroscopic principles to microbial metabolism. *Biotechnol. Bioeng.* **22**, 2457–2514.

Roels, J.A. (1987) Thermodynamics of growth. In: J. Bu'Lock and B. Kristiansen (Eds), *Basic Biotechnology*, pp. 57–74, Academic Press, London.

6

Unsteady-State Material and Energy Balances

An unsteady-state or transient process is one which causes system properties to vary with time. Batch and semi-batch systems are inherently transient; continuous systems are unsteady during start-up and shut-down. Changing from one set of process conditions to another also creates an unsteady state, as does any fluctuation in input or control variables.

The principles of mass and energy balances developed in Chapters 4 and 5 can be applied to unsteady-state processes. Balance equations are used to determine the rate of change of system parameters; solution of these equations generally requires application of calculus. Questions such as what is the concentration of product in the reactor as a function of time?, and how long will it take to reach a particular temperature after flow of steam is started? can be answered using unsteady-state mass and energy balances. In this chapter we will consider some simple unsteady-state problems.

6.1 Unsteady-State Material-Balance Equations

When mass of a system is not constant, we generally need to know how the mass varies as a function of time. To evaluate the *rate of change* of mass in the system, let us first return to the general mass-balance equation introduced in Chapter 4 see p. 52).

$$\left\{\begin{matrix}\text{mass in}\\\text{through}\\\text{system}\\\text{boundaries}\end{matrix}\right\} - \left\{\begin{matrix}\text{mass out}\\\text{through}\\\text{system}\\\text{boundaries}\end{matrix}\right\} + \left\{\begin{matrix}\text{mass}\\\text{generated}\\\text{within}\\\text{system}\end{matrix}\right\} - \left\{\begin{matrix}\text{mass}\\\text{consumed}\\\text{within}\\\text{system}\end{matrix}\right\} = \left\{\begin{matrix}\text{mass}\\\text{accumulated}\\\text{within}\\\text{system}\end{matrix}\right\}.$$

$$(4.1)$$

Consider the flow system of Figure 6.1 in which reactions are taking place. Species A is involved in the process; M is the mass of A in the system. Using the 'hat' symbol ^ to denote rate, let $\hat{M}_i$ be the mass flow rate of A entering the system, and $\hat{M}_o$ the mass flow rate of A leaving. R_G is the mass rate of generation of species A by chemical reaction; R_C is the mass rate of consumption by reaction. The dimensions of $\hat{M}_i$, $\hat{M}_o$, R_G and R_C are MT^{-1}, with units such as g s^{-1}, kg h^{-1}, lb min^{-1}, etc.

All of the variables: $\hat{M}_i$, $\hat{M}_o$, R_G and R_C, may vary with time. However, let us focus on an infinitesimally-small

Figure 6.1 Flow system for an unsteady-state masss balance.

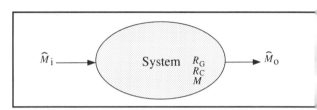

interval of time Δt between times t and $t + \Delta t$. Even though the system variables may be changing, if Δt is sufficiently small we can treat the flow rates $\hat{M}$ and rates of reaction R as if they were constant during this period. Under these circumstances, the terms of the general balance equation (4.1) may be written as follows.

(i)   *Mass in.* During period Δt, the mass of species A transported into the system is $\hat{M}_i \Delta t$. Note that the dimensions of $\hat{M}_i \Delta t$ are M, with units g, kg, lb, etc.

(ii) *Mass out.* Similarly, the mass of species A transported out during time Δt is $\hat{M}_o \Delta t$.

(iii) *Generation.* The mass of A generated during Δt is $R_G \Delta t$.

(iv) *Consumption.* The mass of A consumed during Δt is $R_C \Delta t$.

(v)   *Accumulation.* Let ΔM be the mass of A accumulated in the system during Δt. ΔM may be either positive (accumulation) or negative (depletion).

Entering these terms into the general mass balance Eq. (4.1) with the accumulation term on the left-hand side:

$$\Delta M = \hat{M}_i \Delta t - \hat{M}_o \Delta t + R_G \Delta t - R_C \Delta t.$$

$$(6.1)$$

We can divide both sides of Eq. (6.1) by Δt to give:

$$\frac{\Delta M}{\Delta t} = \hat{M}_i - \hat{M}_o + R_G - R_C.$$

(6.2)

Eq. (6.2) applies when Δt is infinitesimally small. If we take the limit as Δt approaches zero, i.e. as t and $t + \Delta t$ become virtually the same, Eq. (6.2) represents the system at an instant rather than over an interval of time. Mathematical techniques for handling this type of situation are embodied in the rules of calculus. In calculus, the *derivative* of y with respect to x, $\frac{dy}{dx}$, is defined as:

$$\frac{dy}{dx} = \lim_{\Delta x \to 0} \frac{\Delta y}{\Delta x}$$

(6.3)

where

$$\lim_{\Delta x \to 0}$$

represents the limit as Δx approaches zero. As Eq. (6.2) is valid for $\Delta t \to 0$, we can write:

$$\frac{dM}{dt} = \lim_{\Delta t \to 0} \frac{\Delta M}{\Delta t} = \hat{M}_i - \hat{M}_o + R_G - R_C.$$

(6.4)

The derivative $\frac{dM}{dt}$ represents the rate of change of mass with time measured at a particular instant. We have thus derived a differential equation for rate of change of M as a function of system variables $\hat{M}_i$, $\hat{M}_o$, R_G and R_C:

$$\frac{dM}{dt} = \hat{M}_i - \hat{M}_o + R_G - R_C.$$

(6.5)

At steady state there can be no change in mass of the system, so the rate of change, $\frac{dM}{dt}$ must be zero. At steady state, therefore, Eq. (6.5) reduces to the familiar steady-state mass-balance equation (see p. 53):

mass in + mass generated = mass out + mass consumed.

(4.2)

Unsteady-state mass-balance calculations begin with derivation of a differential equation to describe the process. Eq. (6.5) was developed on a mass basis and contains parameters such as mass flow rate $\hat{M}$ and mass rate of reaction R. Another common form of the unsteady-state mass balance is based on volume. The reason for this variation is that reaction rates are usually expressed on a per-volume basis. For example, the rate of a first-order reaction is expressed in terms of the concentration of reactant:

$$r_C = k_1 C_A$$

where r_C is the *volumetric rate of consumption of A by reaction* with units, e.g. g cm^{-3} s^{-1}, k_1 is the first-order reaction-rate constant, and C_A is the concentration of reactant A. This and other reaction-rate equations are described in more detail in Chapter 11. When rate expressions are used in mass- and energy-balance problems, the relationship between mass and volume must enter the analysis. This is illustrated in Example 6.1.

Example 6.1 Unsteady-state material balance for a CSTR

A continuous stirred-tank reactor is operated as shown in Figure 6E1.1. The volume of liquid in the tank is V. Feed enters with volumetric flow rate F_i; product leaves with flow rate F_o. The concentration of reactant A in the feed is C_{Ai}; the concentration of A in the exit stream is C_{Ao}. The density of the feed stream is ρ_i; the density of the product stream is ρ_o. The tank is well mixed. The concentration of A in the tank is C_A and the density of liquid in the tank is ρ. In the reactor, compound A undergoes reaction and is transformed into compound B. The volumetric rate of consumption of A by reaction is give by the expression $r_C = k_1 C_A$.

Figure 6E1.1   Continuous stirred-tank reactor.

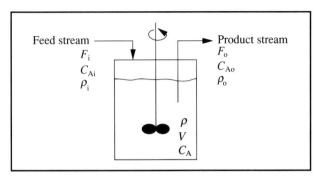

Using unsteady-state balances, derive differential equations for:

(a) total mass; and
(b) mass of component A.

Solution:
The general unsteady-state mass-balance equation is Eq. (6.5):

$$\frac{dM}{dt} = \hat{M}_i - \hat{M}_o + R_G - R_C.$$

(a) For the balance on total mass R_G and R_C are zero; total mass cannot be generated or consumed by chemical reaction. From the definition of density (Section 2.4.1), total mass can be expressed as the product of volume and density. Similarly, mass flow rate can be expressed as the product of volumetric flow rate and density.

> Total mass in the tank: $M = \rho V$; therefore $\dfrac{dM}{dt} = \dfrac{d(\rho V)}{dt}$
> Mass flow rate in: $\hat{M}_i = F_i \rho_i$
> Mass flow rate out: $\hat{M}_o = F_o \rho_o.$

Substituting these terms into Eq. (6.5):

$$\frac{d(\rho V)}{dt} = F_i \rho_i - F_o \rho_o.$$

(6.6)

Eq. (6.6) is a differential equation representing the unsteady-state total-mass balance.

(b) A is not generated in the reaction; therefore $R_G = 0$. The other terms of Eq. (6.5) can be expressed as follows.

> Mass of A in the tank: $M = V C_A$; therefore $\dfrac{dM}{dt} = \dfrac{d(V C_A)}{dt}$
> Mass flow rate of A in: $\hat{M}_i = F_i C_{Ai}$
> Mass flow rate of A out: $\hat{M}_o = F_o C_{Ao}$
> Rate of consumption of A: $R_C = k_1 C_A V.$

Substituting into Eq. (6.5):

$$\frac{\mathrm{d}(VC_A)}{\mathrm{d}t} = F_i C_{Ai} - F_o C_{Ao} - k_1 C_A V$$

(6.7)

Eq. (6.7) is the differential equation representing the unsteady-state mass balance on A.

6.2 Unsteady-State Energy-Balance Equations

In Chapter 5, the law of conservation of energy was represented by the equation (see p. 87):

$$\left\{\begin{array}{c}\text{energy in through}\\\text{system boundaries}\end{array}\right\} - \left\{\begin{array}{c}\text{energy out through}\\\text{system boundaries}\end{array}\right\} = \left\{\begin{array}{c}\text{energy accumulated}\\\text{within the system}\end{array}\right\}.$$

(5.4)

Consider the system shown in Figure 6.2. E is the total energy in the system, $\hat{W}_s$ is the rate at which shaft work is done on the system, $\hat{Q}$ is the rate of heat removal from the system, and $\hat{M}_i$ and $\hat{M}_o$ are mass flow rates to and from the system. All these parameters may vary with time.

Ignoring kinetic and potential energies as discussed in Section 5.2.1, the energy-balance equation can be applied over an infinitesimally-small interval of time Δt, during which we can treat $\hat{W}_s$, $\hat{Q}$, $\hat{M}_i$ and $\hat{M}_o$ as if they were constant. Under these conditions, the terms of the general energy-balance equation are:

(i) *Input.* During time interval Δt, the amount of energy entering the system is $\hat{M}_i h_i \Delta t + \hat{W}_s \Delta t$, where h_i is the specific enthalpy of the incoming flow stream.
(ii) *Output.* Similarly, the amount of energy leaving the system is $\hat{M}_o h_o \Delta t + \hat{Q} \Delta t$.
(iii) *Accumulation.* Let ΔE be the energy accumulated in the system during time Δt. ΔE may be either positive (accumulation) or negative (depletion). In the absence of kinetic and potential energies, E represents the enthalpy of the system.

Entering these terms into Eq. (5.4) with the accumulation term first:

$$\Delta E = \hat{M}_i h_i \Delta t - \hat{M}_o h_o \Delta t - \hat{Q} \Delta t + \hat{W}_s \Delta t.$$

(6.8)

We can divide both sides of Eq. (6.8) by Δt:

$$\frac{\Delta E}{\Delta t} = \hat{M}_i h_i - \hat{M}_o h_o - \hat{Q} + \hat{W}_s.$$

(6.9)

Figure 6.2   Flow system for an unsteady-state energy balance.

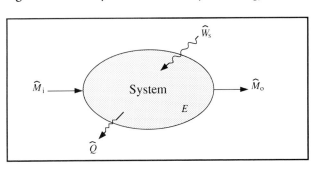

Eq. (6.9) is valid for small Δt. The equation for rate of change of energy at a particular instant in time is determined by taking the limit of Eq. (6.9) as Δt approaches zero:

$$\frac{\mathrm{d}E}{\mathrm{d}t} = \lim_{\Delta t \to 0} \frac{\Delta E}{\Delta t} = \hat{M}_i h_i - \hat{M}_o h_o - \hat{Q} + \hat{W}_s.$$

(6.10)

Eq. (6.10) is the unsteady-state energy-balance equation for a system with only one inlet and one outlet stream. If there are several such streams, all mass flow rates and enthalpies must be added together:

$$\frac{\mathrm{d}E}{\mathrm{d}t} = \underset{\substack{\text{input}\\\text{streams}}}{\Sigma (\hat{M}h)} - \underset{\substack{\text{output}\\\text{streams}}}{\Sigma (\hat{M}h)} - \hat{Q} + \hat{W}_s.$$

(6.11)

Eq. (6.11) can be simplified for fermentation processes using the same arguments as those presented in Section 5.10. If $\Delta \hat{H}_{rxn}$ is the rate at which heat is absorbed or liberated by reaction, and $\hat{M}_v$ is the mass flow rate of evaporated liquid leaving the system, for fermentation processes in which sensible heat changes and heats of mixing can be ignored, the following unsteady-state energy-balance equation applies:

$$\frac{\mathrm{d}E}{\mathrm{d}t} = -\Delta \hat{H}_{rxn} - \hat{M}_v \Delta h_v - \hat{Q} + \hat{W}_s.$$

(6.12)

For exothermic reactions $\Delta \hat{H}_{rxn}$ is negative; for endothermic reactions $\Delta \hat{H}_{rxn}$ is positive.

6.3 Solving Differential Equations

As shown in Sections 6.1 and 6.2, unsteady-state mass and energy balances are represented using differential equations. Once the differential equation for a particular system has been found, the equation must be solved to obtain an expression for mass M or energy E as a function of time. Differential equations are solved by integration. Some simple rules for differentiation and integration are outlined in Appendix D. Of course, there are many more rules of calculus than those included in Appendix D; however those shown are sufficient for handling the unsteady-state problems in this chapter. Further details can be found in any elementary calculus textbook, or in mathematics handbooks written especially for biological scientists, e.g. [1–3].

Before we proceed with solution techniques for unsteady-state mass and energy balances, there are several general points to consider.

(i) *A differential equation can be solved directly only if it contains no more than two variables.* For mass- and energy-balance problems, the differential equation must have the form:

$$\frac{dM}{dt} = f(M, t) \quad \text{or} \quad \frac{dE}{dt} = f(E, t)$$

where $f(M, t)$ represents some function of M and t, and $f(E, t)$ represents some function of E and t. The function may contain constants, but no other variables besides M and t should appear in the expression for dM/dt, and no other variables besides E and t should appear in the expression for dE/dt. Before you attempt to solve these differential equations, check first that all parameters except M and t, or E and t, are constants.

(ii) *Solution of differential equations requires knowledge of boundary conditions.* Boundary conditions contain extra information about the system. The number of boundary conditions required depends on the *order* of the differential equation, which is equal to the order of the highest differential coefficient in the equation. For example, if the equation contains a second derivative, e.g. d^2x/dt^2, the equation is second order. All equations developed in this chapter have been first order; they involve only first order derivatives of the form dx/dt. One boundary condition is required to solve a first-order differential equation; two

boundary conditions are required for a second-order differential equation, and so on. Boundary conditions which apply at the beginning of the process when $t = 0$ are called *initial conditions*.

(iii) *Not all differential equations can be solved algebraically,* even if the equation contains only two variables and the boundary conditions are available. Solution of some differential equations requires application of numerical techniques, preferably using a computer. In this chapter we will be concerned mostly with simple equations that can be solved using elementary calculus.

The easiest way of solving differential equations is to *separate variables* so that each variable appears on only one side of the equation. For example, consider the simple differential equation:

$$\frac{dx}{dt} = a(b - x) \tag{6.13}$$

where a and b are constants. First we must check that the equation contains only two variables x and t, and that all other parameters in the equation are constants. Once this is verified, the equation is separated so that x and t each appear on only one side of the equation. In the case of Eq. (6.13), this is done by dividing each side of the equation by $(b - x)$, and multiplying each side by dt:

$$\frac{1}{(b - x)} dx = a \, dt. \tag{6.14}$$

The equation is now ready for integration:

$$\int \frac{1}{(b - x)} dx = \int a \, dt. \tag{6.15}$$

Using integration rules (D.28) and (D.24) from Appendix D:

$$- \ln(b - x) = at + K. \tag{6.16}$$

Note that the constants of integration from both sides of the equation have been condensed into one constant K; this is valid because a constant $\pm$ a constant = a constant.

6.4 Solving Unsteady-State Mass Balances

Solution of unsteady-state mass balances is sometimes difficult unless certain simplifications are made. Because the aim here is to illustrate application of unsteady-state balances without becoming too involved in integral calculus, the problems presented in this section will be relatively simple. For the majority of problems in this chapter analytical solution is possible.

The following restrictions are common in unsteady-state mass-balance problems.

(i) The system is *well mixed* so that properties of the system do not vary with position. If properties within the system are the same at all points, this includes the point from which any product stream is drawn. Accordingly, when the system is well mixed, properties of the outlet stream are the same as those within the system.

(ii) Expressions for reaction rate involve the concentration of only one reactive species. The mass-balance equation for this species can then be derived and solved; if other species appear in the kinetic expression this introduces extra variables into the differential equation making solution more complex.

The following example illustrates solution of an unsteady-state mass balance without reaction.

Example 6.2 Dilution of salt solution

1.5 kg salt is dissolved in water to make 100 litres. Pure water runs into a tank containing this solution at a rate of 5 l min^{-1}; salt solution overflows at the same rate. The tank is well mixed. How much salt is in the tank at the end of 15 min? Assume the density of salt solution is constant and equal to that of water.

Solution:
(i) *Flowsheet and system boundary.*
These are shown in Figure 6E2.1.

Figure 6E2.1 Well-mixed tank for dilution of salt solution.

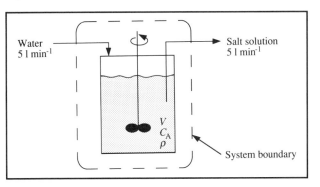

(ii) *Define variables.*
C_A = concentration of salt in the tank; V = volume of solution in the tank; ρ = density of salt solution and water.
(iii) *Assumptions.*
  —no leaks
  —tank is well mixed
  —density of salt solution is the same as water
(iv) *Boundary conditions.*
At the beginning of the process, the salt concentration in the tank is 1.5 kg in 100 l, or 0.015 kg l^{-1}. If we call this initial salt concentration C_{A0}, the initial condition is:

$$\text{at } t = 0 \quad C_A = C_{A0} = 0.015 \text{ kg l}^{-1}.$$

$$(1)$$

We also know that the initial volume of liquid in the tank is 100 l. Another initial condition is:

at $t = 0$ $V = V_0 = 100$ l.

$$(2)$$

(v) *Total mass balance.*
The unsteady-state balance equation for total mass was derived in Example 6.1:

$$\frac{d(\rho V)}{dt} = F_i \rho_i - F_o \rho_o.$$

$$(6.6)$$

In this problem we are told that the volumetric flow rates of inlet and outlet streams are equal; therefore $F_i = F_o$. In addition, the density of the system is constant so that $\rho_i = \rho_o = \rho$. Under these conditions, the terms on the right-hand side of Eq. (6.6) cancel. On the left-hand side, because ρ is constant it can be taken outside of the differential. Therefore:

$$\rho \frac{dV}{dt} = 0$$

or

$$\frac{dV}{dt} = 0.$$

If the derivative of V with respect to t is zero, V must be a constant:

$$V = K$$

where K is the constant of integration. This result means that the volume of the tank is constant and independent of time. Initial condition (2) tells us that $V = 100$ l at $t = 0$; therefore V must equal 100 l at all times. Consequently, the constant of integration K is equal to 100 l, and the volume of liquid in the tank does not vary from 100 l.

(vi) *Mass balance for salt.*
The unsteady mass-balance equation for component A such as salt was derived in Example 6.1:

$$\frac{d(VC_A)}{dt} = F_i C_{Ai} - F_o C_{Ao} - k_1 C_A V.$$

$$(6.7)$$

In the present problem there is no reaction, so k_1 is zero. Also, $F_i = F_o = F = 5$ l min^{-1}. Because the tank is well mixed, the concentration of salt in the outlet stream is equal to that inside the tank, i.e. $C_{Ao} = C_A$. In addition, since the inlet stream does not contain salt, $C_{Ai} = 0$. From the balance on total mass we know that V is constant and can be placed outside of the differential. Taking these factors into consideration, Eq. (6.7) becomes:

$$V \frac{dC_A}{dt} = -F C_A.$$

This differential equation contains only two variables C_A and t; F and V are constants. The variables are easy to separate by dividing both sides by $V C_A$ and multiplying by dt:

$$\frac{\mathrm{d}C_\mathrm{A}}{C_\mathrm{A}} = \frac{-F}{V}\,\mathrm{d}t.$$

The equation is now ready to integrate:

$$\int \frac{\mathrm{d}C_\mathrm{A}}{C_\mathrm{A}} = \int \frac{-F}{V}\,\mathrm{d}t.$$

Using integration rules (D.27) and (D.24) from Appendix D and combining the constants of integration:

$$\ln C_\mathrm{A} = \frac{-F}{V}\,t + K. \tag{3}$$

We have yet to determine the value of K. From initial condition (1), at $t = 0$, $C_\mathrm{A} = C_\mathrm{A0}$. Substituting this information into (3):

$$\ln C_\mathrm{A0} = K.$$

We have thus determined K. Substituting this value for K back into (3):

$$\ln C_\mathrm{A} = \frac{-F}{V}\,t + \ln C_\mathrm{A0}. \tag{4}$$

This is the solution to the mass balance; it gives an expression for concentration of salt in the tank as a function of time. Notice that if we had forgotten to add the constant of integration, the answer would not contain the term $\ln C_\mathrm{A0}$. The equation would then say that at $t = 0$, $\ln C_\mathrm{A} = 0$; i.e. $C_\mathrm{A} = 1$. We know this is not true; instead, at $t = 0$, $C_\mathrm{A} = 0.015$ kg l^{-1}, so the result without the boundary condition is incorrect. It is important to apply boundary conditions every time you integrate.

The solution equation is usually rearranged to give an exponential expression. This is achieved by subtracting $\ln C_\mathrm{A0}$ from both sides of (4):

$$\ln C_\mathrm{A} - \ln C_\mathrm{A0} = \frac{-F}{V}\,t$$

and noting from Eq. (D-9) that $(\ln C_A - \ln C_\mathrm{A0})$ is the same as $\ln {C_\mathrm{A}}/{C_\mathrm{A0}}$:

$$\ln \frac{C_\mathrm{A}}{C_\mathrm{A0}} = \frac{-F}{V}\,t.$$

Taking the anti-logarithm of both sides:

$$\frac{C_\mathrm{A}}{C_\mathrm{A0}} = e^{(-F/V)\,t}$$

or

$$C_\mathrm{A} = C_\mathrm{A0}\,e^{(-F/V)\,t}.$$

We can check that this is the correct solution by taking the derivative of both sides with respect to t and making sure that the original differential equation is recovered.

For $F = 5 \, \mathrm{l \, min^{-1}}$, $V = 100 \, \mathrm{l}$ and $C_{A0} = 0.015 \, \mathrm{kg \, l^{-1}}$, at $t = 15 \, \mathrm{min}$:

$$C_A = (0.015 \, \mathrm{kg \, l^{-1}}) \, e^{(-5 \, 1 \, \mathrm{min^{-1}}/100 \, \mathrm{l})(15 \, \mathrm{min})} = 7.09 \times 10^{-3} \, \mathrm{kg \, l^{-1}}.$$

The salt concentration after 15 min is $7.09 \times 10^{-3} \, \mathrm{kg \, l^{-1}}$. Therefore:

Mass of salt $= C_A \, V = (7.09 \times 10^{-3} \, \mathrm{kg \, l^{-1}}) \, 100 \, \mathrm{l} = 0.71 \, \mathrm{kg}.$

(vii) *Finalise.*
After 15 min, the mass of salt in the tank is 0.71 kg.

In Example 6.2 the density of the system was assumed constant. This simplified the mathematics of the problem so that ρ could be taken outside the differential and cancelled from the total mass balance. The assumption of constant density is justified for dilute solutions because the density does not differ greatly from that of the solvent. The result of the total mass balance makes intuitive sense: for a tank with equal flow rates in and out and constant density, the volume of liquid inside the tank should remain constant.

The effect of reaction on the unsteady-state mass balance is illustrated in Example 6.3.

Example 6.3 Flow reactor

Rework Example 6.2 to include reaction. Assume that a reaction in the tank consumes salt, at a rate given by the first-order equation:

$$r = k_1 \, C_A$$

where k_1 is the first-order reaction constant and C_A is the concentration of salt in the tank. Derive an expression for C_A as a function of time. If $k_1 = 0.02 \, \mathrm{min^{-1}}$, how long does it take for the concentration of salt to fall to a value $^1/_{20}$ the initial level?

Solution:
The flowsheet, boundary conditions and assumptions for this problem are the same as in Example 6.2. The total mass balance is also the same; total mass in the system is unaffected by reaction.

(i) *Mass balance for salt.*
From Example 6.1, the unsteady mass-balance equation for salt is:

$$\frac{d(VC_A)}{dt} = F_i C_{Ai} - F_o C_{Ao} - k_1 C_A V.$$

(6.7)

In this problem $F_i = F_o = F$, $C_{Ai} = 0$, and V is constant. Because the tank is well mixed $C_{Ao} = C_A$. Therefore, Eq. (6.7) becomes:

$$V \frac{dC_A}{dt} = -FC_A - k_1 C_A V.$$

This equation contains only two variables C_A and t; F, V and k_1 are constants. Separate variables by dividing both sides by $V C_A$ and multiplying by dt:

$$\frac{dC_A}{C_A} = \left(\frac{-F}{V} - k_1\right) dt.$$

Integrating both sides gives:

$$\ln C_A = \left(\frac{-F}{V} - k_1\right) t + K.$$

where K is the constant of integration. K is determined from initial condition (1) in Example 6.2: at $t = 0$, $C_A = C_{A0}$. Substituting these values gives:

$$\ln C_{A0} = K.$$

Substituting this value for K back into the answer:

$$\ln C_A = \left(\frac{-F}{V} - k_1\right) t + \ln C_{A0}$$

or

$$\ln \frac{C_A}{C_{A0}} = \left(\frac{-F}{V} - k_1\right) t.$$

For $F = 5 \, \mathrm{l \, min^{-1}}$, $V = 100 \, \mathrm{l}$, $k_1 = 0.02 \, \mathrm{min^{-1}}$ and $C_A / C_{A0} = 1/20$, this equation becomes:

$$\ln \left(\frac{1}{20}\right) = \left(\frac{-5 \, \mathrm{l \, min^{-1}}}{100 \, \mathrm{l}} - 0.02 \, \mathrm{min^{-1}}\right) t$$

or

$$-3.00 = (0.07 \, \mathrm{min^{-1}}) \, t.$$

Solving for t:

$$t = 42.8 \, \mathrm{min}.$$

(ii) *Finalise.*
The concentration of salt in the tank reaches $1/20$ its initial level after 43 min.

6.5 Solving Unsteady-State Energy Balances

Solution of unsteady-state energy-balance problems can be mathematically quite complex. In this chapter only problems with the following characteristics will be treated for ease of mathematical handling.

(i)   The system has at most one input and one output stream; furthermore, these streams have the same mass flow rate.

Under these conditions, total mass of the system is constant.

(ii) The system is well mixed with uniform temperature and composition. Properties of the outlet stream are therefore the same as within the system.

(iii) No chemical reactions or phase changes occur.

(iv) Mixtures and solutions are ideal.

(v)   Heat capacities of the system contents and inlet and outlet streams are independent of composition and temperature, and therefore invariant with time.

(vi) Internal energy U and enthalpy H are independent of pressure.

The principles and equations for unsteady-state energy balances are entirely valid when these conditions are not met; the only difference is that solution of the differential equation is greatly simplified for systems with the above characteristics. The procedure for solution of unsteady-state energy balances is illustrated in Example 6.4.

Example 6.4 Solvent heater

An electric heating-coil is immersed in a stirred tank. Solvent at 15°C with heat capacity 2.1 kJ kg^{-1} °C^{-1} is fed into the tank at a rate of 15 kg h^{-1}. Heated solvent is discharged at the same flow rate. The tank is filled initially with 125 kg cold solvent at 10°C. The rate of heating by the electric coil is 800 W. Calculate the time required for the temperature of the solvent to reach 60°C.

Solution:
(i) *Flowsheet and system boundary.*
These are shown in Figure 6E4.1.

Figure 6E4.1 Continuous process for heating solvent.

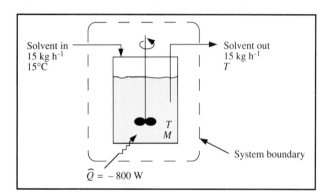

(ii) *Define variables.*
If the tank is well mixed, the temperature of the outlet stream is the same as inside the tank. Let T be the temperature in the tank; T_i is the temperature of the incoming stream, M is the mass of solvent in the tank, and $\hat{M}$ is the mass flow rate of solvent to and from the tank.
(iii) *Assumptions.*
   —no leaks
   —tank is well mixed
   —negligible shaft work
   —heat capacity is independent of temperature
   —no evaporation
   —heat losses to the environment are negligible
(iv) *Reference state.*
$H = 0$ for solvent at 10°C, i.e. $T_{ref} = 10$°C.
(v) *Boundary conditions.*
The initial condition is:

at $t = 0$ $T = T_0 = 10$°C.

$$(1)$$

(vi) *Mass balance.*

Since the mass flow rates of solvent to and from the tank are equal, the mass M of solvent inside the tank is constant and equal to the initial value, 125 kg. The mass balance is therefore complete.

(vii) *Energy balance.*

The unsteady-state energy-balance equation for two flow streams is given by Eq. (6.10):

$$\frac{dE}{dt} = \hat{M}_i h_i - \hat{M}_o h_o - \hat{Q} + \hat{W}_s.$$

In the absence of phase change, reaction and heats of mixing, the enthalpies of input and output streams can be determined from sensible heats only. Similarly, any change in the energy content of the system must be due to change in temperature and sensible heat. With solvent enthalpy defined as zero at T_{ref}, change in E can be calculated from the difference between the system temperature and T_{ref}. Therefore, the terms of the energy-balance equation are:

Accumulation: $\dfrac{dE}{dt} = \dfrac{d}{dt}(MC_p \Delta T) = \dfrac{d}{dt}(MC_p[T - T_{ref}])$

As M, C_p and T_{ref} are constants, $\dfrac{dE}{dt} = MC_p \dfrac{dT}{dt}$

Flow input: $\hat{M}_i h_i = \hat{M}_i C_p(T_i - T_{ref})$

Flow output: $\hat{M}_o h_o = \hat{M}_o C_p(T - T_{ref})$

Shaft work: $\hat{W}_s = 0$.

Substituting these expressions into the energy-balance equation gives:

$$MC_p \frac{dT}{dt} = \hat{M}_i C_p(T_i - T_{ref}) - \hat{M}_o C_p(T - T_{ref}) - \hat{Q}.$$

$M = 125$ kg, $C_p = 2.1$ kJ kg^{-1} °C^{-1}, $\hat{M}_i = \hat{M}_o = 15$ kg h^{-1}, $T_i = 15$°C and $T_{ref} = 10$°C. Converting data for $\hat{Q}$ into consistent units:

$$\hat{Q} = -800 \text{ W} . \left|\frac{1 \text{ J s}^{-1}}{1 \text{ W}}\right| . \left|\frac{1 \text{ kJ}}{1000 \text{ J}}\right| . \left|\frac{3600 \text{ s}}{1 \text{ h}}\right|$$

$$= -2.88 \times 10^3 \text{ kJ h}^{-1}.$$

$\hat{Q}$ is negative because heat flows into the system. Substituting these values into the energy-balance equation:

$$(125 \text{ kg})(2.1 \text{ kJ kg}^{-1}°\text{C}^{-1})\frac{dT}{dt} = (15 \text{ kg h}^{-1})(2.1 \text{ kJ kg}^{-1}°\text{C}^{-1})(15°\text{C} - 10°\text{C})$$
$$- (15 \text{ kg h}^{-1})(2.1 \text{ kJ kg}^{-1}°\text{C}^{-1})(T - 10°\text{C}) - (-2.88 \times 10^3 \text{ kJ h}^{-1}).$$

Grouping terms gives a differential equation for rate of temperature change:

$$\frac{dT}{dt} = 12.77 - 0.12 T$$

where T has units °C and t has units h. Separating variables and integrating:

$$\int \frac{dT}{12.77 - 0.12\,T} = \int dt.$$

Using the integration rule (D.28) from Appendix D:

$$\frac{-1}{0.12} \ln (12.77 - 0.12\,T) = t + K.$$

The initial condition is: at $t = 0$, $T = 15$°C. Therefore, $K = -19.96$ and the solution is:

$$19.96 - \frac{1}{0.12} \ln (12.77 - 0.12\,T) = t.$$

From this equation, at $T = 60$°C, $t = 5.65$ h.
(viii) *Finalise.*
It takes 5.7 h for the temperature to reach 60°C.

6.6 Summary of Chapter 6

At the end of Chapter 6 you should:
(i) know what types of process require unsteady-state analysis;
(ii) be able to derive appropriate *differential equations* for unsteady-state mass and energy balances;
(iii) understand the need for *boundary conditions* to solve differential equations representing actual processes; and
(iv) be able to solve simple unsteady-state mass and energy balances to obtain equations for system parameters as a function of time.

Problems

6.1 Dilution of sewage

In a sewage-treatment plant, a large concrete tank initially contains 440 000 litres liquid and 10 000 kg fine suspended solids. To flush this material out of the tank, water is pumped into the vessel at a rate of 40 000 litres h^{-1}. Liquid containing solids leaves at the same rate. Estimate the concentration of suspended solids in the tank at the end of 5 h.

6.2 Production of fish-protein concentrate

Whole gutted fish are dried to make a protein paste. In a batch drier, the rate at which water is removed from the fish is roughly proportional to the moisture content. If a batch of gutted fish loses half its initial moisture content in the first 20 min, how long will the drier take to remove 95% of the water?

6.3 Contamination of vegetable oil

Vegetable oil is used in a food-processing factory for preparing instant breadcrumbs. A stirred tank is used to hold the oil; during operation of the breadcrumb process, oil is pumped from the tank at a rate of 4.8 l h^{-1}. At 8 p.m. during the night shift, the tank is mistakenly connected to a drum of cod-liver oil which is then pumped into the tank. The volume of vegetable oil in the tank at 8 p.m. is 60 l.

(a) If the flow rate of cod-liver oil into the tank is 7.5 l h^{-1} and the tank has a maximum capacity of 100 l, will the tank overflow before the factory manager arrives at 9 a.m.? Assume that the density of both oils is the same.
(b) If cod-liver oil is pumped into the tank at a rate of 4.8 l h^{-1} instead of 7.5 l h^{-1}, what is the composition of oil in the tank at midnight?

6.4 Batch growth of bacteria

During exponential phase in batch culture, the growth rate of a culture is proportional to the concentration of cells present. When *Streptococcus lactis* bacteria are cultured in milk, the concentration of cells doubles in 45 min. If this rate of growth is maintained for 12 h, what is the final concentration of cells relative to the inoculum level?

6.5 Radioactive decay

A radioactive isotope decays at a rate proportional to the amount of isotope present. If the concentration of isotope is C (mg l^{-1}), its rate of decay is:

$$r_C = k_1 C.$$

(a) A solution of radioactive isotope is prepared at concentration C_0. Show that the half-life of the isotope, i.e. the time required for the isotope concentration to reach half of its original value, is equal to $^{\ln 2}/_{k_1}$.

(b) A solution of the isotope ^{32}P is used to radioactively label DNA for hybridisation studies. The half-life of ^{32}P is 14.3 days. According to institutional safety requirements, the solution cannot be discarded until the activity is 1% its present value. How long will this take?

6.6 Continuous fermentation

A well-mixed fermenter of volume V contains cells initially at concentration x_0. A sterile feed enters the fermenter with volumetric flow rate F; fermentation broth leaves at the same rate. The concentration of substrate in the feed is s_i. The equation for rate of cell growth is:

$$r_X = k_1 x$$

and the expression for rate of substrate consumption is:

$$r_S = k_2 x$$

where k_1 and k_2 are rate constants with dimensions T^{-1}, r_X and r_S have dimensions L^{-3}MT^{-1}, and x is the concentration of cells in the fermenter.

(a) Derive a differential equation for the unsteady-state mass balance of cells.

(b) From this equation, what must be the relationship between F, k_1 and the volume of liquid in the fermenter at steady state?

(c) Solve the differential equation to obtain an expression for cell concentration in the fermenter as a function of time.

(d) Use the following data to calculate how long it takes for the cell concentration in the fermenter to reach 4.0 g l^{-1}:

$$F = 2200 \, \text{l h}^{-1}$$

$$
\begin{aligned}
V &= 10\,000 \, \text{l} \\
x_0 &= 0.5 \, \text{g l}^{-1} \\
k_1 &= 0.33 \, \text{h}^{-1}.
\end{aligned}
$$

V is the volume of liquid in the fermenter.

(e) Set up a differential equation for the mass balance of substrate. Substitute the result for x from (c) to obtain a differential equation in which the only variables are substrate concentration and time. (Do you think you would be able to solve this equation algebraically?)

(f) At steady state, what must be the relationship between s and x?

6.7 Fed-batch fermentation

A feed stream containing glucose enters a fed-batch fermenter at constant flow rate. The initial volume of liquid in the fermenter is V_0. Cells in the fermenter consume glucose at a rate given by:

$$r_S = k_1 s$$

where k_1 is the rate constant (h^{-1}) and s is the concentration of glucose in the fermenter (g l^{-1}).

(a) Assuming constant density, derive an equation for the total mass balance. What is the expression relating volume and time?

(b) Derive the differential equation for rate of change of substrate concentration.

6.8 Plug-flow reactor

When fluid flows through a pipe or channel with sufficiently large Reynolds number, it approximates *plug flow*. Plug flow means that there is no variation of axial velocity over the flow cross-section. When reaction occurs in a plug-flow tubular reactor (PFTR), as reactant is consumed its concentration changes down the length of the tube.

(a) Derive the differential equation for change in reactant concentration with distance at steady state.

(b) What are the boundary conditions?

(c) If the reaction is first order, solve the differential equation to determine an expression for concentration as a function of distance from the front of the tube.

(d) How does this expression compare with that for a well-mixed batch reactor?

Hint: Referring to Figure 6P8.1, consider the accumulation of reactant within a section of the reactor between z and $z + \Delta z$.

In this case, the volumetric flow rate of liquid in and out of the section is Au. Use the following symbols:

A is the reactor cross-sectional area

u is the fluid velocity

z is distance along the tube from the entrance

L is the total length of the reactor

C_{Ai} is the concentration of reactant in the feed stream

C_A is the concentration of reactant in the reactor; C_A is a function of z

r_C is the volumetric rate of consumption of reactant.

6.9 Boiling water

A beaker containing 2 litres water at 18°C is placed on a laboratory hot-plate. The water begins to boil in 11 min.

(a) Neglecting evaporation, write the energy balance for the process.
(b) The hot-plate delivers heat at a constant rate. Assuming that the heat capacity of water is constant, what is that rate?

6.10 Heating glycerol solution

An adiabatic stirred tank is used to heat 100 kg of a 45% glycerol solution in water. An electrical coil delivers 2.5 kW of power to the tank; 88% of the energy delivered by the coil goes into heating the vessel contents. The glycerol solution is initially at 15°C.

(a) Write a differential equation for the energy balance.
(b) Integrate the equation to obtain an expression for temperature as a function of time.
(c) Assuming glycerol and water form an ideal solution, how long will the solution take to reach 90°C?

6.11 Heating molasses

Diluted molasses is heated in a well-stirred steel tank by saturated steam at 40 psia condensing in a jacket on the outside of the tank. The outer walls of the jacket are insulated. 1020 kg h^{-1} molasses solution at 20°C enters the tank, and 1020 kg h^{-1} of heated molasses leaves. The rate of heat transfer from the steam through the jacket and to the molasses is given by the equation:

$$\hat{Q} = UA(T_{steam} - T_{molasses})$$

where $\hat{Q}$ is the rate of heat transfer, U is the overall heat-transfer coefficient, A is the surface area for heat transfer, and T is the temperature. For this system the value of U is 190 kcal m^{-2} h^{-1} °C^{-1}; C_p for the molasses solution is 0.85 kcal kg^{-1} °C^{-1}. The initial mass of molasses solution in the tank is 5000 kg; the initial temperature is 20°C. The surface area for heat transfer between the steam and tank is 1.5 m^2.

(a) Derive the differential equation describing the rate of change of temperature in the tank.
(b) Solve the differential equation to obtain an equation relating temperature and time.
(c) Plot the temperature of molasses leaving the tank as a function of time.
(d) What is the maximum temperature that can be achieved in the tank?
(e) Estimate the time required for this system to reach steady state.
(f) How long does it take for the outlet molasses temperature to rise from 20°C to 40°C?

6.12 Pre-heating culture medium

A glass fermenter used for culture of hybridoma cells contains nutrient medium at 15°C. The fermenter is wrapped in an electrical heating mantle which delivers heat at a rate of 450 W. Before inoculation, the medium and vessel must be at 36°C. The medium is well mixed during heating. Use the following information to determine the time required for medium pre-heating.

Figure 6P8.1   Plug-flow tubular reactor (PFTR).

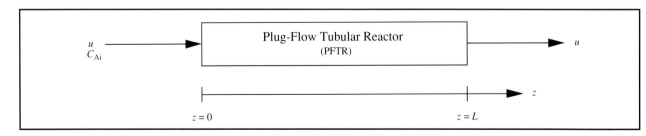

Glass fermenter vessel: mass = 12.75 kg; $C_p = 0.20$ cal g^{-1} °C^{-1}
Nutrient medium: mass = 7.50 kg; $C_p = 0.92$ cal g^{-1} °C^{-1}.

6.13 Water heater

A tank contains 1000 kg water at 24°C. It is planned to heat this water using saturated steam at 130°C in a coil inside the tank. The rate of heat transfer from the steam is given by the equation:

$$\hat{Q} = UA(T_{\text{steam}} - T_{\text{water}})$$

where $\hat{Q}$ is the rate of heat transfer, U is the overall heat-transfer coefficient, A is the surface area for heat transfer, and T is the temperature. The heat-transfer area provided by the coil is 0.3 m^2; the heat-transfer coefficient is 220 kcal m^{-2} h^{-1} °C^{-1}. Condensate leaves the coil saturated.

(a) The tank has a surface area of 0.9 m^2 exposed to the ambient air. The tank exchanges heat through this exposed surface at a rate given by an equation similar to that above. For heat transfer to or from the surrounding air the heat-transfer coefficient is 25 kcal m^{-2} h^{-1} °C^{-1}. If the air temperature is 20°C, calculate the time required to heat the water to 80°C.

(b) What time is saved if the tank is insulated?

Assume the heat capacity of water is constant, and neglect the heat capacity of the tank walls.

References

1. Cornish-Bowden, A. (1981) *Basic Mathematics for Biochemists*, Chapman and Hall, London.
2. Newby, J.C. (1980) *Mathematics for the Biological Sciences*, Oxford University Press, Oxford.
3. Arya, J.C. and R.W. Lardner (1979) *Mathematics for the Biological Sciences*, Prentice-Hall, New Jersey.

Suggestions for Further Reading

Felder, R.M. and R.W. Rousseau (1978) *Elementary Principles of Chemical Processes*, Chapter 11, John Wiley, New York.

Himmelblau, D.M. (1974) *Basic Principles and Calculations in Chemical Engineering*, 3rd edn, Chapter 6, Prentice-Hall, New Jersey.

Shaheen, E.I. (1975) *Basic Practice of Chemical Engineering*, Chapter 4, Houghton Mifflin, Boston, Massachusetts.

Whitwell, J.C. and R.K. Toner (1969) *Conservation of Mass and Energy*, Chapter 9, Blaisdell, Waltham, Massachusetts.

Part 3
Physical
Processes

7

Fluid Flow and Mixing

Fluid mechanics is an important area of engineering science. The nature of flow in pipes, pumps and reactors depends on the power input to the system and the physical characteristics of the fluid. In fermenters, fluid properties affect process energy requirements and the effectiveness of mixing, which can have a dramatic influence on productivity and the success of equipment scale-up. As we shall see in the following chapters, transport of heat and mass is often coupled with fluid flow. To understand the mechanisms of these important transport processes, we must first examine the behaviour of fluid near surfaces and interfaces. Fluids in bioprocessing often contain suspended solids, consist of more than one phase, and have non-Newtonian properties. All of these features complicate analysis of flow behaviour and present many challenges in bioprocess design.

Fluid dynamics accounts for a substantial fraction of the chemical engineering literature. Accordingly, complete treatment of the subject is beyond the scope of this book. Here, we content ourselves with study of those aspects of flow behaviour particularly relevant to fermentation fluids. Further information can be found in the references at the end of the chapter.

7.1 Classification of Fluids

A fluid is a substance which undergoes continuous deformation when subjected to a shearing force. A simple shearing force is one which causes thin parallel plates to slide over each other, as in a pack of cards. Shear can also occur in other geometries; the effect of shear force in planar and rotational systems is illustrated in Figure 7.1. Shear forces in these examples cause *deformation*, which is a change in the relative positions of parts of a body. A shear force must be applied to produce fluid flow.

According to the above definition, fluids can be either gases or liquids. Two physical properties, viscosity and density, are used to classify fluids. If the density of a fluid changes with pressure, the fluid is *compressible*. Gases are generally classed as compressible fluids. The density of liquids is practically independent of pressure; liquids are *incompressible* fluids. Sometimes the distinction between compressible and incompressible fluid is not well defined; for example, a gas may be treated as incompressible if variations of pressure and temperature are small.

Fluids are also classified on the basis of viscosity. Viscosity is the property of fluids responsible for internal friction during flow. An *ideal* or *perfect* fluid is a hypothetical liquid or gas which is incompressible and has zero viscosity. The term *inviscid* applies to fluids with zero viscosity. All *real* fluids have finite viscosity and are therefore called *viscid* or *viscous* fluids.

Fluids can be classified further as *Newtonian* or *non-Newtonian*. This distinction is explained in detail in Sections 7.3 and 7.5.

Figure 7.1 Laminar deformation due to (a) planar shear and (b) rotational shear. (From J.R. van Wazer, J.W. Lyons, K.Y. Kim and R.E. Colwell, 1963, *Viscosity and Flow Measurement*, Interscience, New York.)

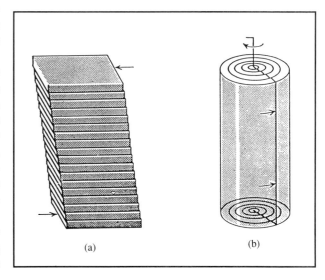

(a)   (b)

7.2 Fluids in Motion

Bioprocesses involve fluids in motion in vessels and pipes. General characteristics of fluid flow are described in the following sections.

7.2.1 Streamlines

When a fluid flows through a pipe or over a solid object, the velocity of the fluid varies depending on position. One way of representing variation in velocity is *streamlines*, which follow the flow path. Constant velocity is shown by equidistant spacing of parallel streamlines as shown in Figure 7.2(a). The velocity profile for slow-moving fluid flowing over a submerged object is shown in Figure 7.2(b); reduced spacing between the streamlines indicates that the velocity at the top and bottom of the object is greater than at the front and back.

Streamlines show only the net effect of fluid motion; although streamlines suggest smooth continuous flow, fluid molecules may actually be moving in an erratic fashion. The slower the flow the more closely the streamlines represent actual motion. Slow fluid flow is therefore called *streamline* or *laminar* flow. In fast motion, fluid particles frequently cross and recross the streamlines. This motion is called *turbulent* flow and is characterised by formation of *eddies*.

7.2.2 Reynolds Number

Transition from laminar to turbulent flow depends not only on the velocity of the fluid, but also on its viscosity and density and the geometry of the flow conduit. A parameter used to characterise fluid flow is the *Reynolds number*. For full flow in pipes with circular cross-section, Reynolds number Re is defined as:

$$Re = \frac{Du\rho}{\mu}$$

(7.1)

where D is pipe diameter, u is average linear velocity of the fluid, ρ is fluid density, and μ is fluid viscosity. For stirred vessels there is another definition of Reynolds number:

$$Re_i = \frac{N_i D_i^2 \rho}{\mu}$$

(7.2)

where Re_i is the *impeller Reynolds number*, N_i is stirrer speed, D_i is impeller diameter, ρ is fluid density and μ is fluid viscosity.

Figure 7.2 Streamlines for (a) constant fluid velocity; (b) steady flow over a submerged object.

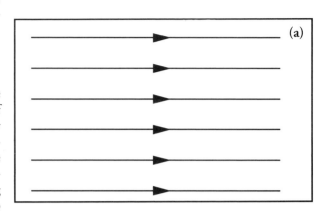

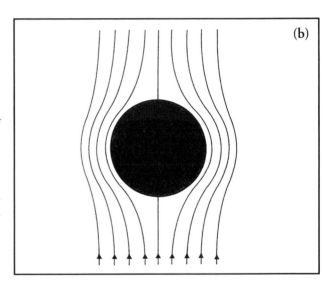

The Reynolds number is a dimensionless variable; the units and dimensions of the parameters in Eqs (7.1) and (7.2) cancel completely.

Reynolds number is named after Osborne Reynolds, who published in 1883 a classical series of papers on the nature of flow in pipes. One of the most significant outcomes of Reynolds' experiments is that there is a *critical Reynolds number* which marks the upper boundary for laminar flow in pipes. In smooth pipes, laminar flow is encountered at Reynolds numbers less than 2100. Under normal conditions, flow is turbulent at Re above about 4000. Between 2100 and 4000 is the *transition region* where flow may be either laminar or turbulent depending on conditions at the entrance of the pipe and other variables. Flow in stirred tanks may also be laminar or turbulent as a function of the impeller Reynolds number.

The value of Re_i marking the transition between these flow regimes depends on the geometry of the impeller and tank; for several commonly-used mixing systems, laminar flow is found at $Re_i \lesssim 10$.

7.2.3 Hydrodynamic Boundary Layers

In most practical applications, fluid flow occurs in the presence of a stationary solid surface, such as the walls of a pipe or tank. That part of the fluid where flow is affected by the solid is called the *boundary layer*. As an example, consider flow of fluid parallel to the flat plate shown in Figure 7.3. Contact between the moving fluid and the plate causes formation of a boundary layer beginning at the leading edge and developing on both top and bottom of the plate. Figure 7.3 shows only the upper stream; fluid motion below the plate will be a mirror image of that above.

As indicated by the arrows in Figure 7.3(a), the bulk fluid velocity in front of the plate is uniform and of magnitude u_B. The extent of the boundary layer is indicated by the broken

line. Above the boundary layer, fluid motion is the same as if the plate were not there. The boundary layer grows in thickness from the leading edge until it develops its full size. Final thickness of the boundary layer depends on the Reynolds number for bulk flow.

When fluid flows over a stationary object, a thin film of fluid in contact with the surface adheres to it to prevent slippage over the surface. Fluid velocity at the surface of the plate in Figure 7.3 is therefore zero. When part of a flowing fluid has been brought to rest, the flow of adjacent fluid layers is slowed down by the action of *viscous drag*. This phenomenon is illustrated in Figure 7.3(b). Velocity of fluid within the boundary layer, u, is represented by arrows; u is zero at the surface of the plate. Viscous drag forces are transmitted upwards through the fluid from the stationary layer at the surface. The fluid layer just above the surface moves at a slow but finite velocity; layers further above move at increasing velocity as the drag forces associated with the stationary layer decrease. At the edge of the boundary layer, fluid is unaffected by the presence of the plate and the velocity is close to that of the bulk flow, u_B. The magnitude of u at various points in the boundary layer is indicated in Figure 7.3(b) by the length of the arrows in the direction of flow. The line connecting the heads of the velocity arrows shows the *velocity profile* in the fluid. A *velocity gradient*, i.e. a change in velocity with distance from the plate, is thus established in a direction perpendicular to the direction of flow. The velocity gradient forms as the drag force resulting from retardation of fluid at the surface is transmitted through the fluid.

Formation of boundary layers is important not only in determining characteristics of fluid flow, but also for transfer of heat and mass between phases. These topics are discussed further in Chapters 8 and 9.

7.2.4 Boundary-Layer Separation

What happens when contact is broken between a fluid and a solid immersed in the flow path? As an example, consider a flat plate aligned perpendicular to the direction of fluid flow, as shown in Figure 7.4. Fluid impinges on the surface of the plate, and forms a boundary layer as it flows either up or down the object. When fluid reaches the top or bottom of the plate its momentum prevents it from making the sharp turn around the edge. As a result, fluid separates from the plate and proceeds outwards into the bulk fluid. Directly behind the plate is a zone of highly decelerating fluid in which large eddies or *vortices* are formed. This zone is called the *wake*. Eddies in the wake are kept in rotational motion by the force of bordering currents.

Figure 7.3 Fluid boundary layer for flow over a flat plate. (a) The boundary layer forms at the leading edge. (b) Compared with velocity u_B in the bulk fluid, velocity in the boundary layer is zero at the plate surface but increases with distance from the plate to reach u_B near the outer limit of the boundary layer.

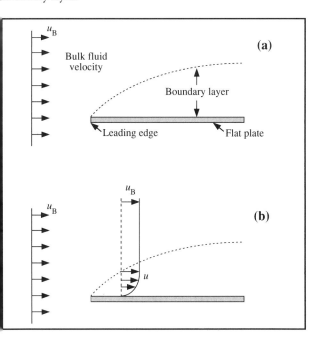

Figure 7.4 Flow around a flat plate aligned perpendicular to the direction of flow. (From W.L. McCabe and J.C. Smith, 1976, *Unit Operations of Chemical Engineering*, 3rd edn, McGraw-Hill, Tokyo.)

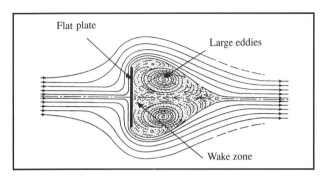

Figure 7.5 Velocity profile for Couette flow between parallel plates.

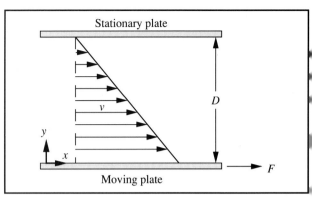

Boundary-layer separation such as that shown in Figure 7.4 occurs whenever an abrupt change in either magnitude or direction of fluid velocity is too great for the fluid to keep to a solid surface. It occurs in sudden contractions, expansions or bends in the flow channel, or when an object is placed across the flow path. Considerable energy is associated with the wake; this energy is taken from the bulk flow. Formation of wakes should be minimised if large pressure losses in the fluid are to be avoided; however, for some purposes such as promotion of mixing and heat transfer, boundary-layer separation may be desirable.

7.3 Viscosity

Viscosity is the most important property affecting flow behaviour of a fluid; viscosity is related to the fluid's resistance to motion. Viscosity has a marked effect on pumping, mixing, mass transfer, heat transfer and aeration of fluids; these in turn exert a major influence on bioprocess design and economics. Viscosity of fermentation fluids is affected by the presence of cells, substrates, products and air.

Viscosity is an important aspect of *rheology*, the science of deformation and flow. Viscosity is determined by relating the velocity gradient in fluids to the shear force causing flow to occur. This relationship can be explained by considering the development of laminar flow between parallel plates, as shown in Figure 7.5. The plates are a relatively short distance apart and, initially, the fluid between them is stationary. The lower plate is then moved steadily to the right with shear force F, while the upper plate remains fixed.

A thin film of fluid adheres to the surface of each plate. Therefore as the lower plate moves, fluid moves with it, while at the surface of the stationary plate the fluid velocity is zero. Due to viscous drag, fluid just above the moving plate is set

into motion, but with reduced speed. Layers further above also move; however, as we get closer to the top plate, the fluid is affected by viscous drag from the stationary film attached to the upper plate surface. As a consequence, fluid velocity between the plates decreases from that of the moving plate at $y = 0$, to zero at $y = D$. The velocity at different levels between the plates is indicated in Figure 7.5 by the arrows marked v. Laminar flow due to a moving surface as shown in Figure 7.5 is called *Couette flow*.

When steady Couette flow is attained in simple fluids, the velocity profile is as indicated in Figure 7.5; the slope of the line connecting all the velocity arrows is constant and proportional to the shear force F responsible for motion of the plate. The slope of the line connecting the velocity arrows is the velocity gradient, dv/dy. When the magnitude of the velocity gradient is directly proportional to F, we can write:

$$\frac{dv}{dy} \propto F.$$

(7.3)

If we define τ as the *shear stress*, equal to the shear force per unit area of plate:

$$\tau = \frac{F}{A}$$

(7.4)

it follows from Eq. (7.3) that:

$$\tau \propto \frac{dv}{dy}$$

(7.5)

This proportionality is represented by the equation:

$$\tau = -\mu \frac{\mathrm{d}v}{\mathrm{d}y}$$

(7.6)

where μ is the proportionality constant. Eq. (7.6) is called *Newton's law of viscosity*, and μ is the viscosity. The minus sign is necessary in Eq. (7.6) because the velocity gradient is always negative if the direction of F, and therefore τ, is considered positive. $-\mathrm{d}v/\mathrm{d}y$ is called the *shear rate*, and is usually denoted by the symbol $\dot{\gamma}$.

Viscosity as defined in Eq. (7.6) is sometimes called *dynamic viscosity*. As τ has dimensions $L^{-1}MT^{-2}$ and $\dot{\gamma}$ has dimensions T^{-1}, μ must therefore have dimensions $L^{-1}MT^{-1}$. The SI unit of viscosity is the Pascal second (Pa s), which is equal to 1 N s m^{-2} or $1 \text{ kg m}^{-1}\text{ s}^{-1}$. Other units include centipoise, cP. Direct conversion factors for viscosity units are given in Table A.9 in Appendix A. The viscosity of water at 20°C is approximately 1 cP or 10^{-3} Pa s. A modified form of viscosity is the *kinematic viscosity*, defined as μ/ρ where ρ is fluid density; kinematic viscosity is usually given the symbol ν.

Fluids which obey Eq. (7.6) with constant μ are known as *Newtonian fluids*. The *flow curve* or *rheogram* for a Newtonian fluid is shown in Figure 7.6; the slope of a plot of τ versus $\dot{\gamma}$ is constant and equal to μ. The viscosity of Newtonian fluids remains constant despite changes in shear stress (force applied) or shear rate (velocity gradient). This does not imply that the viscosity is invariant; viscosity depends on many parameters such as temperature, pressure and fluid composition. However, under a given set of these conditions, viscosity of Newtonian fluids is independent of shear stress and shear rate. On the other hand, the ratio between shear stress and shear rate is not constant for *non-Newtonian* fluids, but depends on the shear force exerted on the fluid. Accordingly, μ in Eq. (7.6) is not a constant, and the velocity profile during Couette flow is not as simple as that shown in Figure 7.5.

7.4 Momentum Transfer

Viscous drag forces responsible for the velocity gradient in Figure 7.5 are the instrument of *momentum transfer* in fluids. At $y = 0$ the fluid acquires momentum in the x-direction due to motion of the lower plate. This fluid imparts some of its momentum to the adjacent layer of fluid above the plate, causing it also to move in the x-direction. Momentum in the x-direction is thus transmitted through the fluid in the y-direction.

Momentum transfer in fluids is represented by Eq. (7.6).

Figure 7.6 Flow curve for a Newtonian fluid.

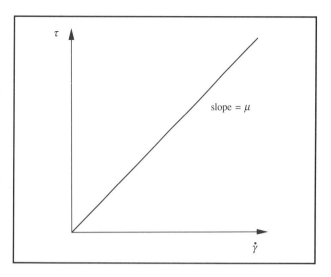

To interpret this equation in terms of momentum transfer, shear stress τ is considered as the *flux* of x-momentum in the y-direction. The validity of this definition can be verified by checking the dimensions of momentum flux and shear stress.

Momentum is given by the expression Mv where M is mass and v is velocity; momentum has dimensions LMT^{-1}. *Flux means rate per unit area*; therefore momentum flux has dimensions $L^{-1}MT^{-2}$, which are also the dimensions of τ. So with τ representing momentum flux, according to Eq. (7.6), flux of momentum is directly proportional to the velocity gradient $\mathrm{d}v/\mathrm{d}y$. The negative sign in Eq. (7.6) means that momentum is transferred from regions of high velocity to regions of low velocity, i.e. in a direction opposite to the direction of increasing velocity. The magnitude of the velocity gradient $\mathrm{d}v/\mathrm{d}y$ determines the magnitude of the momentum flux; $\mathrm{d}v/\mathrm{d}y$ thus acts as the 'driving force' for momentum transfer.

Interpretation of fluid flow as momentum transfer perpendicular to the direction of flow may seem peculiar at first. The reason it is mentioned here is that there are many parallels between momentum transfer, heat transfer and mass transfer in terms of mechanism and equations. The analogy between these physical processes will be discussed further in Chapters 8 and 9.

7.5 Non-Newtonian Fluids

Most slurries, suspensions and dispersions are non-Newtonian, as are homogeneous solutions of long-chain polymers and other large molecules. Many fermentation processes involve materials which exhibit non-Newtonian behaviour, such as starches, extracellular polysaccharides, and culture broths containing cell suspensions or pellets. Examples of non-Newtonian fluids are listed in Table 7.1.

Classification of non-Newtonian fluids depends on the relationship between the shear stress imposed on the fluid and the shear rate developed. Common types of non-Newtonian fluid include *pseudoplastic, dilatant, Bingham plastic* and *Casson plastic*; flow curves for these materials are shown in Figure 7.7. In each case, the ratio between shear stress and shear rate is not constant; nevertheless, this ratio for non-Newtonian fluids is often called the *apparent viscosity, μ_a*. Apparent viscosity is not a physical property of the fluid in the same way as Newtonian viscosity; it is dependent on the shear force exerted on the fluid. It is therefore meaningless to specify the apparent viscosity of a non-Newtonian fluid without noting the shear stress or shear rate at which it was measured.

7.5.1 Two-Parameter Models

Pseudoplastic and dilatant fluids obey the *Ostwald–de Waele* or *power law*:

$$\tau = K \dot{\gamma}^n$$

$$(7.7)$$

where τ is shear stress, K is the *consistency index*, $\dot{\gamma}$ is shear rate,

and n is the *flow behaviour index*. The parameters K and n characterise the rheology of power-law fluids. The flow behaviour index n is dimensionless; the dimensions of K, $L^{-1}MT^{n-2}$, depend on n. As indicated in Figure 7.7, when $n < 1$ the fluid exhibits pseudoplastic behaviour; when $n > 1$ the fluid is dilatant. $n = 1$ corresponds to Newtonian behaviour. For power-law fluids, apparent viscosity μ_a is expressed as:

$$\mu_a = \frac{\tau}{\dot{\gamma}} = K \dot{\gamma}^{n-1}.$$

$$(7.8)$$

For pseudoplastic fluids $n < 1$ and the apparent viscosity decreases with increasing shear rate; these fluids are said to exhibit *shear thinning*. On the other hand, apparent viscosity increases with shear rate for dilatant or *shear thickening* fluids.

Also included in Figure 7.7 are flow curves for plastic flow. Some fluids do not produce motion until some finite *yield stress* has been applied. For *Bingham plastic* fluids:

$$\tau = \tau_0 + K_p \dot{\gamma}$$

$$(7.9)$$

where τ_0 is the yield stress. Once the yield stress is exceeded and flow initiated, Bingham plastics behave like Newtonian fluids; a constant ratio K_p exists between change in shear stress

Table 7.1   Common non-Newtonian fluids

(Adapted from B. Atkinson and F. Mavituna, 1991, Biochemical Engineering and Biotechnology Handbook, *2nd edn, Macmillan, Basingstoke)*

Fluid type	Examples
Newtonian	All gases, water, dispersions of gas in water, low-molecular-weight liquids, aqueous solutions of low-molecular-weight compounds
Non-Newtonian	
Pseudoplastic	Rubber solutions, adhesives, polymer solutions, some greases, starch suspensions, cellulose acetate, mayonnaise, some soap and detergent slurries, some paper pulps, paints, wallpaper paste, biological fluids
Dilatant	Some cornflour and sugar solutions, starch, quicksand, wet beach sand, iron powder dispersed in low-viscosity liquids, wet cement aggregates
Bingham plastic	Some plastic melts, margarine, cooking fats, some greases, toothpaste, some soap and detergent slurries, some paper pulps
Casson plastic	Blood, tomato sauce, orange juice, melted chocolate, printing ink

Figure 7.7 Classification of fluids according to their rheological behaviour. (From B. Atkinson and F. Mavituna, 1991, *Biochemical Engineering and Biotechnology Handbook*, 2nd edn, Macmillan, Basingstoke.)

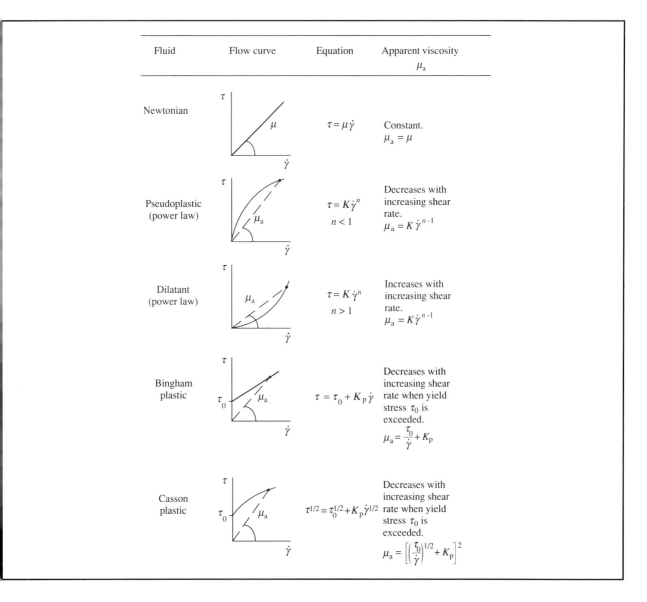

and change in shear rate. Another common plastic behaviour is described by the *Casson* equation:

$$\tau^{1/2} = \tau_0^{1/2} + K_p \dot{\gamma}^{1/2}.$$

(7.10)

Once the yield stress is exceeded, the behaviour of Casson fluids is pseudoplastic. Several other equations describing non-Newtonian flow have also been developed [1].

7.5.2 Time-Dependent Viscosity

When a shear force is exerted on some fluids, the apparent viscosity either increases or decreases with duration of the force. If apparent viscosity increases with time, the fluid is said to be *rheopectic*; rheopectic fluids are relatively rare in occurrence. If apparent viscosity decreases with time the fluid is *thixotropic*. Thixotropic behaviour is not uncommon in cultures containing fungal mycelia or extracellular microbial polysaccharides, and appears to be related to reversible 'structure' effects associated with the orientation of cells and macromolecules in the fluid. Rheological properties vary during application of the shear force because it takes time for equilibrium to be established between structure breakdown and re-development.

7.5.3 Viscoelasticity

Viscoelastic fluids, such as some polymer solutions, exhibit an elastic response to changes in shear stress. When shear forces are removed from a moving viscoelastic fluid, the direction of flow may be reversed due to elastic forces developed during flow. Most viscoelastic fluids are also pseudoplastic and may exhibit other rheological characteristics such as yield stress. Mathematical analysis of viscoelasticity is therefore quite complex.

7.6 Viscosity Measurement

Many different instruments or *viscometers* are available for measurement of rheological properties. Space does not permit a detailed discussion of viscosity measurement in this text; further information can be found elsewhere [1–5]. Specifications for commercial viscometers are also available [2, 3, 6].

The objective of any viscosity measurement system is to create a controlled flow situation where easily measured parameters can be related to the shear stress τ and shear rate $\dot{\gamma}$. Usually the fluid is set in rotational motion and the parameters measured are torque M and angular velocity Ω. These quantities are used to calculate τ and $\dot{\gamma}$ using approximate formulae which depend on the geometry of the apparatus. Once obtained, τ and $\dot{\gamma}$ are applied for evaluation of viscosity in Newtonian fluids, or viscosity parameters such as K, n, and τ_0 for non-Newtonian fluids. Equations for particular viscometers can be found in other texts [2, 3, 6]. Most modern viscometers use microprocessors to provide automatic read-out of parameters such as shear stress, shear rate and apparent viscosity.

Three types of viscometer commonly used in bioprocessing applications are the cone-and-plate viscometer, the coaxial-cylinder rotary viscometer, and the impeller viscometer.

7.6.1 Cone-and-Plate Viscometer

The cone-and-plate viscometer consists of a flat horizontal plate and an inverted cone, the apex of which is near contact with the plate as shown in Figure 7.8. The angle ϕ between the plate and cone is very small, usually less than 3°, and the fluid to be measured is located in this small gap. Large cone angles are not used for routine work for a variety of reasons, the most important being that analysis of the results for non-Newtonian fluids would be complex or impossible. The cone is rotated in the fluid, and the angular velocity Ω and torque M measured. It

Figure 7.8 Cone-and-plate viscometer.

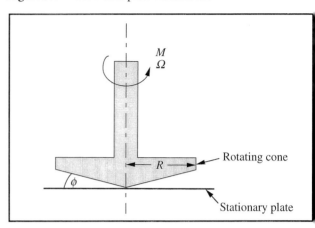

is generally assumed that the fluid undergoes streamline flow in concentric circles about the axis of rotation of the cone. This assumption is not always valid; however for ϕ less than about 3°, the error is small. Temperature can be controlled by circulating water from a constant-temperature bath beneath the plate; this is effective provided the speed of rotation is not too high. Limitations of the cone-and-plate method for measurement of flow properties, including corrections for edge and temperature effects and turbulence, are discussed elsewhere [3].

7.6.2 Coaxial-Cylinder Rotary Viscometer

The coaxial-cylinder viscometer is a popular rotational device for measuring rheological properties. A typical coaxial-cylinder instrument is shown in Figure 7.9. This device is designed to shear fluid located in the annulus between two

Figure 7.9 Coaxial-cylinder viscometer.

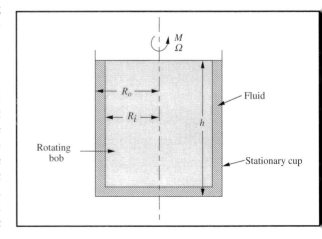

concentric cylinders, one of which is held stationary while the other rotates. A cylindrical bob of radius R_i is suspended in sample fluid held in a stationary cylindrical cup of radius R_o. Liquid covers the bob to a height h from the bottom of the outer cup. As the inner cylinder rotates, the angular velocity Ω and torque M are measured. In some designs the outer cylinder rather than the inner bob rotates; in any case the motion is relative with magnitude Ω.

Coaxial-cylinder viscometers are used with Newtonian or non-Newtonian fluids. When flow is non-Newtonian, shear rate is not related simply to rotational speed and geometric factors and the calculations can be somewhat complicated. Limitations of the coaxial-cylinder method, including corrections for end effects, slippage, temperature variation and turbulence, are discussed elsewhere [2, 3, 6].

7.6.3 Impeller Viscometer

Because of difficulties (discussed in Section 7.6.4) associated with standard rotational viscometers, modified apparatus employing turbine and other impellers have been developed for rheological study of fermentation fluids [7, 8]. Instead of the rotating inner cylinder of Figure 7.9, a small impeller on a stirring shaft is used to shear the fluid sample. As the impeller rotates slowly in the fluid, accurate measurements of torque M and rotational speed N_i are made. For a turbine impeller under laminar-flow conditions, the following relationships apply [8]:

$$\dot{\gamma} = k N_i \tag{7.11}$$

and

$$\tau = \frac{2\pi M k}{64 D_i^3} \tag{7.12}$$

where D_i is the impeller diameter and k is a constant which depends on the geometry of the impeller (see Section 7.13). The relationship of Eq. (7.11) is experimentally derived; for turbine impellers k is approximately 10. The exact value of k for a particular apparatus is evaluated using liquid with a known viscosity–shear rate relationship.

Because Eqs (7.11) and (7.12) are valid only for laminar flow, viscosity measurements using the impeller method must be carried out under laminar flow conditions. Accordingly, if a turbine impeller is used, Re_i should not be greater than about 10. As Re_i is directly proportional to N_i which, from Eq. (7.11), determines the value of $\dot{\gamma}$, the necessity for laminar flow limits the range of shear rates that may be investigated. This range can be extended if anchor or helical agitators are used instead of the conventional disc turbine (see Figure 7.15 for illustrations of these impellers). Laminar flow is maintained at higher Re_i with anchor and helical impellers; the value of k in Eq. (7.11) is also greater so that higher shear rates can be tested. As Eq. (7.12) is valid only for turbine impellers, the relationship between τ and M must be modified if alternative impellers are used. Application of anchor and helical impellers for viscosity measurement is described in the literature [9, 10].

Because the flow patterns in stirred fluids are relatively complex, analysis of data from impeller viscometers is not absolutely rigorous from a rheological point of view. However, the procedure is based on well-proven and widely-accepted empirical correlations and is considered the most reliable technique for mycelial broths. As discussed below, the method eliminates many of the operating problems associated with conventional viscometers for study of fermentation fluids.

7.6.4 Use of Viscometers With Fermentation Broths

Measurement of rheological properties is difficult when the fluid contains suspended solids such as cells. Viscosity of fermentation broths often appears time-dependent due to artifacts associated with the measuring device. With viscometers such as the cone-and-plate and coaxial cylinder, the following problems can arise:

(i) the suspension is effectively centrifuged in the viscometer so that a region with lower cell density is formed near the rotating surface;
(ii) solids settle out of suspension during measurement;
(iii) large particles about the same size as the gap in the coaxial viscometer, or about the same size as the cone angle in the cone-and-plate, interfere with accurate measurement;
(iv) the measurement will depend somewhat on the orientation of particles in the flow field;
(v) some types of particle will begin to flocculate or de-flocculate when the shear field is applied; and
(vi) particles can be destroyed during measurement.

The first problem is particularly troublesome because it is hard to detect and can give viscosity results which are too small by a factor of up to 100. For suspensions containing solids, the impeller method offers significant advantages compared with other measurement procedures. Stirring by the impeller prevents sedimentation, promotes uniform distribution of solids through the fluid, and reduces time-dependent changes in suspension composition. The method has proved very useful for rheological measurements on microbial suspensions [5].

Table 7.2 Rheological properties of microbial and plant-cell suspensions

(Adapted from M. Charles, 1978, Technical aspects of the rheological properties of microbial cultures. Adv. Biochem. Eng. 8, 1–62)

Culture	Shear rate (s^{-1})	Viscometer	Comments	Reference
Saccharomyces cerevisiae (pressed cake diluted with water)	2–100	rotating spindle	Newtonian below 10% solids ($\mu < 4$–5 cP); pseudoplastic above 10% solids	[11]
Aspergillus niger (washed cells in buffer)	0–21.6	rotating spindle (guard removed)	pseudoplastic	[12]
Penicillium chrysogenum (whole broth)	1–15	turbine impeller	Casson plastic	[8]
Penicillium chrysogenum (whole broth)	not given	coaxial cylinder	Bingham plastic	[13]
Penicillium chrysogenum (whole broth)	not given	coaxial cylinder	pseudoplastic; K and n vary with CO_2 content of inlet gas	[14]
Endomyces sp. (whole broth)	not given	coaxial cylinder	pseudoplastic; K and n vary over course of batch culture	[15]
Streptomyces noursei (whole broth)	4–28	rotating spindle (guard removed)	Newtonian in batch culture; viscosity 40 cP after 96 h	[16]
Streptomyces aureofaciens (whole broth)	2–58	rotating spindle/ coaxial cylinder	initially Bingham plastic due to high starch concentration in medium; changes to Newtonian as starch is broken down; increasingly pseudoplastic as mycelium concentration increases	[17]
Aureobasidium pullulans (whole broth)	10.2–1020	coaxial cylinder	Newtonian at the beginning of culture; increasingly pseudoplastic as concentration of product (exopolysaccharide) increases	[18]
Xanthomonas campestris	0.0035–100	cone-and-plate	pseudoplastic; K increases continually; n levels off when xanthan concentration reaches 0.5%; cell mass (max 0.6%) has relatively little effect on viscosity	[4]

contd.

Culture	Shear rate (s⁻¹)	Viscometer	Comments	Reference
Cellulomonas uda (whole broth)	0.8–100	anchor impeller	shredded newspaper used as substrate; broth pseudoplastic with constant n until end of cellulose degradation; Newtonian thereafter	[10]
Nicotiana tabacum (whole broth)	not given	rotating spindle	pseudoplastic	[19]
Datura stramonium (whole broth)	0–1000	rotating spindle/ parallel-plate	pseudoplastic and viscoelastic, with yield stress	[20]

7.7 Rheological Properties of Fermentation Broths

Rheological data have been reported for a range of fermentation fluids. This information has been obtained using various viscometers and measurement techniques; however, operating problems such as particle settling and broth centrifugation have been ignored in many cases. Most mycelial suspensions have been modelled as pseudoplastic fluids or, if there is a yield stress, Bingham or Casson plastic. On the other hand, the rheology of dilute broths and cultures of yeast and non-chain-forming bacteria is usually Newtonian. Rheological properties of some microbial and plant-cell suspensions are listed in Table 7.2. In most cases, the results are valid over only a limited range of shear conditions which is largely dictated by the choice of viscometer. When the fermentation produces extracellular polymers such as in microbial production of pullulan and xanthan, the rheological characteristics of the broth depend strongly on the properties and concentration of these materials.

7.8 Factors Affecting Broth Viscosity

The rheology of fermentation broths often changes throughout batch culture. For broths obeying the power law, the flow behaviour index n and consistency index K can vary substantially depending on culture time. As an example, Figure 7.10 shows changes in n and K during batch culture of *Endomyces* [15]; the culture starts off Newtonian ($n = 1$) but quickly becomes pseudoplastic ($n < 1$). K rises steadily throughout most of the batch period; this gives a direct indication of the increase in apparent viscosity since, as indicated in Eq. (7.8), μ_a is directly proportional to K.

Changes in rheology of fermentation broths are caused by variation of one or more of the following properties:

Figure 7.10 Variation of rheological parameters in *Endomyces* fermentation. (From H. Taguchi and S. Miyamoto, 1966, Power requirement in non-Newtonian fermentation broth. *Biotechnol. Bioeng.* **8**, 43–54.)

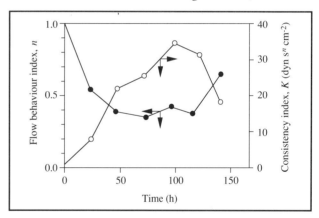

(i)   cell concentration;
(ii) cell morphology, including size, shape and mass;
(iii) flexibility and deformability of cells;
(iv) osmotic pressure of the suspending fluid;
(v)   concentration of polymeric substrate;
(vi) concentration of polymeric product; and
(vii) rate of shear.

Some of these parameters are considered below.

7.8.1 Cell Concentration

The viscosity of a suspension of spheres in Newtonian liquid can be predicted using the *Vand equation*:

$$\mu = \mu_L (1 + 2.5\psi + 7.25\psi^2) \tag{7.13}$$

where μ_L is the viscosity of the suspending liquid and ψ is the volume fraction of solids. Eq. (7.13) has been found to hold for yeast and spore suspensions up to 14 vol% solids [21]. Many other cell suspensions do not obey Eq. (7.13); cell concentration can have a much stronger influence on rheological properties than is predicted by the Vand equation. As an example, Figure 7.11 shows how cell concentration affects the apparent viscosity of various pseudoplastic plant-cell suspensions [22]; a doubling in cell concentration causes the apparent viscosity to increase by a factor of up to 90. Similar results have been found for mould pellets in liquid culture [23]. When viscosity is so strongly dependent on cell concentration, a steep drop in viscosity can be achieved by diluting the broth with water or medium. Periodic removal of part of the culture and refilling with fresh medium reduces the viscosity and improves fluid flow in viscous fermentations.

7.8.2 Cell Morphology

Morphological characteristics exert a profound influence on broth rheology. Disperse filamentous growth produces 'structure' in the broth, resulting in pseudoplasticity, yield-stress behaviour, or both. On the other hand, broths containing pelleted cells tends to be more Newtonian, depending on how

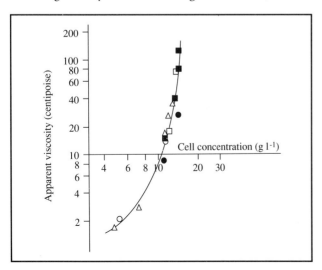

Figure 7.11 Relationship between apparent viscosity and cell concentration for plant-cell suspensions forming aggregates of various size. (○) *Cudrania tricuspidata* 44–149 μm; (●) *C. tricuspidata* 149–297 μm; (□) *Vinca rosea* 44–149 μm; (■) *V. rosea* 149–297 μm; (△) *Nicotiana tabacum* 150–800 μm. (From H. Tanaka, 1982, Oxygen transfer in broths of plant cells at high density. *Biotechnol. Bioeng.* **24**, 425–442.)

readily the pellets are deformed during flow. The extent of branching of hyphal cells can also affect rheology; cells with a high branching frequency are generally less flexible than non-branching cells and produce higher viscosities.

Sample rheological data for pseudoplastic mycelial broths are shown in Figure 7.12 [9]. Pelleted mycelia are more closely Newtonian in behaviour than filamentous cells; the flow behaviour index *n* for pellets is closer to unity. As indicated in Figure 7.12(b), the consistency index, and therefore the apparent viscosity, can differ by several orders of magnitude depending on cell morphology.

7.8.3 Osmotic Pressure

Osmotic pressure of the culture medium affects cell turgor pressure. This in turn affects the hyphal flexibility of filamentous cells; increased osmotic pressure gives a lower turgor pressure making the hyphae more flexible. Improved hyphal flexibility reduces broth viscosity, and can also have a marked effect on yield stress.

7.8.4 Product and Substrate Concentrations

When the product of fermentation is a polymer, continued excretion in batch culture raises the broth viscosity. For example, during production of exopolysaccharide by *Aureobasidium pullulans*, apparent viscosity measured at a shear rate of 1 s^{-1} can reach as high as 24 000 cP [18]. Cell concentration usually has a negligible effect on overall viscosity in these fermentations; the rheological properties of the fluid are dominated by the dissolved polymer. Other products having a similar effect on culture rheology include dextran, alginate and xanthan gum.

In contrast, when the fermentation medium contains polymeric substrate such as starch, apparent viscosity will decrease as the fermentation progresses and the polymer is broken down. There could also be a progressive change from non-Newtonian to Newtonian behaviour. In mycelial fermentations this change is usually short lived; as the cells grow and develop a structured filamentous network, the broth becomes increasingly pseudoplastic and viscous.

7.9 Mixing

Mixing is a physical operation which reduces non-uniformities in fluid by eliminating gradients of concentration, temperature and other properties. Mixing is accomplished by interchanging material between different locations to produce a mingling of components. If a system is perfectly

Figure 7.12   Effect of morphology on the rheology of mycelial broths. (From J.H. Kim, J.M. Lebeault and M. Reuss, 1983, Comparative study on rheological properties of mycelial broth in filamentous and pelleted forms. *Eur. J. Appl. Microbiol. Biotechnol.* **18**, 11–16.)

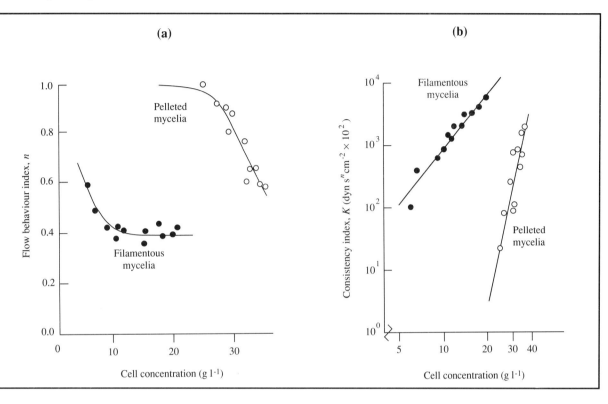

mixed, there is a random homogeneous distribution of system properties. Mixing involves:

i)   blending soluble components of the medium such as sugars;

ii) dispersing gases such as air through the liquid in the form of small bubbles;

iii) maintaining suspension of solid particles such as cells;

iv) where necessary, dispersing immiscible liquids to form an emulsion or suspension of fine drops; and

v)   promoting heat transfer to or from the liquid.

Mixing is one of the most important operations in bioprocessing. To create the optimal environment for fermentation, bioreactors must provide the cells access to all substrates, including oxygen in aerobic culture. It is not enough to just fill the fermenter with nutrient-rich medium; unless the culture is mixed, zones of nutrient depletion will develop as the cells rapidly consume the materials they need. This problem is heightened if mixing does not maintain a uniform suspension of biomass; substrate concentrations can quickly drop to zero in areas where cells settle out of suspension. Another import-

ant function of mixing is heat transfer. Bioreactors must be capable of transferring heat to or from the broth rapidly enough so that the desired temperature is maintained. Cooling water is used to take up excess heat from fermentations; the rate of heat transfer from the broth through the walls of the vessel to the cooling water depends on mixing conditions in the vessel. The effectiveness of mixing depends in turn on the rheological properties of the culture fluid.

Mixing can be achieved in many different ways. In this chapter we will concentrate on the most common mixing technique in bioprocessing: mechanical agitation using an impeller.

7.9.1 Mixing Equipment

Mixing is usually carried out in a stirred tank, such as that shown in Figure 7.13. Stirred tanks are usually cylindrical in shape. If possible, the base of the tank is rounded at the edges rather than angled; this eliminates sharp corners and pockets into which fluid currents may not penetrate and discourages formation of stagnant regions. Mixing is achieved using an *impeller* mounted in the tank; for use with Newtonian fluids, the ratio of tank diameter to impeller diameter is normally about 3:1. The impeller is usually positioned overhead on a centrally-located *stirrer shaft*. Sometimes, stirrer shafts are designed to enter at the bottom of the vessel; the disadvantage of this arrangement is that leaks can develop if the seal between the shaft and the tank floor is not perfect. The stirrer shaft is driven rapidly by the stirrer motor; the effect of the rotating impeller is to pump the liquid and create a regular flow pattern. Liquid is forced away from the impeller, circulates through the vessel, and periodically returns to the impeller region. For efficient mixing with a single impeller, the depth of liquid in the tank should be no more than 1.0–1.25 times the tank diameter.

Baffles, which are vertical strips of metal mounted against the wall of the tank, are installed to reduce vortexing and swirling of the liquid. Baffles are attached to the tank by means of welded brackets; four equally-spaced baffles are usually sufficient to prevent vortex formation. The optimum baffle width depends on the impeller design and fluid viscosity but is of the order $^1/_{10}$–$^1/_{12}$ the tank diameter. For low-viscosity liquids, baffles are usually attached perpendicular to the wall as illustrated in Figure 7.14(a). Alternatively, as shown in Figures 7.14(b) and (c), baffles can be mounted away from the wall with a clearance of about $^1/_{50}$ the tank diameter, or set at an angle. These arrangements prevent sedimentation and development of stagnant zones at the inner edge of the baffle during mixing of viscous cell suspensions.

Many impeller designs are available for mixing applications; a small selection is illustrated in Figure 7.15. Further details and descriptions of impellers can be found elsewhere [24, 25]. Some impellers have flat blades; in others such as the

propeller and helical screw, the slope of the individual blades varies continuously. Specification of the *pitch* of a propeller blade refers to its properties as a segment of a screw; pitch is the advance per revolution. Choice of impeller depends on several factors, including viscosity of the liquid to be mixed and sensitivity of the system to mechanical shear. The recommended viscosity ranges for a number of common impellers are indicated in Figure 7.16. For low-to-medium-viscosity liquids, propellers and flat-blade turbines are recommended. The most frequently-used impeller in the fermentation industry is the 6-flat-blade disc-mounted turbine shown in Figure 7.15; this impeller is also known as the *Rushton turbine*.

Figure 7.14 Baffle arrangements. (a) Baffles attached to the wall for low-viscosity liquids. (b) Baffles set away from the wall for moderate-viscosity liquids. (c) Baffles set away from the wall and at an angle for high-viscosity liquids. (From F.A. Holland and F.S. Chapman, 1966, *Liquid Mixing and Processing in Stirred Tanks*, Reinhold, New York.)

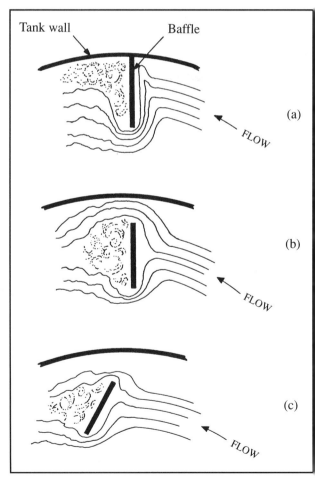

Figure 7.13 Typical configuration of a stirred tank.

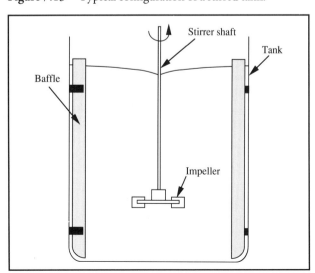

Figure 7.15 Impeller designs.

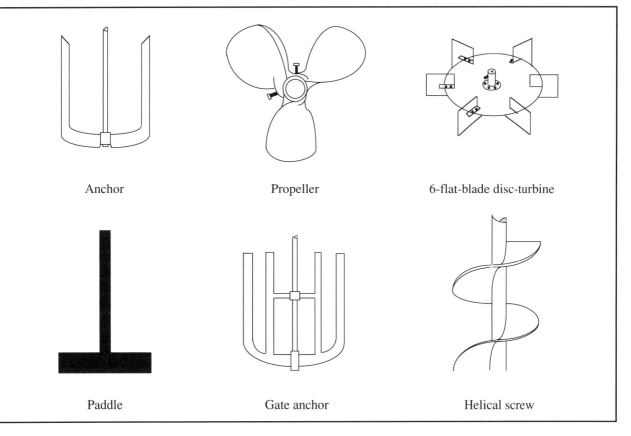

Anchor   Propeller   6-flat-blade disc-turbine

Paddle   Gate anchor   Helical screw

Figure 7.16 Viscosity ranges for different impellers. (From F.A. Holland and F.S. Chapman, 1966, *Liquid Mixing and Processing in Stirred Tanks*, Reinhold, New York.)

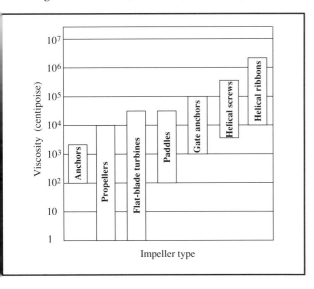

7.9.2 Flow Patterns in Agitated Tanks

The flow pattern in an agitated tank depends on the impeller design, the properties of the fluid, and the size and geometric proportions of the vessel, baffles and agitator.

Although most stirrers are rotational in action, simple circular flow of liquid around the shaft such as that illustrated in Figure 7.17(a) is generally disadvantageous and should be avoided. In circular flow, liquid moves in a streamline fashion and there is little mixing between fluid at different heights in the tank. Circular flow also leads to vortex development as shown in Figure 7.17(b). At high impeller speeds, the vortex may reach down to the impeller so that gas from the surrounding atmosphere is drawn into the liquid; this is generally undesirable as it produces very high mechanical stresses in the stirrer shaft, bearings and seal. Prevention of circular flow has a high priority in design of mixing systems, and is usually achieved by installing baffles which interrupt the circular flow pattern and create turbulence in the fluid.

As well as circular flow, motion of fluid occurs in the radial direction (i.e. from the stirrer out to the sides of the tank and back again) and in the axial direction (i.e. up and down the height of the tank). Axial and radial flows are generated at the impeller, and it is these components of fluid motion which are primarily responsible for bulk mixing. Impellers are broadly classified as *axial flow* or *radial flow* depending on the direction of liquid leaving the impeller; some impellers have both axial- and radial-flow characteristics.

7.9.2.1 *Radial-flow impellers*

Radial-flow impellers have blades which are parallel to the vertical axis of the stirrer shaft and tank; the six-flat-blade disc turbine of Figure 7.15 is an example. The flow pattern set up by high-speed rotation of a radial-flow impeller is illustrated in

Figure 7.17 (a) Circular flow in an unbaffled stirred tank viewed from above. (b) Vortex formation during circular flow. (From J.H. Rushton, E.W. Costich and H.J. Everett, 1950, Power characteristics of mixing impellers: Part I. *Chem. Eng. Prog.* **46**, 395–404.)

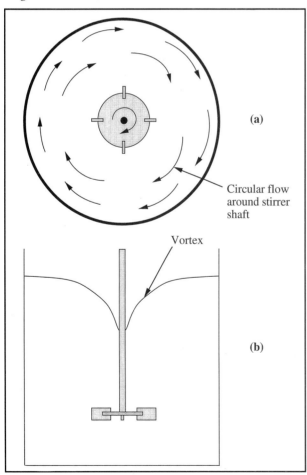

Figure 7.18. Liquid is driven radially from the impeller against the walls of the tank where it divides into two streams, one flowing up to the top of the tank and the other flowing down to the bottom. These streams eventually reach the central axis of the tank and are drawn back to the impeller. Radial-flow impellers also set up circular flow which must be reduced by baffles.

7.9.2.2 *Axial-flow impellers*

In general, axial-flow impellers have blades which make an angle of less than 90° to the plane of rotation and promote axial top-to-bottom motion. Propellers are axial-flow devices, as are pitched-blade turbines such as that shown in Figure 7.19. The flow pattern set up by a typical axial-flow impeller is illustrated in Figure 7.20. Fluid leaving the impeller is driven downward until it is deflected from the floor of the vessel. It then spreads out over the floor and flows up along the wall before being drawn back to the impeller. Axial-flow impellers are particularly useful when strong vertical currents are required. For example, if the fluid contains solids, a strong axial flow of liquid leaving the impeller will discourage settling at the bottom of the tank.

7.9.3 Mechanism of Mixing

As illustrated in Figures 7.18 and 7.20, large liquid-circulation loops develop in stirred vessels. For mixing to be effective, fluid circulated by the impeller must sweep the entire vessel in a reasonable time. In addition, the velocity of fluid leaving the impeller must be sufficient to carry material into the most remote parts of the tank. Turbulence must also be developed in the fluid; mixing is certain to be poor unless flow in the tank is turbulent. All these factors are important in mixing, which can be described as a combination of three physical processes:

(i)   distribution;
(ii) dispersion; and
(iii) diffusion.

Distribution is sometimes called *macromixing*; diffusion is also called *micromixing*. Dispersion can be classified as either micro- or macromixing depending on the scale of fluid motion.

   The pattern of bulk fluid flow in a vessel stirred by a radial-flow impeller is shown in detail in Figure 7.21. Near the impeller there is a region of high turbulence where fluid currents converge and exchange material. Away from the impeller flow is slower and largely streamline; mixing in these regions is much less intense than near the impeller. The contents of the vessel are recirculated through the mixing zone in a very regular manner due to the periodic pumping action of the impeller.

   Let us consider what happens when a small amount of liquid dye is dropped onto the top of the fluid in Figure 7.21.

Figure 7.18   Flow pattern produced by a radial-flow impeller in a baffled tank. (a) side view; (b) bottom view. (From J.H. Rushton, E.W. Costich and H.J. Everett, 1950, Power characteristics of mixing impellers. Part I. *Chem. Eng. Prog.* **46**, 395–404.)

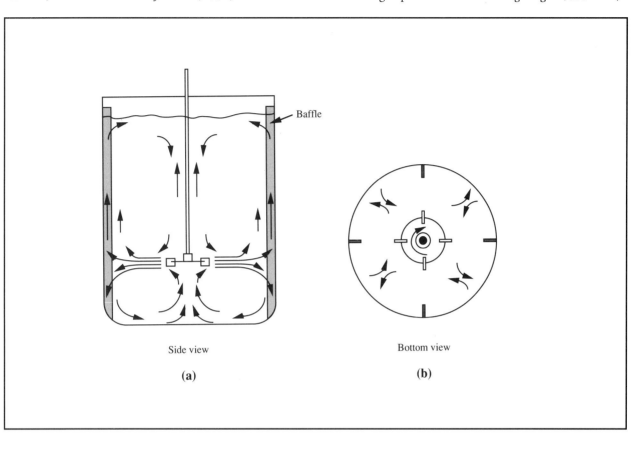

Side view

(a)

Bottom view

(b)

Figure 7.19   Pitched-blade turbine.

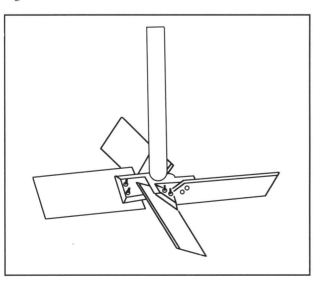

First, the dye is swept by circulating currents down to the impeller. At the impeller there is vigorous and turbulent motion of fluid; the dye is mechanically dispersed into smaller volumes and distributed between the large circulation loops. These smaller parcels of dye are then carried around the tank, dispersing all the while into those parts of the system not yet containing dye. Returning again to the impeller, the dye aliquots are broken up into even smaller volumes for further distribution. After a time, dye is homogeneously distributed throughout the tank with uniform concentration.

The process whereby dye is transported to all regions of the vessel by bulk circulation currents is called *distribution*. Distribution is an important process in mixing, but can be relatively slow. In large tanks, the size of the circulation paths is also large and the time taken to traverse them is long; this, together with the regularity of fluid pumping at the impeller, inhibits rapid mixing. Accordingly, *distribution is often the slowest step in the mixing process*. If the rotational speed of the impeller is sufficiently high, superimposed on the distribution

Figure 7.20 Flow pattern produced by an axial-flow impeller in a baffled tank. (a) side view; (b) bottom view. (From J.H. Rushton, E.W. Costich and H.J. Everett, 1950, Power characteristics of mixing impellers. Part I. *Chem. Eng. Prog.* **46**, 395–404.)

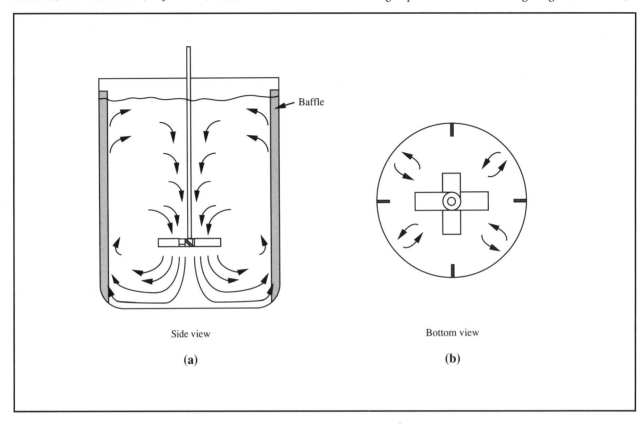

Side view

(a)

Bottom view

(b)

process is turbulence. Turbulent flow occurs when fluid no longer travels along streamlines but moves erratically in the form of cross-currents. The kinetic energy of turbulent fluid is directed into regions of rotational flow called *eddies*; masses of eddies of various size coexist during turbulent flow. Large eddies are continuously formed by action of the stirrer; these break down into small eddies which produce even smaller eddies. Eddies, like spinning tops, possess kinetic energy which is transferred to eddies of decreasing size. When the eddies become very small they can no longer sustain rotational motion and their kinetic energy is dissipated as heat. At steady state in a mixed tank, most of the energy from the stirrer is dissipated through eddies as heat; energy lost as fluid collides with the tank walls is generally negligible.

The process of breaking up bulk flow into smaller and smaller eddies is called *dispersion*; dispersion facilitates rapid transfer of material throughout the vessel. The degree of homogeneity possible as a result of dispersion is limited by the size of the smallest eddies which may be formed in a particular fluid. This size is given approximately as the *Kolmogorov scale of mixing*, or *scale of turbulence*, λ:

$$\lambda = \left(\frac{\nu^3}{\varepsilon}\right)^{1/4}$$

$$(7.14)$$

where λ is the characteristic dimension of the smallest eddies, ν is the kinematic viscosity of the fluid, and ε is the local rate of turbulent energy dissipation per unit mass of liquid. At steady state, the rate of energy dissipation by turbulence is equal to the power supplied by the impeller. According to Eq. (7.14), the greater the power input to the fluid, the smaller are the eddies. λ is also dependent on viscosity; at a given power input, smaller eddies are produced in low-viscosity fluids. For low-viscosity liquids such as water, λ is usually in the range 30–100 µm. For such fluids, this is the smallest scale of mixing achievable by dispersion.

Within eddies there is little mixing because rotational flow occurs in streamlines. Therefore, to achieve mixing on a scale smaller than the Kolmogorov scale, we must rely on *diffusion*. Molecular diffusion is generally regarded as a slow process; however, over small distances it can be accomplished quite rapidly. Within eddies of 30–100 µm diameter, homogeneity is achieved in about 1 s for low-viscosity fluids. Consequently,

Figure 7.21   Flow pattern developed by a centrally-positioned radial-flow impeller. (From R.M. Voncken, J.W. Rotte and A.Th. ten Houten, Circulation model for continuous-flow, turbine-stirred, baffled tanks. In: *Mixing—Theory Related to Practice,* Proc. Symp. 10, AIChE–IChE Joint Meeting, London, June 1965.)

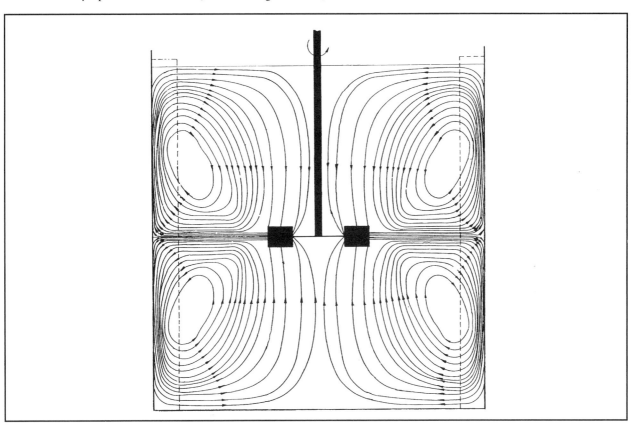

if power input to a stirred vessel produces eddies of this dimension, mixing on a molecular scale is accomplished virtually simultaneously.

7.9.4 Assessing Mixing Effectiveness

As explained in the above section, to achieve rapid mixing in a stirred tank the agitator must provide good bulk circulation or macromixing. Micromixing at or near the molecular scale is also important, but usually occurs relatively quickly compared with macromixing. Assessment of mixing effectiveness can therefore be reduced in most cases to measuring the rate of bulk flow.

Mixing time is a useful parameter for assessing mixing efficiency and is applied to characterise bulk flow in fermenters and reactors. The mixing time t_m is the time required to achieve a given degree of homogeneity starting from the completely segregated state. It can be measured by injecting a tracer into the vessel and following its concentration at a fixed point in the tank. Tracers in common use include acids, bases

and concentrated salt solutions; corresponding detectors are pH probes and conductivity cells. Mixing time can also be determined by measuring the temperature response after addition of a small quantity of heated liquid.

Let us assume that a small pulse of tracer is added to fluid in a stirred tank already containing tracer material at concentration C_i. When flow in the system is circulatory, the tracer concentration measured at some fixed point in the tank will follow a pattern similar to that shown in Figure 7.22. Before mixing is complete, a relatively high concentration will be detected every time the bulk flow brings tracer to the measurement point. The peaks in concentration will be separated by a period approximately equal to the average time taken for fluid to traverse one bulk circulation loop. In stirred vessels this period is called the *circulation time*, t_c. After several circulations the desired degree of homogeneity is reached.

Definition of the mixing time t_m depends on the degree of homogeneity required. Usually, mixing time is defined as the time after which the concentration of tracer differs from the

Figure 7.22 Concentration response after dye is injected into a stirred tank.

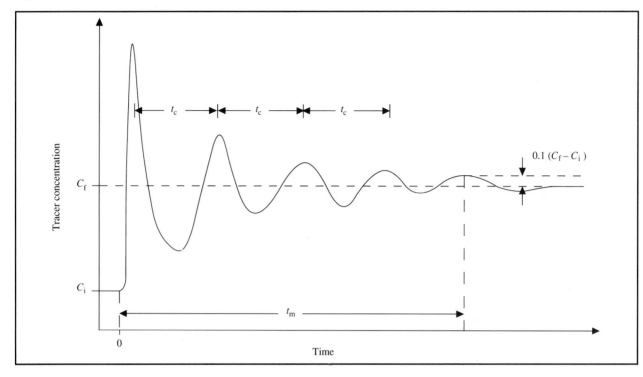

Figure 7.23 Variation of mixing time with Reynolds number for a six-blade Rushton turbine in a baffled tank. The impeller is located one-third the tank diameter off the floor of the vessel; the impeller diameter is one-third the tank diameter. The liquid height is equal to the tank diameter; the tank has four baffles of width one-tenth the tank diameter. Several measurement techniques and tank sizes were used: (●) thermal method, 1.8-m diameter vessel; (○) thermal method, 0.24-m vessel; (△) decoloration method, 0.24-m vessel. (Reprinted from C.J. Hoogendoorn and A.P. den Hartog, Model studies on mixers in the viscous flow region, *Chem. Eng. Sci.* **22**, 1689–1699. Copyright 1967, with permission from Pergamon Press Ltd, Oxford.)

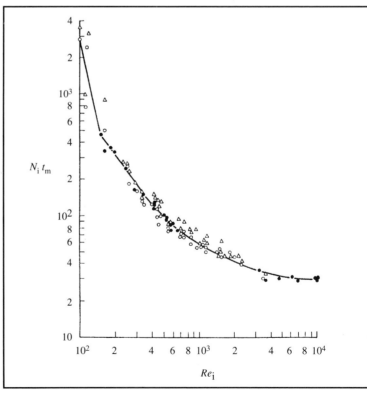

final concentration C_f by less than 10% of the total concentration difference $(C_f - C_i)$. However, there is no single, universally-applied definition of mixing time; sometimes deviations greater or less than 10% are specified. Nevertheless, at t_m the tracer concentration is relatively steady and the fluid composition approaches uniformity. For a single-phase liquid in a stirred tank with several baffles and small impeller, there is an approximate relationship between mixing time and circulation time [26]:

$$t_m = 4t_c.$$
(7.15)

Industrial-scale stirred vessels with working volumes between 1 and 100 m³ have mixing times between about 30 and 120 s, depending on conditions.

Intuitively, we can predict that mixing time in stirred tanks will depend on variables such as the size of the tank and impeller, fluid properties such as viscosity, and stirrer speed. The relationship between mixing time and several of these variables has been determined experimentally for different impellers [27]; results for a Rushton turbine in a baffled tank are shown in Figure 7.23. The dimensionless product $N_i t_m$ is plotted as a function of the impeller Reynolds number Re_i defined in Eq. (7.2); t_m is the mixing time based on a 10% deviation from final conditions, and N_i is rotational speed of the stirrer. $N_i t_m$ represents the number of stirrer rotations required to homogenise the liquid. At low Reynolds number, $N_i t_m$ increases significantly with decreasing Re_i. However, as Reynolds number is increased above about 5×10^3, $N_i t_m$ approaches a constant value which persists at high Re_i. For Rushton turbines, this constant value can be estimated using the following relationship [28]:

$$N_i t_m = \frac{1.54 V}{D_i^3} \quad \text{at high } Re_i$$
(7.16)

where V is liquid volume and D_i is impeller diameter. Thus, $N_i t_m$ at high Reynolds number depends only on the size of the tank and stirrer. The relationship between $N_i t_m$ and Re_i for most impellers is similar to that shown in Figure 7.23 for Rushton turbines. In some systems a slight increase in $N_i t_m$ with increasing Re_i has been reported; however, in practice we can assume $N_i t_m$ reaches a constant value [29]. With $N_i t_m$ constant, mixing time reduces in direct proportion to increase in stirrer speed.

Example 7.1 Estimation of mixing time

A fermentation broth with viscosity 10^{-2} Pa s and density 1000 kg m⁻³ is agitated in a 2.7 m³ baffled tank using a Rushton turbine with diameter 0.5 m and stirrer speed 1 s⁻¹. Estimate the mixing time.

Solution:
From Eq. (7.2):

$$Re_i = \frac{1 \text{ s}^{-1} (0.5 \text{ m})^2 \, 1000 \text{ kg m}^{-3}}{10^{-2} \text{ kg m}^{-1} \text{ s}^{-1}} = 2.5 \times 10^4.$$

$Re_i > 5 \times 10^3$; therefore $N_i t_m$ is constant and can be calculated from Eq. (7.16):

$$N_i t_m = \frac{1.54 \, (2.7 \text{ m}^3)}{(0.5 \text{ m})^3} = 33.3.$$

Therefore:

$$t_m = \frac{33.3}{1 \text{ s}^{-1}} = 33.3 \text{ s}.$$

The mixing time is about 33 s.

For rapid and effective mixing, t_m should be as small as possible. From Eq. (7.16) we can conclude that, in a tank of fixed volume, mixing time is reduced if we use a large impeller and high stirring speed. However, as the power requirements for mixing are also dependent on impeller diameter and stirrer speed, it is not always possible to achieve small mixing times without consuming enormous amounts of energy, especially in large vessels. Relationships between power requirements, mixing time, tank size, fluid properties and other operating variables are explored further in the following section.

Figure 7.24   Correlation between power number and Reynolds number for Rushton turbine, paddle and marine propeller without sparging. (From J.H. Rushton, E.W. Costich and H.J. Everett, 1950, Power characteristics of mixing impellers. Parts I and II. *Chem. Eng. Prog.* **46**, 395–404, 467–476.)

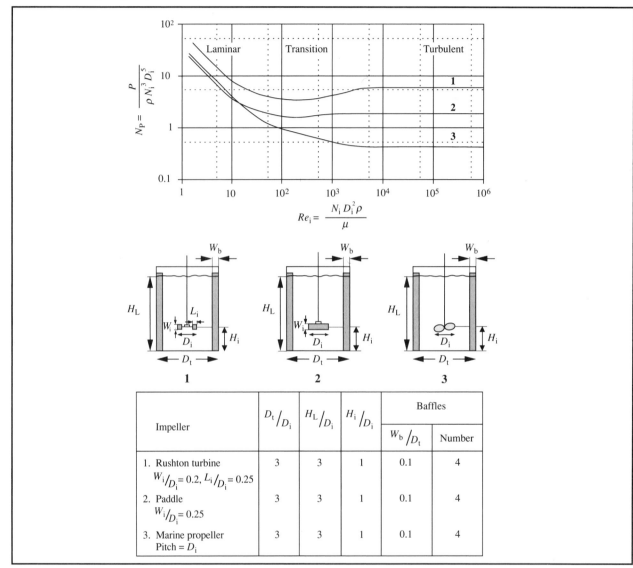

7.10 Power Requirements for Mixing

Usually, electrical power is used to drive impellers in stirred vessels. For a given stirrer speed, the power required depends on the resistance offered by the fluid to rotation of the impeller. Average power consumption per unit volume for industrial bioreactors ranges from $10\ \mathrm{kW\ m^{-3}}$ for small vessels (ca. $0.1\ \mathrm{m^3}$), to 1–$2\ \mathrm{kW\ m^{-3}}$ for large vessels (ca. $100\ \mathrm{m^3}$). Friction in the stirrer motor gearbox and seals reduces the energy transmitted to the fluid; therefore, the electrical power

consumed by stirrer motors is always greater than the mixing power by an amount depending on the efficiency of the drive. Energy costs for operation of stirrers in bioreactors are an important consideration in process economics. General guidelines for calculating power requirements are discussed below.

7.10.1 Ungassed Newtonian Fluids

Mixing power for non-aerated fluids depends on the stirrer speed, the impeller diameter and geometry, and properties of the fluid such as density and viscosity. The relationship between these variables is usually expressed in terms of dimensionless numbers such as the impeller Reynolds number Re_i and the *power number* N_P. N_P is defined as:

$$N_P = \frac{P}{\rho N_i^3 D_i^5}$$

(7.17)

where P is power, ρ is fluid density, N_i is stirrer speed and D_i is impeller diameter. The relationship between Re_i and N_P has been determined experimentally for a range of impeller and tank configurations. The results for five impeller designs: Rushton turbine, paddle, marine propeller, anchor and helical ribbon, are shown in Figures 7.24 and 7.25 [29–31]. Once the value of N_P is known, the power required is calculated from Eq. (7.17) as:

$$P = N_P \rho N_i^3 D_i^5.$$

(7.18)

For a given impeller, the general relationship between power number and Reynolds number depends on the flow regime in the tank. Three flow regimes can be identified in Figures 7.24 and 7.25:

(i) *Laminar regime.* The laminar regime corresponds to $Re_i < 10$ for many impellers; for stirrers with very small wall-clearance such as the anchor and helical-ribbon mixer, laminar flow persists until $Re_i = 100$ or greater. In the laminar regime:

$$N_P \propto \frac{1}{Re_i} \quad \text{or} \quad P = k_1 \mu N_i^2 D_i^3$$

(7.19)

where k_1 is a proportionality constant. Values of k_1 for the impellers illustrated in Figures 7.24 and 7.25 are listed in Table 7.3 [29]. Power required for laminar flow is independent of the density of the fluid but directly proportional to fluid viscosity.

(ii) *Turbulent regime.* Power number is independent of Reynolds number in turbulent flow. Therefore:

$$P = N_P' \rho N_i^3 D_i^5$$

(7.20)

where N_P' is the constant value of the power number in the turbulent regime. Approximate values of N_P' for the impellers of Figures 7.24 and 7.25 are listed in Table 7.3 [29]. N_P' for turbines is significantly higher than for most other impellers, indicating that turbines transmit more

power to the fluid than other designs. Power required for turbulent flow is independent of the viscosity of the fluid but proportional to fluid density. The turbulent regime is fully developed at $Re_i > 10^3$ or 10^4 for most small impellers in baffled vessels. For the same impellers in vessels without baffles, the power curves are somewhat different from those shown in Figure 7.24. Without baffles, turbulence is not fully developed until $Re_i > 10^5$; even then the value of N_P' is reduced to between $1/2$ and $1/10$ that with baffles [29–31].

(iii) *Transition regime.* Between laminar and turbulent flow lies the transition regime. Both density and viscosity affect power requirements in this regime. There is usually a gradual transition from laminar to fully-developed turbulent flow in stirred tanks; the flow pattern and Reynolds-number range for transition depend on system geometry.

Eqs (7.19) and (7.20) express the strong dependence of power consumption on stirrer diameter and, to a lesser extent, stirrer speed. Small changes in impeller size have a large effect on power requirements, as would be expected from dependency on impeller diameter raised to the third or fifth power. In the turbulent regime, a 10% increase in impeller diameter increases the power required by more than 60%; a 10% increase in stirrer speed raises the power required by over 30%.

Frictional drag, and therefore the power required for stirring, depend on the geometry of the impeller and configuration of the tank. The curves of Figures 7.24 and 7.25 refer to the particular geometries specified and will change if the number or size of baffles, the number, length, width, pitch or angle of blades on the impeller, the height of impeller from the bottom of the tank, etc. are changed. For a Rushton turbine in a baffled tank under fully turbulent conditions ($Re_i > 10^4$), the power number lies between about 2 and 10 depending on these parameters [25, 34]. For propellers, impeller pitch has a significant effect on power number in the turbulent regime [25].

Table 7.3 Constants in Eqs (7.19) and (7.20)

Impeller type	k_1 ($Re_i = 1$)	N_P' ($Re_i = 10^5$)
Rushton turbine	70	5–6
Paddle	35	2
Marine propeller	40	0.35
Anchor	420	0.35
Helical ribbon	1000	0.35

Figure 7.25 Correlation between power number and Reynolds number for anchor and helical-ribbon impellers without sparging. (From M. Zlokarnik and H. Judat, 1988, Stirring. In: W. Gerhartz, Ed, *Ullmann's Encyclopedia of Industrial Chemistry*, vol. B2, pp. 25-1–25-33, VCH, Weinheim.)

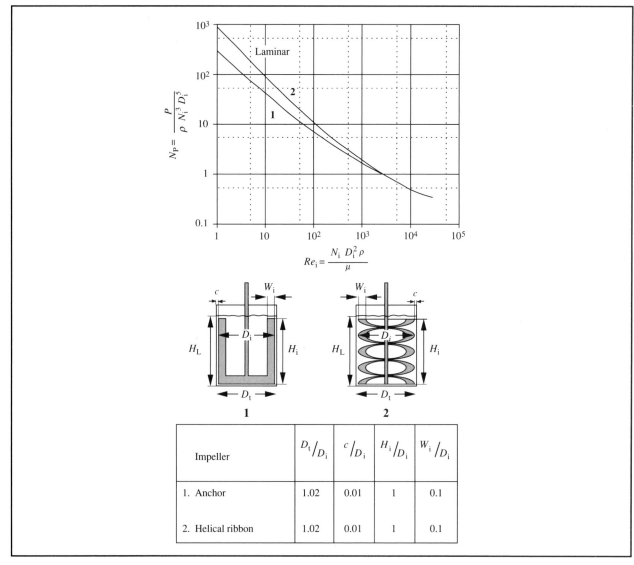

Impeller	D_t/D_i	c/D_i	H_i/D_i	W_i/D_i
1. Anchor	1.02	0.01	1	0.1
2. Helical ribbon	1.02	0.01	1	0.1

Example 7.2 Calculation of power requirements

A fermentation broth with viscosity 10^{-2} Pa s and density 1000 kg m^{-3} is agitated in a 50 m^3 baffled tank using a marine propeller 1.3 m in diameter. The tank geometry is as specified in Figure 7.24. Calculate the power required for a stirrer speed of 4 s^{-1}.

Solution:
From Eq. (7.2):

$$Re_i = \frac{4\,\text{s}^{-1}\,(1.3\,\text{m})^2\,1000\,\text{kg m}^{-3}}{10^{-2}\,\text{kg m}^{-1}\,\text{s}^{-1}} = 6.76 \times 10^5.$$

From Figure 7.24, flow at this Re_i is fully turbulent. From Table 7.3, N_P' is 0.35; therefore:

$$P = (0.35)\ 1000\ \text{kg m}^{-3}\ (4\ \text{s}^{-1})^3\ (1.3\ \text{m})^5 = 8.3 \times 10^4\ \text{kg m}^2\ \text{s}^{-3}$$

$$P = 83\ \text{kW}.$$

7.10.2 Ungassed Non-Newtonian Fluids

Estimation of power requirements for non-Newtonian fluids is more difficult. It may be impossible with highly viscous fluids to achieve fully-developed turbulence so that N_P is always dependent on Re_i. In addition, because the viscosity of non-Newtonian liquids varies with shear conditions, the impeller Reynolds number used to correlate power requirements must be re-defined. Some power correlations have been developed using an impeller Reynolds number based on the apparent viscosity μ_a:

$$Re_i = \frac{N_i D_i^2 \rho}{\mu_a} \tag{7.21}$$

so that, from Eq. (7.8) for power-law fluids:

$$Re_i = \frac{N_i D_i^2 \rho}{K\ \dot{\gamma}^{\,n-1}} \tag{7.22}$$

where n is the flow behaviour index and K is the consistency index. A problem with application of Eq. (7.22) is evaluation of $\dot{\gamma}$. For stirred tanks, an approximate relation for pseudoplastic fluids is often used:

$$\dot{\gamma} = k\,N_i \tag{7.11}$$

where k is a constant with magnitude dependent on the geometry of the impeller. The relationship of Eq. (7.11) is discussed further in Section 7.13; for turbine impellers k is about 10. Substituting Eq. (7.11) into Eq. (7.22) gives an appropriate Reynolds number for pseudoplastic fluids:

$$Re_i = \frac{N_i^{2-n} D_i^2 \rho}{k^{n-1} K}. \tag{7.23}$$

The relationship between power number and Reynolds number for a Rushton turbine in a baffled tank containing pseudoplastic non-Newtonian fluid is shown in Figure 7.26 [32, 35]. The upper line was measured using Newtonian fluid for which Re_i is defined by Eq. (7.2); this line corresponds to part of the curve already shown in Figure 7.24. The lower line gives the N_P–Re_i relationship for pseudoplastic fluid with Re_i

Figure 7.26 Correlation between power number and Reynolds number for a Rushton turbine in unaerated non-Newtonian fluid in a baffled tank. (From A.B. Metzner, R.H. Feehs, H. Lopez Ramos, R.E. Otto and J.D. Tuthill, 1961, Agitation of viscous Newtonian and non-Newtonian fluids. *AIChE J.* 7, 3–9.)

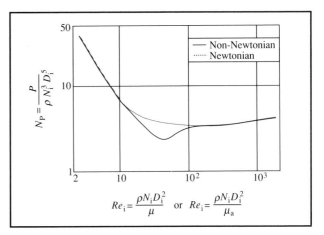

defined by Eq. (7.23). The laminar region extends to higher Reynolds numbers in pseudoplastic fluids than in Newtonian systems. At Re_i below 10 and above 200 the results for Newtonian and non-Newtonian fluids are essentially the same; in the intermediate range, pseudoplastic liquids consume less power than Newtonian fluids.

There are several practical difficulties with application of Figure 7.26 for design of bioreactors. As discussed further in Section 7.13, flow patterns in pseudoplastic and Newtonian fluids differ significantly. Even when there is high turbulence near the impeller in pseudoplastic systems, the bulk liquid may be moving very slowly and consuming relatively little power. Another problem is that, as illustrated in Figure 7.10, the non-Newtonian parameters K and n, and therefore μ_a, can vary substantially during fermentation.

7.10.3 Gassed Fluids

Liquids into which gas is sparged have reduced power requirements. Gas bubbles decrease the density of the fluid; however, the influence of density on power requirements as expressed by Eq. (7.20) does not adequately explain all the power characteristics of gas–liquid systems. The presence of bubbles also affects the hydrodynamic behaviour of fluid around the impeller. Large gas-filled cavities develop behind the stirrer blades in aerated liquids; these cavities reduce the resistance to fluid flow and decrease the drag coefficient of the impeller. Typical gas cavities are shown in Figure 7.27; this photograph taken through the base of a baffled tank shows a nine-blade disc-turbine with sparger positioned just below the impeller [33].

All of the changes in hydrodynamic behaviour due to gassing are not completely understood. Power consumption is strongly controlled by gas-cavity formation; because this process is discontinuous and appears somewhat random, reduction in power consumption is typically non-uniform. The random nature of gas dispersion in agitated tanks means that it is difficult to obtain an accurate prediction of power requirements. However, an expression for the ratio of gassed to ungassed power as a function of operating conditions has been obtained [34]:

$$\frac{P_g}{P_0} = 0.10 \left(\frac{F_g}{N_i V}\right)^{-0.25} \left(\frac{N_i^2 D_i^4}{g\, W_i\, V^{2/3}}\right)^{-0.20} \tag{7.24}$$

where P_g is power consumption with sparging, P_0 is power consumption without sparging, F_g is volumetric gas flow rate, N_i is stirrer speed, V is liquid volume, D_i is impeller diameter, g is gravitational acceleration, and W_i is impeller blade width. The average deviation of experimental values from Eq. (7.24)

Figure 7.27 Gas cavities formed behind the blades of a 7.6-cm nine-blade flat-disc turbine in water sparged with air. The stirrer speed was 720 rpm. (From W. Bruijn, K. van't Riet and J.M. Smith, 1974, Power consumption with aerated Rushton turbines. *Trans. IChE* 52, 88–104.)

is about 12%. With sparging, the power consumed could be reduced to as little as half the ungassed value, depending on gas flow rate [33].

7.11 Scale-Up of Mixing Systems

Design of industrial-scale bioprocesses is usually based on the performance of small-scale prototypes. Determining optimum operating conditions at production scale is expensive and time-consuming; accordingly, it is always better to know whether a particular process will work properly before it is constructed in full size. Ideally, scale-up should be carried out so that conditions in the large vessel are as close as possible to those producing good results in the small vessel. As mixing is an important function of bioreactors, it would seem desirable to keep the mixing time constant on scale-up. Unfortunately, as explained below, the relationship between mixing time and power consumption makes this rarely possible in practice.

As the volume of mixing vessels is increased, so too are the lengths of the flow paths for bulk circulation. To keep the mixing time constant, the velocity of fluid in the tank must be increased in proportion to the size. As a rough guide, under turbulent conditions the power per unit volume is proportional to the fluid velocity squared:

$$P/_V \propto v^2 \tag{7.25}$$

where P is power, V is liquid volume, and v is fluid linear velocity. The effect of this relationship on power requirements is illustrated in the following example. Suppose a cylindrical 1 m³ pilot-scale stirred tank is scaled up to 100 m³. If the tanks are geometrically similar, the length of the flow path in the large tank is about 4.5 times that in the small tank. Therefore, to keep the same mixing time, fluid velocity in the large tank must be approximately 4.5 times faster. From Eq. (7.25) this would entail a $(4.5)^2$ or 20-fold increase in power per unit volume. So, if the power input to the 1 m³ pilot-scale vessel is P, the power required for the same mixing time in the 100 m³ tank is about 2000P. This represents an extremely large increase in power, much greater than is economically or technically feasible with most equipment used for stirring. Because the criterion of constant mixing time can hardly ever be applied for scale-up, it is inevitable that mixing times increase with scale. If instead of mixing time, $P/_V$ is kept constant during scale-up, mixing time can be expected to increase in proportion to vessel diameter raised to the power 0.67 [29].

Reduced productivity and performance often accompany scale-up of bioreactors as a result of lower mixing efficiency and subsequent alteration of the physical environment. One way of improving the design procedure is to use *scale-down methods*.

The general idea behind scale-down is that small-scale experiments to determine operating parameters are carried out under conditions that can actually be realised, physically and economically, at production scale. For example, if we decide that power input to a large-scale vessel cannot exceed a certain limit, we can calculate the corresponding mixing time and use an appropriate power input to a small-scale reactor to simulate mixing conditions in the large-scale system. Using this approach, as long as the flow regime is the same in the small- and large-scale fermenters, there is a better chance that results achieved in the small-scale unit will be reproducible in the larger system.

7.12 Improving Mixing in Fermenters

Sometimes, for the reasons outlined in the previous section, it is not possible to reduce mixing times by simply raising the power input to the stirrer. So, while increasing the stirrer speed is an obvious way of improving fluid circulation, other techniques may be required.

Mixing can sometimes be improved by changing the configuration of the system. Baffles should be installed; this is routine for stirred fermenters and produces greater turbulence. For efficient mixing the impeller should be mounted below the geometric centre of the vessel. In standard designs the impeller is located about one impeller diameter, or one-third the tank diameter, above the bottom of the tank. Mixing is facilitated when circulation currents below the impeller are smaller than those above; fluid particles leaving the impeller at the same instant then take different periods of time to return and exchange material. Rate of distribution throughout the vessel is increased when upper and lower circulation loops are asynchronous.

Another device for improving mixing is multiple impellers, although this requires an increase in power input. Typical bioreactors used for aerobic culture are tall cylindrical vessels with liquid depths significantly greater than the tank diameter. This design produces a higher hydrostatic pressure at the bottom of the vessel, and gives rising air bubbles a longer contact time with the liquid. Effective mixing in tall fermenters requires more than one impeller, as illustrated in Figure 7.28. Each impeller generates its own circulation currents. The distance between impellers should be 1.0 to 1.5 impeller diameters. If the impellers are spaced too far apart, unagitated zones develop between them; conversely, impellers located too close together produce flow streams which interfere with each other and disrupt circulation to the far reaches of the vessel. In ungassed systems with spacing between impellers of at least one impeller diameter, the power dissipated by multiple impellers is approximated by the following relationship:

$$(P)_n = n(P)_1 \tag{7.26}$$

Figure 7.28   Multiple impellers in a tall fermenter.

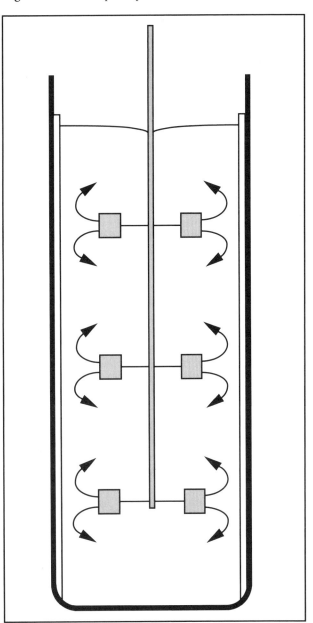

where $(P)_n$ is the power required by n impellers and $(P)_1$ is the power required by a single impeller. Two turbines spaced less than one impeller diameter apart can draw as much as 2.4 times the power of a single turbine. When the vessel is sparged with gas the power relationship may not be so simple. As described in Section 7.10.3, the presence of gas bubbles reduces the power required for the lowest impeller. However the quantity of gas passing through the upper impellers is often much smaller; when this occurs the power drawn by each additional impeller is large compared with a single-impeller system [36].

Additional mixing problems can occur in fermenters when material is fed into the system during operation. If bulk distribution is slow, fermenters operated continuously or in fed-batch mode develop high localised concentrations of substrate near the feed point. This has been observed particularly in large-scale processes for production of single-cell-protein from methanol. Because high levels of methanol are toxic to cells, biomass yields decrease significantly when mixing of feed material into the broth is slow. Problems like this can be alleviated by using multiple injection points to aid distribution of substrate. It is much less expensive to do this than to increase the fluid velocity and power input.

7.13 Effect of Rheological Properties on Mixing

For effective mixing there must be turbulent conditions in the mixing vessel. Intensity of turbulence is represented by the impeller Reynolds number Re_i. As shown in Figure 7.23 for a baffled tank with turbine impeller, once Re_i falls below about 5×10^3 turbulence is damped and mixing time increases significantly. Re_i as defined in Eq. (7.2) decreases in direct proportion to increase in viscosity. Accordingly, non-turbulent conditions and poor mixing are likely to occur during agitation of highly viscous fluids. Increasing the impeller speed is an obvious solution but, as discussed in Section 7.11, this requires considerable increase in power consumption and therefore may not be feasible.

Most non-Newtonian fluids in bioprocessing are pseudoplastic. Because the apparent viscosity of these fluids depends on the shear rate, the rheological behaviour of many culture broths depends on shear conditions in the fermenter. Metzner and Otto [32] have proposed that the average shear rate in a stirred vessel is a linear function of stirrer speed:

$$\dot{\gamma}_{av} = k N_i \qquad (7.27)$$

where $\dot{\gamma}_{av}$ is the average shear rate, k is a constant dependent on impeller design and N_i is stirrer speed. Experimental values of k are summarised in Table 7.4. The validity of Eq. (7.27) was established in studies by Metzner *et al.* [35]. However, shear rate in stirred vessels is far from uniform, being strongly dependent on distance from the impeller. Figure 7.29 indicates the rapid decline in shear rate in pseudoplastic fluid with increasing radial distance from the tip of a flat-blade turbine impeller [37]. The maximum shear rate close to the impeller is much higher than the average calculated from Eq. (7.27).

Pseudoplastic fluids are shear thinning, i.e. their apparent viscosity decreases with increasing shear. Accordingly, in stirred vessels, pseudoplastic fluids have relatively low apparent viscosity in the high-shear zone near the impeller, and relatively high apparent viscosity when the fluid is away from the impeller. As a result, flow patterns similar to that illustrated in Figure 7.30 can develop; a small circulating pool of highly sheared fluid surrounds the impeller while the bulk liquid scarcely moves at all. In bioreactors containing non-Newtonian broths, this can lead to development of stagnant zones away from the impeller.

The effects of local fluid thinning in pseudoplastic fluids can be countered by modifying the geometry of the system or impeller design. Stirrers of larger diameter are recommended. For turbine impellers, instead of the conventional tank-to-impeller diameter ratio of 3:1 used with low viscosity fluids, this ratio is reduced to between 1.6 and 2. Different impeller designs which sweep the entire volume of the vessel are also recommended. The most common types used for viscous mixing are helical impellers and gate- and paddle-anchors mounted with small clearance between the impeller and tank wall. Mixing with these stirrers is accomplished at low speed without high-velocity streams. Helical agitators have been used to reduce shear damage and improve mixing in viscous cell suspensions [38].

Alternative impeller designs such as the helical ribbon and anchor improve mixing in viscous fluids; however their application in fermenters is only possible when oxygen demand in the culture is relatively low. Although large-diameter impellers operating at relatively slow speed give superior bulk mixing, high-shear systems with small, high-speed impellers are preferable for breaking up gas bubbles and promoting oxygen transfer to the liquid. In design of fermenters for viscous cultures, a compromise is usually required between mixing effectiveness and adequate mass transfer.

7.14 Role of Shear in Stirred Fermenters

Mixing in bioreactors must provide the shear conditions necessary to disperse bubbles, droplets and cell flocs. Dispersion of

Table 7.4 Observed values of k in Eq. (7.27)

(From S. Nagata, 1975, Mixing: Principles and Applications, *Kodansha, Tokyo)*

Impeller type	k
Rushton turbine	10–13
Paddle	10–13
Curved-blade paddle	7.1
Propeller	10
Anchor	20–25
Helical ribbon	30

Figure 7.29   Shear rate in pseudoplastic fluid as a function of stirrer speed and radial distance from the impeller: (○) impeller tip; (▲) 0.10 in; (■) 0.20 in; (▽) 0.34 in; (●) 0.50 in; (□) 1.00 in. The impeller diameter is 4 in. (From A.B. Metzner and J.S. Taylor, 1960, Flow patterns in agitated vessels. *AIChE J.* **6**, 109–114.)

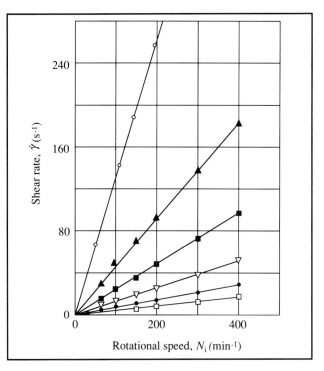

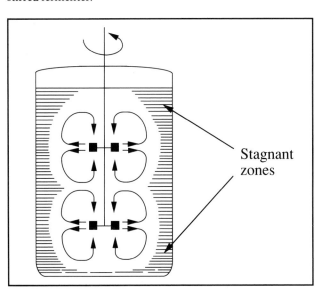

Figure 7.30   Mixing pattern for pseudoplastic liquid in a stirred fermenter.

gas bubbles by agitation involves a balance between opposing forces. Shear forces in turbulent eddies stretch and distort the bubbles and break them into smaller sizes; at the same time, surface tension at the gas–liquid interface tends to restore the bubbles to their spherical shape. In the case of solid material such as cell flocs or aggregates, shear forces in turbulent flow are resisted by the mechanical strength of the particles.

While bubble break-up is required in fermenters to facilitate oxygen transfer, disruption of cells is undesirable. Different cell types display different levels of *shear sensitivity*; insect, mammalian and plant cells are known to be particularly sensitive to mechanical forces. Bioreactors used for culture of these cells must limit the intensity of shear while still providing adequate mixing and mass transfer. At the present time, the effects of shear on cells are not well understood. Cell disruption is an obvious outcome of high shear forces; however more subtle changes such as retardation of growth and product synthesis, denaturation of extracellular proteins, change in morphology, and thickening of the cell wall, may also occur.

Because there is significant spatial variation in shear intensity in stirred vessels, the precise shear conditions experienced by cells are poorly defined. There have been many publications in recent years addressing the problem of shear damage, especially in insect- and mammalian-cell cultures. Several mechanisms have been considered in terms of their contribution to cell damage:

(i)   interaction between cells and turbulent eddies;
(ii)   collisions between cells, collision of cells with the impeller, and collision of cells with stationary surfaces in the vessel;
(iii) generation of shear forces in the boundary layers and wakes near solid objects in the reactor, especially the impeller;
(iv)   generation of shear forces as bubbles rise through liquid; and
(v)   bursting of bubbles at the liquid surface.

Detailed discussion of these effects can be found elsewhere [39–50]. In general, when gas bubbles are not present in the liquid, interactions between cells and turbulent eddies are considered most likely to damage cells. However, if the vessel is sparged with air, shear damage can occur at much lower impeller speeds due to shear effects associated with bubbles [44].

7.14.1 Interaction Between Cells and Turbulent Eddies

Hydrodynamic effects have been studied mainly with animal cells because shear damage is a significant problem in large-scale culture. Many animal cells used in bioprocessing are *anchorage-dependent*; this means that the cells must be attached to a solid

Figure 7.31 Chinese hamster ovary (CHO) cells attached to microcarrier beads; magnification × 85. (Photograph courtesy of J. Crowley.)

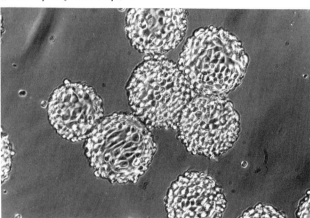

surface for survival. In bioreactors, the surface area required for cell attachment is provided very effectively by *microcarrier beads*, which range in diameter from 80 to 200 μm. As shown in Figure 7.31, cells cover the surface of the beads which are then suspended in nutrient medium. There are many benefits associated with use of microcarriers; however, a disadvantage is that cells attached to microcarriers cannot easily change position or rotate in response to shear forces in the fluid. This, coupled with the lack of a protective cell wall, make animal cells on microcarriers especially susceptible to shear damage.

Interactions between microcarriers and eddies in turbulent flow have the potential to cause mechanical damage to cells.

The intensity of shear associated with these interactions is dependent on the relative sizes of the eddies and microcarrier particles. If the particles are small relative to the eddies, they tend to be captured or entrained in the eddies as shown in Figure 7.32(a). As fluid motion within eddies is laminar, if the density of the microcarriers is about the same as the suspending fluid, there is little relative motion of the particles. Accordingly, the velocity difference between the fluid streamlines and the microcarriers is small, except for brief periods of acceleration when the bead enters a new eddy. On average, therefore, if the particles are smaller than the eddies, the shear effects of eddy–cell interactions are minimal.

If the stirrer speed is increased and the average eddy size reduced, interactions between eddies and microcarriers can occur in two possible ways. A single eddy that cannot fully engulf the particle will act on part of its surface and cause the particle to rotate in the fluid; this will result in a relatively low level of shear at the surface of the bead. However, much higher shear stresses result when several eddies with opposing rotation interact with the particle simultaneously, as illustrated in Figure 7.32(b). It has been found experimentally that detrimental effects start to occur when the Kolmogorov scale (Eq. 7.14) for eddy size drops below $^2/_3$–$^1/_2$ the diameter of the microcarrier beads [41, 42, 49]. Excessive agitation leads to formation of eddies with size small enough and of sufficient energy to cause damage to the cells. These findings for cells on microcarriers apply also to freely suspended cells; however, because cells are smaller than microcarriers, eddy sizes causing shear damage are also smaller.

Example 7.3 Operating conditions for turbulent shear damage

Microcarrier beads 120 μm in diameter are used to culture recombinant CHO cells for production of growth hormone. It is proposed to use a 6-cm turbine impeller to mix the culture in a 3.5-litre stirred tank. Air and carbon dioxide are supplied by flow through the reactor headspace. The microcarrier suspension has a density of approximately 1010 kg m^{-3} and a viscosity of 1.3×10^{-3} Pa s. Estimate the maximum allowable stirrer speed which avoids turbulent shear damage of the cells.

Solution:

Damage due to eddies is avoided if the Kolmogorov scale remains greater than $^2/_3$–$^1/_2$ the diameter of the beads. Let us determine the stirrer speed required to create eddies with size $\lambda = {}^2/_3$ (120 μm) = 80 μm = 8×10^{-5} m. The stirrer power producing eddies of this dimension can be estimated using Eq. (7.14) and the properties of the fluid.

$$\text{Kinematic viscosity, } v = \frac{\mu}{\rho} = \frac{1.3 \times 10^{-3} \text{ kg m}^{-1} \text{ s}^{-1}}{1010 \text{ kg m}^{-3}} = 1.29 \times 10^{-6} \text{ m}^2 \text{ s}^{-1}.$$

Raising both sides of Eq. (7.14) to the fourth power:

$$\lambda^4 = \frac{v^3}{\varepsilon}$$

so that

Figure 7.32 Eddy–microcarrier interactions. (a) Microcarriers are captured in large eddies and move within the streamline flow. (b) When several eddies with opposing rotation interact with the microcarrier simultaneously, high levels of shear develop on the bead surface. (From R.S. Cherry and E.T. Papoutsakis, 1986, Hydrodynamic effects on cells in agitated tissue culture reactors. *Bioprocess Eng.* **1**, 29–41.)

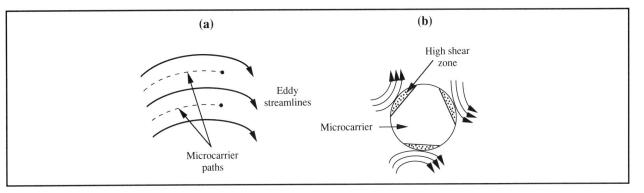

$$\varepsilon = \frac{v^3}{\lambda^4}.$$

Therefore:

$$\varepsilon = \frac{(1.29 \times 10^{-6})^3 \, \text{m}^6 \, \text{s}^{-3}}{(8 \times 10^{-5})^4 \, \text{m}^4} = 0.052 \, \text{m}^2 \, \text{s}^{-3}.$$

ε represents the power dissipated per unit mass of fluid. The mass of fluid in which turbulent power is dissipated may be taken as the entire contents of the tank; however power dissipation occurs unevenly throughout the vessel with highest levels in the vicinity of the impeller. ε based on the mass of fluid near the impeller is considered the more appropriate value in these calculations [50]; fluid mass in the impeller zone is roughly equal to $\rho \, D_i^3$ where ρ is fluid density and D_i is impeller diameter. Therefore, the stirrer power P is equal to ε multiplied by $\rho \, D_i^3$:

$$P = (0.052 \, \text{m}^2 \, \text{s}^{-3}) \, (1010 \, \text{kg m}^{-3}) \, (6 \times 10^{-2} \, \text{m})^3$$

$$P = 1.13 \times 10^{-2} \, \text{kg m}^2 \, \text{s}^{-3} = 1.13 \times 10^{-2} \, \text{W}.$$

From Figure 7.24 and Table 7.3, N_p' is about 5 for a turbine impeller operating in the turbulent regime, depending on tank geometry. The stirrer speed corresponding to these conditions can be calculated from Eq. (7.20):

$$N_i^3 = \frac{P}{N_p' \rho \, D_i^5}$$

$$N_i^3 = \frac{1.13 \times 10^{-2} \, \text{kg m}^2 \, \text{s}^{-3}}{(5) \, (1010 \, \text{kg m}^{-3}) \, (6 \times 10^{-2} \, \text{m})^5} = 2.89 \, \text{s}^{-3}.$$

Taking the cube root:

$$N_i = 1.42 \, \text{s}^{-1} = 85.5 \, \text{rpm}.$$

Flow is just turbulent with $Re_i = 4 \times 10^3$. This analysis indicates that shear damage from turbulent eddies is not expected until the stirrer speed exceeds about 85 rpm. If the culture were sparged with gas, it is possible that shear damage would occur due to other mechanisms, e.g. bursting bubbles.

As indicated by Eq. (7.14), if the viscosity of the fluid is increased, the size of the smallest eddies also increases. Increasing the fluid viscosity should, therefore, reduce shear damage in bioreactors. This effect has been demonstrated by addition of thickening agents to animal-cell growth medium; moderate increases in viscosity have been shown to significantly reduce turbulent cell death [42].

7.14.2 Bubble Shear

When liquid containing shear-sensitive cells is sparged with air, other damaging mechanisms come into play. From experiments conducted so far, these appear to be associated primarily with bubbles bursting at the surface of the liquid. Breakage of the thin bubble film and rapid flow from the bubble rim back into the liquid generate high shear forces capable of damaging certain types of cell. Further discussion of these effects can be found in the literature [43–47].

7.15 Summary of Chapter 7

Chapter 7 covers a wide range of topics in fluid dynamics, rheology and mixing. At the end of Chapter 7 you should:

(i)   understand the difference between *laminar* and *turbulent* flow;
(ii) be able to describe how fluid *boundary layers* develop in terms of *viscous drag*;
(iii) be able to define *viscosity* in terms of *Newton's law*;
(iv) know what *Newtonian* and *non-Newtonian* fluids are, and the difference between viscosity for Newtonian fluids and *apparent viscosity* for non-Newtonian fluids;
(v)   be familiar with equipment used for mixing in stirred vessels;
(vi) be able to describe the mechanisms of mixing and their effect on *mixing time*;
(vii) understand the effects of scale-up on mixing;
(viii) know how liquid properties, gas sparging, impeller size and stirrer speed affect power consumption in stirred vessels; and
(ix) understand how cells can be damaged by shear in stirred fermenters.

Problems

7.1 Rheology of fermentation broth

The fungus *Aureobasidium pullulans* is used to produce an extracellular polysaccharide by fermentation of sucrose. After 120 h fermentation, the following measurements of shear stress and shear rate were made with a rotating-cylinder viscometer.

Shear stress (dyn cm^{-2})	Shear rate (s^{-1})
44.1	10.2
235.3	170
357.1	340
457.1	510
636.8	1020

(a) Plot the rheogram for this fluid.
(b) Determine the appropriate non-Newtonian parameters.
(c) What is the apparent viscosity at shear rates of:
   (i)   15 s^{-1}; and
   (ii)   200 s^{-1}?

7.2 Rheology of yeast suspensions

Apparent viscosities for pseudoplastic cell suspensions at varying cell concentrations are measured using a coaxial-cylinder rotary viscometer. The results are:

Cell concentration (%)	Shear rate (s^{-1})	Apparent viscosity (cP)
1.5	10	1.5
	100	1.5
3	10	2.0
	100	2.0
6	20	2.5
	45	2.4
10.5	10	4.7
	20	4.0
	50	4.1
	100	3.8
12	1.8	40
	4.0	30
	7.0	22
	20	15
	40	12
18	1.8	140
	7.0	85
	20	62
	40	55
21	1.8	710
	4.0	630
	7.0	480
	40	330
	70	290

Show on an appropriate plot how K and n vary with cell concentration.

7.3 Impeller viscometer

The rheology of a *Penicillium chrysogenum* broth is examined using an impeller viscometer. The density of the cell suspension is approximately 1000 kg m^{-3}. Samples of broth are poured into a glass beaker of diameter 15 cm and stirred slowly using a Rushton turbine of diameter 4 cm. The average shear rate generated by this impeller is greater than the stirrer speed by a factor of about 10.2. When the stirrer shaft is attached to a device for measuring torque and rotational speed, the following results are recorded.

Stirrer speed (s^{-1})	Torque (N m)
0.185	3.57×10^{-6}
0.163	3.45×10^{-6}
0.126	3.31×10^{-6}
0.111	3.20×10^{-6}

(a) Can the rheology be described using a power-law model? If so, evaluate K and n.

(b) Viscosity measurements using impeller viscometers must be carried out under laminar flow conditions. Check that flow in this experiment is laminar.

(c) Use of turbines for impeller viscometry restricts the range of shear rates that can be tested. How is the situation improved with a helical-ribbon impeller?

7.4 Particle suspension and gas dispersion

Cells in fermenters must be kept from settling out of suspension. The minimum stirrer speed required to keep the bottom of the tank free of cells can be estimated roughly using a relation given by Zwietering [51]:

$$N_i^* = C v_L^{0.1} D_p^{0.2} \left| \frac{g(\rho_p - \rho_L)}{\rho_L} \right|^{0.45} D_i^{-0.85} x^{0.13}$$

where:

N_i^*	=	minimum stirrer speed for suspension of solids, s^{-1};
C	=	a constant (= 7.7 for a turbine impeller with diameter one-third of that of the tank);
v_L	=	liquid kinematic viscosity, m^2 s^{-1};
D_p	=	mean cell diameter, m;
g	=	gravitational acceleration, m s^{-2};
ρ_p	=	density of the cells, kg m^{-3};
ρ_L	=	density of the suspending liquid, kg m^{-3};
D_i	=	impeller diameter, m; and
x	=	cell concentration, wt%.

A certain minimum stirrer speed is also required in aerobic systems for proper dispersion of air bubbles. From the data of Westerterp *et al.* [52], the minimum stirrer tip speed (tip speed = $\pi N_i D_i$) for this purpose can be assumed to lie between 1.5 and 2.5 m s^{-1}, depending on the surface tension between the gas and liquid, the fluid density, and the tank-to-impeller diameter ratio.

A fermentation broth contains 40 wt% cells of average dimension 10 μm and density 1.04 g cm^{-3}. The diameter of the impeller in the fermenter is 30 cm. Assuming that the density and viscosity of the suspending medium are the same as water, determine which takes more power to achieve, cell suspension or bubble dispersion.

7.5 Scale-up of mixing system

To ensure turbulent conditions and minimum mixing time during agitation with a turbine impeller, the Reynolds number must be at least 10^4.

(a) A stirred laboratory-scale fermenter with a turbine impeller 5 cm in diameter is operated at 800 rpm. If the density of broth being stirred is close to that of water, what is the upper limit for viscosity of the suspension if adequate mixing is to be maintained?

(b) The mixing system is scaled up so the tank and impeller are 15 times the diameter of the laboratory equipment. The stirrer in the large vessel is operated so that the stirrer tip speed (tip speed = $\pi N_i D_i$) is the same as in the laboratory apparatus. How does scale-up affect the maximum viscosity allowable for maintenance of turbulent mixing conditions?

7.6 Effect of viscosity on power requirements

A cylindrical bioreactor of diameter 3 m has four baffles. A Rushton turbine mounted in the reactor has a diameter one-third the tank diameter and is operated at a speed of 90 rpm. The density of the fluid is approximately 1 g cm^{-3}. The reactor is used to culture an anaerobic organism that does not require gas sparging. The broth can be assumed Newtonian. As the cells grow, the viscosity of the broth increases.

(a) Compare power requirements when the viscosity is:
  (i)   approximately that of water;
  (ii) 100 times greater than water; and
  (iii) 10^4 times greater than water.

(b) When the viscosity is 1000 times greater than water, estimate the power required to achieve turbulence.

7.7 Electrical power required for mixing

Laboratory-scale fermenters are usually mixed using small stirrers with electric motors rated between 100 and 500 W. One such motor is used to drive a 7-cm turbine impeller in a small reactor containing fluid with the properties of water. The stirrer speed is 900 rpm. Estimate the power requirements for this process. How do you explain the difference between the amount of electrical power consumed by the motor and the power dissipated by the stirrer?

7.8 Mixing time with aeration

A cylindrical stirred bioreactor of diameter and height 2 m has a Rushton turbine one-third the tank diameter in size. The bioreactor contains Newtonian culture broth with the same density as water and with viscosity 4 cP.

(a) If the specific power consumption must not exceed 1.5 kW m^{-3}, determine the maximum allowable stirrer speed. What is the mixing time under these conditions?
(b) The tank is now aerated. In the presence of gas bubbles, the approximate relationship between ungassed power number $(N_p)_0$ and gassed power number $(N_p)_g$ is: $(N_p)_g = 0.5 (N_p)_0$. What maximum stirrer speed is now possible in the sparged reactor? Estimate the mixing time.

References

1. Skelland, A.H.P. (1967) *Non-Newtonian Flow and Heat Transfer*, John Wiley, New York.
2. van Wazer, J.R., J.W. Lyons, K.Y. Kim and R.E. Colwell (1963) *Viscosity and Flow Measurement*, Interscience, New York.
3. Whorlow, R.W. (1980) *Rheological Techniques*, Ellis Horwood, Chichester.
4. Charles, M. (1978) Technical aspects of the rheological properties of microbial cultures. *Adv. Biochem. Eng.* **8**, 1–62.
5. Metz, B., N.W.F. Kossen and J.C. van Suijdam (1979) The rheology of mould suspensions. *Adv. Biochem. Eng.* **11**, 103–156.
6. Sherman, P. (1970) *Industrial Rheology*, Academic Press, London.
7. Bongenaar, J.J.T.M., N.W.F. Kossen, B. Metz and F.W. Meijboom (1973) A method for characterizing the rheological properties of viscous fermentation broths. *Biotechnol. Bioeng.* **15**, 201–206.
8. Roels, J.A., J. van den Berg and R.M. Voncken (1974) The rheology of mycelial broths. *Biotechnol. Bioeng.* **16**, 181–208.
9. Kim, J.H., J.M. Lebeault and M. Reuss (1983) Comparative study on rheological properties of mycelial broth in filamentous and pelleted forms. *Eur. J. Appl. Microbiol. Biotechnol.* **18**, 11–16.
10. Rapp, P., H. Reng, D.-C. Hempel and F. Wagner (1984) Cellulose degradation and monitoring of viscosity decrease in cultures of *Cellulomonas uda* grown on printed newspaper. *Biotechnol. Bioeng.* **26**, 1167–1175.
11. Labuza, T.P., D. Barrera Santos and R.N. Roop (1970) Engineering factors in single-cell protein production. I. Fluid properties and concentration of yeast by evaporation. *Biotechnol. Bioeng.* **12**, 123–134.
12. Berkman-Dik, T., M. Özilgen and T.F. Bozoğlu (1992) Salt, EDTA, and pH effects on rheological behavior of mold suspensions. *Enzyme Microbiol. Technol.* **14**, 944–948.
13. Deindoerfer, F.H. and E.L. Gaden (1955) Effects of liquid physical properties on oxygen transfer in penicillin fermentation. *Appl. Microbiol.* **3**, 253–257.
14. Ju, L.-K., C.S. Ho and J.F. Shanahan (1991) Effects of carbon dioxide on the rheological behavior and oxygen transfer in submerged penicillin fermentations. *Biotechnol. Bioeng.* **38**, 1223–1232.
15. Taguchi, H. and S. Miyamoto (1966) Power requirement in non-Newtonian fermentation broth. *Biotechnol. Bioeng.* **8**, 43–54.
16. Deindoerfer, F.H. and J.M. West (1960) Rheological examination of some fermentation broths. *J. Biochem. Microbiol. Technol. Eng.* **2**, 165–175.
17. Tuffile, C.M. and F. Pinho (1970) Determination of oxygen-transfer coefficients in viscous streptomycete fermentations. *Biotechnol. Bioeng.* **12**, 849–871.
18. LeDuy, A., A.A. Marsan and B. Coupal (1974) A study of the rheological properties of a non-Newtonian fermentation broth. *Biotechnol. Bioeng.* **16**, 61–76.
19. Kato, A., S. Kawazoe and Y. Soh (1978) Viscosity of the broth of tobacco cells in suspension culture. *J. Ferment. Technol.* **56**, 224–228.
20. Ballica, R., D.D.Y. Ryu, R.L. Powell and D. Owen (1992) Rheological properties of plant cell suspensions. *Biotechnol. Prog.* **8**, 413–420.
21. Deindoerfer, F.H. and J.M. West (1960) Rheological properties of fermentation broths. *Adv. Appl. Microbiol.* **2**, 265–273.
22. Tanaka, H. (1982) Oxygen transfer in broths of plant cells at high density. *Biotechnol. Bioeng.* **24**, 425–442.
23. Laine, J. and R. Kuoppamäki (1979) Development of the design of large-scale fermentors. *Ind. Eng. Chem. Process Des. Dev.* **18**, 501–506.
24. Holland, F.A. and F.S. Chapman (1966) *Liquid Mixing and Processing in Stirred Tanks*, Reinhold, New York.

25. Bates, R.L., P.L. Fondy and J.G. Fenic (1966) Impeller characteristics and power. In: V.W. Uhl and J.B. Gray (Eds), *Mixing: Theory and Practice*, vol. 1, pp. 111–178, Academic Press, New York.

26. Nagata, S. (1975) *Mixing: Principles and Applications*, Kodansha, Tokyo.

27. Hoogendoorn, C.J. and A.P. den Hartog (1967) Model studies on mixers in the viscous flow region. *Chem. Eng. Sci.* 22, 1689–1699.

28. Kossen, N.W.F. and N.M.G. Oosterhuis (1985) Modelling and scaling-up of bioreactors. In: H.-J. Rehm and G. Reed (Eds), *Biotechnology*, vol. 2, pp. 571–605, VCH, Weinheim.

29. Zlokarnik, M. and H. Judat (1988) Stirring. In: W. Gerhartz (Ed), *Ullmann's Encyclopedia of Industrial Chemistry*, vol. B2, pp. 25-1–25-33, VCH, Weinheim.

30. Rushton, J.H., E.W. Costich and H.J. Everett (1950) Power characteristics of mixing impellers. Part I. *Chem. Eng. Prog.* 46, 395–404.

31. Rushton, J.H., E.W. Costich and H.J. Everett (1950) Power characteristics of mixing impellers. Part II. *Chem. Eng. Prog.* 46, 467–476.

32. Metzner, A.B. and R.E. Otto (1957) Agitation of non-Newtonian fluids. *AIChE J.* 3, 3–10.

33. Bruijn, W., K. van't Riet and J.M. Smith (1974) Power consumption with aerated Rushton turbines. *Trans. IChE* 52, 88–104.

34. Hughmark, G.A. (1980) Power requirements and interfacial area in gas–liquid turbine agitated systems. *Ind. Eng. Chem. Process Des. Dev.* 19, 638–641.

35. Metzner, A.B., R.H. Feehs, H. Lopez Ramos, R.E. Otto and J.D. Tuthill (1961) Agitation of viscous Newtonian and non-Newtonian fluids. *AIChE J.* 7, 3–9.

36. Nienow, A.W. and M.D. Lilly (1979) Power drawn by multiple impellers in sparged agitated vessels. *Biotechnol. Bioeng.* 21, 2341–2345.

37. Metzner, A.B. and J.S. Taylor (1960) Flow patterns in agitated vessels. *AIChE J.* 6, 109–114.

38. Jolicoeur, M., C. Chavarie, P.J. Carreau and J. Archambault (1992) Development of a helical-ribbon impeller bioreactor for high-density plant cell suspension culture. *Biotechnol. Bioeng.* 39, 511–521.

39. Cherry, R.S. and E.T. Papoutsakis (1986) Hydrodynamic effects on cells in agitated tissue culture reactors. *Bioprocess Eng.* 1, 29–41.

40. Tramper, J., D. Joustra and J.M. Vlak (1987) Bioreactor design for growth of shear-sensitive insect cells. In: C. Webb and F. Mavituna (Eds), *Plant and Animal Cells: Process Possibilities*, pp. 125–136, Ellis Horwood, Chichester.

41. Croughan, M.S., J.-F. Hamel and D.I.C. Wang (1987) Hydrodynamic effects on animal cells grown in microcarrier cultures. *Biotechnol. Bioeng.* 29, 130–141.

42. Croughan, M.S., E.S. Sayre and D.I.C. Wang (1989) Viscous reduction of turbulent damage in animal cell culture. *Biotechnol. Bioeng.* 33, 862–872.

43. Handa-Corrigan, A., A.N. Emery and R.E. Spier (1989) Effect of gas–liquid interfaces on the growth of suspended mammalian cells: mechanisms of cell damage by bubbles. *Enzyme Microb. Technol.* 11, 230–235.

44. Kunas, K.T. and E.T. Papoutsakis (1990) Damage mechanisms of suspended animal cells in agitated bioreactors with and without bubble entrainment. *Biotechnol. Bioeng.* 36, 476–483.

45. Jöbses, I., D. Martens and J. Tramper (1991) Lethal events during gas sparging in animal cell culture. *Biotechnol. Bioeng.* 37, 484–490.

46. Chalmers, J.J. and Bavarian, F. (1991) Microscopic visualization of insect cell-bubble interactions. II. The bubble film and bubble rupture. *Biotechnol. Prog.* 7, 151–158.

47. Cherry, R.S. and C.T. Hulle (1992) Cell death in the thin films of bursting bubbles. *Biotechnol. Prog.* 8, 11–18.

48. van't Riet, K. and J. Tramper (1991) *Basic Bioreactor Design*, Chapter 8, Marcel Dekker, New York.

49. McQueen, A., E. Meilhoc and J.E. Bailey (1987) Flow effects on the viability and lysis of suspended mammalian cells. *Biotechnol. Lett.* 9, 831–836.

50. Cherry, R.S. and E.T. Papoutsakis (1988) Physical mechanisms of cell damage in microcarrier cell culture bioreactors. *Biotechnol. Bioeng.* 32, 1001–1014.

51. Zwietering, Th.N. (1958) Suspending of solid particles in liquid by agitators. *Chem. Eng. Sci.* 8, 244–253.

52. Westerterp, K.R., L.L. van Dierendonck and J.A. de Kraa (1963) Interfacial areas in agitated gas–liquid contactors. *Chem. Eng. Sci.* 18, 157–176.

Suggestions for Further Reading

1. Atkinson, B. and F. Mavituna (1991) *Biochemical Engineering and Biotechnology Handbook*, 2nd edn, Chapter 11, Macmillan, Basingstoke.

2. Oldshue, J.Y. (1983) *Fluid Mixing Technology*, McGraw-Hill, New York.

3. Thomas, C.R. (1990) Problems of shear in biotechnology. In: M.A. Winkler (Ed), *Chemical Engineering Problems in Biotechnology*, pp. 23–93, Elsevier Applied Science, Barking.

8

Heat Transfer

In this chapter we are concerned with the process of heat flow between hot and cold systems. The rate at which heat is transferred depends directly on two variables: the temperature difference between the hot and cold bodies, and the surface area available for heat exchange. It is also influenced by many other factors, such as the geometry and physical properties of the system and, if fluid is present, the flow conditions. Fluids are often heated or cooled in bioprocessing. Typical examples are removal of heat during fermenter operation using cooling water, and heating of raw medium to sterilisation temperature by steam.

As shown in Chapters 5 and 6, energy balances allow us to determine the heating and cooling requirements of fermenters and enzyme reactors. Once the rate of heat transfer for a particular purpose is known, the surface area required to achieve this rate can be calculated using design equations. Estimating the heat-transfer surface area is a central objective in design as this parameter determines the size of heat-exchange equipment. In this chapter the principles governing heat transfer are outlined with applications in bioprocess design.

First, let us look at the types of equipment used for industrial heat-exchange.

8.1 Heat-Transfer Equipment

In bioprocessing, heat exchange occurs most frequently between fluids. Equipment is provided to allow transfer of heat while preventing the fluids from actually coming into contact with each other. In most heat exchangers, heat is transferred through a solid metal wall which separates the fluid streams. Sufficient surface area is provided so that the desired rate of heat transfer can be achieved. Heat transfer is facilitated by agitation and turbulent flow of the fluids.

8.1.1 Bioreactors

Two applications of heat transfer are common in bioreactor operation. The first is *in situ* batch sterilisation of liquid medium. In this process, the fermenter vessel containing medium is heated using steam and held at the sterilisation temperature for a period of time; cooling water is then used to bring the temperature back to normal operating conditions. Sterilisation is discussed in more detail in Chapter 13. The

other application of heat transfer is for temperature control during reactor operation. Metabolic activity of cells generates a substantial amount of heat in fermenters; this heat must be removed to avoid temperature increases. Most fermentations take place in the range 30–37°C; in large-scale operations, cooling water is used to maintain the temperature usually to within 1°C. Small-scale fermenters have different heat-exchange requirements; because the external surface area to volume ratio is much greater and heat losses through the wall of the vessel more significant, laboratory-scale units often require heating rather than cooling. Many enzyme reactions also require heating to maintain optimum temperature.

Equipment used for heat exchange in bioreactors usually takes one of the forms illustrated in Figure 8.1. The fermenter may have an external jacket (Figure 8.1a) or coil (Figure 8.1b) through which steam or cooling water is circulated. Alternatively, helical (Figure 8.1c) or baffle (Figure 8.1d) coils may be located internally. Another method is to pump liquid from the reactor through a separate heat-exchange unit as shown in Figure 8.1(e).

The surface area available for heat transfer is lower in the external jacket and coil designs of Figures 8.1(a) and (b) than when internal coils are completely submerged in the reactor contents. External jackets on bioreactors provide sufficient heat-transfer area for laboratory and other small-scale systems; however they are likely to be inadequate for large-scale fermentations. Internal coils are frequently used in production vessels; the coil can be operated with high liquid velocity and the entire tube surface is exposed to the reactor contents providing a relatively large heat-transfer area. There are some disadvantages with internal structures: they interfere with mixing in the vessel and make cleaning of the reactor difficult;

Figure 8.1   Heat-transfer configurations for bioreactors: (a) jacketed vessel; (b) external coil; (c) internal helical coil; (d) internal baffle-type coil; (e) external heat exchanger.

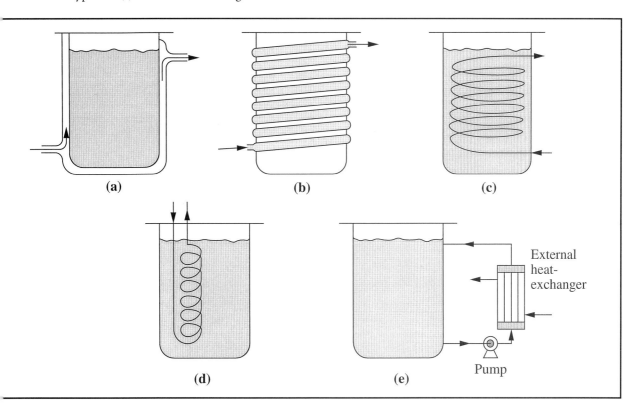

(a)   **(b)**   **(c)**

(d)   **(e)**

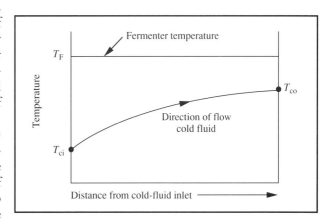

another problem is film growth of cells on the heat-transfer surface. In contrast, the external heat exchange unit shown in Figure 8.1(e) is independent of the reactor, easy to scale-up, and can provide better heat-transfer capabilities than any of the other configurations. However, conditions of sterility must be met, the cells must be able to withstand the shear forces imposed during pumping, and, in aerobic fermentations, the residence time in the heat exchanger must be small enough to ensure the medium does not become depleted of oxygen.

When internal coils such as those in Figures 8.1(c) and (d) are used to carry cooling water for removal of heat from a fermenter, the variation of water temperature with distance through the coil is as shown in Figure 8.2. The temperature of the cooling water rises as it flows through the tube and takes up heat from the fermenter contents. The water temperature increases steadily from its inlet temperature T_{ci} to the outlet temperature T_{co}. On the other hand, if the fermenter contents are well mixed, temperature gradients in the bulk fluid are negligible and the temperature is uniform at T_F.

Figure 8.2   Temperature changes for control of fermentation temperature using cooling water.

8.1.2 General Equipment For Heat Transfer

Many types of general-purpose equipment are used industrially for heat-exchange operations. The simplest form of heat-transfer equipment is the double-pipe heat exchanger; for larger capacities, more elaborate shell-and-tube units containing hundreds of square metres of heat-exchange area are required. These devices are described below.

8.1.2.1 *Double-pipe heat exchanger*

A *double-pipe heat exchanger* consists of two metal pipes, one inside the other as shown in Figure 8.3. One fluid flows through the inner tube while the other fluid flows in the annular space between the pipe walls. When one of the fluids is hotter than the other, heat flows from it through the wall of the inner tube into the other fluid. As a result, the hot fluid becomes cooler and the cold fluid becomes warmer.

Double-pipe heat exchangers can be operated with countercurrent or cocurrent flow of fluid. If, as indicated in Figure 8.3, the two fluids enter at opposite ends of the device and pass in opposite directions through the pipes, the flow is *countercurrent*. Cold fluid entering the device meets the hot fluid just as it is leaving, i.e. cold fluid at its lowest temperature is placed in thermal contact with hot fluid also at its lowest temperature. Changes in temperature of the two fluids as they flow countercurrently through the length of the pipe are shown in Figure 8.4. The four terminal temperatures are as follows: T_{hi} is the inlet temperature of the hot fluid, T_{ho} is the outlet temperature of the hot fluid, T_{ci} is the inlet temperature of the cold fluid, and T_{co} is the outlet temperature of the cold fluid leaving the system. A sign of efficient operation is T_{co} close to T_{hi}, or T_{ho} close to T_{ci}.

The alternative to countercurrent flow is *cocurrent* or *parallel* flow. In this mode of operation, both fluids enter their respective tubes at the same end of the exchanger and flow in the same direction to the other end. The temperature curves for cocurrent flow are given in Figure 8.5. Cocurrent operation

Figure 8.3 Double-pipe heat exchanger. (From A.S. Foust, L.A. Wenzel, C.W. Clump, L. Maus and L.B. Andersen, 1960, *Principles of Unit Operations*, John Wiley, New York.)

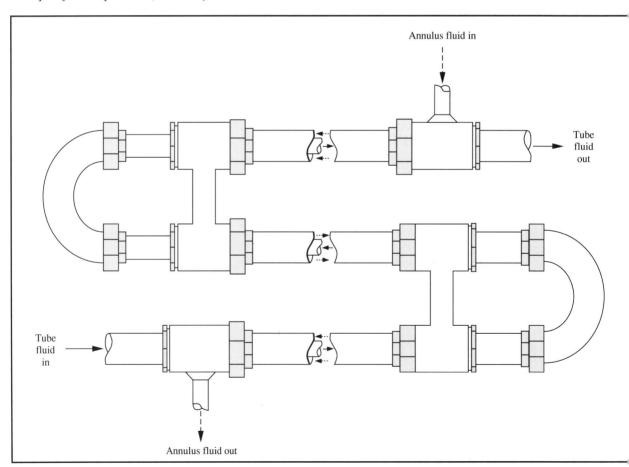

Figure 8.4 Temperature changes for countercurrent flow in a double-pipe heat exchanger.

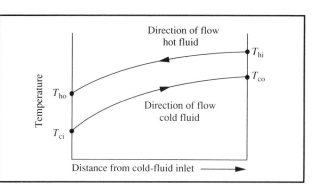

Figure 8.5 Temperature changes for cocurrent flow in a double-pipe heat exchanger.

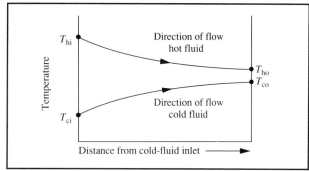

is not as effective as countercurrent; it is not possible using cocurrent flow to bring the exit temperature of one fluid close to the entrance temperature of the other. Instead, the exit temperatures of both streams lie between the two entrance temperatures. Less heat can be transferred in parallel flow than in countercurrent flow; consequently, parallel flow is applied less frequently.

Double-pipe heat exchangers can be extended to several passes arranged in a vertical stack, as illustrated in Figure 8.3. However, when large surface areas are needed to achieve the desired rate of heat transfer, the weight of the outer pipe becomes so great that an alternative design, the shell-and-tube

heat exchanger, is a better and more economical choice. As a rule of thumb, if the heat-transfer area between the fluids must be more than 10–15 m^2, a shell-and-tube exchanger is required.

8.1.2.2 *Shell-and-tube heat exchangers*

Shell-and-tube heat exchangers are used for heating and cooling all types of fluid. They have the advantage of containing very large surface areas in a relatively small volume. The simplest form, called a *single-pass shell-and-tube heat exchanger*, is shown in Figure 8.6.

Figure 8.6 Single-pass shell-and-tube heat exchanger. (From A.S. Foust, L.A. Wenzel, C.W. Clump, L. Maus and L.B. Andersen, 1960, *Principles of Unit Operations*, John Wiley, New York.)

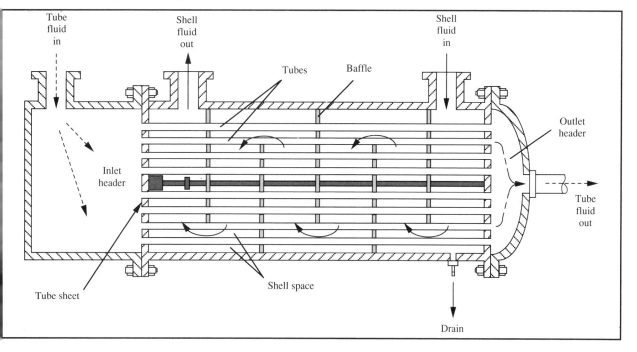

Consider the device of Figure 8.6 for exchange of sensible heat from one fluid to another. The heat-transfer system is divided into two sections: a *tube bundle* containing pipes through which one fluid flows, and a *shell* or cavity where the other fluid flows. Hot or cold fluid may be put into either the tubes or the shell. In a single-pass exchanger, the shell and tube fluids pass down the length of the equipment only once. The fluid which is to travel in the tubes enters at the inlet header. The header is divided from the rest of the apparatus by a *tube sheet*. Open tubes are fitted into the tube sheet; fluid in the header cannot enter the main cavity of the exchanger but must pass into the tubes. The tube-side fluid leaves the exchanger through another header at the outlet. Shell-side fluid enters the internal cavity of the exchanger and flows around the outsides of the tubes in a direction which is largely countercurrent to the tube fluid. Heat is exchanged across the tube walls from hot fluid to cold fluid. As shown in Figure 8.6, *baffles* are often installed in the shell to decrease the cross-sectional area for flow and divert the shell fluid so it flows mainly across rather than parallel to the tubes. Both these effects promote turbulence in the shell fluid which improves the rate of heat transfer. As well as directing the flow of shell fluid, baffles also

support the tube bundle and keep the tubes from sagging.

The length of tubes in a single-pass heat exchanger determines the surface area available for heat transfer, and therefore the rate at which heat can be exchanged. However, there are practical and economic limits to the maximum length of single-pass tubes; if greater heat-transfer capacity is required *multiple-pass heat exchangers* are employed. Heat exchangers containing more that one tube pass are used routinely; Figure 8.7 shows the structure of a heat exchanger with one shell pass and a double tube pass. In this device, the header for the tube fluid is divided into two sections. Fluid entering the header is channelled into the lower half of the tubes and flows to the other end of the exchanger. The header at the other end diverts the fluid into the upper tubes; the tube-side fluid therefore leaves the exchanger at the same end it entered. On the shell side, the configuration is the same as for the single-pass structure of Figure 8.6; fluid enters one end of the shell and flows around several baffles to the other end.

In a double-tube pass exchanger, flow of tube and shell fluids is mainly countercurrent for one tube pass and mainly cocurrent for the other; however, because of the action of the baffles, cross-flow of shell-side fluid normal to the tubes also

Figure 8.7 Double tube-pass heat exchanger.

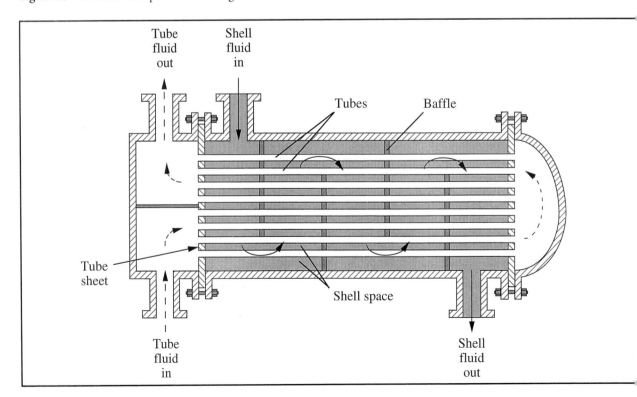

occurs. Temperature curves for the exchanger depend on the location of the shell-side entry nozzle; this is illustrated in Figure 8.8 for hot fluid flowing in the shell. In the temperature profile of Figure 8.8(b), a *temperature cross* occurs where, at some point in the exchanger, the temperature of the cold fluid equals the temperature of the hot fluid. This situation should be avoided because, after the cross, the cold fluid is actually cooled rather than heated. The solution is an increased number of shell passes or, more practically, provision of another heat exchanger in series with the first.

Heat exchangers with multiple shell-passes can also be used. However, in comparison with multiple-tube pass equipment, multiple-shell exchangers are complex in construction and are normally applied only in very large installations.

8.2 Mechanisms of Heat Transfer

Heat transfer occurs by one or more of the following three mechanisms.

(i) *Conduction.* Heat conduction occurs by transfer of vibrational energy between molecules, or movement of free electrons. Conduction is particularly important with metals and occurs without observable movement of matter.

(ii) *Convection.* Convection requires movement on a macroscopic scale; it is therefore confined to gases and liquids. *Natural convection* occurs when temperature gradients in the system generate localised density differences which result in flow currents. In *forced convection*, flow currents are set in motion by an external agent such as a stirrer or pump and are independent of density gradients. Higher rates of heat transfer are possible with forced convection compared with natural convection.

(iii) *Radiation.* Energy is radiated from all materials in the form of waves; when this radiation is absorbed by matter it appears as heat. Because radiation is important at much higher temperatures than those normally encountered in biological processing, it will not be mentioned further.

Figure 8.8 Temperature changes for a double tube-pass heat exchanger. (From J.M. Coulson and J.F. Richardson, 1977, *Chemical Engineering*, 3rd edn, Pergamon Press, Oxford.)

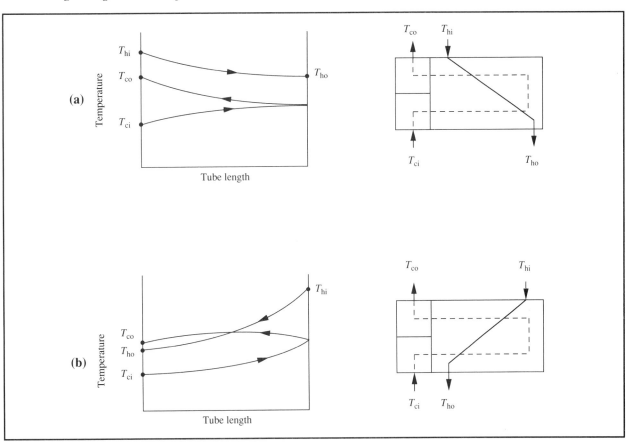

8.3 Conduction

In most heat-transfer equipment, heat is exchanged between fluids separated by a solid wall. Heat transfer through the wall occurs by conduction. In this section we consider equations describing rate of conduction as a function of operating variables.

Conduction of heat through a homogeneous solid wall is depicted in Figure 8.9. The wall has thickness B; on one side of the wall the temperature is T_1, on the other side the temperature is T_2. The area of wall exposed to each temperature is A. The rate of heat conduction through the wall is given by *Fourier's law*:

$$\hat{Q} = -kA\frac{dT}{dy}$$

$$(8.1)$$

where $\hat{Q}$ is rate of heat transfer, k is the *thermal conductivity* of the wall, A is surface area perpendicular to the direction of heat flow, T is temperature, and y is distance measured normal to A. dT/dy is the *temperature gradient*, or change of temperature with distance through the wall. The negative sign in Eq. (8.1) indicates that heat always flows from hot to cold irrespective of whether dT/dy is positive or negative. To illustrate this, when

Figure 8.9 Heat conduction through a flat wall.

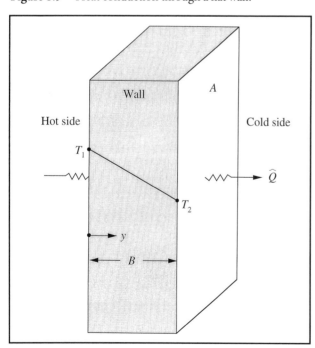

the temperature gradient is negative relative to co-ordinate y (as shown in Figure 8.9), heat flows in the positive y-direction; conversely, if the gradient were positive (i.e. $T_1 < T_2$), heat would flow in the negative y-direction.

Fourier's law can also be expressed in terms of *heat flux*, $\hat{q}$. Heat flux is defined as the rate of heat transfer per unit area normal to the direction of heat flow. Therefore, from Eq. (8.1):

$$\hat{q} = -k\frac{dT}{dy}.$$

$$(8.2)$$

Rate of heat transfer $\hat{Q}$ is also known as *power*. The SI unit for $\hat{Q}$ is the watt (W); in imperial units $\hat{Q}$ is measured in Btu h^{-1}. Corresponding units of $\hat{q}$ are W m^{-2} and Btu h^{-1} ft^{-2}.

Thermal conductivity is a transport property of materials; values can be found in handbooks. The dimensions of k are $LMT^{-3}\Theta^{-1}$; units include W m^{-1} K^{-1} and Btu h^{-1} ft^{-1} °F^{-1}. The magnitude of k in Eqs (8.1) and (8.2) reflects the ease with which heat is conducted; the higher the value of k the faster is the heat transfer. Table 8.1 lists thermal conductivities for some common materials; metals generally have higher thermal conductivities than other substances. Solids with low k values are used as *insulators* to minimise the rate of conduction, for example, around steam pipes or in buildings. Thermal conductivity varies somewhat with temperature; however for small ranges of temperature, k can be considered constant.

8.3.1 Analogy Between Heat and Momentum Transfer

An analogy exists between the equations for heat and momentum transfer. Newton's law of viscosity given by Eq. (7.6):

$$\tau = -\mu\frac{dv}{dy}$$

$$(7.6)$$

has the same mathematical form as Eq. (8.2). In heat transfer, the temperature gradient dT/dy is the driving force for heat flow; in momentum transfer the driving force is the velocity gradient dv/dy. Both heat flux and momentum flux are directly proportional to the driving force, and the proportionality constant, μ or k, is a physical property of the material. As we shall see in Chapter 9, the analogy between heat and momentum transfer can be extended to include mass transfer.

Table 8.1   Thermal conductivities

(From J.M. Coulson and J.F. Richardson, 1977, Chemical Engineering, *vol. 1, 3rd edn, Pergamon Press, Oxford)*

Material	*Temperature*	*k*	
	(K)	(W m^{-1} K^{-1})	(Btu h^{-1} ft^{-1} °F^{-1})
Solids: Metals			
Aluminium	573	230	133
Bronze	–	189	109
Copper	373	377	218
Iron (wrought)	291	61	35
Iron (cast)	326	48	27.6
Lead	373	33	19
Stainless steel	293	16	9.2
Steel (1% C)	291	45	26
Solids: Non-metals			
Asbestos	273	0.16	0.09
Bricks (building)	293	0.69	0.4
Cotton wool	303	0.050	0.029
Glass	303	1.09	0.63
Rubber (hard)	273	0.15	0.087
Cork	303	0.043	0.025
Glass wool	–	0.041	0.024
Liquids			
Acetic acid (50%)	293	0.35	0.20
Ethanol (80%)	293	0.24	0.137
Glycerol (40%)	293	0.45	0.26
Water	303	0.62	0.356
Water	333	0.66	0.381
Gases			
Air	273	0.024	0.014
Air	373	0.031	0.018
Carbon dioxide	273	0.015	0.0085
Oxygen	273	0.024	0.0141
Water vapour	373	0.025	0.0145

8.3.2 Steady-State Conduction

Consider again the conduction of heat through the wall shown in Figure 8.9. At steady state there can be neither accumulation nor depletion of heat within the wall; this means that the rate of heat flow $\hat{Q}$ must be the same at each point in the wall. If k is largely independent of temperature and A is also constant, the only variables in Eq. (8.1) are temperature T and distance y. We can therefore integrate Eq. (8.1) to obtain an

expression for rate of conduction as a function of the temperature difference across the wall.

Separating variables in Eq. (8.1) gives:

$$\hat{Q}\,dy = -kA\,dT.$$

$$(8.3)$$

Both sides of Eq. (8.3) can be integrated after taking the constants $\hat{Q}$, k and A outside of the integral signs:

$$\hat{Q} \int dy = -kA \int dT.$$

(8.4)

From the rules of integration given in Appendix D:

$$\hat{Q} y = -kAT + K$$

(8.5)

where K is the integration constant. K is evaluated by applying a single boundary condition; in this case we can use the boundary condition: $T = T_1$ at $y = 0$. Substituting this information into Eq. (8.5) gives:

$$K = kAT_1.$$

(8.6)

Thus eliminating K from Eq. (8.5) gives:

$$\hat{Q} y = -kA(T_1 - T)$$

(8.7)

or

$$\hat{Q} = \frac{kA}{y}(T_1 - T).$$

(8.8)

Because $\hat{Q}$ at steady state is the same at all points in the wall, Eq. (8.8) holds for all values of y including at $y = B$ where $T = T_2$. Substituting these values into Eq. (8.8) gives the expression:

$$\hat{Q} = \frac{kA}{B}(T_1 - T_2)$$

(8.9)

or

$$\hat{Q} = \frac{kA}{B} \Delta T.$$

(8.10)

Eq. (8.10) allows us to calculate $\hat{Q}$ if we know the heat-transfer area A and the total temperature drop across the slab ΔT. Eq. (8.10) can also be written in the form:

$$\hat{Q} = \frac{\Delta T}{R_w}$$

(8.11)

where R_w is the *thermal resistance* to heat transfer offered by the wall:

$$R_w = \frac{B}{kA}.$$

(8.12)

In heat transfer, the ΔT responsible for flow of heat is known as the *temperature-difference driving force*. Eq. (8.11) is an example of the general rate principle which equates rate of transfer to the ratio of driving force and resistance. Eq. (8.12) can be interpreted as follows: the wall would pose more of a resistance to heat transfer if its thickness were increased; on the other hand, resistance is reduced if the surface area is increased, or the material in the wall is replaced with a substance of higher thermal conductivity.

8.3.3 Combining Thermal Resistances in Series

When a system contains several different heat-transfer resistances in series, *the overall resistance is equal to the sum of the individual resistances*. For example, if the wall shown in Figure 8.9 were constructed of several layers of different material, each layer would represent a separate resistance to heat transfer. Consider the three-layer system illustrated in Figure 8.10 with surface area A, layer thicknesses B_1, B_2 and B_3, thermal conductivities k_1, k_2 and k_3, and temperature drops across the layers ΔT_1, ΔT_2 and ΔT_3. If the layers are in perfect thermal contact so that there is no temperature drop across the interfaces, the temperature change across the entire structure is:

$$\Delta T = \Delta T_1 + \Delta T_2 + \Delta T_3.$$

(8.13)

Rate of heat conduction in this system is given by Eq. (8.11), with the overall resistance R_w equal to the sum of the individual resistances:

$$\hat{Q} = \frac{\Delta T}{R_w} = \frac{\Delta T}{(R_1 + R_2 + R_3)}$$

(8.14)

where R_1, R_2 and R_3 are the thermal resistances of the individual layers:

$$R_1 = \frac{B_1}{k_1 A} \qquad R_2 = \frac{B_2}{k_2 A} \qquad \text{and} \qquad R_3 = \frac{B_3}{k_3 A}.$$

(8.15)

Figure 8.10 Heat conduction through three resistances in series.

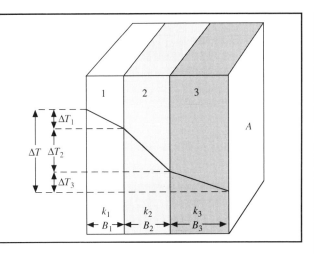

Figure 8.11 Heat transfer between fluids separated by a solid wall.

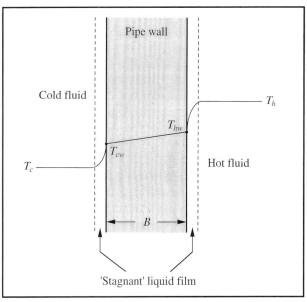

Eq. (8.14) represents the important principle of *additivity of resistances.* We shall use this principle later for analysis of convective heat transfer in pipes and stirred vessels.

8.4 Heat Transfer Between Fluids

Convection and conduction both play important roles in heat transfer in fluids. In agitated, single-phase systems, convective heat transfer in the bulk fluid is linked directly to mixing and turbulence and is generally quite rapid. However, in the heat-exchange devices described in Section 8.1, additional resistances are encountered.

8.4.1 Thermal Boundary Layers

Figure 8.11 depicts the heat-transfer situation at any point on the pipe wall of a heat exchanger. Hot and cold fluids flow on either side of the wall; we assume that both fluids are in turbulent flow. The bulk temperature of the hot fluid away from the wall is T_h; T_c is the bulk temperature of the cold fluid. T_{hw} and T_{cw} are the respective temperatures of hot and cold fluids at the wall.

As explained in Section 7.2.3, when fluid contacts a solid, a fluid boundary layer develops at the surface as a result of viscous drag. Therefore, fluids such as those represented in Figure 8.11 consist of a turbulent core which accounts for the bulk of the fluid, and a thin sublayer or film near the wall where the velocity is relatively low. In the turbulent part of the fluid, rapidly moving eddies transfer heat quickly so that any temperature gradients in the bulk fluid can be neglected. The film

of liquid at the wall is called the *thermal boundary layer* or *stagnant film*, although the fluid in it is not actually stationary. This viscous sublayer has an important effect on the rate of heat transfer. Most of the resistance to heat transfer to or from the fluid is contained in the film; the reason for this is that heat flow through it must occur mainly by conduction rather than convection because of the reduced velocity of the fluid. The width of the film indicated by the broken lines in Figure 8.11 is the approximate distance from the wall where the temperature reaches the bulk-fluid temperature, either T_h or T_c. The thickness of the thermal boundary layer in most heat-transfer situations is less than the hydrodynamic boundary layer described in Section 7.2.3. In other words, as we move away from the wall, the temperature normally reaches that of the bulk fluid before the velocity reaches that of the bulk flow stream.

8.4.2 Individual Heat-Transfer Coefficients

Heat exchanged between the fluids in Figure 8.11 encounters three major resistances in series: the hot-fluid film resistance at the wall, resistance due to the wall itself, and the cold-fluid film resistance. Equations for rate of conduction through the wall have already been developed in Section 8.3.2. Rate of heat transfer through each thermal boundary layer is given by an equation somewhat analogous to Eq. (8.10) for steady-state conduction:

$$\hat{Q} = hA\Delta T \tag{8.16}$$

where h is the *individual heat-transfer coefficient*, A is the area for heat transfer normal to the direction of heat flow, and ΔT is the temperature difference between the wall and the bulk stream. $\Delta T = T_h - T_{hw}$ for the hot-fluid film; $\Delta T = T_{cw} - T_c$ for the cold-fluid film. Eq. (8.16) does not contain a separate term for the thickness of the boundary layer; this thickness is difficult to measure and depends strongly on the prevailing flow conditions. Instead, the effect of film thickness is included in the value of h so that, unlike thermal conductivity, values for h cannot be found in handbooks. The heat-transfer coefficient h is an empirical parameter incorporating the effects of system geometry, flow conditions and fluid properties. Because it involves fluid flow, convective heat transfer is a more complex process than conduction. Consequently, there is little theoretical basis for calculation of h; h must be determined experimentally or evaluated using published correlations based on experimental data. Suitable correlations for heat-transfer coefficients are presented in Section 8.5.3. SI units for h are $W\ m^{-2}\ K^{-1}$; in the imperial system h is expressed as $Btu\ h^{-1}\ ft^{-2}\ {}^\circ F^{-1}$. Magnitudes of h vary greatly; some typical values are listed in Table 8.2.

Rate of heat transfer $\hat{Q}$ in each fluid boundary layer can be written as the ratio of the driving force ΔT and the resistance. Therefore, from Eq. (8.16) the two resistances on either side of the pipe wall are as follows:

$$R_h = \frac{1}{h_h A} \tag{8.17}$$

and

$$R_c = \frac{1}{h_c A} \tag{8.18}$$

where R_h is the resistance to heat transfer in the hot fluid, R_c is the resistance to heat transfer in the cold fluid, h_h is the individual heat-transfer coefficient for the hot fluid, h_c is the individual heat-transfer coefficient for the cold fluid and A is the surface area for heat transfer.

8.4.3 Overall Heat-Transfer Coefficient

Use of Eq. (8.16) to calculate rate of heat transfer in each boundary layer requires knowledge of ΔT for each fluid; this is usually difficult because of lack of information about T_{hw} and T_{cw}. It is easier and more accurate to measure the bulk temperatures of fluids rather than wall temperatures. This problem is removed by introduction of the *overall heat-transfer coefficient*, U, for the total heat-flow process through both fluids and the wall. U is defined by the equation:

$$\hat{Q} = UA\Delta T = UA(T_h - T_c). \tag{8.19}$$

The units of U are the same as h, e.g. $W\ m^{-2}\ K^{-1}$ or $Btu\ h^{-1}\ ft^{-2}\ {}^\circ F^{-1}$. Eq. (8.19) written in terms of the ratio of driving force (ΔT) and resistance yields an expression for the total resistance to heat flow, R_T:

$$R_T = \frac{1}{UA} . \tag{8.20}$$

In Section 8.3.3 it was noted that when there are thermal

Table 8.2 Individual heat-transfer coefficients

(From W.H. McAdams, 1954, Heat Transmission, *3rd edn, McGraw-Hill, New York)*

Process	Range of values of h	
	($W\ m^{-2}\ K^{-1}$)	($Btu\ ft^{-2}\ h^{-1}\ {}^\circ F^{-1}$)
Condensing steam	6000–115 000	1000–20 000
Boiling water	1700–50 000	300–9000
Condensing organic vapour	1100–2200	200–400
Heating or cooling water	300–17 000	50–3000
Heating or cooling oil	60–1700	10–300
Superheating steam	30–110	5–20
Heating or cooling air	1–60	0.2–10

resistances in series, the total resistance is the sum of the individual resistances. Applying this now to the situation of heat exchange between fluids, R_T is equal to the sum of R_h, R_w and R_c:

$$R_T = R_h + R_w + R_c. \tag{8.21}$$

Combining Eqs (8.12), (8.17), (8.18), (8.20) and (8.21) gives:

$$\frac{1}{UA} = \frac{1}{h_h A} + \frac{B}{k A} + \frac{1}{h_c A}. \tag{8.22}$$

In Eq. (8.22), the surface area A appears in each term. When fluids are separated by a flat wall, the surface area for heat transfer through each boundary layer and the wall is the same, so that A can be cancelled from the equation. However, a minor complication arises with cylindrical geometry such as pipes. Let us assume that hot fluid is flowing inside a pipe while cold fluid flows outside, as shown in Figure 8.12. Because the inside diameter of the pipe is smaller than the outside diameter, the surface areas for heat transfer between the fluid and the pipe wall are different for the two fluids. The surface area of a cylinder is equal to its circumference multiplied by length, i.e. $A = 2\pi R L$ where R is the radius of the cylinder and L is its length. Therefore, the heat-transfer area at the hot-fluid boundary layer inside the tube is $A_i = 2\pi R_i L$; the heat-transfer area at the cold-fluid film outside the tube is

$A_o = 2\pi R_o L$. The surface area available for conduction through the wall varies between A_i and A_o.

The variation of heat-transfer area in cylindrical systems depends on the thickness of the pipe wall; for thin walls the variation will be relatively small. In engineering design, these variations in surface area are incorporated into the equations for heat transfer. However, for the sake of simplicity, in this chapter we will ignore any differences in surface area; we will assume in effect that the pipes are thin-walled. Accordingly, for cylindrical as well as flat geometry, we can cancel A from Eq. (8.22), and write a simplified equation for U:

$$\frac{1}{U} = \frac{1}{h_h} + \frac{B}{k} + \frac{1}{h_c}. \tag{8.23}$$

The overall heat-transfer coefficient characterises the operating conditions used for heat transfer. Small U for a particular process means that the system has limited capacity for heat exchange; U can be improved by manipulating operating conditions such as fluid velocity in shell-and-tube equipment or stirrer speed in bioreactors. U is independent of A. Heat transfer in an exchanger with small U can also be improved by increasing the heat-transfer area and size of the unit; however increasing A raises the cost of the equipment. If U is large, the heat exchanger is well designed and is operating under conditions which enhance heat transfer.

8.4.4 Fouling Factors

Heat-transfer equipment in service does not remain clean. Dirt and scale deposit on one or both sides of the pipes, providing additional resistance to heat flow and reducing the overall heat-transfer coefficient. Resistances to heat transfer when fouling affects both sides of the heat-transfer surface are represented in Figure 8.13. Five resistances are present in series: the thermal boundary layer on the hot-fluid side, a fouling layer on the hot-fluid side, the pipe wall, a fouling layer on the cold-fluid side, and the cold-fluid boundary layer.

Each fouling layer has associated with it a heat-transfer coefficient; for scale and dirt the coefficient is called a *fouling factor*. Let h_{fh} be the fouling factor on the hot-fluid side, and h_{fc} be the fouling factor on the cold-fluid side. When these additional resistances are present, they must be included in the expression for the overall heat-transfer coefficient, U. Eq. (8.23) becomes:

$$\frac{1}{U} = \frac{1}{h_{fh}} + \frac{1}{h_h} + \frac{B}{k} + \frac{1}{h_c} + \frac{1}{h_{fc}}. \tag{8.24}$$

Figure 8.12 Effect of pipe wall thickness on surface area for heat transfer.

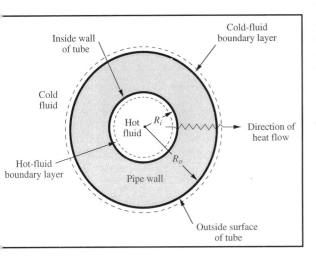

Figure 8.13 Resistances to heat transfer with fouling deposits on both surfaces.

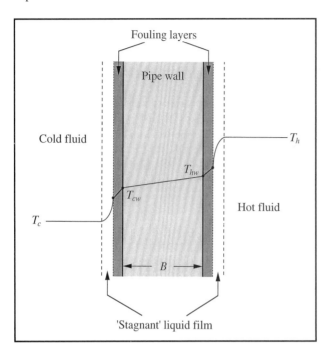

Adding fouling factors in Eq. (8.24) increases $1/U$, thu decreasing the value of U.

Accurate estimation of fouling factors is very difficult. Th chemical nature of the deposit and its thermal conductivi depend on the fluid in the tube and the temperature; foulin thickness can also vary between cleanings. Typical values o fouling factors for various fluids are listed in Table 8.3.

8.5 Design Equations For Heat-Transfer Systems

The basic equation for design of heat exchangers is Eq. (8.19) If $\hat{Q}$, U and ΔT are known, this equation allows us to calculat A. Specification of A is a major objective of heat-exchang design; the surface area required dictates the configuration an size of the equipment and its cost. In the following sections, w will consider procedures for determining $\hat{Q}$, U and ΔT for us in Eq. (8.19).

8.5.1 Energy Balance

In heat-exchanger design, energy balances are applied to deter mine $\hat{Q}$ and all inlet and outlet temperatures used to specif

Table 8.3 Fouling factors for scale deposits

(Data from J.M. Coulson and J.F. Richardson, 1977, Chemical Engineering, vol. 1, 3rd edn, Pergamon Press, Oxford)

Source of deposit	Fouling factor	
	$(\mathrm{W\,m^{-2}\,K^{-1}})$	$(\mathrm{Btu\,ft^{-2}\,h^{-1}\,{}^{\circ}F^{-1}})$
Water*		
Distilled	11 000	2 000
Sea	11 000	2 000
Clear river	4 800	800
Untreated cooling tower	1 700	300
Hard well	1 700	300
Steam		
Good quality, oil free	19 000	3 000
Liquids		
Treated brine	3 700	700
Organics	5 600	1 000
Fuel oils	1 000	200
Gases		
Air	2 000–4 000	300–700
Solvent vapour	7 000	1 300

* Velocity 1 m s^{-1}; temperature less than 320K

ΔT. These energy balances are based on general equations for flow systems derived in Chapters 5 and 6.

Let us first consider the equations for double-pipe or shell-and-tube heat-exchangers. From Eq. (6.10), under steady-state conditions $dE/dt = 0$ and in the absence of shaft work ($\hat{W}_s = 0$), the energy-balance equation is:

$$\hat{M}_i\, h_i - \hat{M}_o\, h_o - \hat{Q} = 0 \tag{8.25}$$

where $\hat{M}_i$ is mass flow rate in, $\hat{M}_o$ is mass flow rate out, h_i is specific enthalpy of the incoming stream, h_o is specific enthalpy of the outgoing stream and $\hat{Q}$ is rate of heat removal from the system. Unfortunately, the conventional symbols for individual heat-transfer coefficient and specific enthalpy are the same: h. In this section, h in Eqs (8.25)–(8.30) and (8.33) denotes specific enthalpy; otherwise in this chapter, h represents the individual heat-transfer coefficient.

Eq. (8.25) can be applied separately to each fluid in the heat exchanger. As the mass flow rate is the same at the inlet as at the outlet, for the hot fluid:

$$\hat{M}_h\, (h_{hi} - h_{ho}) - \hat{Q}_h = 0 \tag{8.26}$$

$$\hat{M}_h\, (h_{hi} - h_{ho}) = \hat{Q}_h \tag{8.27}$$

where subscript h denotes hot fluid and $\hat{Q}_h$ is the rate of heat transfer from that fluid. Equations similar to Eqs (8.26) and (8.27) can be derived for the cold fluid:

$$\hat{M}_c\, (h_{ci} - h_{co}) + \hat{Q}_c = 0 \tag{8.28}$$

or

$$\hat{M}_c\, (h_{co} - h_{ci}) = \hat{Q}_c \tag{8.29}$$

where subscript c refers to cold fluid. $\hat{Q}_c$ is the rate of heat flow into the cold fluid; therefore $\hat{Q}_c$ is added rather than subtracted in Eq. (8.28).

When there are no heat losses from the exchanger, all heat removed from the hot stream is taken up by the cold stream. We can therefore equate $\hat{Q}$ terms in Eqs (8.27) and (8.29): $\hat{Q}_h = \hat{Q}_c = \hat{Q}$. Therefore:

$$\hat{M}_h\, (h_{hi} - h_{ho}) = \hat{M}_c\, (h_{co} - h_{ci}) = \hat{Q}. \tag{8.30}$$

When sensible heat is exchanged between fluids, the enthalpy differences in Eq. (8.30) can be expressed in terms of the heat capacity C_p and the temperature change for each fluid. If we assume C_p is constant over the temperature range in the exchanger, Eq. (8.30) becomes:

$$\hat{M}_h\, C_{ph}\, (T_{hi} - T_{ho}) = \hat{M}_c\, C_{pc}\, (T_{co} - T_{ci}) = \hat{Q} \tag{8.31}$$

where C_{ph} is the heat capacity of the hot fluid, C_{pc} is the heat capacity of the cold fluid, T_{hi} is the inlet temperature of the hot fluid, T_{ho} is the outlet temperature of the hot fluid, T_{ci} is the inlet temperature of the cold fluid, and T_{co} is the outlet temperature of the cold fluid.

In heat-exchanger design, Eq. (8.31) is used to determine $\hat{Q}$ and the inlet and outlet conditions of the fluid streams. This is illustrated in Example 8.1.

Example 8.1 Heat exchanger

Hot, freshly-sterilised nutrient medium is cooled in a double-pipe heat exchanger before being used in a fermentation. Medium leaving the steriliser at 121°C enters the exchanger at a flow rate of 10 m³ h⁻¹; the desired outlet temperature is 30°C. Heat from the medium is used to raise the temperature of 25 m³ h⁻¹ water initially at 15°C. The system operates at steady state. Assume that nutrient medium has the properties of water.

(a) What rate of heat transfer is required?
(b) Calculate the final temperature of the cooling water as it leaves the heat exchanger.

Solution:
The density of water and medium is 1000 kg m⁻³. Therefore:

$$\hat{M}_h = 10 \text{ m}^3 \text{ h}^{-1} \cdot \left| \frac{1 \text{ h}}{3600 \text{ s}} \right| \cdot \left| \frac{1000 \text{ kg}}{1 \text{ m}^3} \right| = 2.78 \text{ kg s}^{-1}$$

$$\hat{M}_c = 25 \text{ m}^3 \text{ h}^{-1} \cdot \left| \frac{1 \text{ h}}{3600 \text{ s}} \right| \cdot \left| \frac{1000 \text{ kg}}{1 \text{ m}^3} \right| = 6.94 \text{ kg s}^{-1}.$$

The heat capacity of water can be taken as 75.4 J gmol^{-1} °C^{-1} for most of the temperature range of interest (Table B.3). Therefore:

$$C_{ph} = C_{pc} = 75.4 \text{ J gmol}^{-1} \text{°C}^{-1} \cdot \left| \frac{1 \text{ gmol}}{18 \text{ g}} \right| \cdot \left| \frac{1000 \text{ kg}}{1 \text{ kg}} \right|$$

$$= 4.19 \times 10^3 \text{ J kg}^{-1} \text{°C}^{-1}.$$

(a) From Eq. (8.31) for the hot fluid:

$$\hat{Q} = (2.78 \text{ kg s}^{-1}) (4.19 \times 10^3 \text{ J kg}^{-1} \text{°C}^{-1}) (121 - 30)\text{°C}$$

$$= 1.06 \times 10^6 \text{ J s}^{-1} = 1060 \text{ kW}.$$

(b) For the cold fluid, from Eq. (8.31):

$$T_{co} = T_{ci} + \frac{\hat{Q}}{\hat{M}_c C_{pc}}$$

$$T_{co} = 15\text{°C} + \frac{1.06 \times 10^6 \text{ J s}^{-1}}{(6.94 \text{ kg s}^{-1}) (4.19 \times 10^3 \text{ J kg}^{-1} \text{°C})} = 51.5\text{°C}.$$

The exit water temperature is 52°C.

Eq. (8.31) can also be applied to heat removal from a reactor for the purpose of temperature control. At steady state, the temperature of the hot fluid, e.g. fermentation broth, does not change; therefore the left-hand side of Eq. (8.31) is zero. If energy is absorbed by the cold fluid as sensible heat, the energy-balance equation becomes:

$$\hat{M}_c C_{pc} (T_{co} - T_{ci}) = \hat{Q}. \tag{8.32}$$

To use Eq. (8.32) for bioreactor design we must know $\hat{Q}$. $\hat{Q}$ is found by considering all significant heat sources and sinks in the system; an expression involving $\hat{Q}$ for fermentation systems was presented in Chapter 6 based on relationships derived in Chapter 5:

$$\frac{dE}{dt} = -\Delta\hat{H}_{rxn} - \hat{M}_v \Delta h_v - \hat{Q} + \hat{W}_s \tag{6.12}$$

where $\Delta\hat{H}_{rxn}$ is the rate of heat absorption or evolution due to metabolic reaction, $\hat{M}_v$ is the mass flow rate of evaporated liquid leaving the system, Δh_v is the latent heat of evaporation, and $\hat{W}_s$ is the rate of shaft work done on the system. For exothermic reactions $\Delta\hat{H}_{rxn}$ is negative, for endothermic reactions $\Delta\hat{H}_{rxn}$ is positive. In most fermentation systems the only source of shaft work is the stirrer; therefore $\hat{W}_s$ is the power P dissipated by the impeller. Methods for estimating P are described in Section 7.10. Eq. (6.12) represents a considerable simplification of the energy balance. It is applicable to systems in which heat of reaction dominates the energy balance so that contributions from sensible heat and heats of solution can be ignored. In large insulated fermenters, heat produced by metabolic activity is by far the dominant source of heat; energy dissipated by stirring may also be worth considering. The other heat sources and sinks are relatively minor and can generally be neglected.

At steady state $dE/dt = 0$ and Eq. (6.12) becomes:

$$\hat{Q} = -\Delta\hat{H}_{rxn} - \hat{M}_v\Delta h_v + \hat{W}_s.$$

$$(8.33)$$

Application of Eq. (8.33) to determine $\hat{Q}$ is illustrated in

Examples 5.7 and 5.8. Once $\hat{Q}$ has been estimated, Eq. (8.32) is used to evaluate unknown operating conditions as shown in Example 8.2.

Example 8.2 Cooling coil

A 150 m³ bioreactor is operated at 35°C to produce fungal biomass from glucose. The rate of oxygen uptake by the culture is 1.5 kg m⁻³ h⁻¹; the agitator dissipates heat at a rate of 1 kW m⁻³. 60 m³ h⁻¹ cooling water available from a nearby river at 10°C is passed through an internal coil in the fermentation tank. If the system operates at steady state, what is the exit temperature of the cooling water?

Solution:

Rate of heat generation by aerobic cultures is calculated directly from the oxygen demand. As outlined in Section 5.9.2, approximately 460 kJ heat is released for each gmol oxygen consumed. Therefore, the metabolic heat load is:

$$\Delta\hat{H}_{rxn} = \frac{-460\text{ kJ}}{\text{gmol}} \cdot \left|\frac{1\text{ gmol}}{32\text{ g}}\right| \cdot \left|\frac{1000\text{ g}}{1\text{ kg}}\right| \cdot (1.5\text{ kg m}^{-3}\text{h}^{-1}) \cdot \left|\frac{1\text{ h}}{3600\text{ s}}\right| \cdot 150\text{ m}^3$$

$$= -898\text{ kJ s}^{-1}$$

$$= -898\text{ kW.}$$

$\Delta\hat{H}_{rxn}$ is negative because fermentation is exothermic. The rate of heat dissipation by the agitator is:

$$(1\text{ kW m}^{-3})\ 150\text{ m}^3 = 150\text{ kW.}$$

We can now calculate $\hat{Q}$ from Eq. (8.33):

$$\hat{Q} = (898 + 150)\text{ kW} = 1048\text{ kW.}$$

The density of the cooling water is 1000 kg m⁻³; therefore:

$$\hat{M}_c = 60\text{ m}^3\text{h}^{-1} \cdot \left|\frac{1\text{ h}}{3600\text{ s}}\right| \cdot \left|\frac{1000\text{ kg}}{1\text{ m}^3}\right| = 16.7\text{ kg s}^{-1}.$$

The heat capacity of water is 75.4 J gmol⁻¹ °C⁻¹ (Table B.3). Therefore:

$$C_{pc} = 75.4\text{ J gmol}^{-1}°C^{-1} \cdot \left|\frac{1\text{ gmol}}{18\text{ g}}\right| \cdot \left|\frac{1000\text{ g}}{1\text{ kg}}\right| = 4.19 \times 10^3\text{ J kg}^{-1}°C^{-1}.$$

We can now apply Eq. (8.32) by rearranging and solving for T_{co}:

$$T_{co} = T_{ci} + \frac{\hat{Q}}{\hat{M}_c C_{pc}}$$

$$T_{co} = 10°C + \frac{1048 \times 10^3\text{ J s}^{-1}}{(16.7\text{ kg s}^{-1})(4.19 \times 10^3\text{ J kg}^{-1}°C^{-1})} = 25.0°C.$$

The water outlet temperature is 25°C.

8.5.2 Logarithmic- and Arithmetic-Mean Temperature Differences

Application of the heat-exchanger design equation, Eq. (8.19), requires knowledge of the temperature-difference driving force for heat transfer, ΔT. ΔT is equal to the difference in temperature between hot and cold fluids. However, as we have seen in Figures 8.2, 8.4, 8.5 and 8.8, fluid temperatures vary with position in heat exchangers; for example, the temperature difference between hot and cold fluids at one end of the exchanger may be more or less than at the other end. The driving force for heat transfer therefore varies from point to point in the system. For application of Eq. (8.19), this difficulty is overcome by use of an average ΔT.

If the temperature varies in both fluids in either countercurrent or cocurrent flow, the *logarithmic-mean temperature difference* ΔT_L is used:

$$\Delta T_L = \frac{\Delta T_2 - \Delta T_1}{\ln (\Delta T_2/\Delta T_1)} = \frac{\Delta T_2 - \Delta T_1}{2.303 \log (\Delta T_2/\Delta T_1)} \tag{8.34}$$

where ΔT_1 and ΔT_2 are the temperature differences betwee hot and cold fluids at the ends of the equipment. ΔT_1 and ΔT are calculated using the values for T_{hi}, T_{ho}, T_{ci} and T_c obtained from the energy balance. For convenience and t eliminate negative numbers and their logarithms, subscripts and 2 can refer to either end of the exchanger. Eq. (8.34) ha been derived using the following assumptions:

(i)   the overall heat-transfer coefficient U is constant;
(ii) the specific heats of the hot and cold fluids are constant;
(iii) heat losses from the system are negligible; and
(iv) the system is at steady-state in either countercurrent o cocurrent flow.

The most questionable of these assumptions is that of constan U, since this coefficient varies with temperature of the fluids However, because the change with temperature is gradua when temperature differences in the system are moderate th assumption is not seriously in error. Other details of the deri vation of Eq. (8.34) can be found elsewhere [1, 2].

Example 8.3 Log-mean temperature difference

A liquid stream is cooled from 70°C to 32°C in a double-pipe heat exchanger. Fluid flowing countercurrently with this stream i heated from 20°C to 44°C. Calculate the log-mean temperature difference.

Solution:
The heat-exchanger configuration is shown in Figure 8E3.1. At the left-hand end of the equipment, $\Delta T_1 = (32 - 20)°C = 12°C$ At the other end, $\Delta T_2 = (70 - 44)°C = 26°C$. From Eq. (8.34):

$$\Delta T_L = \frac{(26 - 12)°C}{\ln (26 / 12)} = 18.1°C.$$

Figure 8E3.1   Flow configuration for heat exchanger.

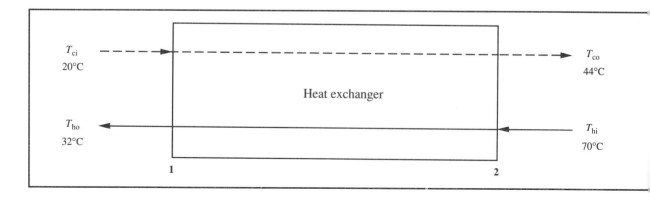

As noted above, the log-mean temperature difference is applicable to systems with cocurrent or countercurrent flow. In multiple-pass shell-and-tube heat exchangers, flow is neither countercurrent nor cocurrent. In these units the flow pattern is complex, with cocurrent, countercurrent and cross-flow all present. For shell-and-tube heat exchangers with more than a single tube or shell pass, the log-mean temperature difference must be used with a suitable correction factor to account for the geometry of the exchanger. Correction factors for cross-flow are available in other references [1–3].

When one fluid in the heat-exchange system remains at a constant temperature such as in a fermenter, the *arithmetic-mean temperature difference* ΔT_A is the appropriate ΔT to use in heat-exchanger design:

$$\Delta T_A = \frac{2 T_F - (T_1 + T_2)}{2}$$

(8.35)

where T_F is the temperature of fluid in the fermenter and T_1 and T_2 are the inlet and exit temperatures of the other fluid.

8.5.3 Calculation of Heat-Transfer Coefficients

As described in Sections 8.4.3 and 8.4.4, U in Eq. (8.19) can be determined as a combination of individual heat-transfer coefficients, properties of the separating wall, and, if applicable, fouling factors. Values of the individual heat-transfer coefficients h_h and h_c depend on the thickness of the fluid boundary layers, which is in turn strongly dependent on flow velocity and fluid properties such as viscosity and thermal conductivity. Increasing the level of turbulence and decreasing the viscosity will reduce the thickness of the liquid film, and hence increase the heat-transfer coefficient. Individual heat-transfer coefficients for flow in pipes or stirred vessels are usually evaluated using empirical correlations expressed in terms of dimensionless numbers.

The general form of correlations for heat-transfer coefficients is:

$$Nu = f\left(Re \text{ or } Re_i, Pr, Gr, \frac{D}{L}, \frac{\mu_b}{\mu_w}\right)$$

(8.36)

where f means 'some function of', and:

$$Nu = \text{Nusselt number} = \frac{hD}{k_{fb}}$$

(8.37)

$$Re = \text{Reynolds number for pipe flow} = \frac{Du\rho}{\mu_b}$$

(8.38)

$$Re_i = \text{impeller Reynolds number} = \frac{N_i D_i^2 \rho}{\mu_b}$$

(8.39)

$$Pr = \text{Prandtl number} = \frac{C_p \mu_b}{k_{fb}}$$

(8.40)

and

$$Gr = \text{Grashof number for heat transfer} = \frac{D^3 g \rho^2 \beta \Delta T}{\mu_b^2}.$$

(8.41)

Parameters in the above equations are as follows: h is the individual heat-transfer coefficient, D is the pipe or tank diameter, k_{fb} is the thermal conductivity of the bulk fluid, u is the linear velocity of fluid in the pipe, ρ is the average density of the fluid, μ_b is the viscosity of the bulk fluid, N_i is the rotational speed of the impeller, D_i is the impeller diameter, C_p is the average heat capacity of the fluid, g is gravitational acceleration, β is the coefficient of thermal expansion of the fluid, ΔT is the variation of fluid temperature in the system, L is pipe length, and μ_w is the viscosity of fluid at the wall.

The Nusselt number contains the heat-transfer coefficient h, and represents the ratio of rates of convective and conductive heat transfer. The Prandtl number represents the ratio of momentum and heat transfer; Pr contains physical constants which, for Newtonian fluids, are independent of flow conditions. The Grashof number represents the ratio of gravitational to viscous forces, and appears in correlations only when the fluid is not well mixed. Under these conditions the fluid density is no longer uniform and natural convection becomes an important heat-transfer mechanism. In most industrial applications, heat transfer occurs between turbulent fluids in pipes or in stirred vessels; forced convection in these systems is therefore more important than natural convection and the Grashof number is not of concern. The form of the correlation used to evaluate Nu and therefore h depends on the configuration of the heat-transfer equipment, the flow conditions and other factors.

A wide variety of heat-transfer situations is met in practice and there are many correlations available to biochemical engineers designing heat-exchange equipment. The most common heat-transfer applications are as follows:

(i) heat flow to or from fluids inside tubes, without phase change;
(ii) heat flow to or from fluids outside tubes, without phase change;
(iii) heat flow from condensing fluids; and
(iv) heat flow to boiling liquids.

Different equations are generally required to evaluate h_h and h_c depending on the flow geometry of the hot and cold fluids. Examples of correlations for heat-transfer coefficients in bioprocessing are given in the next section. Others can be found in the references listed at the end of this chapter.

8.5.3.1 *Flow in tubes without phase change*

There are several widely-accepted correlations for forced convection in tubes. The heat-transfer coefficient for fluid flowing inside a tube can be calculated from the following equation [2]:

$$Nu = 0.023\,Re^{0.8}\,Pr^{0.4}.$$

(8.42)

Eq. (8.42) is valid for either heating or cooling of liquids with viscosity close to water, and applies under the following conditions: $10^4 \leqslant Re \leqslant 1.2 \times 10^5$ (turbulent flow), $0.7 \leqslant Pr \leqslant 120$ and $L/D \geqslant 60$. Application of Eq. (8.42) to evaluate the tube-side heat-transfer coefficient in a heat exchanger is illustrated in Example 8.4.

Example 8.4 Tube-side heat-transfer coefficient

A single-pass shell-and-tube heat exchanger is used to heat a dilute salt solution used in large-scale protein chromatography. $25.5\ \mathrm{m^3\ h^{-1}}$ solution passes through 42 parallel tubes inside the heat exchanger; the internal diameter of the tubes is 1.5 cm and the tube length is 4 m. The viscosity of the bulk salt solution is $10^{-3}\ \mathrm{kg\ m^{-1}\ s^{-1}}$, the density is $1010\ \mathrm{kg\ m^{-3}}$, the average heat capacity is $4\ \mathrm{kJ\ kg^{-1}\,^{\circ}C^{-1}}$ and the thermal conductivity is $0.64\ \mathrm{W\ m^{-1}\,^{\circ}C^{-1}}$. Calculate the heat-transfer coefficient.

Solution:

First we must evaluate Re and Pr. All parameter values for calculation of these dimensionless groups are known except u, the linear fluid velocity. u is obtained by dividing the volumetric flow rate of the fluid by the total flow cross-sectional area.

Total flow cross-sectional area = (cross-sectional area of each tube) (number of tubes)

$$= 42\,(\pi R^2)$$

$$= 42\,\pi\left(\frac{1.5 \times 10^{-2}\,\mathrm{m}}{2}\right)^2$$

$$= 7.42 \times 10^{-3}\,\mathrm{m^2}.$$

Therefore:

$$u = \frac{25.5\ \mathrm{m^3\ h^{-1}}}{7.42 \times 10^{-3}\,\mathrm{m^2}} \cdot \left|\frac{1\mathrm{h}}{3600\ \mathrm{s}}\right| = 0.95\ \mathrm{m\ s^{-1}}.$$

From Eqs (8.38) and (8.40):

$$Re = \frac{(1.5 \times 10^{-2}\,\mathrm{m})\,(0.95\ \mathrm{m\ s^{-1}})\,(1010\ \mathrm{kg\ m^{-3}})}{10^{-3}\ \mathrm{kg\ m^{-1}\ s^{-1}}} = 1.44 \times 10^4$$

and

$$Pr = \frac{(4 \times 10^{-3}\ \mathrm{J\ kg^{-1}\,^{\circ}C^{-1}})\,(10^{-3}\ \mathrm{kg\ m^{-1}\ s^{-1}})}{0.64\ \mathrm{J\ s^{-1}\ m^{-1}\,^{\circ}C^{-1}}} = 6.25.$$

Also, $L/D = 267$. As $10^4 \leqslant Re \leqslant 1.2 \times 10^5$, $0.7 \leqslant Pr \leqslant 120$ and $L/D \geqslant 60$, Eq. (8.42) is valid.

$$Nu = 0.023\,(1.44 \times 10^4)^{0.8}\,(6.25)^{0.4} = 101.6.$$

Calculating h from this value of Nu:

$$h = \frac{101.6\,(0.64 \text{ W m}^{-1}\,{}^\circ\text{C}^{-1})}{1.5 \times 10^{-2}\text{ m}} = 4335 \text{ W m}^{-2}\,{}^\circ\text{C}^{-1}.$$

The heat-transfer coefficient is $4.3 \text{ kW m}^{-2}\,{}^\circ\text{C}^{-1}$.

For very viscous liquids, because of the temperature variation across the thermal boundary layer, there may be a marked difference between the viscosity of fluid in bulk flow and the viscosity of fluid adjacent to the wall. A modified form of Eq. (8.42) includes a viscosity correction term [3, 4]:

$$Nu = 0.027\,Re^{0.8}\,Pr^{0.33}\left(\frac{\mu_b}{\mu_w}\right)^{0.14}. \tag{8.43}$$

When flow in pipes is laminar rather than turbulent, Eqs (8.42) and (8.43) do not apply. In liquids with $Re < 2100$, fluid buoyancy and natural convection play an important role in heat transfer. Heat-transfer correlations for laminar flow in tubes can be found in other texts, e.g. [2].

8.5.3.2 Flow outside tubes without phase change

In the shell section of shell-and-tube heat exchangers, fluid flows around the outside of the tubes. The degree of turbulence, and therefore the external heat-transfer coefficient for the shell fluid, depend in part on the geometric arrangement of the tubes and baffles. Tubes in the exchanger may be arranged 'in line' as shown in Figure 8.14(a), or 'staggered' as shown in Figure 8.14(b). The degree of turbulence is considerably less for tubes in line than for staggered tubes. Because the area for flow through a bank of staggered tubes is continually changing, this is a difficult system to analyse.

The following correlation has been proposed for flow of fluid at right-angles to a bank of tubes more than 10 rows deep [5, 6]:

$$Nu = C\,Re_{max}^{0.6}\,Pr^{0.33}. \tag{8.44}$$

In Eq. (8.44) $C = 0.33$ for staggered tubes and $C = 0.26$ for in-line tubes. Re_{max} is the Reynolds number evaluated using Eq.

Figure 8.14 Configuration of tubes in shell-and-tube heat exchanger: (a) tubes 'in line'; (b) 'staggered' tubes.

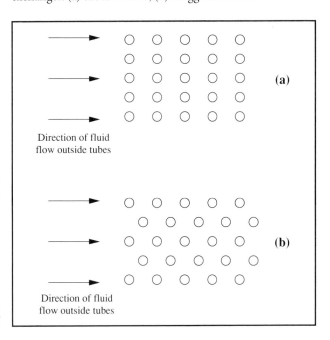

Direction of fluid flow outside tubes

Direction of fluid flow outside tubes

(8.38) with D equal to the outside tube diameter and u equal to the maximum fluid velocity based on the minimum free area available for fluid flow. For flow outside tubes, D in Nu is usually taken as the outside tube diameter. Eq. (8.44) is valid for $Re_{max} > 6 \times 10^3$ [6].

8.5.3.3 Stirred liquids

The heat-transfer coefficient in stirred vessels depends on the degree of agitation and the fluid properties. When heat is transferred to or from a helical coil in the vessel, h can be determined using the following equation [7]:

$$Nu = 0.87 \, Re_i^{0.62} \, Pr^{0.33} \left(\frac{\mu_b}{\mu_w}\right)^{0.14}$$

$$(8.45)$$

where Re_i is given by Eq. (8.39), and D in Nu refers to the inside diameter of the tank. For low-viscosity fluids such as water, the viscosity at the wall μ_w can usually be assumed equal to the bulk

viscosity μ_b. When heat is transferred to or from a jacket rather than a coil, the correlation is slightly modified [7]:

$$Nu = 0.36 \, Re_i^{0.67} \, Pr^{0.33} \left(\frac{\mu_b}{\mu_w}\right)^{0.14}.$$

$$(8.46)$$

Example 8.5 Heat-transfer coefficient for stirred vessel

A stirred fermenter of diameter 5 m contains an internal helical coil for heat transfer. The fermenter is mixed using a turbine impeller 1.8 m in diameter operated at 60 rpm. The fermentation broth has the following properties: $\mu_b = 5 \times 10^{-3}$ Pa s; $\rho = 1000$ kg m^{-3}; $C_p = 4.2$ kJ kg^{-1} °C^{-1}; $k_{fb} = 0.70$ W m^{-1} °C^{-1}. Neglecting viscosity changes at the wall of the coil, calculate the heat-transfer coefficient.

Solution:
From Eqs (8.39) and (8.40):

$$Re_i = \frac{60 \text{ min}^{-1}. \left|\frac{1 \text{ min}}{60 \text{ s}}\right|. (1.8 \text{ m})^2 (1000 \text{ kg m}^{-3})}{5 \times 10^{-3} \text{ kg m}^{-1} \text{ s}^{-1}} = 6.48 \times 10^5$$

$$Pr = \frac{(4.2 \times 10^3 \text{ J kg}^{-1} \text{°C}^{-1}) (5 \times 10^{-3} \text{ kg m}^{-1} \text{ s}^{-1})}{0.70 \text{ J s}^{-1} \text{ m}^{-1} \text{°C}^{-1}} = 30.$$

These values can be substituted into Eq. (8.45) to evaluate Nu:

$$Nu = 0.87 \, (6.48 \times 10^5)^{0.62} \, (30)^{0.33} = 1.07 \times 10^4.$$

Calculating h from this value of Nu:

$$h = \frac{(1.07 \times 10^4) (0.70 \text{ W m}^{-1} \text{°C}^{-1})}{5 \text{ m}} = 1501 \text{ W m}^{-2} \text{°C}^{-1}.$$

The heat-transfer coefficient is 1.5 kW m^{-2} °C^{-1}.

8.6 Application of the Design Equations

The equations of Sections 8.4 and 8.5 provide the essential elements for design of heat-transfer systems. Eq. (8.19) is used as the design equation, with $\hat{Q}$ and the inlet and outlet temperatures available from energy-balance calculations as described in Section 8.5.1. The overall heat-transfer coefficient is evaluated from correlations such as those given in Section 8.5.3; additional terms are included if the heat-transfer surfaces are fouled. With these parameters at hand, the area required and the size and capital cost of the equipment can be determined. Because metabolic rates, and therefore heat-production rates,

vary during fermentation, the rate at which heat is removed from the bioreactor must also be varied to maintain constant temperature. Heat-transfer design is based on the maximum heat load for the system. When the rate of heat generation drops, operating conditions can be changed to reduce the rate of heat removal; the simplest way of achieving this is to decrease the cooling-water flow rate.

The procedures outlined in this chapter represent the simplest and most direct approach to heat-transfer design. If several independent variables remain unfixed prior to design calculations, many different design outcomes are possible. When variables such as type of fluid, mass flow rates and erminal temperatures are unspecified, they can be manipulated to produce an *optimum design*, e.g. the design yielding the lowest total cost per year of operation. Computer packages which optimise heat-exchanger design are available commercially.

Calculation of equipment requirements for heating or cooling of fermenters can sometimes be simplified by considering the relative importance of each heat-transfer resistance. For large fermentation vessels containing cooling coils, the fluid velocity in the vessel is generally much slower than in the coils; accordingly, the tube-side thermal boundary layer is relatively thin and most of the heat-transfer resistance is located in the fermentation medium. Especially when there is no fouling present in the tubes, the heat-transfer coefficient for the cooling water can often be omitted when calculating U. Likewise, the wall resistance can sometimes be ignored as conduction of heat through metal is generally very rapid. An exception is stainless steel which is used widely in the fermentation industry; the low thermal conductivity of this material means that

wall resistance should be taken into account if the wall thickness is greater than about 5 mm.

The influence of sparging on heat transfer in stirred bioreactors is not clear; however, as the effect is minor compared with other parameters, correlations developed for ungassed systems are applied to aerobic fermentations. For non-Newtonian broths, apparent viscosity can be substituted for μ_b in the dimensionless groups Re, Re_i and Pr; this substitution is not straightforward when rheological parameters such as n, K and τ_0 change during the culture period. Apparent viscosity also depends on the shear rate in the fermenter which varies greatly throughout the vessel. Correlations for heat-transfer coefficients such as those presented in this chapter were not developed for fermentation systems and must not be considered to give exact values. The actual flow behaviour in many bioprocesses is poorly characterised due to the effects of non-Newtonian rheology and the presence of cells and air bubbles.

Application of heat-exchanger design equations to specify a fermenter cooling-system is illustrated in Example 8.6.

Example 8.6 Cooling-coil length in fermenter design

A fermenter used for antibiotic production must be kept at 27°C. After considering the oxygen demand of the organism and the heat dissipation from the stirrer, the maximum heat-transfer rate required is estimated as 550 kW. Cooling water is available at 10°C; the exit temperature of the cooling water is calculated using an energy balance as 25°C. The heat-transfer coefficient for the fermentation broth is estimated from Eq. (8.45) as 2150 W m^{-2}°C^{-1}. The heat-transfer coefficient for the cooling water is calculated as 14 000 W m^{-2}°C^{-1}. It is proposed to install a helical cooling coil inside the fermenter; the outer diameter of the coil pipe is 8 cm, the pipe thickness is 5 mm and the thermal conductivity of the steel is 60 W m^{-1}°C^{-1}. An average internal fouling-factor of 8500 W m^{-2}°C^{-1} is expected; the fermenter side surface of the coil is kept relatively clean. What length of cooling coil is required?

Solution:

$\hat{Q} = 550 \times 10^3$ W. ΔT is calculated as the arithmetic-average temperature difference from Eq. (8.35):

$$\Delta T = \frac{2(27°C) - (10 + 25)°C}{2} = 9.5°C.$$

U is calculated using Eq. (8.24) after omitting h_{fh} as there is no fouling layer on the hot side of the system:

$$\frac{1}{U} = \left(\frac{1}{2150} + \frac{5 \times 10^{-3}}{60} + \frac{1}{14\,000} + \frac{1}{8500} \right) \text{m}^2\,°C\,W^{-1}$$

$$= (4.65 \times 10^{-4} + 8.33 \times 10^{-5} + 7.14 \times 10^{-5} + 1.18 \times 10^{-4}) \text{m}^2\,°C\,W^{-1}$$

$$= 7.38 \times 10^{-4}\,\text{m}^2\,°C\,W^{-1}$$

$$U = 1355\,\text{W m}^{-2}\,°C^{-1}.$$

Note the relative magnitudes of the four contributions to U: the cooling-water coefficient and the wall resistance make a comparatively minor contribution and can often be neglected in design calculations.

We can now apply Eq. (8.19) to evaluate the required surface area A.

$$A = \frac{550 \times 10^3 \text{ W}}{(9.5°\text{C})(1355 \text{ W m}^{-2}\text{ °C}^{-1})} = 42.7 \text{ m}^2.$$

The heat-transfer area is equal to the surface area of the pipe:

$$A = 2\pi RL$$

where R is the pipe radius and L is the pipe length. Therefore:

$$L = \frac{42.7 \text{ m}^2}{2\pi\left(\dfrac{8 \times 10^{-2}}{2}\right)\text{ m}} = 169.9 \text{ m}.$$

The length of coil required is 170 m. The cost of such a length of pipe is a significant factor in the overall cost of the fermenter.

8.6.1 Relationship Between Heat Transfer, Cell Concentration and Stirring Conditions

The design equation (8.19) and the energy-balance equation (8.33) allow us to derive some important relationships for fermenter operation. Because cell metabolism is usually the largest source of heat in fermenters, the capacity of the system for heat removal can be linked directly to the maximum cell concentration in the reactor. Assuming that heat dissipated from the stirrer and the cooling effects of evaporation are negligible compared with the heat of reaction, Eq. (8.33) becomes:

$$\hat{Q} = -\Delta\hat{H}_{\text{rxn}}. \tag{8.47}$$

In aerobic fermentation, heat of reaction is related to the rate of oxygen consumption by the cells. As outlined in Section 5.9.2, approximately 460 kJ heat is released for each gmol oxygen consumed. If Q_O is the rate of oxygen uptake per unit volume in the fermenter:

$$\Delta\hat{H}_{\text{rxn}} = (-460 \text{ kJ gmol}^{-1})\, Q_O V \tag{8.48}$$

where V is the reactor volume. Typical units for Q_O are gmol m^{-3} s^{-1}. $\Delta\hat{H}_{\text{rxn}}$ in Eq. (8.48) is negative because the reaction is exothermic. Substituting this equation in Eq. (8.47):

$$\hat{Q} = (460 \text{ kJ gmol}^{-1})\, Q_O V. \tag{8.49}$$

If q_O is the *specific oxygen-uptake rate*, or rate of oxygen consumption per unit cell, $Q_O = q_O x$ where x is cell concentration. Typical units for q_O are gmol g^{-1} s^{-1}. Therefore:

$$\hat{Q} = (460 \text{ kJ gmol}^{-1})\, q_O x V. \tag{8.50}$$

Substituting this into Eq. (8.19) gives:

$$(460 \text{ kJ gmol}^{-1})\, q_O x V = UA\Delta T. \tag{8.51}$$

The fastest rate of heat transfer occurs when the temperature difference between fermenter contents and cooling water is maximum. Hypothetically, this occurs when the cooling water remains at its inlet temperature, i.e. $\Delta T = (T_F - T_{ci})$ where T_F is the fermentation temperature and T_{ci} is the water inlet temperature. Therefore, we can determine from Eq. (8.51) the maximum cell concentration supported by the heat-transfer system:

$$x_{\max} = \frac{UA(T_F - T_{ci})}{(460 \text{ kJ gmol}^{-1})\, q_O V}. \tag{8.52}$$

It is undesirable for biomass concentration in fermenters to be limited by heat-transfer capacity. Therefore, if the maximum cell concentration estimated using Eq. (8.52) is lower than that desired in the process, the heat-transfer facilities must be improved. For example, area A could be increased by installing a longer cooling coil, or the overall heat-transfer coefficient could be improved by increasing the stirrer speed. Eq. (8.52) was derived for fermenters in which shaft work could be ignored; if stirrer power adds significantly to the total heat load, x_{max} will be somewhat smaller than that calculated.

It should be evident from Chapters 7 and 8 that mixing and heat transfer are not independent functions in bioreactors. Agitation rate and stirrer size affect the value of the heat-transfer coefficient in the fermentation fluid; turbulence in the reactor decreases the thickness of the thermal boundary layer and facilitates rapid heat transfer. However, stirring also generates heat and contributes to the total heat load that must be removed from the reactor to maintain constant temperature. Heat removal can be a severe problem in bioreactors if the fluid is viscous; turbulence and high heat-transfer coefficients are difficult to achieve in highly viscous liquids without enormous power input which itself generates an extra heat load. These relationships are explored further in Problem 8.8.

8.7 Summary of Chapter 8

After studying Chapter 8, you should:

(i) understand the mechanisms of *conduction* and *convection* in heat transfer;

(ii) know *Fourier's law* of conduction in terms of the *thermal conductivity* of materials;

(iii) be able to describe equipment used for heat exchange in industrial bioprocesses;

(iv) understand the importance of the *thermal boundary layer* in heat transfer;

(v) know the *heat-transfer design equation* and the meaning of the *overall heat-transfer coefficient U*;

(vi) understand how the overall heat-transfer coefficient can be expressed in terms of individual resistances to heat transfer;

(vii) know how individual heat-transfer coefficients are estimated;

(viii) know how to incorporate fouling factors into heat-transfer analysis;

(ix) for heat exchange to or from fermentation vessels, know what are the major resistances to heat transfer; and

(x) be able to carry out simple calculations for design of heat-transfer systems.

Problems

8.1 Rate of conduction

(a) A furnace wall is constructed of firebrick 15-cm thick. The temperature inside the wall is 700°C; the temperature outside is 80°C. If the thermal conductivity of the brick under these conditions is 0.3 W m^{-1} K^{-1}, what is the rate of heat loss through 1.5 m^2 of wall surface?

(b) The 1.5 m^2 area in part (a) is insulated with 4-cm thick asbestos with thermal conductivity 0.1 W m^{-1} K^{-1}. What is the rate of heat loss now?

8.2 Overall heat-transfer coefficient

Heat is transferred from one fluid to a second fluid across a metal wall. The film coefficients are 1.2 and 1.7 kW m^{-2} K^{-1}. The metal is 6-mm thick and has a thermal conductivity of 19 W m^{-1} K^{-1}. On one side of the wall there is a scale deposit with a fouling factor estimated at 830 W m^{-2} K^{-1}. What is the overall heat-transfer coefficient?

8.3 Effect of cooling-coil length on coolant requirements

A fermenter is maintained at 35°C by water circulating at a rate of 0.5 kg s^{-1} in a cooling coil inside the vessel. The inlet and outlet temperatures of the water are 8°C and 15°C, respectively.

The length of the cooling coil is increased by 50%. In order to maintain the same fermentation temperature, the rate of heat removal must be kept the same. Determine the new cooling-water flow rate and outlet temperature by carrying out the following calculations. The heat capacity of the cooling water can be taken as 4.18 kJ kg^{-1} °C^{-1}.

(a) From a steady-state energy balance on the cooling water, calculate the rate of cooling with the original coil.

(b) Determine the mean temperature difference with the original coil.

(c) Evaluate UA for the original coil.

(d) If the length of the coil is increased by 50%, the area available for heat transfer, A', also increases by 50% so that $A' = 1.5 A$. The value of the overall heat-transfer coefficient is not expected to change very much. For the new coil, what is the value of UA'?

(e) Evaluate the new cooling-water outlet temperature.

(f) By how much are the cooling-water requirements reduced after the new coil is installed?

8.4 Calculation of heat-transfer area in fermenter design

A 100 m^3 fermenter of diameter 5 m is stirred using a turbine impeller 1.7 m in diameter at a speed of 80 rpm. The culture fluid inside the fermenter has the following properties:

C_p = 4.2 kJ kg^{-1}°C^{-1}
k_{fb} = 0.6 W m^{-1}°C^{-1}
ρ = 10^3 kg m^{-3}
μ_b = 10^{-3} N s m^{-2}.

Assume that the viscosity at the wall is equal to the bulk-fluid viscosity.

Heat is generated by the fermentation at a rate of 2500 kW. This heat is removed to cooling water flowing in a helical stainless-steel coil inside the vessel. The coil wall thickness is 6 mm and the thermal conductivity of the metal is 20 W m^{-1}°C^{-1}. There are no fouling layers present, and the heat-transfer coefficient for the cooling water can be neglected. The fermentation temperature is 30°C; cooling water enters the coils at 10°C.

(a) Calculate the fermenter-side heat-transfer coefficient.
(b) Calculate the overall heat-transfer coefficient U. What proportion of the total resistance to heat transfer is due to the pipe wall?
(c) The surface area needed for cooling depends on the cooling-water flow rate. Prepare a graph showing the outlet cooling-water temperature and the area required for heat transfer as functions of coolant flow-rate for 1.2 × 10^5 kg h^{-1} ≤ $\hat{M}_c$ ≤ 2 × 10^6 kg h^{-1}.
(d) For a cooling-water flow rate of 5 × 10^5 kg h^{-1}, estimate the length of cooling-coil needed if the diameter is 10 cm.

8.5 Effect of fouling on heat-transfer resistance

In current service, 20 kg s^{-1} cooling water at 12°C must be circulated through a coil inside a fermenter to maintain the temperature at 37°C. The coil is 150 m long with pipe diameter 12 cm; the exit water temperature is 28°C. After the inner and outer surfaces of the coil are cleaned it is found that only 13 kg s^{-1} cooling water is required to control the fermentation temperature.

(a) Calculate the overall heat-transfer coefficient before cleaning.
(b) What is the outlet water temperature after cleaning?
(c) What fraction of the total resistance to heat transfer before cleaning was due to fouling deposits?

8.6 Pre-heating of nutrient medium

Nutrient medium is to be heated from 10°C to 28°C in a single-pass countercurrent shell-and-tube heat exchanger before being pumped into a fed-batch fermenter. Medium passes through the tubes of the exchanger; the shell-side fluid is water which enters with flow rate 3 × 10^4 kg h^{-1} and temperature 60°C. Pre-heated medium is required at a rate of 50 m^3 h^{-1}. The density, viscosity and heat capacity of the medium are the same as water; the thermal conductivity of the medium is 0.54 W m^{-1}°C^{-1}.

It is proposed to use 30 steel tubes with inner diameter 5 cm; the tubes will be arranged in line. The pipe wall is 5-mm thick; the thermal conductivity of the metal is 50 W m^{-1}°C^{-1}. The maximum linear shell-side fluid velocity is estimated as 0.15 m s^{-1}. Estimate the tube length required by carrying out the following calculations.

(a) What is the rate of heat transfer?
(b) Calculate individual heat-transfer coefficients for the tube- and shell-side fluids.
(c) Calculate the overall heat-transfer coefficient.
(d) Calculate the log-mean temperature difference.
(e) Determine the heat-transfer area.
(f) What tube length is required?

8.7 Suitability of an existing cooling-coil

An enzyme manufacturer in the same industrial park as your antibiotic factory has a re-conditioned 20 m^3 fermenter for sale. You are in the market for a cheap 20 m^3 fermenter; however the vessel on offer is fitted with a 45-m steel helical cooling-coil with inner pipe-diameter 7.5 cm. You propose to use the fermenter for your newest production organism which is known to have a maximum oxygen demand of 90 mol m^{-3} h^{-1} at its optimum culture temperature of 28°C. You consider that the 3-m diameter vessel should be stirred with a 1-m diameter turbine-impeller operated at an average speed of 50 rpm. The fermentation fluid can be assumed to have the properties of water. If 20 m^3 h^{-1} cooling water at 12°C is available, should you make an offer for the second-hand fermenter and cooling coil?

8.8 Optimum stirring speed for removal of heat from viscous broth

The viscosity of a fermentation broth containing exopolysaccharide is about 10 000 cP. The broth is stirred in an aerated 10 m^3 fermenter of diameter 2.3 m using a single 0.78-m diameter turbine impeller. Other properties of the broth are as follows:

$$C_p = 2 \, \text{kJ kg}^{-1} \, ^\circ\text{C}^{-1}$$
$$k_{\text{fb}} = 2 \, \text{W m}^{-1} \, ^\circ\text{C}^{-1}$$
$$\rho = 10^3 \, \text{kg m}^{-3}.$$

The fermenter is equipped with an internal cooling-coil which provides a heat-transfer area of 14 m². Cooling water is provided; the average temperature difference for heat transfer is 20°C. Neglect any variation of viscosity at the wall of the coil. Assume that the power dissipated in aerated broth is 40% lower than in ungassed liquid.

(a) Using logarithmic coordinates, plot $\hat{Q}$ as a function of stirrer speed between 0.5 and 10 s⁻¹.

(b) From equations presented in Section 7.10, calculate $\hat{W}_s$, the power dissipated from the stirrer, as a function of stirrer speed. Plot these values on the same graph as $\hat{Q}$.

(c) If evaporation, heat losses and other factors have only a negligible effect on heat load, the difference between $\hat{Q}$ and $\hat{W}_s$ is equal to the rate of metabolic heat removal from the fermenter, $\Delta\hat{H}_{\text{rxn}}$. Plot $\Delta\hat{H}_{\text{rxn}}$ as a function of stirrer speed.

(d) At what stirrer speed is removal of metabolic heat most rapid?

(e) The specific rate of oxygen consumption is 6 mmol g⁻¹ h⁻¹. If the fermenter is operated at the stirrer speed identified in (d), what is the maximum cell concentration?

(f) How do you interpret the intersection of the curves for $\hat{Q}$ and $\hat{W}_s$ at high stirrer speed in terms of the capacity of the system to handle exothermic reactions?

References

1. McCabe, W.L. and J.C. Smith (1976) *Unit Operations of Chemical Engineering*, 3rd edn, Section 3, McGraw-Hill, New York.

2. McAdams, W.H. (1954) *Heat Transmission*, 3rd edn, McGraw-Hill, New York.

3. Coulson, J.M. and J.F. Richardson (1977) *Chemical Engineering*, vol. 1, 3rd edn, Chapter 7, Pergamon Press, Oxford.

4. Sieder, E.N. and G.E. Tate (1936) Heat transfer and pressure drop of liquids in tubes. *Ind. Eng. Chem.* **28**, 1429–1435.

5. Colburn, A.P. (1933) A method of correlating forced convection heat transfer data and a comparison with fluid friction. *Trans. AIChE* **29**, 174–210.

6. Rohsenow, W.M. and H.Y. Choi (1961) *Heat, Mass, and Momentum Transfer*, Prentice-Hall, New Jersey.

7. Chilton, T.H., T.B. Drew and R.H. Jebens (1944) Heat transfer coefficients in agitated vessels. *Ind. Eng. Chem.* **36**, 510–516.

Suggestions for Further Reading

Heat-Transfer Theory (see also refs 1, 2, 4 and 6)

Kern, D.Q. (1950) *Process Heat Transfer*, McGraw-Hill, Tokyo.

Heat-Transfer Equipment

Perry, R.H., D.W. Green and J.O. Maloney (Eds) (1984) *Chemical Engineers' Handbook*, 6th edn, Section 11, McGraw-Hill, New York.

Heat Transfer in Bioprocessing

Atkinson, B. and F. Mavituna (1991) *Biochemical Engineering and Biotechnology Handbook*, 2nd edn, Chapter 14, Macmillan, Basingstoke.

Brain, T.J.S. and K.L. Man (1989) Heat transfer in stirred tank bioreactors. *Chem. Eng. Prog.* **85**(7), 76–80.

Swartz, J.R. (1985) Heat management in fermentation processes. In: M. Moo-Young (Ed), *Comprehensive Biotechnology*, vol. 2, pp. 299–303, Pergamon Press, Oxford.

Mass Transfer

Mass transfer occurs in mixtures containing local concentration variations. For example, when dye is dropped into a pail of water, mass-transfer processes are responsible for movement of dye molecules through the water until equilibrium is established and the concentration is uniform. Mass is transferred from one location to another under the influence of a concentration difference or concentration gradient in the system.

There are many situations in bioprocessing where concentrations of compounds are not uniform; we rely on mechanisms of mass transfer to transport material from regions of high concentration to regions where the concentration is low. An example is the supply of oxygen in fermenters for aerobic culture. Concentration of oxygen at the surface of air bubbles is high compared with the rest of the fluid; this concentration gradient promotes oxygen transfer from the bubbles into the medium. Another example of mass transfer is extraction of penicillin from fermentation liquor using organic solvents such as butyl acetate. When solvent is added to the broth, the relatively low concentration of penicillin in the organic phase causes mass transfer of penicillin into the solvent. Solvent extraction is an efficient downstream-processing technique as it selectively removes the desired product from the rest of the fermentation fluid.

Mass transfer plays a vital role in many reaction systems. As distance between the reactants and site of reaction becomes greater, rate of mass transfer is more likely to influence or control the conversion rate. Taking again the example of oxygen in aerobic culture, if mass transfer of oxygen from the bubbles is slow, the rate of cell metabolism will become dependent on the rate of oxygen supply from the gas phase. Because oxygen is a critical component of aerobic fermentations and is so sparingly soluble in aqueous solutions, much of our interest in mass transfer lies with the transfer of oxygen across gas–liquid interfaces. However, liquid–solid mass transfer can also be important in systems containing clumps, pellets, flocs or films of cells or enzymes; in these cases, nutrients in the liquid phase must be transported into the solid before they can be utilised in reaction. Unless mass transfer is rapid, supply of nutrients will limit the rate of biological conversion.

In a solid or quiescent fluid, mass transfer occurs as a result of molecular diffusion. However, most mass-transfer systems contain moving fluid; in turbulent flow, mass transfer by molecular motion is supplemented by convective transfer.

There is an enormous variety of circumstances in which convective mass transfer takes place. In this chapter, we will consider the theory of mass transfer with applications relevant to the bioprocessing industry.

9.1 Molecular Diffusion

Molecular diffusion is the movement of component molecules in a mixture under the influence of a concentration difference in the system. Diffusion of molecules occurs in the direction required to destroy the concentration gradient. If the gradient is maintained by constantly supplying material to the region of high concentration and removing it from the region of low concentration, diffusion will be continuous. This situation is often exploited in mass-transfer operations and reaction systems.

9.1.1 Diffusion Theory

In this text, we confine our discussion of diffusion to *binary mixtures*, i.e. mixtures or solutions containing only two components. Consider a system containing molecular components A and B. Initially, the concentration of A in the system is not uniform; as indicated in Figure 9.1, concentration C_A varies from C_{A1} to C_{A2} is a function of distance y. In response to this concentration gradient, molecules of A will diffuse away from the region of high concentration until eventually the whole system acquires uniform composition. If there is no large-scale fluid motion in the system, e.g. due to stirring, mixing occurs solely by random molecular movement.

Assume that mass transfer of A occurs across area a perpendicular to the direction of diffusion. In single-phase systems, the rate of mass transfer due to molecular diffusion is given by *Fick's law of diffusion*, which states that mass flux is proportional to the concentration gradient:

Figure 9.1 Concentration gradient of component A inducing mass transfer across area *a*.

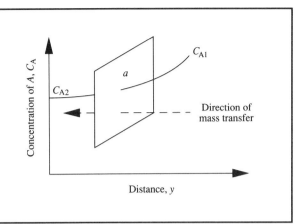

$$J_A = \frac{N_A}{a} = -\mathscr{D}_{AB}\frac{dC_A}{dy}.$$

(9.1)

In Eq. (9.1), J_A is the *mass flux* of component A, N_A is the *rate of mass transfer* of component A, *a* is the area across which mass transfer occurs, $\mathscr{D}_{AB}$ is the *binary diffusion coefficient* or *diffusivity* of component A in a mixture of A and B, C_A is the concentration of component A, and *y* is distance. dC_A/dy is the concentration gradient, or change in concentration of A with distance. As indicated in Eq. (9.1), mass flux is defined as the rate of mass transfer per unit area perpendicular to the direction of movement; J_A has units of, e.g. gmol s^{-1} m^{-2}. Corresponding units for N_A are gmol s^{-1}, for C_A gmol m^{-3}, and for $\mathscr{D}_{AB}$ m^2 s^{-1}. Mass rather than mole units may be used for J_A, N_A and C_A; Eq. (9.1) holds in either case. Eq. (9.1) indicates that rate of diffusion can be enhanced by increasing the area available for mass transfer, the concentration gradient in the system, and the magnitude of the diffusion coefficient. The negative sign in Eq. (9.1) indicates that the direction of mass transfer is always from high concentration to low concentration, opposite to that of the concentration gradient. In other words, if the slope of C_A versus *y* is positive as in Figure 9.1, the direction of mass transfer is in the negative *y*-direction, and vice versa.

The diffusion coefficient $\mathscr{D}_{AB}$ is a property of materials; values can be found in handbooks. $\mathscr{D}_{AB}$ reflects the ease with which diffusion takes place. Its value depends on both components of the mixture; for example, the diffusivity of carbon dioxide in water will be different from the diffusivity of carbon dioxide in another solvent such as ethanol. The value of $\mathscr{D}_{AB}$ is also dependent on temperature. Diffusivity of gases varies with pressure; for liquids there is an approximate linear dependence

on concentration. Diffusivity values in liquids are several orders of magnitude smaller than in gases. As examples, $\mathscr{D}_{AB}$ for oxygen in air at 0°C and 1 atm is 1.78×10^{-5} m^2 s^{-1}; $\mathscr{D}_{AB}$ is 2.5×10^{-9} m^2 s^{-1} for oxygen in water at 25°C and 1 atm, and 6.9×10^{-10} m^2 s^{-1} for glucose in dilute solution in water at 25°C.

When diffusivity values are not available for the exact temperature and pressure of interest, $\mathscr{D}_{AB}$ can be estimated using equations. Relationships for calculating diffusivities are available from other references [1–3]. The theory of diffusion in liquids is not as well advanced as with gases; there are also fewer experimental data available for liquid systems.

9.1.2 Analogy Between Mass, Heat and Momentum Transfer

There is a close similarity between the processes of mass, heat and momentum transfer occurring as a result of molecular motion. This is suggested by the form of the equations for mass, heat and momentum fluxes:

$$J_A = -\mathscr{D}_{AB}\frac{dC_A}{dy}$$

(9.2)

$$\hat{q} = -k\frac{dT}{dy}$$

(8.2)

and

$$\tau = -\mu\frac{dv}{dy}.$$

(7.6)

The three processes represented above are quite different on the molecular level, but the basic equations have the same form. In each case, flux in the *y*-direction is directly proportional to the driving force (either dC_A/dy, dT/dy or dv/dy), with the proportionality constant ($\mathscr{D}_{AB}$, k or μ) a physical property of the material. The negative signs in Eqs (9.2), (8.2) and (7.6) indicate that transfer of mass, heat or momentum is always in the direction opposite to that of increasing concentration, temperature or velocity. The similarity in the form of the three rate equations makes it possible in some situations to apply analysis of one process to either of the other two.

The analogy of Eqs (9.2), (8.2) and (7.6) is valid for transport of mass, heat and momentum resulting from motion or vibration of molecules. Extension of the analogy to turbulent flow is generally valid for heat and mass transfer; however the analogy with momentum transfer presents a number of difficulties. Several analogy theories have been proposed in the chemical engineering literature to describe simultaneous transport phenomena in turbulent systems. Details are presented elsewhere [2, 4, 5].

9.2 Role of Diffusion in Bioprocessing

Fluid mixing is carried out in most industrial processes where mass transfer takes place. Bulk fluid motion causes more rapid large-scale mixing than molecular diffusion; why then is diffusive transport still important? Areas of bioprocessing in which diffusion plays a major role are described below.

(i) *Scale of mixing.* As discussed in Section 7.9.3, turbulence in fluids produces bulk mixing on a scale equal to the smallest eddy size. Within the smallest eddies, flow is largely streamline so that further mixing must occur by diffusion of fluid components. Mixing on a molecular scale therefore relies on diffusion as the final step in the mixing process.

(ii) *Solid-phase reaction.* In biological systems, reactions are sometimes mediated by catalysts in solid form, e.g. clumps, flocs and films of cells and immobilised-enzyme and -cell particles. When cells or enzyme molecules are clumped together into a solid particle, substrates must be transported into the solid before reaction can take place. Mass transfer within solid particles is usually unassisted by bulk fluid convection; the only mechanism for intraparticle mass transfer is molecular diffusion. As the reaction proceeds, diffusion is also responsible for removal of product molecules away from the site of reaction. As discussed more fully in Chapter 12, when reaction is coupled with diffusion, the overall reaction rate can be significantly reduced if diffusion is slow.

(iii) *Mass transfer across a phase boundary.* Mass transfer between phases occurs often in bioprocessing. Oxygen transfer from gas bubbles to fermentation broth, penicillin recovery from aqueous to organic liquid, and glucose transfer from liquid medium into mould pellets are typical examples. When different phases come into contact, fluid velocity near the phase interface is significantly decreased and diffusion becomes crucial for mass transfer across the phase interface. This is discussed further in the next section.

9.3 Film Theory

The *two-film theory* is a useful model for mass transfer between phases. Mass transfer of solute from one phase to another involves transport from the bulk of one phase to the *phase*

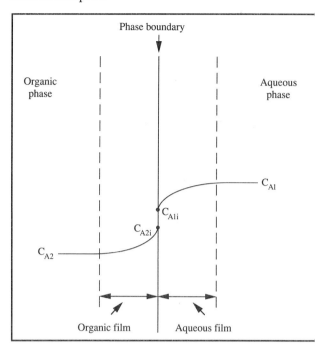

Figure 9.2   Film resistance to mass transfer between two immiscible liquids.

boundary or *interface*, and then from the interface to the bulk of the second phase. The film theory is based on the idea that a fluid film or *mass-transfer boundary layer* forms wherever there is contact between two phases.

Let us consider mass transfer of component A across the phase interface represented in Figure 9.2. Assume that the phases are two immiscible liquids such as water and chloroform, and that A is initially at higher concentration in the aqueous phase than in the organic phase. Each phase is well mixed and in turbulent flow. The concentration of A in the bulk aqueous phase is C_{A1}; the concentration of A in the bulk organic phase is C_{A2}.

According to the film theory, turbulence in each fluid dies out at the phase boundary. A thin film of relatively stagnant fluid exists on either side of the interface; mass transfer through this film is effected solely by molecular diffusion. The concentration of A changes near the interface as indicated in Figure 9.2; C_{A1i} is the interfacial concentration of A in the aqueous phase; C_{A2i} is the interfacial concentration of A in the organic phase. Most of the resistance to mass transfer resides in the liquid films rather than in the bulk liquid. For practical purposes it is generally assumed that there is negligible resistance to transport at the interface itself; this is equivalent to assuming that the phases are in equilibrium at the plane of contact. The difference between C_{A1i} and C_{A2i} at the interface accounts for the possibility that, at equilibrium, A may be more soluble in one phase than in the other. For example, if A were acetic acid in contact at the interface with both water and chloroform, the equilibrium concentration in water would be greater than that in chloroform by a factor of between 5 and 10. C_{A1i} would then be significantly higher than C_{A2i}.

Even though the bulk liquids in Figure 9.2 may be well mixed, diffusion of component A is crucial in effecting mass transfer because the local fluid velocities approach zero at the interface. The film theory as described above is applied extensively in analysis of mass transfer, although it is a greatly simplified representation. There are other models of mass transfer in fluids which lead to more realistic mathematical outcomes than the film theory [1, 4]. Nevertheless, irrespective of how mass transfer is visualised, diffusion is always an important mechanism of mass transfer close to the interface between fluids.

9.4 Convective Mass Transfer

The term *convective mass transfer* refers to mass transfer occurring in the presence of bulk fluid motion. Molecular diffusion will occur whenever there is a concentration gradient; however if the bulk fluid is also moving, the overall rate of mass transfer will be higher due to the contribution of convective currents. Analysis of mass transfer is most important in multi-phase systems where interfacial boundary layers provide significant mass-transfer resistance. Let us develop an expression for rate of mass transfer which is applicable to mass-transfer boundary layers.

Rate of mass transfer is directly proportional to the driving force for transfer, and the area available for the transfer process to take place. This can be expressed as:

Transfer rate $\propto$ (transfer area) $\times$ (driving force).

The proportionality coefficient in this equation is called the *mass-transfer coefficient*, so that:

Transfer rate = (mass-transfer coefficient) $\times$ (transfer area) $\times$ (driving force).

For each fluid on either side of a phase boundary, the driving force for mass transfer can be expressed in terms of a concentration difference. Therefore, rate of mass transfer to a phase boundary is given by the equation:

$$N_A = k\,a\,\Delta C_A = k\,a\,(C_{Ao} - C_{Ai})$$
(9.3)

where N_A is the rate of mass transfer of component A, k is the mass-transfer coefficient, a is the area available for mass transfer, C_{Ao} is the bulk concentration of component A away from the phase boundary, and C_{Ai} is the concentration of A at the interface. Eq. (9.3) is usually used to represent the *volumetric rate of mass transfer*, so units of N_A are, for example, gmol m^{-3} s^{-1}. Consistent with this representation, a is the interfacial area per unit volume with dimensions L^{-1} and units of, for example, m^2 m^{-3} or m^{-1}. The dimensions of the mass-transfer coefficient are LT^{-1}; SI units are m s^{-1}. Eq. (9.3) indicates that the rate of convective mass transfer can be enhanced by increasing the area available for mass transfer, the concentration difference between the bulk fluid and the interface, and the magnitude of the mass-transfer coefficient. By analogy with Eq. (8.11) for heat transfer, Eq. (9.3) can also be written in the form:

$$N_A = \frac{\Delta C_A}{R_m}$$
(9.4)

where R_m is the resistance to mass transfer:

$$R_m = \frac{1}{k\,a}.$$
(9.5)

Mass transfer coupled with fluid flow is a more complicated process than diffusive mass transfer. The value of the mass-transfer coefficient reflects the contribution to mass transfer from all the processes in the system that affect the boundary layer. Like the heat-transfer coefficient in Chapter 8, k depends on the combined effects of flow velocity, geometry of the mass-transfer system, and fluid properties such as viscosity and diffusivity. Because the hydrodynamics of most practical systems are not easily characterised, k cannot be calculated reliably from first principles. Instead, it is measured experimentally or estimated using correlations available from the literature. In general, reducing the thickness of the boundary layer or improving the diffusion coefficient in the film will result in enhancement of k and improvement in the rate of mass transfer.

Three mass-transfer situations which occur in bioprocessing are *liquid–solid mass transfer*, *liquid–liquid mass transfer* between immiscible solvents and *gas–liquid mass transfer*. Use of Eq. (9.3) to determine the rate of mass transfer in these systems is discussed in the following sections.

9.4.1 Liquid–Solid Mass Transfer

Mass transfer between a moving liquid and a solid is important in biological processing in a variety of applications. Transport of substrates to solid-phase cell or enzyme catalysts has already been mentioned. Adsorption of molecules onto surfaces, such as in chromatography, requires transport from liquid phase to solid; liquid–solid mass transfer is also important in crystallisation as molecules move from the liquid to the face of the growing crystal. Conversely, the process of dissolving a solid in liquid requires liquid–solid mass transfer directed away from the solid surface.

Let us assume that component A is required for reaction at the surface of a solid. The situation at the interface between flowing liquid containing A and the solid is illustrated in Figure 9.3. Near the interface, the fluid velocity is reduced and a boundary layer develops. As A is consumed by reaction, the local concentration of A at the surface decreases and a concentration gradient is established through the film. The concentration difference between the bulk liquid and the phase interface drives mass transfer of A from the liquid to the solid, allowing the reaction to continue. If the solid is non-porous, A does not penetrate further than the surface. The concentration of A at the phase boundary is C_{Ai}; the concentration of A in the bulk liquid outside the film is C_{Ao}. If a is the liquid–solid interfacial area per unit volume, the volumetric rate of mass transfer can be determined from Eq. (9.3) as:

$$N_A = k_L a (C_{Ao} - C_{Ai})$$

$$(9.6)$$

where k_L is the *liquid-phase mass-transfer coefficient.*

Application of Eq. (9.6) requires knowledge of the mass-transfer coefficient, the interfacial area between the phases, the bulk concentration of A, and the concentration of A at the interface. Bulk compositions are generally easy to measure; in simple cases, area a can be determined from the shape and size of the solid. The value of the mass-transfer coefficient is either measured or calculated using published correlations. Estimation of the interfacial concentration C_{Ai} is more difficult; measuring compositions at phase boundaries is not easy experimentally. To overcome this problem, we must consider the processes in the system which are linked to mass transfer of A. In the example of Figure 9.3, transport of A is linked to reaction at the surface of the solid, so that the value of C_{Ai} will depend on the rate of consumption of A at the interface. In practical terms, we can therefore calculate the rate of mass transfer of A only if we have information about the rate of reaction at the solid surface. Simultaneous reaction and mass transfer occurs in many bioprocesses; Chapter 12 treats solid-phase reaction coupled with mass transfer in more detail.

Figure 9.3 Concentration gradient for liquid–solid mass transfer.

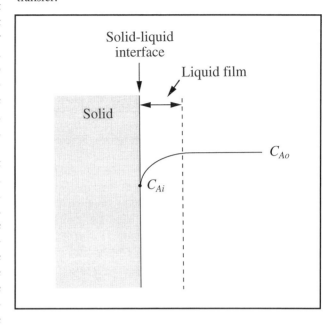

9.4.2 Liquid–Liquid Mass Transfer

Liquid–liquid mass transfer between immiscible solvents is most often encountered in the product-recovery stages of bioprocessing. Organic solvents are used to isolate antibiotics, steroids and alkaloids from fermentation broths; two-phase aqueous systems are used in protein purification. Liquid–liquid mass transfer is also important when hydrocarbons are used as substrates in fermentation, e.g. in production of microbial biomass for single-cell protein.

The situation at the interface between two immiscible liquids is shown in Figure 9.2. Component A is present at bulk concentration C_{A1} in one liquid phase; this concentration falls to C_{A1i} at the interface. In the other liquid, the concentration of A falls from C_{A2i} at the interface to C_{A2} in the bulk. The rate of mass transfer N_A in each liquid phase can be obtained from Eq. (9.3):

$$N_{A1} = k_{L1} a (C_{A1} - C_{A1i})$$

$$(9.7)$$

and

$$N_{A2} = k_{L2} a (C_{A2i} - C_{A2})$$

$$(9.8)$$

where k_L is the liquid-phase mass-transfer coefficient, and subcripts 1 and 2 refer to the two liquid phases. As mentioned already in Section 9.4.1, application of Eqs (9.7) and (9.8) is difficult because we cannot easily measure interfacial concentrations. However, in this case these terms can be eliminated by considering the physical situation at the interface and manipulating the equations.

First, let us recognise that at steady state, because there can be no accumulation of A at the interface or anywhere else in the system, any A transported through liquid 1 must also be transported through liquid 2. This means that N_{A1} in Eq. (9.7) must be equal to N_{A2} in Eq. (9.8); we will call $N_{A1} = N_{A2} = N_A$. We can then rearrange Eqs (9.7) and (9.8):

$$\frac{N_A}{k_{L1}a} = C_{A1} - C_{A1i}$$

$$(9.9)$$

and

$$\frac{N_A}{k_{L2}a} = C_{A2i} - C_{A2}.$$

$$(9.10)$$

Normally, it can be assumed that there is negligible resistance to mass transfer at the actual interface, i.e. within distances corresponding to molecular free paths on either side of the phase boundary. This is equivalent to assuming that the phases are in equilibrium at the interface; therefore, C_{A1i} and C_{A2i} are equilibrium concentrations. The assumption of phase-boundary equilibrium has been subjected to many tests. As a result it is known that there are special situations, such as when there is adsorption of material at the interface, where the assumption is invalid. However, in ordinary situations, the evidence is that equilibrium does exist at the interface between phases. Note that we are not proposing to relate bulk concentrations C_{A1} and C_{A2} using equilibrium relationships, only C_{A1i} and C_{A2i}. If the bulk liquids were in equilibrium, no nett mass transfer would take place.

A typical equilibrium curve relating concentrations of solute A in two immiscible liquid phases is shown in Figure 9.4. The points making up the curve are obtained readily from experiments; alternatively equilibrium data can be found in handbooks. Equilibrium distribution of one solute between two phases is conveniently described in terms of the *distribution law*. At equilibrium, the ratio of solute concentrations in the two phases is given by the *distribution coefficient* or *partition coefficient, m*. As shown in Figure 9.4, when the concentration of A is low, the equilibrium curve is approxi-

Figure 9.4 Equilibrium curve for solute A in two immiscible solvents 1 and 2.

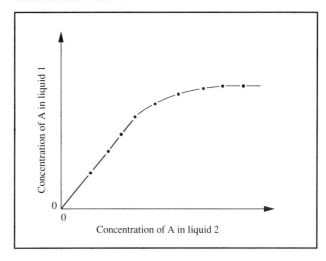

mately a straight line, so that m is constant. The distribution law is accurate only if both solvents are immiscible and there is no chemical reaction.

Therefore if C_{A1i} and C_{A2i} are equilibrium concentrations, they can be related using the distribution coefficient m.

$$m = \frac{C_{A1i}}{C_{A2i}}$$

$$(9.11)$$

i.e.

$$C_{A1i} = mC_{A2i}$$

$$(9.12)$$

or, alternatively:

$$C_{A2i} = \frac{C_{A1i}}{m}.$$

$$(9.13)$$

Eqs (9.12) and (9.13) can now be used to eliminate the interfacial concentrations from Eqs (9.9) and (9.10). First, we make a direct substitution:

$$\frac{N_A}{k_{L1}a} = C_{A1} - mC_{A2i}$$

$$(9.14)$$

and

$$\frac{N_A}{k_{L2} a} = \frac{C_{A1i}}{m} - C_{A2}.$$

(9.15)

If we now multiply Eq. (9.10) by m:

$$\frac{m N_A}{k_{L2} a} = m C_{A2i} - m C_{A2}$$

(9.16)

and divide Eq. (9.9) by m:

$$\frac{N_A}{m k_{L1} a} = \frac{C_{A1}}{m} - \frac{C_{A1i}}{m}$$

(9.17)

and add Eq. (9.14) to Eq. (9.16), and Eq. (9.15) to Eq. (9.17), we eliminate the interfacial-concentration terms completely:

$$N_A \left(\frac{1}{k_{L1} a} + \frac{m}{k_{L2} a} \right) = C_{A1} - m C_{A2}$$

(9.18)

$$N_A \left(\frac{1}{m k_{L1} a} + \frac{1}{k_{L2} a} \right) = \frac{C_{A1}}{m} - C_{A2}.$$

(9.19)

Eqs (9.18) and (9.19) combine mass-transfer resistances in the two liquid films, and relate the rate of mass transfer N_A to the bulk fluid concentrations C_{A1} and C_{A2}. The bracketed terms for the combined mass-transfer coefficients are used to define the *overall liquid-phase mass-transfer coefficient, K_L*. Depending on the form used for the concentration difference, we can define two overall mass-transfer coefficients:

$$\frac{1}{K_{L1} a} = \left(\frac{1}{k_{L1} a} + \frac{m}{k_{L2} a} \right)$$

(9.20)

and

$$\frac{1}{K_{L2} a} = \left(\frac{1}{m k_{L1} a} + \frac{1}{k_{L2} a} \right)$$

(9.21)

where K_{L1} is the overall mass-transfer coefficient based on the bulk concentration in liquid 1, and K_{L2} is the overall mass-transfer coefficient based on the bulk concentration in liquid 2.

We can now summarise the results to obtain two equations for the mass-transfer rate in liquid–liquid systems:

$$N_A = K_{L1} a (C_{A1} - m C_{A2})$$

(9.22)

and

$$N_A = K_{L2} a \left(\frac{C_{A1}}{m} - C_{A2} \right)$$

(9.23)

where K_{L1} and K_{L2} are given by Eqs (9.20) and (9.21). Use of either of these two equations requires knowledge of the concentrations of A in the bulk fluids, the partition coefficient m, the interfacial area a between the two liquid phases, and the value of either K_{L1} or K_{L2}. C_{A1} and C_{A2} are generally easy to measure. m can also be measured, or is found in handbooks of physical properties. The overall mass-transfer coefficients can be measured experimentally, or are estimated from published correlations for k_{L1} and k_{L2} in the literature. The only remaining parameter is the interfacial area, a. In many applications of liquid–liquid mass transfer it is difficult to know how much interfacial area is available between phases. For example, liquid–liquid extraction is often carried out in stirred tanks where an impeller is used to disperse and mix droplets of one phase through the other. The interfacial area under these circumstances will depend on the size, shape and number of the droplets, which will in turn depend on the intensity of agitation and properties of the fluid. Because these factors also affect the value of k_L, correlations for mass-transfer coefficients are often given in terms of $k_L a$ as a combined parameter.

Eqs (9.22) and (9.23) indicate that the rate of mass transfer between two phases is not dependent simply on the concentration difference; the equilibrium relationship is also an important factor. According to Eq. (9.22), the driving force for transfer of A out of liquid 1 is the difference between the bulk concentration C_{A1} *and the concentration of A in liquid 1 which would be in equilibrium with concentration C_{A2} in liquid 2*. Similarly, the driving force for mass transfer according to Eq. (9.23) is the difference between C_{A2} and the concentration of A in liquid 2 which would be in equilibrium with C_{A1} in liquid 1.

9.4.3 Gas–Liquid Mass Transfer

Gas–liquid mass transfer is of paramount importance in bioprocessing because of the requirement for oxygen in aerobic fermentations. Transfer of a solute such as oxygen from gas to

Figure 9.5 Concentration gradients for gas–liquid mass transfer.

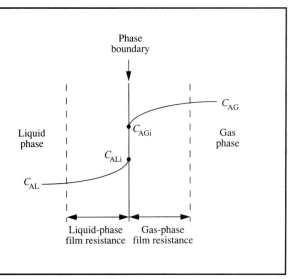

liquid is analysed in a similar way to liquid–liquid and liquid–solid mass transfer.

Figure 9.5 shows the situation at an interface between gas and liquid phases containing component A. Let us assume that A is transferred from the gas phase into the liquid. The concentration of A in the liquid is C_{AL} in the bulk and C_{ALi} at the interface. In the gas, the concentration is C_{AG} in the bulk and C_{AGi} at the interface.

From Eq. (9.3), the rate of mass transfer of A through the gas boundary layer is:

$$N_{AG} = k_G\, a\,(C_{AG} - C_{AGi})$$
(9.24)

and the rate of mass transfer of A through the liquid boundary layer is:

$$N_{AL} = k_L\, a\,(C_{ALi} - C_{AL})$$
(9.25)

where k_G is the gas-phase mass-transfer coefficient and k_L is the liquid-phase mass-transfer coefficient. To eliminate C_{AGi} and C_{ALi}, we must manipulate the equations as in Section 9.4.2.

If we assume that equilibrium exists at the interface, C_{AGi} and C_{ALi} can be related. For dilute concentrations of most gases and for a wide range of concentration for some gases, equilibrium concentration in the gas phase is a linear function of liquid concentration. Therefore, we can write:

$$C_{AGi} = mC_{ALi}$$
(9.26)

or, alternatively:

$$C_{ALi} = \frac{C_{AGi}}{m}$$
(9.27)

where m is the distribution factor. These equilibrium relationships can be incorporated into Eqs (9.24) and (9.25) at steady state using procedures which parallel those already used for liquid–liquid mass transfer. The results are also similar:

$$N_A \left(\frac{1}{k_G a} + \frac{m}{k_L a} \right) = C_{AG} - mC_{AL}$$
(9.28)

$$N_A \left(\frac{1}{m\, k_G a} + \frac{1}{k_L a} \right) = \frac{C_{AG}}{m} - C_{AL}.$$
(9.29)

The combined mass-transfer coefficients in Eqs (9.28) and (9.29) can be used to define overall mass-transfer coefficients. The *overall gas-phase mass-transfer coefficient* K_G is defined by the equation:

$$\frac{1}{K_G a} = \frac{1}{k_G a} + \frac{m}{k_L a}$$
(9.30)

and the *overall liquid-phase mass-transfer coefficient* K_L is defined as

$$\frac{1}{K_L a} = \frac{1}{m\, k_G a} + \frac{1}{k_L a}.$$
(9.31)

The rate of mass transfer in gas–liquid systems can therefore be expressed using either of two equations:

$$N_A = K_G\, a\,(C_{AG} - mC_{AL})$$
(9.32)

or

$$N_A = K_L\, a\left(\frac{C_{AG}}{m} - C_{AL} \right).$$
(9.33)

Eqs (9.32) and (9.33) are usually expressed using equilibrium concentrations. mC_{AL} is equal to C_{AG}^* , the gas-phase concentration of A in equilibrium with C_{AL}, and (C_{AG}/m) is equal to C_{AL}^*, the liquid-phase concentration of A in equilibrium with C_{AG}. Eqs (9.32) and (9.33) become:

$$N_A = K_G a (C_{AG} - C_{AG}^*) \tag{9.34}$$

and

$$N_A = K_L a (C_{AL}^* - C_{AL}). \tag{9.35}$$

In real mass-transfer systems, obtaining the values of C_{AG}, C_{AL} and m in Eqs (9.32) and (9.33) is reasonably straightforward. However, as in liquid–liquid mass-transfer systems, it is generally difficult to evaluate the interfacial area a. When gas is sparged through a liquid the interfacial area will depend on the size and number of bubbles present, which in turn depend on many other factors such as medium composition, stirrer speed and gas flow rate. Mass-transfer coefficients are measured experimentally or estimated using empirical correlations from the literature.

Eqs (9.34) and (9.35) can be simplified for systems in which most of the resistance to mass transfer lies in either the gas-phase interfacial film or the liquid-phase film. When solute A is very soluble in the liquid, for example in transfer of ammonia to water, the liquid-side resistance is small compared with that posed by the gas interfacial film. Therefore, from Eq. (9.5), if the liquid-side resistance is small $k_L a$ must be relatively large; from Eq. (9.30), $K_G a$ is then approximately equal to $k_G a$. Using this result in Eq. (9.34) gives:

$$N_A = k_G a (C_{AG} - C_{AG}^*). \tag{9.36}$$

Conversely, if A is poorly soluble in the liquid, e.g. oxygen in aqueous solution, the liquid-phase mass-transfer resistance dominates and $k_G a$ is much larger than $k_L a$. From Eq. (9.31), this means that $K_L a$ is approximately equal to $k_L a$, and Eq. (9.35) can be simplified to:

$$N_A = k_L a (C_{AL}^* - C_{AL}). \tag{9.37}$$

9.5 Oxygen Uptake in Cell Cultures

Cells in aerobic culture take up oxygen from the liquid. The rate of oxygen transfer from gas to liquid is therefore of prime importance, especially at high cell densities when cell growth is likely to be limited by availability of oxygen in the medium. An expression for rate of oxygen transfer from gas to liquid is given by Eq. (9.37); N_A is the rate of oxygen transfer per unit volume of fluid (gmol m^{-3} s^{-1}), k_L is the liquid-phase mass-transfer coefficient (m s^{-1}), a is the gas–liquid interfacial area per unit volume of fluid (m^2 m^{-3}), C_{AL} is the oxygen concentration in the broth (gmol m^{-3}), and C_{AL}^* is the oxygen concentration in the broth in equilibrium with the gas phase (gmol m^{-3}). The equilibrium concentration C_{AL}^* is also known as the *solubility* of oxygen in the broth. The difference $(C_{AL}^* - C_{AL})$ between the maximum possible and actual oxygen concentrations in the liquid represents the *concentration-difference driving force* for mass transfer.

The solubility of oxygen in aqueous solutions at ambient temperature and pressure is only about 10 ppm. This amount of oxygen is quickly consumed in aerobic cultures and must be constantly replaced by sparging. For an actively respiring yeast population with a cell density of 10^9 cells per ml, it can be calculated that the oxygen content of the broth must be replaced about 12 times per minute to keep up with cellular oxygen demand [6]. This is not an easy task because the low solubility of oxygen guarantees that the concentration difference $(C_{AL}^* - C_{AL})$ is always very small. Design of fermenters for aerobic operation must take these factors into account and provide optimum mass-transfer conditions.

9.5.1 Factors Affecting Cellular Oxygen Demand

The rate at which oxygen is consumed by cells in fermenters determines the rate at which it must be transferred from gas to liquid. Many factors influence oxygen demand; the most important of these are cell species, culture growth phase, and nature of the carbon source in the medium. In batch culture rate of oxygen uptake varies with time. The reasons for this are twofold. First, the concentration of cells increases during the course of batch culture and the total rate of oxygen uptake is proportional to the number of cells present. In addition, the rate of oxygen consumption per cell, known as the *specific oxygen uptake rate*, also varies. Typically, specific oxygen demand passes through a maximum in early exponential phase as illustrated in Figure 9.6, even though the cell concentration is relatively small at that time. If Q_O is the oxygen uptake rate per volume of broth and q_O is the specific oxygen uptake rate:

$$Q_O = q_O x \tag{9.38}$$

where x is cell concentration. Typical units for q_O are g g^{-1} s^{-1}, and for Q_O, g l^{-1} s^{-1}.

Figure 9.6 Variation in specific rate of oxygen consumption and biomass concentration during batch culture. (From R.T. Darby and D.R. Goddard, 1950, Studies of the respiration of the mycelium of the fungus *Myrothecium verrucaria*. *Am. J. Bot.* **37**, 379–387.)

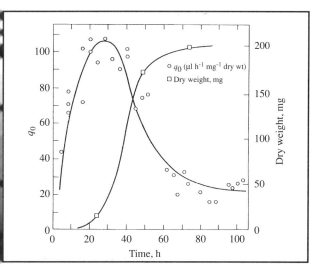

Figure 9.7 Relationship between specific rate of oxygen consumption by cells and dissolved-oxygen concentration.

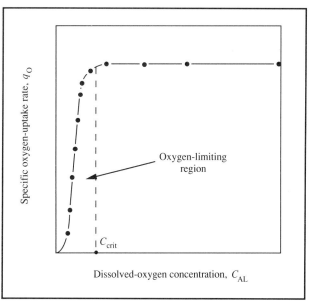

The inherent demand of an organism for oxygen (q_O) depends primarily on the biochemical nature of the cell and its nutritional environment. However, when the level of dissolved oxygen in the medium falls below a certain point, the specific rate of oxygen uptake is also dependent on the oxygen concentration in the liquid. The dependence of q_O on C_{AL} is shown in Figure 9.7. If C_{AL} is above the *critical oxygen concentration* C_{crit}, q_O is a constant maximum and independent of C_{AL}. If C_{AL} is below C_{crit}, q_O is approximately linearly dependent on oxygen concentration.

To eliminate oxygen limitations and allow cell metabolism to function at its fastest, the dissolved-oxygen concentration at every point in the fermenter must be above C_{crit}. The exact value of C_{crit} depends on the organism, but under average operating conditions usually falls between 5% and 10% of air saturation. For cells with relatively high C_{crit} levels, the task of transferring sufficient oxygen to maintain $C_{AL} > C_{crit}$ is always more challenging than for cultures with low C_{crit} values.

Choice of substrate for the fermentation can also significantly affect oxygen demand. Because glucose is generally consumed more rapidly than other sugars or carbon-containing substrates, rates of oxygen demand are higher when glucose is used. For example, maximum oxygen-consumption rates of 5.5, 6.1 and 12 mmol l^{-1} h^{-1} have been observed for *Penicillium* mould growing on lactose, sucrose and glucose, respectively [7].

9.5.2 Oxygen Transfer From Gas Bubble to Cell

In aerobic fermentation, oxygen molecules must overcome a series of transport resistances before being utilised by the cells. Eight mass-transfer steps involved in transport of oxygen from the interior of gas bubbles to the site of intracellular reaction are represented diagrammatically in Figure 9.8. They are:

(i) transfer from the interior of the bubble to the gas–liquid interface;

(ii) movement across the gas–liquid interface;

(iii) diffusion through the relatively stagnant liquid film surrounding the bubble;

(iv) transport through the bulk liquid;

(v) diffusion through the relatively stagnant liquid film surrounding the cells;

(vi) movement across the liquid–cell interface;

(vii) if the cells are in a floc, clump or solid particle, diffusion through the solid to the individual cell; and

(viii) transport through the cytoplasm to the site of reaction.

Note that resistance due to the gas boundary layer on the inside of the bubble has been neglected; because of the low solubility of oxygen in aqueous solutions, we can assume that the liquid-film resistance dominates gas–liquid mass transfer (see

Figure 9.8   Steps for transfer of oxygen from gas bubble to cell.

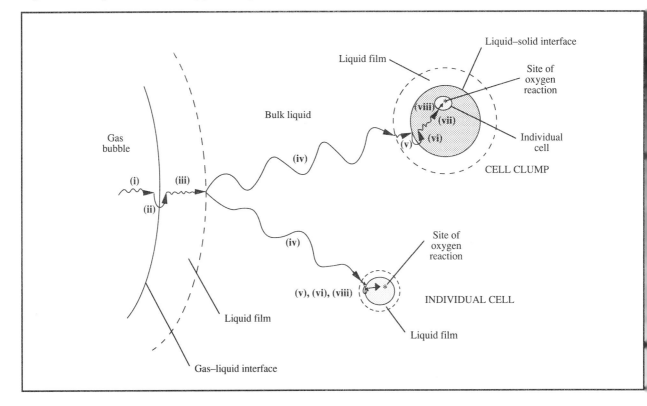

Section 9.4.3). If the cells are individually suspended in liquid rather than in a clump, step (vii) disappears.

The relative magnitudes of the various mass-transfer resistances depend on the composition and rheological properties of the liquid, mixing intensity, bubble size, cell-clump size, interfacial adsorption characteristics and other factors. For most bioreactors the following analysis is valid.

(i)   Transfer through the bulk gas phase in the bubble is relatively fast.

(ii)   The gas–liquid interface itself contributes negligible resistance.

(iii)   The liquid film around the bubbles is a major resistance to oxygen transfer.

(iv)   In a well-mixed fermenter, concentration gradients in the bulk liquid are minimised and mass-transfer resistance in this region is small. However, rapid mixing can be difficult to achieve in viscous fermentation broths; if this is the case, oxygen-transfer resistance in the bulk liquid may be important.

(v)   Because single cells are much smaller than gas bubbles, the liquid film surrounding each cell is much thinner than that around the bubbles and its effect on mass

transfer can generally be neglected. On the other hand, if the cells form large clumps, liquid-film resistance can be significant.

(vi)   Resistance at the cell–liquid interface is generally neglected.

(vii)   When the cells are in clumps, intraparticle resistance is likely to be significant as oxygen has to diffuse through the solid pellet to reach the interior cells. The magnitude of this resistance depends on the size of the clumps.

(viii)   Intracellular oxygen-transfer resistance is negligible because of the small distances involved.

When cells are dispersed in the liquid and the bulk fermentation broth is well mixed, *the major resistance to oxygen transfer is the liquid film surrounding the gas bubbles.* Transport through this film becomes the rate-limiting step in the complete process, and controls the overall mass-transfer rate. Consequently, the rate of oxygen transfer from the bubble all the way to the cell is dominated by the rate of step (iii). The mass-transfer rate for this step can be calculated using Eq. (9.37).

At steady state there can be no accumulation of oxygen at any location in the fermenter; therefore, the rate of oxygen transfer from the bubbles must be equal to the rate of oxygen onsumption by the cells. If we make N_A in Eq. (9.37) equal to Q_O in Eq. (9.38) we obtain the following equation:

$$k_L a \, (C_{AL}^* - C_{AL}) = q_O x.$$

$$(9.39)$$

$_L a$ is used to characterise the oxygen mass-transfer capability of fermenters. If $k_L a$ for a particular system is small, the ability of the reactor to deliver oxygen to the cells is limited. We can predict the response of the fermenter to changes in mass-transfer operating conditions using Eq. (9.39). For example if the rate of cell metabolism remains unchanged but $k_L a$ is increased, e.g. by raising the stirrer speed to reduce the thickness of the boundary layer around the bubbles, the dissolved-oxygen concentration C_{AL} must rise in order for the left-hand side of Eq. (9.39) to remain equal to the right-hand side. Similarly, if the rate of oxygen consumption by the cells accelerates while $k_L a$ is unaffected, C_{AL} must decrease.

We can use Eq. (9.39) to deduce some important relationships for fermenters. First, let us estimate the maximum cell concentration that can be supported by the fermenter's oxygen-transfer system. For a given set of operating conditions, the maximum rate of oxygen transfer occurs when the concentration-difference driving force $(C_{AL}^* - C_{AL})$ is highest, i.e. when the concentration of dissolved oxygen C_{AL} is zero. Therefore from Eq. (9.39), the maximum cell concentration

that can be supported by the mass-transfer functions of the reactor is:

$$x_{max} = \frac{k_L a \, C_{AL}^*}{q_O}.$$

$$(9.40)$$

If x_{max} estimated using Eq. (9.40) is lower than the cell concentration required in the fermentation process, $k_L a$ must be improved. It is generally undesirable for cell density to be limited by rate of mass transfer. Comparison of x_{max} values evaluated using Eqs (8.52) and (9.40) can be used to gauge the relative effectiveness of heat and mass transfer in aerobic fermentation. For example, if x_{max} from Eq. (9.40) were small while x_{max} calculated from heat-transfer considerations were large, we would know that mass-transfer operations are more likely to limit biomass growth. If both x_{max} values are greater than that desired for the process, heat and mass transfer are adequate.

Another important parameter is the minimum $k_L a$ required to maintain $C_{AL} > C_{crit}$ in the fermenter. This can be determined from Eq. (9.39) as:

$$(k_L a)_{crit} = \frac{q_O x}{(C_{AL}^* - C_{crit})}.$$

$$(9.41)$$

Example 9.1 Cell concentration in aerobic culture

A strain of *Azotobacter vinelandii* is cultured in a 15 m³ stirred fermenter for alginate production. Under current operating conditions $k_L a$ is 0.17 s⁻¹. Oxygen solubility in the broth is approximately 8×10^{-3} kg m⁻³.

a) The specific rate of oxygen uptake is 12.5 mmol g⁻¹ h⁻¹. What is the maximum possible cell concentration?
b) The bacteria suffer growth inhibition after copper sulphate is accidently added to the fermentation broth. This causes a reduction in oxygen uptake rate to 3 mmol g⁻¹ h⁻¹. What maximum cell concentration can now be supported by the fermenter?

Solution:

a) From Eq. (9.40):

$$x_{max} = \frac{(0.17 \text{ s}^{-1})(8 \times 10^{-3} \text{ kg m}^{-3})}{\dfrac{12.5 \text{ mmol}}{\text{g h}}} \cdot \left|\frac{1 \text{ h}}{3600 \text{ s}}\right| \cdot \left|\frac{1 \text{ gmol}}{1000 \text{ mmol}}\right| \cdot \left|\frac{32 \text{ g}}{1 \text{ gmol}}\right| \cdot \left|\frac{1 \text{ kg}}{1000 \text{ g}}\right|$$

$$= 1.2 \times 10^4 \text{ g m}^{-3} = 12 \text{ g l}^{-1}.$$

b) Assume that addition of copper sulphate does not affect C_{AL}^* or $k_L a$. If q_O is reduced by a factor of $12.5/3 = 4.167$, x_{max} is increased to:

$$x_{max} = 4.167 \, (12 \text{ g l}^{-1}) = 50 \text{ g l}^{-1}.$$

To achieve the calculated cell densities all other conditions must be favourable, e.g. sufficient substrate and time must be provided.

9.6 Oxygen Transfer in Fermenters

The rate of oxygen transfer in fermentation broths is influenced by several physical and chemical factors that change either the value of k_L or the value of a, or the driving force for mass transfer, $(C_{AL}^* - C_{AL})$. As a general rule of thumb, k_L in fermentation liquids is about 3–4×10^{-4} m s^{-1} for bubbles greater than 2–3 mm diameter; this can be reduced to 1×10^{-4} m s^{-1} for smaller bubbles depending on bubble rigidity. Once the bubbles are above 2–3 mm in size, k_L is relatively constant and insensitive to conditions. If substantial improvement in mass-transfer rates is required, it is usually more productive to focus on increasing the interfacial area a. Operating values in bioreactors for the combined coefficient $k_L a$ span a wide range over about three orders of magnitude; this is due mainly to the large variation in a. In production-scale fermenters, the value of $k_L a$ is typically in the range 0.02 s^{-1} to 0.25 s^{-1}.

In this section, several aspects of fermenter design and operation are discussed in terms of their effect on oxygen mass transfer.

9.6.1 Bubbles

The efficiency of gas–liquid mass transfer depends to a large extent on the characteristics of bubbles in the liquid medium. Bubble behaviour strongly affects the value of $k_L a$; some properties of bubbles affect mainly the magnitude of k_L, whereas others change the interfacial area a. The important aspects of bubble behaviour in fermenters are described below.

Stirred fermenters are used most commonly for aerobic culture. In these vessels, oxygen is supplied to the medium by sparging swarms of air bubbles underneath the impeller. The action of the impeller then creates a dispersion of gas throughout the vessel. In small laboratory-scale fermenters all of the liquid is close to the impeller; bubbles in these systems are frequently subjected to severe distortions as they interact with turbulent liquid currents in the vessel. In contrast, bubbles in most industrial stirred tanks spend a large proportion of their time floating free and unimpeded through the liquid after initial dispersion at the impeller. Liquid in large fermenters away from the impeller does not possess sufficient energy for continuous break-up of bubbles. This is a consequence of scale; most laboratory fermenters operate with stirrer power between 10 and 20 kW m^{-3}, whereas large agitated vessels operate at 0.5–5 kW m^{-3}. The result is that virtually all large commercial-size stirred-tank reactors operate mostly in the free bubble-rise regime [8].

The most important property of air bubbles in fermenters is their size. For a given volume of gas, more interfacial area a is provided if the gas is dispersed into many small bubbles rather than a few large ones; therefore a major goal in bioreactor design is a high level of gas dispersion. However, there are other important benefits associated with small bubbles. Small bubbles have correspondingly slow bubble-rise velocities; consequently they stay in the liquid longer, allowing more time for the oxygen to dissolve. Small bubbles therefore create high *gas hold-up*, defined as the fraction of the fluid volume in the reactor occupied by gas:

$$\varepsilon = \frac{V_G}{V_L + V_G}$$

$$(9.42)$$

where ε is the gas hold-up, V_G is the volume of gas bubbles in the reactor, and V_L is the volume of liquid. Because the total interfacial area for oxygen transfer depends on the total volume of gas in the system as well as on the average bubble size, high mass-transfer rates are achieved at high gas hold-ups. Gas hold-up values are very difficult to predict and may be anything from very low (0.01) up to a maximum in commercial-scale stirred fermenters of about 0.2. Under normal operating conditions, a significant fraction of the oxygen in fermentation vessels is contained in the gas hold-up. For example if the culture is sparged with air and the broth saturated with dissolved oxygen, for an air hold-up of only 0.03 about half the total oxygen in the system is in the gas phase.

While it is desirable to have small bubbles, there are practical limits. Bubbles << 1 mm diameter can become a nuisance in bioreactors. Oxygen concentration in these bubbles equilibrates with that in the medium within seconds, so that the gas hold-up no longer reflects the capacity of the system for mass transfer [9]. Problems with very small bubbles are exacerbated in viscous non-Newtonian broths; tiny bubbles remain lodged in these fluids for long periods of time because their velocity of rise is reduced. As a rule of thumb, relatively large bubbles must be employed in viscous cultures.

Bubble size also affects the value of k_L. In most fermentation broths, if the bubbles have diameters less than 2–3 mm, surface tension effects dominate the behaviour of the bubble surface. As a result, the bubbles behave as rigid spheres with immobile surfaces and no internal gas circulation. A rigid bubble surface gives lower k_L values; k_L decreases with decreasing bubble diameter below 2–3 mm. On the other hand, bubbles in fermentation media with sizes greater than about 3 mm develop internal circulation and relatively mobile surfaces, depending on liquid properties. Bubbles with mobile surfaces are able to wobble and move in spirals during free rise;

Figure 9.9 Flow pattern in stirred aerated bioreactors as a function of impeller speed N_i and gas flow rate F_g. (From A.W. Nienow, D.J. Wisdom and J.C. Middleton, 1978, The effect of scale and geometry on flooding, recirculation, and power in gassed stirred vessels. *Proc. 2nd Eur. Conf. on Mixing*, Cambridge, England, 1977, pp. F1-1–F1-16, BHRA Fluid Engineering, Cranfield.)

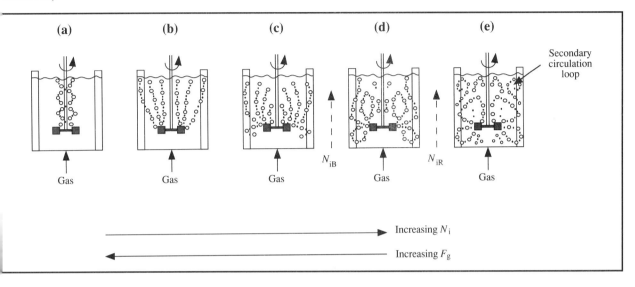

this behaviour has a marked beneficial effect on k_L and rate of mass transfer.

To summarise the influence of bubble size on oxygen mass transfer, small bubbles are generally beneficial because of the increased gas hold-up and larger interfacial surface-area. However, k_L for bubbles less than about 3 mm diameter is reduced due to surface effects. Very small bubbles << 1 mm should be avoided, especially in viscous broth.

9.6.2 Sparging, Stirring and Medium Properties

Because bubble size is such a critical parameter affecting oxygen transfer, it is useful to consider the physical processes in fermenters which determine bubble size. These processes include *bubble formation*, *gas dispersion* and *coalescence*.

Air bubbles are formed at the sparger. There exists a large variety of sparger designs, including simple open pipes, perforated tubes, porous diffusers, and complex two-phase injector devices. Bubbles leaving the sparger usually fall within a relatively narrow size range depending on the sparger type. This size range is a significant parameter in design of air-driven fermenters such as bubble and airlift columns because there is no other mechanism for bubble dispersion in these reactors. However in stirred vessels, design of the sparger and the mechanics of bubble formation are of secondary importance compared with the effects of the impeller. As a result of contin-

ual bubble break-up and dispersion by the impeller and coalescence from bubble collisions, bubble sizes in stirred reactors often bear little relationship to those formed at the sparger. Coalescence of small bubbles into bigger bubbles is generally undesirable because it reduces the total interfacial area and gas hold-up. Frequency of coalescence depends mainly on the liquid properties. In a coalescing liquid, a large fraction of bubble collisions results in the formation of bigger bubbles, while in non-coalescing liquids colliding bubbles do not coalesce readily. Salts act to suppress coalescence; therefore, fermentation media are usually non-coalescing to some extent depending on composition. This is an advantage for oxygen mass transfer.

Flow patterns set up in stirred vessels in the absence of aeration have been described in Chapter 7. As illustrated in Figure 9.9, when air is sparged different gas flow patterns develop depending on the relative rates of gas input and stirring. As shown in Figure 9.9(a), if the agitator speed N_i is low and the gas feed rate F_g is high, gas envelopes the impeller without dispersion and the flow pattern is dominated by air flow up the stirrer shaft. *Impeller flooding* is said to occur; this means that the gas-handling capacity of the stirrer is smaller than the amount introduced. Flooding should be avoided because an impeller surrounded by gas no longer contacts the liquid properly, resulting in poor mixing and gas dispersion. As the impeller speed increases, gas is captured behind the agitator blades and is dispersed into the liquid. N_{iB} is the minimum stirrer speed required to just completely disperse the gas,

Figure 9.10 Dependence of $k_L a$ on stirrer speed N_i in an agitated tank with 0.381 cm s^{-1} superficial gas velocity. Symbols represent various concentrations of sodium polyacrylate and sodium carboxyl methyl cellulose in aqueous solution. (From H. Yagi and F. Yoshida, 1975, Gas absorption by Newtonian and non-Newtonian fluids in sparged agitated vessels. *Ind. Eng. Chem. Process Des. Dev.* **14**, 488–493.)

Figure 9.11 Effect on $k_L a$ of number of impellers used to mix a viscous mycelial suspension. The ratio of liquid depth to tank diameter is 2.41. P_g is power consumption with gas sparging. (○) Two impellers, apparent viscosity = 500 cP, impeller spacing/impeller diameter = 2.06; (□) three impellers, apparent viscosity = 500 cP, impeller spacing/impeller diameter = 1.37; (●) two impellers, apparent viscosity = 700 cP, impeller spacing/impeller diameter = 2.06; (■) three impellers, apparent viscosity = 700 cP, impeller spacing/impeller diameter = 1.37. (From H. Taguchi, 1971, The nature of fermentation fluids. *Adv. Biochem. Eng.* **1**, 1–30.)

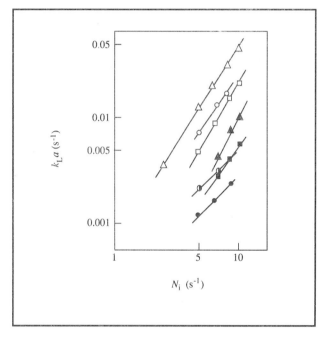

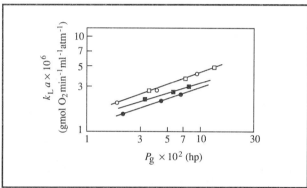

including below the impeller. The minimum agitator tip speed for dispersion of air bubbles even at very low gas flow rates has been estimated by Westerterp *et al.* [10] as roughly 1.5–2.5 m s^{-1}; tip speed $= \pi N_i D_i$ where N_i is impeller speed and D_i is impeller diameter. With further increases in stirrer speed, small recirculation patterns start to emerge as indicated in Figure 9.9(e); N_{iR} is the speed at which gross recirculation of gas back to the agitator starts to occur.

Gas dispersion in stirred vessels takes place mainly in the immediate vicinity of the impeller. As shown in Figure 7.27 (see p. 154), when gas is present in stirred liquids it is drawn into low-pressure cavities behind the stirrer blades. Gas from the sparger together with a large fraction of the recirculating gas in the system is entrained in these cavities. Gas contacting the impeller blades leads to a decrease in the friction or drag coefficient associated with impeller rotation and a concomitant reduction in power consumption. As the impeller blades rotate at high speed, small gas bubbles are thrown out from the back of the cavities into the bulk liquid under the influence of

dispersion processes that are not yet completely understood. In a non-coalescing liquid, the bubbles remain close to the size produced at the back of the cavities. Because bubbles formed at the sparger are immediately drawn into the impeller zone, dispersion of gas in stirred vessels is largely independent of sparger design; when the sparger is located under the stirrer, it has been shown that sparger type does not significantly affect mass transfer.

Under typical fermenter operating conditions, increasing the stirrer speed improves the value of $k_L a$, as shown in Figure 9.10. In contrast, except at very low sparging rates, increasing the gas flow is generally considered to exert only a minor influence on $k_L a$. As indicated in Figure 9.11, increasing the number of impellers on the stirrer shaft does not necessarily improve $k_L a$ even though the power consumption is increased. The quantity of gas passing through the upper impellers is small compared with the lower impeller, so that any additional gas dispersion is not significant.

9.6.3 Antifoam Agents

Most cell cultures produce a variety of foam-producing and foam-stabilising agents, such as proteins, polysaccharides and fatty acids. Foam build-up in fermenters is very common, particularly in aerobic systems. Foaming causes a range of reactor operating problems; foam control is therefore an important consideration in fermentation design. Excessive foam overflowing from the top of the fermenter provides a route for entry of contaminating organisms and causes blockage of outlet gas lines. Liquid and cells trapped in the foam represent a loss of bioreactor volume; conditions in the foam may not be favourable for metabolic activity. In addition, fragile cells can be damaged by collapsing foam.

Addition of special antifoam compounds to the medium is the most common method of reducing foam build-up in fermenters. However, antifoam compounds affect the surface chemistry of bubbles and their tendency to coalesce, and have a significant effect on $k_L a$. Most antifoam agents are strong surface tension-lowering substances. Decrease in surface tension reduces the average bubble diameter, thus producing higher values of a. However, this is countered by a reduction in mobility of the gas–liquid interface which lowers the value of k_L. With most silicon-based antifoams, the decrease in k_L is generally larger than the increase in a so that, overall, $k_L a$ is reduced [11, 12]. The resulting decrease in rate of oxygen transfer can be dramatic, by up to a factor of 10.

In order to maintain the non-coalescing character of the medium and high $k_L a$ values, mechanical rather than chemical methods of disrupting foam are preferred because the liquid properties are not changed. Mechanical foam breakers, such as high-speed discs rotating at the top of the vessel and centrifugal foam destroyers, are suitable when foam development is moderate. However, some of these devices need large quantities of power to operate in commercial-scale vessels; their limited foam-destroying capacity is also a problem with highly-foaming cultures. In many cases, use of chemical antifoam agents is unavoidable.

9.6.4 Temperature

The temperature of aerobic fermentations affects both the solubility of oxygen C_{AL}^* and the mass-transfer coefficient k_L. Increasing temperature causes C_{AL}^* to drop, so that the driving force for mass transfer $(C_{AL}^* - C_{AL})$ is reduced. At the same time, diffusivity of oxygen in the liquid film surrounding the bubbles is increased, resulting in an increase in k_L. The net effect of temperature on oxygen transfer depends on the range of temperature considered. For temperatures between 10°C and 40°C, increase in temperature is more likely to increase the rate of oxygen transfer. Above 40°C the solubility of oxygen drops significantly, adversely affecting the driving force and rate of mass transfer.

9.6.5 Gas Pressure and Oxygen Partial Pressure

Pressure and oxygen partial pressure of the gas used to aerate fermenters affect the value of C_{AL}^*. The equilibrium relationship between these parameters for dilute liquid solutions is given by *Henry's law*:

$$p_{AG} = p_T \, y_{AG} = H \, C_{AL}^*$$

$$(9.43)$$

where p_{AG} is the partial pressure of component A in the gas, p_T is total gas pressure, y_{AG} is the mole fraction of A in the gas, H is *Henry's constant* which is a function of temperature, and C_{AL}^* is the solubility of component A in the liquid. From Eq. (9.43), if the total gas pressure p_T or concentration of oxygen in the gas y_{AG} is increased at constant temperature, C_{AL}^* and therefore the mass-transfer driving force $(C_{AL}^* - C_{AL})$ also increase.

In some fermentations, oxygen-enriched air or pure oxygen is used to improve mass transfer. Alternatively, oxygen solubility is increased by sparging compressed air at high pressure. Both these strategies increase the operating cost of the fermenter; it is also possible in some cases that the culture will suffer inhibitory effects from exposure to very high oxygen partial pressures.

9.6.6 Presence of Cells

Oxygen transfer is influenced by the presence of cells in fermentation broths; the nature of the effect depends on the species of organism, its morphology and concentration. Cells with complex morphology generally lead to lower transfer rates. Cells interfere with bubble break-up and coalescence; cells, proteins and other molecules which adsorb at gas–liquid interfaces also cause *interfacial blanketing* which reduces the contact area between gas and liquid. The quantitative effect of interfacial blanketing is highly system-specific. Because concentrations of cells, substrates and products change throughout batch fermentation, the value of $k_L a$ can also vary. An example of change in $k_L a$ due to these factors is shown in Figure 9.12.

9.7 Measuring Dissolved-Oxygen Concentrations

The concentration of dissolved oxygen C_{AL} in fermenters is normally measured using a *dissolved-oxygen electrode*. There are two types in common use: *galvanic electrodes* and *polarographic*

Figure 9.12 Variation in $k_L a$ during a batch 300-l streptomycete fermentation. (From C.M. Tuffile and F. Pinho, 1970, Determination of oxygen-transfer coefficients in viscous streptomycete fermentations. *Biotechnol. Bioeng.* **12**, 849–871.)

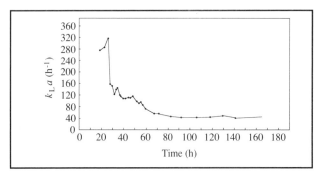

Figure 9.13 Diffusion of oxygen from bulk liquid to the cathode of an oxygen electrode.

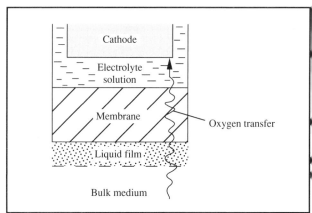

electrodes. Details of the construction and operating principles of these probes can be found in other references [13, 14]. In both designs, a membrane which is permeable to oxygen separates the fermentation fluid from the electrode. As illustrated in Figure 9.13, oxygen diffuses through the membrane to the cathode, where it reacts to produce a current between anode and cathode proportional to the oxygen partial pressure in the fermentation broth. An electrolyte solution in the electrode supplies ions which take part in the reactions, and must be replenished at regular intervals. Depending on flow conditions in the bulk medium, liquid properties, and the rate of oxygen utilisation by the probe, a liquid boundary layer develops at the interface between the solid probe and the liquid.

As indicated in Figure 9.13, supply of oxygen molecules from the bulk medium to the cathode is in itself a mass-transfer process. Because there is no bulk fluid motion in the membrane or electrolyte solution and little motion in the liquid film at the membrane interface, operation of the probe relies on diffusion of oxygen across these thicknesses. This takes time, so the response of an electrode to sudden changes in dissolved-oxygen level is subject to delay. The *electrode response time* can be measured by quickly transferring the probe from a beaker containing medium saturated with nitrogen to one saturated with air. The response time is defined as the time taken for the probe to indicate 63% of the total change in dissolved-oxygen level. For commercially-available steam-sterilisable electrodes, response times are usually 10–100 s. Electrode response can usually be improved if the bulk liquid is stirred rapidly; this decreases the thickness of the liquid film at the membrane surface. Micro-probes for dissolved-oxygen measurement are also available. The smaller cathode size and lower rate of oxygen consumption by these instruments means that their response is quicker; so much so that micro-probes can be used to measure dissolved oxygen levels in unagitated systems. Steam-sterilisable probes are inserted directly into fermentation vessels for on-line monitoring of dissolved oxygen. Repeated calibration of dissolved-oxygen probes is usually necessary; fouling by cells attaching to the membrane surface, electronic noise due to air bubbles passing close to the membrane, and signal drift are the main operating problems.

Both galvanic and polarographic electrodes measure the partial pressure of dissolved oxygen or *oxygen tension* in the fermentation broth, not the dissolved-oxygen concentration. To convert this to dissolved-oxygen concentration, it is necessary to know the solubility of oxygen in the liquid at the temperature and pressure of measurement.

9.8 Estimating Oxygen Solubility

The concentration difference $(C_{AL}^* - C_{AL})$ is the driving force for oxygen mass transfer. Because this difference is usually very small, it is important that the solubility C_{AL}^* be known accurately; small errors in C_{AL}^* will result to large errors in $(C_{AL}^* - C_{AL})$. Experimentally-determined values for solubility of oxygen in water can be found in many literature sources, e.g. [15–17]; Table 9.1 shows the solubility of oxygen in water under 1 atm oxygen pressure at various temperatures. However, fermentations are not carried out using pure water and pure oxygen. Because oxygen partial pressure in the gas phase and presence of dissolved material in the liquid are major factors affecting oxygen solubility, the values given in Table 9.1 cannot be applied directly to bioprocessing systems.

Table 9.1   Solubility and Henry's constant for oxygen in pure water under 1 atm oxygen pressure

(Calculated from data in International Critical Tables, *1928, vol. III, p. 257. McGraw-Hill, New York)*

Temperature (°C)	Oxygen solubility ($kg\,m^{-3}$)	Henry's constant ($atm\,m^3\,kg^{-1}$)
	7.03×10^{-2}	14.2
	5.49×10^{-2}	18.2
	4.95×10^{-2}	20.2
	4.50×10^{-2}	22.2
	4.14×10^{-2}	24.2
	4.07×10^{-2}	24.6
	4.01×10^{-2}	24.9
	3.95×10^{-2}	25.3
	3.89×10^{-2}	25.7
	3.84×10^{-2}	26.1
	3.58×10^{-2}	27.9
	3.37×10^{-2}	29.7

Table 9.2   Solubility of oxygen in water under 1 atm air pressure

(Calculated from data in Table 9.1 and Henry's law)

Temperature (°C)	Oxygen solubility ($kg\,m^{-3}$)
	1.48×10^{-2}
	1.15×10^{-2}
	1.04×10^{-2}
	9.45×10^{-3}
	8.69×10^{-3}
	8.55×10^{-3}
	8.42×10^{-3}
	8.29×10^{-3}
	8.17×10^{-3}
	8.05×10^{-3}
	7.52×10^{-3}
	7.07×10^{-3}

9.8.1 Effect of Oxygen Partial Pressure

According to the *International Critical Tables* [18], the mole fraction of oxygen in air is 0.2099, so the partial pressure of oxygen at 1 atm air pressure is 0.2099 atm. At a given temperature, the effect of gas-phase oxygen partial pressure on solubility is given by Henry's law, Eq. (9.43). Therefore, the solubility of oxygen in water under 1 atm air pressure is 0.2099

times that under 1 atm pure oxygen. Values for solubility of oxygen in water sparged with air are given in Table 9.2.

9.8.2 Effect of Temperature

The variation of oxygen solubility with temperature is shown in Tables 9.1 and 9.2 for water in the range 0–40°C. Solubility falls with increasing temperature. Oxygen solubility in pure water between 0° and 36°C has been correlated by the following equation [15]:

$$C_{AL}^* = 14.161 - 0.3943\,T + 0.007714\,T^2 - 0.0000646\,T^3 \tag{9.44}$$

where C_{AL}^* is oxygen solubility in units of $mg\,l^{-1}$, and T is temperature in °C.

9.8.3 Effect of Solutes

Presence of solutes such as salts, acids and sugars has a significant effect on oxygen solubility in water, as indicated in Tables 9.3 and 9.4. These data indicate that oxygen solubility is decreased by the ions and sugars normally added to fermentation media. The effect on oxygen solubility of ionic and non-ionic solutes such as molasses, corn-steep liquor, protein and antifoam agents is reported in several publications [19–24]. Quicker *et al.* [23] have developed an empirical correlation to correct values of oxygen solubility in water for the effects of cations, anions and sugars:

Table 9.3 Solubility of oxygen in aqueous solutions at 25°C under 1 atm oxygen pressure

(Calculated from data in International Critical Tables, *1928, vol. III, p. 271. McGraw-Hill, New York)*

Concentration	Oxygen solubility (kg m^{-3})		
(M)	HCl	$\frac{1}{2}\,H_2SO_4$	NaCl
0	4.14×10^{-2}	4.14×10^{-2}	4.14×10^{-2}
0.5	3.87×10^{-2}	3.77×10^{-2}	3.43×10^{-2}
1.0	3.75×10^{-2}	3.60×10^{-2}	2.91×10^{-2}
2.0	3.50×10^{-2}	3.28×10^{-2}	2.07×10^{-2}

Table 9.4 Solubility of oxygen in aqueous solutions of sugars under 1 atm oxygen pressure

(Calculated from data in International Critical Tables, *1928, vol. III, p. 272. McGraw-Hill, New York)*

Sugar	Concentration (gmol per kg H_2O)	Temperature (°C)	Oxygen solubility (kg m^{-3})
Glucose	0	20	4.50×10^{-2}
	0.7	20	3.81×10^{-2}
	1.5	20	3.18×10^{-2}
	3.0	20	2.54×10^{-2}
Sucrose	0	15	4.95×10^{-2}
	0.4	15	4.25×10^{-2}
	0.9	15	3.47×10^{-2}
	1.2	15	3.08×10^{-2}

$$\log_{10}\left(\frac{C_{AL0}^*}{C_{AL}^*}\right) = 0.5 \sum_i H_i z_i^2 \, C_{iL} + \sum_j K_j \, C_{jL}$$

$$(9.45)$$

where:

C_{AL0}^* = oxygen solubility at zero solute concentration (mol m^{-3})

C_{AL}^* = oxygen solubility (mol m^{-3})

H_i = constant for ionic component i $(\text{m}^3\,\text{mol}^{-1})$

z_i = valency of ionic component i

C_{iL} = concentration of ionic component i in the liquid (mol m^{-3})

K_j = constant for non-ionic component j $(\text{m}^3\,\text{mol}^{-1})$

C_{jL} = concentration of non-ionic component j in the liquid (mol m^{-3})

Values of H_i and K_j for use in Eq. (9.45) are listed in Table 9.5. In a typical fermentation medium, oxygen solubility is between 5% and 25% lower than in water as a result of solu effects.

9.9 Mass-Transfer Correlations

In general, there are two approaches to evaluating k_L and calculation using empirical correlations, and experimen measurement. In both cases, separate determination of k_L an a is laborious and sometimes impossible. It is convenie therefore to directly evaluate the product $k_L a$; the combin term $k_L a$ is often referred to as the mass-transfer coefficie rather than just k_L. In this section we consider methods for ca culating $k_L a$ using published correlations.

In aerobic fermenters, k_L and a are dependent on the hydr dynamic conditions around the gas bubbles. Relationship between $k_L a$ and parameters such as bubble diameter, liqu velocity, density, viscosity and oxygen diffusivity have be investigated extensively, and empirical correlations betwe

Table 9.5 Values of H_i and K_j in Eq. (9.45) at 25°C

(From A. Schumpe, I. Adler and W.-D. Deckwer, 1978, Solubility of oxygen in electrolyte solutions, Biotechnol. Bioeng. 20, 145–150; and G. Quicker, A. Schumpe, B. König and W.-D. Deckwer, 1981, Comparison of measured and calculated oxygen solubilities in fermentation media, Biotechnol. Bioeng. 23, 635–650)

Cation	$H_i \times 10^3$ $(m^3\,mol^{-1})$	Anion	$H_i \times 10^3$ $(m^3\,mol^{-1})$	Sugar	$K_j \times 10^3$ $(m^3\,mol^{-1})$
H^+	−0.774	OH^-	0.941	Glucose	0.119
K^+	−0.596	Cl^-	0.844	Lactose	0.197
Na^+	−0.550	CO_3^{2-}	0.485	Sucrose	0.149*
NH_4^+	−0.720	SO_4^{2-}	0.453		
NEt_4^+	−0.912	NO_3^-	0.802		
Mg^{2+}	−0.314	HCO_3^-	1.058		
Ca^{2+}	−0.303	$H_2PO_4^-$	1.037		
Mn^{2+}	−0.311	HPO_4^{2-}	0.485		
		PO_4^{3-}	0.320		

* Approximately valid for sucrose concentrations up to about $200\,g\,l^{-1}$.

mass-transfer coefficients and important operating variables have been developed. Theoretically, these correlations allow prediction of mass-transfer coefficients based on information gathered from a large number of previous experiments. In practice, however, the accuracy of published correlations applied to biological systems is generally poor. The main reason is that mass transfer is strongly affected by the additives usually present in fermentation media. Because fermentation liquids contain varying levels of substrates, products, salts, surface-active agents and cells, the surface chemistry of bubbles and therefore the mass-transfer situation become very complex. Most available correlations for oxygen mass-transfer coefficients were determined using pure air in water, and it is very difficult to correct these correlations for different liquid compositions. The effective mass-transfer area a is also influenced by interfacial blanketing of the bubble–liquid surface by cells and other components of the broth. Prediction of $k_L a$ under these conditions is problematic.

When mass-transfer coefficients are required for large-scale equipment, another factor related to hydrodynamic conditions limits the applicability of published correlations. Most studies of oxygen mass-transfer have been carried out in laboratory-scale stirred reactors, which are characterised by high turbulence throughout most of the vessel. The gas phase in small-scale agitated tanks is well dispersed; break-up and coalescence of bubbles occur constantly due to the high level of turbulence and frequent bubble collisions. In contrast,

bubbles in industrial-scale stirred tanks are mostly in free rise. The result is that, due to the different hydrodynamic regimes present in small- and large-scale vessels, mass-transfer correlations for stirred tanks developed in the laboratory tend to overestimate the oxygen-transfer capacity of commercial-scale systems. Better results can be achieved with a two-compartment model of large fermenters, applying different correlations for the mixed zone close to the impeller and the bubble zone away from the impeller. Application of this technique for calculation of oxygen transfer-rate is described by Oosterhuis and Kossen [25].

In this section we will consider one empirical correlation for oxygen mass transfer; its application in fermentation systems is subject to the problems mentioned above. A widely-used correlation for stirred vessels relates $k_L a$ directly to gas velocity and power input to the stirrer; all the effects of flow and turbulence on bubble dispersion and the mass-transfer boundary layer are represented by the power term. An expression for stirred fermenters containing non-coalescing non-viscous media is [26]:

$$k_L a = 2.0 \times 10^{-3} \left(\frac{P}{V} \right)^{0.7} u_G^{0.2} .$$

$$(9.46)$$

In Eq. (9.46), $k_L a$ is the combined mass-transfer coefficient in units of s^{-1}, P is the power dissipated by the stirrer in W, and

V is the fluid volume in m^3. u_G is the *superficial gas velocity* in m s^{-1}; superficial gas velocity is defined as the volumetric gas flow rate divided by the cross-sectional area of the fermenter. Eq. (9.46) was obtained with water containing ions in vessel volumes $2 \times 10^{-3} < V < 4.4$ m^3, and $500 < P/V < 10\,000$ W m^{-3}.

Experimental results for $k_L a$ are reported to agree with Eq. (9.46) to within 20–40%; however application of this correlation to production-size vessels up to 25 m^3 in volume has been found to overestimate rate of oxygen transfer by about 100% [25]. Note that this correlation does not depend on the sparger or stirrer design; the power dissipated by the impeller determines $k_L a$ independent of stirrer type. Like most published correlations, Eq. (9.46) does not take into account the non-Newtonian behaviour of many culture fluids, the effect of added sugars and antifoam agents, and the presence of solids such as cells.

Eq. (9.46) suggests that $k_L a$ can be increased by raising the superficial gas velocity in the reactor. However, the exponent on u_G in Eq. (9.46) is much less than unity, so the effect of gas flow rate is relatively minor. There is usually limited scope for increasing u_G depending on vessel configuration; at high u_G the liquid contents can be blown out of the fermenter. The maximum operating value for u_G also depends on the stirrer speed; as discussed in Section 9.6.2, unless the impeller is able to disperse all the air impinging on it, impeller flooding will occur. Because the exponent on P in Eq. (9.46) is also less than one, increasing $k_L a$ by raising either the air flow rate or power input becomes progressively less efficient and more costly as the inputs increase.

9.10 Measurement of $k_L a$

Because of the difficulty in predicting $k_L a$ in bioreactors using correlations, mass-transfer coefficients for oxygen are usually determined experimentally. This is not without its own problems however, as discussed below. Whatever method is used to measure $k_L a$, the measurement conditions should match those in the fermenter during normal operation. Techniques for measuring $k_L a$ have been reviewed by van't Riet [26].

9.10.1 Oxygen-Balance Method

This technique is based on the equation for gas–liquid mass transfer, Eq. (9.37). In the experiment, the oxygen content of gas streams flowing to and from the fermenter are measured. From a mass balance at steady state, the difference in oxygen

flow between inlet and outlet must be equal to the rate of oxygen transfer from gas to liquid:

$$N_A = \frac{1}{V_L}\left[(F_g\,C_{AG})_i - (F_g\,C_{AG})_o\right]$$

(9.47)

where V_L is the volume of liquid in the fermenter, F_g is the volumetric gas flow rate, C_{AG} is the gas-phase concentration of oxygen, and subscripts i and o refer to inlet and outlet gas streams, respectively. The first term on the right-hand side of Eq. (9.47) represents the rate at which oxygen enters the fermenter in the inlet-gas stream; the second term is the rate at which oxygen leaves. The difference between them is the rate at which oxygen is transferred out of the gas into the liquid, N_A. Because gas concentrations are generally measured as partial pressures, the ideal gas law Eq. (2.32) can be incorporated into Eq. (9.47) to obtain an alternative expression:

$$N_A = \frac{1}{R\,V_L}\left[\left(\frac{F_g p_{AG}}{T}\right)_i - \left(\frac{F_g p_{AG}}{T}\right)_o\right]$$

(9.48)

where R is the universal gas constant (see Table 2.5 on p. 20), p_{AG} is the oxygen partial pressure in the gas and T is absolute temperature. Because oxygen partial pressures in the inlet and exit gas streams are usually not very different during operation of fermenters, they must be measured very accurately, e.g. using mass spectrometry. The temperature and flow rate of the gases must also be measured carefully to ensure an accurate value of N_A is determined. Once N_A is known and C_{AL} and C^*_{AL} found using the methods described in Sections 9.7 and 9.8, $k_L a$ can be calculated from Eq. (9.37).

The steady-state oxygen-balance method is the most reliable procedure for measuring $k_L a$, and allows determination from a single-point measurement. An important advantage is that the method can be applied to fermenters during normal operation. It depends, however, on accurate measurement of gas composition, flow rate, pressure and temperature; large errors as high as $\pm$ 100% can be introduced if measurement techniques are inadequate. Considerations for design of laboratory equipment to ensure accurate oxygen uptake measurements are described by Brooks *et al.* [27].

9.10.2 Dynamic Method

This method for measuring $k_L a$ is based on an unsteady-state mass balance for oxygen. The main advantage of the dynamic

Figure 9.14 Variation of oxygen tension for dynamic measurement of $k_L a$.

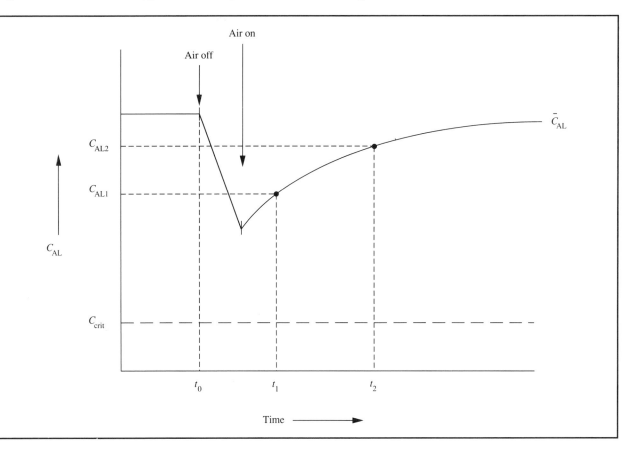

method over the steady-state technique is the low cost of the equipment needed.

There are several different versions of the dynamic method; only one will be described here. Initially, the fermenter contains cells in batch culture. As shown in Figure 9.14, at some time t_0 the broth is de-oxygenated either by sparging nitrogen into the vessel or by stopping the air flow if the culture is oxygen-consuming. Dissolved-oxygen concentration C_{AL} drops during this period. Air is then pumped into the broth at a constant flow-rate and the increase in C_{AL} monitored as a function of time. It is important that the oxygen concentration remains above C_{crit} so that the rate of oxygen uptake by the cells is independent of oxygen level. Assuming re-oxygenation of the broth is fast relative to cell growth, the dissolved-oxygen level will soon reach a steady state value $\bar{C}_{AL}$ which reflects a balance between oxygen supply and oxygen consumption in the system. C_{AL1} and C_{AL2} are two oxygen concentrations measured during re-oxygenation at times t_1 and t_2, respectively. We can develop an equation for $k_L a$ in terms of these experimental data.

During the re-oxygenation step, the system is not at steady state. The rate of change in dissolved-oxygen concentration during this period is equal to the rate of oxygen transfer from gas to liquid, minus the rate of oxygen uptake by the cells:

$$\frac{dC_{AL}}{dt} = k_L a (C_{AL}^* - C_{AL}) - q_O x$$

$$(9.49)$$

where $q_O x$ is the rate of oxygen consumption. We can determine an expression for $q_O x$ by considering the final steady dissolved-oxygen concentration, $\bar{C}_{AL}$. When $C_{AL} = \bar{C}_{AL}$, $dC_{AL}/dt = 0$ because there is no change in C_{AL} with time. Therefore, from Eq. (9.49):

$$q_O x = k_L a (C_{AL}^* - \bar{C}_{AL}).$$

$$(9.50)$$

Substituting this result into Eq. (9.49) and cancelling the

$k_L a C_{AL}^*$ terms gives:

$$\frac{dC_{AL}}{dt} = k_L a (\bar{C}_{AL} - C_{AL}).$$

(9.51)

Assuming $k_L a$ is constant with time, we can integrate Eq. (9.51) between t_1 and t_2 using the integration rules described in Appendix D. The resulting equation for $k_L a$ is:

$$k_L a = \frac{\ln\left(\dfrac{\bar{C}_{AL} - C_{AL1}}{\bar{C}_{AL} - C_{AL2}}\right)}{t_2 - t_1}.$$

(9.52)

$k_L a$ can be estimated using two points from Figure 9.14 or, more accurately, from several values of $(C_{AL1},\ t_1)$ and (C_{AL2}, t_2). When

$$\ln\left(\frac{\bar{C}_{AL} - C_{AL1}}{\bar{C}_{AL} - C_{AL2}}\right)$$

is plotted against $(t_2 - t_1)$ as shown in Figure 9.15, the slope is

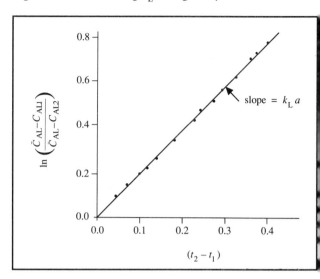

Figure 9.15 Evaluating $k_L a$ using the dynamic method.

$k_L a$. Eq. (9.52) can be applied to actively respiring cultures, or to systems without oxygen uptake. In the latter case, $\bar{C}_{AL} = C_{AL}^*$.

Example 9.2 Estimating $k_L a$ using the dynamic method

A 20-l stirred fermenter containing a *Bacillus thuringiensis* culture at 30°C is used for production of microbial insecticide. $k_L a$ is determined using the dynamic method. Air flow is shut off for a few minutes and the dissolved-oxygen level drops; the air supply is then re-connected. When steady state is established, the dissolved-oxygen tension is 78% air saturation. The following results are obtained.

Time (s)	5	15
Oxygen tension (% air saturation)	50	66

(a) Estimate $k_L a$.
(b) An error is made determining the steady-state oxygen level which, instead of 78%, is taken as 70%. What is the percentage error in $k_L a$ resulting from this 10% error in $\bar{C}_{AL}$?

Solution:

(a) $\bar{C}_{AL} = 78\%$ air saturation. Let us define $t_1 = 5$ s, $C_{AL1} = 50\%$, $t_2 = 15$ s and $C_{AL2} = 66\%$. From Eq. (9.52):

$$k_L a = \frac{\ln\left(\dfrac{78 - 50}{78 - 66}\right)}{(15 - 5)\,\text{s}} = 0.085\ \text{s}^{-1}.$$

(b) If $\bar{C}_{AL}$ is taken to be 70% air saturation:

$$k_L a = \frac{\ln\left(\dfrac{70 - 50}{70 - 66}\right)}{(15 - 5)\,\text{s}} = 0.16\,\text{s}^{-1}.$$

The error in $k_L a$ is almost 100%. This example illustrates how important it is to obtain accurate data for $k_L a$ determination.

An oxygen probe with fast response time is required for measurement of C_{AL}, otherwise the dynamic method will not give accurate results. Frequently, however, the response time of the electrode is similar in magnitude to the time for mass transfer. When this is the case, values of C_{AL} measured during the experiment do not reflect the instantaneous oxygen concentration. Electrode response time should always be measured in conjunction with the dynamic method for $k_L a$ determination. In principle, the probe response time (Section 9.7) should be much smaller than the mass-transfer response time, $1/k_L a$. Van't Riet [26] has estimated that acceptable results are obtained using commercial electrodes with response times between 2 and 3 s for $k_L a$ values up to 0.1 s^{-1}. When the response time is closer to $1/k_L a$, the measurements should be corrected for electrode dynamics. Further discussion and analysis of this problem can be found in the literature [26, 28–34].

Another factor affecting accuracy of the dynamic method is the average residence time of gas in the system when de-oxygenation with nitrogen is followed by a switch to aeration at the beginning of the measurement. Because there is a nitrogen gas hold-up in the vessel when air is re-introduced, measurement of C_{AL} does not reflect the kinetics of simple oxygen transfer until a hold-up of air is established. This takes longer in large vessels; dynamic methods for $k_L a$ measurement are therefore restricted to vessels with height less than 1 m [35]. Dynamic methods are inappropriate for viscous broths for similar reasons; the large residence times of bubbles in viscous broths affect the accuracy of measurement [9]. Corrections to account for gas-phase dynamics in small vessels are described by Dunn and Einsele [29] and Dang *et al.* [32]. When the gas velocity is low and the hold-up high, such large corrections are required that the original measurements become meaningless.

9.10.3 Sulphite Oxidation

This method is based on oxidation of sodium sulphite to sulphate in the presence of a catalyst such as Cu^{2+}. Although the sulphite method has been used extensively, the results appear to depend on operating conditions in an unknown way, and usually give higher $k_L a$ values than other techniques. Accordingly, its application is discouraged [26].

9.11 Oxygen Transfer in Large Vessels

Special difficulties are associated with measurement of oxygen transfer in large fermenters. Problems of residual gas hold-up with the dynamic method have already been mentioned. Implicit in application of the experimental techniques described in Section 9.10 is the assumption that $k_L a$ is constant throughout the entire reactor. This requires that the gas and liquid phases be perfectly mixed with uniform turbulence. In commercial-size reactors (> 10 m^3), perfect mixing is difficult to achieve; as a result the calculated value of $k_L a$ may depend on the position in the tank where the measurements of C_{AL} are made. Even when mixing is good, variation in composition of the gas phase is inevitable as changing static pressure and continuing dissolution of oxygen reduce the oxygen partial pressure in the bubbles as they rise.

Significant variation between inlet and outlet oxygen partial pressures affects the value for C_{AL}^* used in mass-transfer calculations. Allowance can be made for this by determining an average concentration driving force $(C_{AL}^* - C_{AL})$ across the system. A suitable average is the *logarithmic-mean concentration difference*, $(C_{AL}^* - C_{AL})_L$:

$$(C_{AL}^* - C_{AL})_L = \frac{(C_{AL}^* - C_{AL})_o - (C_{AL}^* - C_{AL})_i}{\ln\left[\dfrac{(C_{AL}^* - C_{AL})_o}{(C_{AL}^* - C_{AL})_i}\right]}.$$

$$(9.53)$$

In Eq. (9.53), subscripts i and o represent the gas inlet and outlet ends of the vessel, respectively.

9.12 Summary of Chapter 9

At the end of Chapter 9 you should:

(i) be able to describe the *two-film theory* of mass transfer between phases;

(ii)   know *Fick's law* in terms of the *diffusion coefficient*, $\mathscr{D}_{AB}$;
(iii) be able to describe in simple terms the mathematical analogy between mass, heat and momentum transfer;
(iv)   know the empirical equation for rate of mass transfer in terms of the *mass-transfer coefficient* and *concentration-difference driving force*;
(v)   know the importance of the *critical oxygen concentration*;
(vi)   be able to identify which steps are most likely to present major resistances to oxygen mass transfer from bubbles to cells;
(vii) understand how oxygen mass-transfer and $k_L a$ can limit the biomass density in fermenters;
(viii) understand the mechanisms of *gas dispersion* and *coalescence* in stirred fermenters and the importance of bubble size in determining *gas hold-up* and $k_L a$;
(ix)   know how temperature, total pressure, oxygen partial pressure and presence of dissolved and suspended material in the medium affect oxygen solubility and rates of oxygen mass transfer in fermenters; and
(x)   know the techniques and limitations of *steady-state* and *dynamic methods* for experimental determination of $k_L a$ for oxygen transfer.

Problems

9.1 Rate-controlling processes in fermentation

Serratia marcescens bacteria are used for production of threonine. The maximum specific oxygen uptake rate of *S. marcescens* in batch culture is 5 mmol O_2 g^{-1} h^{-1}. The bacteria are grown in a stirred fermenter to a cell density of 40 g l^{-1}; $k_L a$ under these circumstances is 0.15 s^{-1}. At the fermenter operating temperature and pressure, the solubility of oxygen in the culture liquid is 8×10^{-3} kg m^{-3}. Is the rate of cell metabolism limited by mass-transfer, or dependent solely on metabolic kinetics?

9.2 $k_L a$ required to maintain critical oxygen concentration

A genetically-engineered strain of yeast is cultured in a bioreactor at 30°C for production of heterologous protein. The oxygen requirement is 80 mmol l^{-1} h^{-1}; the critical oxygen concentration is 0.004 mM. The solubility of oxygen in the fermentation broth is estimated to be 10% lower than in water due to solute effects.

(a) What is the minimum mass-transfer coefficient necessary to sustain this culture if the reactor is sparged with air at approximately 1 atm pressure?

(b) What mass-transfer coefficient is required if pure oxygen is used instead of air?

9.3 Single-point $k_L a$ determination using the oxygen-balance method

A 200-litre stirred fermenter contains a batch culture of *Bacillus subtilis* bacteria at 28°C. Air at 20°C is pumped into the vessel at a rate of 1 vvm; (vvm stands for volume of gas per volume of liquid per minute). The average pressure in the fermenter is 1 atm. The volumetric flow rate of off-gas from the fermenter is measured as 189 l min^{-1}. The exit gas stream is analysed for oxygen and is found to contain 20.1% O_2. The dissolved-oxygen concentration in the broth is measured using an oxygen electrode as 52% air saturation. The solubility of oxygen in the fermentation broth at 28°C and 1 atm air pressure is 7.8×10^{-3} kg m^{-3}.

(a) Calculate the oxygen transfer rate.
(b) Determine the value of $k_L a$ for the system.
(c) The oxygen analyser used to measure the exit gas composition has been incorrectly calibrated. If the oxygen content has been overestimated by 10%, what error is associated with the result for $k_L a$?

9.4 $k_L a$ measurement

Escherichia coli bacteria are cultured at 35°C in the following medium:

Component	g l^{-1}
glucose	20
sucrose	8.5
$CaCO_3$	1.3
$(NH_4)_2SO_4$	1.3
Na_2HPO_4	0.09
KH_2PO_4	0.12

The stirred fermenter used for this culture has an operating volume of 20 m^3 and a liquid height of 3.5 m. Air at 25°C is sparged into the bottom of the vessel at a rate of 25 m^3 min^{-1}. Oxygen tension in the fermenter is measured using polarographic electrodes located at the top and bottom of the vessel. At the top the reading is 50% air saturation; the reading at the bottom is 65%. The gas flow rate leaving the fermenter is measured with a rotary gas meter and is found to be 407 l s^{-1}. The oxygen content of the off-gas is 20.15%.

(a) Calculate the pressure at the sparger. (Static pressure p_s due to the height of liquid is given by the equation: $p_s = \rho g h$, where ρ is liquid density, g is gravitational acceleration, and h is liquid height.)
(b) Estimate the solubility of oxygen in the fermentation broth at 35°C and 1 atm air pressure using the correlation of Eq. (9.45).
(c) Estimate the solubility of oxygen at the bottom of the tank.
(d) Calculate the logarithmic-mean concentration driving force, $(C_{AL}^* - C_{AL})_L$.
(e) What is the oxygen transfer rate?
(f) Determine the value of $k_L a$.
(g) What is the maximum cell concentration that can be supported in this fermenter if the oxygen demand of the organism is 7.4 mmol $g^{-1} h^{-1}$?

9.5 Dynamic $k_L a$ measurement

The dynamic method is used to measure $k_L a$ in a fermenter operated at 30°C. Data for dissolved-oxygen concentration as a function of time during the re-oxygenation step is as follows:

Time (s)	C_{AL} (% air saturation)
10	43.5
15	53.5
20	60.0
30	67.5
40	70.5
50	72.0
70	73.0
100	73.5
130	73.5

The equilibrium concentration of oxygen in the broth is 7.9×10^{-3} kg m^{-3}. Determine $k_L a$.

9.6 Measurement of $k_L a$ as a function of stirrer speed: the oxygen-balance method of Mukhopadhyay and Ghose

Combining Eqs (9.37) and (9.48) gives:

$$k_L a\,(C_{AL}^* - C_{AL}) = \frac{1}{R V_L}\left[\left(\frac{F_g p_{AG}}{T}\right)_i - \left(\frac{F_g p_{AG}}{T}\right)_o\right].$$
(9P6.1)

Let us make the following assumptions.

(i) the volumetric flow rate of exit gas is equal to that of the inlet, i.e. $(F_g)_i = (F_g)_o = F_g$; and
(ii) inlet and outlet gases have the same density, i.e.

$$\left(\frac{p_T}{RT}\right)_i = \left(\frac{p_T}{RT}\right)_o = \frac{p_T}{RT}$$

where p_T is total gas pressure.

Applying these assumptions, Eq. (9P6.1) can be written as:

$$k_L a\,(C_{AL}^* - C_{AL}) = \frac{p_T F_g}{R T V_L}\,[\,(y_{AG})_i - (y_{AG})_o]$$
(9P6.2)

where y_{AG} is the mole fraction of oxygen in the gas. Rearranging gives an expression for C_{AL}:

$$C_{AL} = C_{AL}^* - \frac{p_T F_g}{k_L a R T V_L}\,[\,(y_{AG})_i - (y_{AG})_o]\,.$$
(9P6.3)

In Eq. (9P6.3) the only variables are C_{AL} and $(y_{AG})_o$; C_{AL}^*, p_T, F_g, $k_L a$, R, T, V_L and $(y_{AG})_i$ are assumed to be constant. Accordingly, a linear plot of C_{AL} versus $[\,(y_{AG})_i - (y_{AG})_o]$ should give a straight line with slope

$$\frac{-p_T F_g}{k_L a R T V_L}$$

and intercept C_{AL}^*.

Data collected during fermentation of *Pseudomonas ovalis* B1486 at varying stirrer speeds are given by Mukhopadhyay and Ghose [36]. The fermenter volume was 3 litres. Air flow into the vessel was maintained at 1 vvm (vvm means volume of gas per volume of liquid per minute). The air pressure was 3 atm and the temperature 29°C. The following data were measured.

Fermentation time (h)	Agitator speed					
	300 rpm		500 rpm		700 rpm	
	C_{AL} (ppm)	$(y_{AG})_o$	C_{AL} (ppm)	$(y_{AG})_o$	C_{AL} (ppm)	$(y_{AG})_o$
0	5.9	0.210	5.9	0.210	5.9	0.210
4	–	–	5.6	0.209	5.7	0.209
5	5.3	0.209	–	–	–	–
6	–	–	5.2	0.208	5.4	0.208
7	4.7	0.208	4.9	0.207	5.1	0.207
8	4.1	0.207	4.4	0.206	4.7	0.206
9	3.4	0.206	4.0	0.205	4.1	0.204
10	3.4	0.206	4.0	0.205	4.1	0.204
11	3.5	0.207	4.2	0.206	4.2	0.205

(a) Determine $k_L a$ for each stirrer speed.
(b) What is the solubility of oxygen in the fermentation broth?
(c) By considering Eq. (9.39) and Eq. (9P6.2) together, calculate the maximum oxygen uptake rate at each stirrer speed.

References

1. Treybal, R.E. (1968) *Mass-Transfer Operations*, 2nd edn, McGraw-Hill, Tokyo.
2. Sherwood, T.K., R.L. Pigford and C.R. Wilke (1975) *Mass Transfer*, McGraw-Hill, New York.
3. Perry, R.H., D.W. Green and J.O. Maloney (Eds) (1984) *Chemical Engineers' Handbook*, 6th edn, McGraw-Hill, New York.
4. Coulson, J.M. and J.F. Richardson (1977) *Chemical Engineering*, vol. 1, 3rd edn, Chapters 8 and 10, Pergamon Press, Oxford.
5. Brauer, H. (1985) Analogy of momentum, heat and mass transfer. In: H.-J. Rehm and G. Reed (Eds), *Biotechnology*, vol. 2, pp. 153–157, VCH, Weinheim.
6. Bailey, J.E. and D.F. Ollis (1986) *Biochemical Engineering Fundamentals*, 2nd edn, McGraw-Hill, New York.
7. Johnson, M.J. (1946) Metabolism of penicillin-producing molds. *Ann. N.Y. Acad. Sci.* **48**, 57–66.
8. Andrew, S.P.S. (1982) Gas–liquid mass transfer in microbiological reactors. *Trans. IChE.* **60**, 3–13.
9. Heijnen, J.J., K. van't Riet and A.J. Wolthuis (1980) Influence of very small bubbles on the dynamic $k_L A$ measurement in viscous gas–liquid systems. *Biotechnol. Bioeng.* **22**, 1945–1956.
10. Westerterp, K.R., L.L. van Dierendonck and J.A. de Kraa (1963) Interfacial areas in agitated gas–liquid contactors. *Chem. Eng. Sci.* **18**, 157–176.
11. Kawase, Y. and M. Moo-Young (1990) The effect of antifoam agents on mass transfer in bioreactors. *Bioprocess Eng.* **5**, 169–173.
12. Prins, A. and K. van't Riet (1987) Proteins and surface effects in fermentation: foam, antifoam and mass transfer. *Trends in Biotechnol.* **5**, 296–301.
13. Lee, Y.H. and G.T. Tsao (1979) Dissolved oxygen electrodes. *Adv. Biochem. Eng.* **13**, 35–86.
14. Atkinson, B. and F. Mavituna (1991) *Biochemical Engineering and Biotechnology Handbook*, 2nd edn, Chapter 18, Macmillan, Basingstoke.
15. Truesdale, G.A., A.L. Downing and G.F. Lowden (1955) The solubility of oxygen in pure water and seawater. *J. Appl. Chem.* **5**, 53–62.
16. Battino, R. and H.L. Clever (1966) The solubility of gases in liquids. *Chem. Rev.* **66**, 395–463.
17. Stephen, H. and T. Stephen (Eds) (1963) *Solubilities of Inorganic and Organic Compounds*, vol. 1, pp. 87–88, Pergamon Press, Oxford.
18. *International Critical Tables* (1926) McGraw-Hill, New York.
19. Hikita, H., S. Asai and Y. Azuma (1978) Solubility and diffusivity of oxygen in aqueous sucrose solutions. *Can. J. Chem. Eng.* **56**, 371–374.
20. Schumpe, A., I. Adler and W.-D. Deckwer (1978)

Solubility of oxygen in electrolyte solutions. *Biotechnol. Bioeng.* **20**, 145–150.

21. Schumpe, A. and W.-D. Deckwer (1979) Estimation of O_2 and CO_2 solubilities in fermentation media. *Biotechnol. Bioeng.* **21**, 1075–1078.

22. Baburin, L.A., J.E. Shvinka and U.E. Viesturs (1981) Equilibrium oxygen concentration in fermentation fluids. *Eur. J. Appl. Microbiol. Biotechnol.* **13**, 15–18.

23. Quicker, G., A. Schumpe, B. König and W.-D. Deckwer (1981) Comparison of measured and calculated oxygen solubilities in fermentation media. *Biotechnol. Bioeng.* **23**, 635–650.

24. Ju, L.-K., C.S. Ho and R.F. Baddour (1988) Simultaneous measurements of oxygen diffusion coefficients and solubilities in fermentation media with polarographic oxygen electrodes. *Biotechnol. Bioeng.* **31**, 995–1005.

25. Oosterhuis, N.M.G. and N.W.F. Kossen (1983) Oxygen transfer in a production scale bioreactor. *Chem. Eng. Res. Des.* **61**, 308–312.

26. van't Riet, K. (1979) Review of measuring methods and results in nonviscous gas–liquid mass transfer in stirred vessels. *Ind. Eng. Chem. Process Des. Dev.* **18**, 357–364.

27. Brooks, J.D., D.G. Maclennan, J.P. Barford and R.J. Hall (1982) Design of laboratory continuous-culture equipment for accurate gaseous metabolism measurements. *Biotechnol. Bioeng.* **24**, 847–856.

28. Wernau, W.C. and C.R. Wilke (1973) New method for evaluation of dissolved oxygen probe response for $K_L a$ determination. *Biotechnol. Bioeng.* **15**, 571–578.

29. Dunn, I.J. and A. Einsele (1975) Oxygen transfer coefficients by the dynamic method. *J. Appl. Chem. Biotechnol.* **25**, 707–720.

30. Linek, V. and V. Vacek (1976) Oxygen electrode response lag induced by liquid film resistance against oxygen transfer. *Biotechnol. Bioeng.* **18**, 1537–1555.

31. Linek, V. and V. Vacek (1977) Dynamic measurement of the volumetric mass transfer coefficient in agitated vessels: effect of the start-up period on the response of an oxygen electrode. *Biotechnol. Bioeng.* **19**, 983–1008.

32. Dang, N.D.P., D.A. Karrer and I.J. Dunn (1977) Oxygen transfer coefficients by dynamic model moment analysis. *Biotechnol. Bioeng.* **19**, 853–865.

33. Philichi, T.L. and M.K. Stenstrom (1989) Effects of dissolved oxygen probe lag on oxygen transfer parameter estimation. *J. Water Poll. Contr. Fed.* **61**, 83–86.

34. Merchuk, J.C., S. Yona, M.H. Siegel and A. Ben Zvi (1990) On the first-order approximation to the response of dissolved oxygen electrodes for dynamic $K_L a$ estimation. *Biotechnol. Bioeng.* **35**, 1161–1163.

35. van't Riet, K. and J. Tramper (1991) *Basic Bioreactor Design*, Chapter 11, Marcel Dekker, New York.

36. Mukhopadhyay, S.N. and T.K. Ghose (1976) A simple dynamic method of $k_L a$ determination in laboratory fermenter. *J. Ferment. Technol.* **54**, 406–419.

Suggestions for Further Reading

Mass-transfer theory (see also refs 1, 2 and 4)

McCabe, W.L. and J.C. Smith (1976) *Unit Operations of Chemical Engineering*, 3rd edn, Section 4, McGraw-Hill, Tokyo.

Oxygen transfer in fermenters

Bell, G.H. and M. Gallo (1971) Effect of impurities on oxygen transfer. *Process Biochem.* **6** (April), 33–35.

van't Riet, K. (1983) Mass transfer in fermentation. *Trends in Biotechnol.* **1**, 113–119.

10

Unit Operations

Bioprocesses treat raw materials and generate useful products. Individual operations or steps within the process that change or separate components are called unit operations. Although the specific objectives of bioprocesses vary from factory to factory, each processing scheme can be viewed as a series of component operations which appear again and again in different systems. For example, most bioprocesses involve one or more of the following unit operations: centrifugation, chromatography, cooling, crystallisation, dialysis, distillation, drying, evaporation, filtration, heating, humidification, membrane separation, milling, mixing, precipitation, solids handling, solvent extraction. A particular sequence of unit operations used for manufacture of enzymes is shown in the flow sheet of Figure 10.1. Although the same operations are involved in other processes, the order in which they are carried out, the conditions used and the actual materials handled account for the differences in final results. Engineering principles for design of unit operations are independent of specific industries or applications.

In a typical fermentation process, raw materials are altered most significantly by reactions occurring in the fermenter. However physical changes before and after fermentation are also important to prepare the substrates for reaction and to extract and purify the desired product from the culture broth. The term 'unit operation' usually refers to the physical steps in processes; chemical or biochemical transformations are the subject of reaction engineering which is considered in detail in Chapters 11–13.

Fermentation broths are complex mixtures of components containing products in dilute solution. In bioprocessing, any treatment of the culture broth after fermentation is known as *downstream processing*. The purpose of downstream processing is to concentrate and purify the product for sale; in most cases this requires only physical modification. Although each recovery scheme will be different, downstream processing follows a general sequence of steps.

(i) *Cell removal.* A common first step in product recovery is removal of cells from the fermentation liquor. This is necessary if the biomass itself is the desired product, e.g. bakers' yeast, or if the product is contained within the cells. Removal of cells can also assist recovery of product from the liquid phase. Filtration and centrifugation are typical unit operations for cell removal.

(ii) *Primary isolation.* A wide variety of techniques is available for primary isolation of fermentation products from cells or cell-free broth. The method used depends on the physical and chemical properties of the product and surrounding material. The aim of primary isolation is to remove components with properties significantly different from those of the product. Typically, processes for primary isolation treat large volumes of material and are relatively non-selective; however significant increases in product quality and concentration can be accomplished. Unit operations such as adsorption, liquid extraction and precipitation are used for primary isolation.

(iii) *Purification.* Processes for purification are highly selective and separate the product from impurities with similar properties. Typical unit operations are chromatography, ultrafiltration and fractional precipitation.

(iv) *Final isolation.* The final purity required depends on the product application. Crystallisation, followed by centrifugation or filtration and drying, are typical operations used for high-quality products such as pharmaceuticals.

A typical profile of product quality through the various stages of downstream processing is given in Table 10.1.

Downstream processing can account for a substantial part of the total production cost of a fermentation product. For example, the ratio of fermentation cost to cost of product recovery is approximately 60:40 for antibiotics such as penicillin. For newer antibiotics this ratio is reversed; product recovery is more costly than fermentation. Many modern products of biotechnology such as recombinant proteins and monoclonal antibodies require expensive downstream processing which accounts for 80–90% of process costs [1]. Starting product levels before recovery have a strong influence on cost;

Figure 10.1 Unit operations used in manufacture of enzymes. (From B. Atkinson and F. Mavituna, 1991, *Biochemical Engineering and Biotechnology Handbook*, 2nd edn, Macmillan, Basingstoke; and W.T. Faith, C.E. Neubeck and E.T. Reese, 1971, Production and applications of enzymes, *Adv. Biochem. Eng.* **1**, 77–111.)

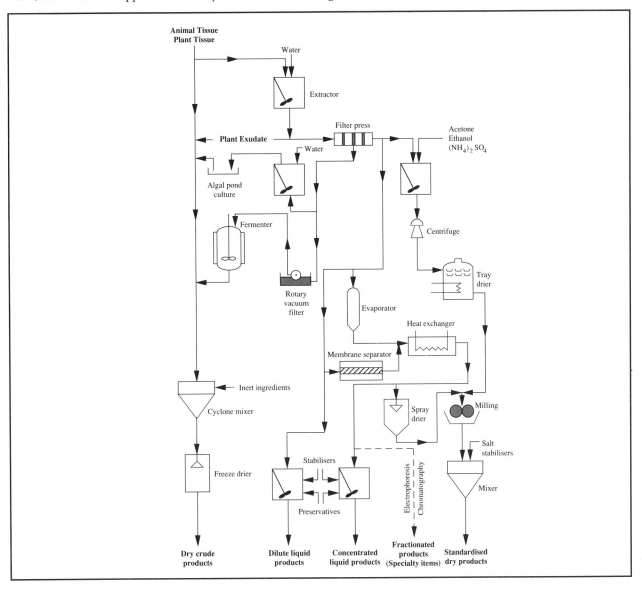

purification is more expensive when the concentration of product in the biomass or fermentation broth is low. As illustrated in Figure 10.2, the higher the starting concentration the cheaper is the final product. Each downstream-processing step involves some product loss; these losses can be substantial for multi-step procedures. For example, if 80% of the product is retained at each purification step, after a five-step process the overall product recovery is only about one-third. If the starting concentration is very low, more recovery stages are required

with higher attendant losses and costs. This situation can be improved by either enhancing product synthesis during fermentation or developing better downstream-processing techniques which minimise product loss.

There is an extensive literature on downstream processing, much of it dealing with recent advances. To thoroughly cover all unit operations used in bioprocessing is beyond the scope of this book; at least one entire separate volume would be required. Rather than attempt such a treatise, this chapter considers the

Table 10.1 Typical profile of product quality during downstream processing

(From P.A. Belter, E.L. Cussler and W.-S. Hu, 1988, Bioseparations: Downstream Processing For Biotechnology, *John Wiley, New York)*

Step	Typical unit operation	Product concentration (g l^{-1})	Product quality (%)
Harvest broth	–	0.1–5	0.1–1
Cell removal	Filtration	1–5	0.2–2
Primary isolation	Extraction	5–50	1–10
Purification	Chromatography	50–200	50–80
Final isolation	Crystallisation	50–200	90–100

engineering principles of a small selection of unit operations commonly applied for recovery of fermentation products. Information about other equally-important unit operations can be found in references listed at the end of the chapter.

10.1 Filtration

In filtration, solid particles are separated from a fluid–solid mixture by forcing the fluid through a *filter medium* or *filter cloth* which retains the particles. Solids are deposited on the filter and, as the deposit or *filter cake* increases in depth, pose a resistance to further filtration. Filtration can be performed using either vacuum or positive-pressure equipment. The pressure difference exerted across the filter to separate fluid from the solids is called the filtration *pressure drop.*

Ease of filtration depends on the properties of the solid and fluid; filtration of crystalline, incompressible solids in low-viscosity liquids is relatively straightforward. In contrast, fermentation broths can be difficult to filter because of the small size and gelatinous nature of the cells and the viscous non-Newtonian behaviour of the broth. Most microbial filter cakes are *compressible*, i.e. the porosity of the cake declines as pressure drop across the filter increases. This can be a major problem causing reduced filtration rates and greater loss of product. Filtration of fermentation broths is usually carried out under non-aseptic conditions; the process must therefore be efficient and reliable to avoid undue contamination and degradation of labile products.

10.1.1 Filter Aids

Filter aids such as diatomaceous earth have found widespread use in the fermentation industry to improve the efficiency of filtration. Diatomaceous earth, also known as kieselguhr, is the fused skeletal remains of diatoms. Packed beds of granulated kieselguhr have very high porosity; only about 15% of the total volume of packed kieselguhr is solid, the rest is empty space. Such high porosity facilitates liquid flow around the particles and improves rate of filtration through the bed.

Filter aids are applied in two ways. As shown in Figure 10.3, filter aid can be used as a pre-coat on the filter medium to prevent blockage or 'blinding' of the filter by solids which would otherwise wedge themselves into the pores of the cloth. Filter aid can also be added to the fermentation broth to increase the porosity of the cake as it forms. This is only recommended when the fermentation product is extracellular; for intracellular products requiring further processing of the cells, severe handling problems can arise if the cake is contaminated with filter aid. Filter aid adds to the cost of filtration; the minimum quantity needed to achieve the desired result must be established experimentally. Kieselguhr absorbs liquid; therefore, if the fermentation product is in the liquid phase, some will be lost. Another disadvantage is that as filtration rate increases with the assistance of filter aid, filtrate clarity is reduced. Disposal of waste cell material is more difficult if it contains kieselguhr; for example, biomass cannot be used as animal feed unless the filter aid is removed.

Fermentation broths can be pre-treated to improve filtration characteristics. Heating to denature proteins enhances the filterability of mycelial broths such as in penicillin production. Alternatively, electrolytes may be added to promote coagulation of colloids into larger, denser particles which are easier to filter. Ease of filtration is also affected by the duration of the fermentation; this affects the composition and viscosity of the medium and properties of the cell cake.

Figure 10.2 Relationship between selling price and concentration before downstream processing for several fermentation products. (From J.L. Dwyer, 1984, Scaling up bio-product separation with high performance liquid chromatography, *Bio/Technology* 2, 957–964; and J. van Brunt, 1988, How big is big enough? *Bio/Technology* 6, 479–485.)

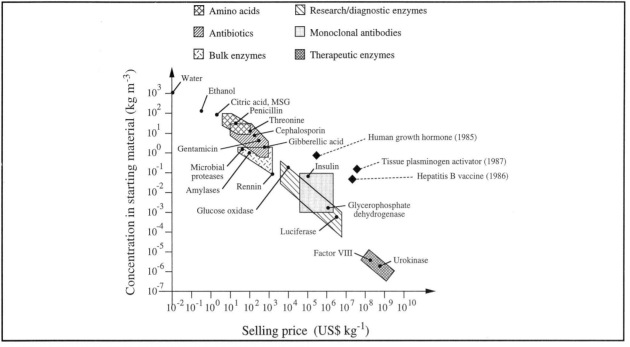

Figure 10.3 Use of filter aid in filtration of fermentation broth.

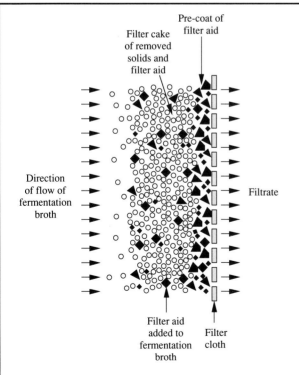

10.1.2 Filtration Equipment

Plate filters are suitable for filtration of small fermentation batches; this type of filter gradually accumulates biomass and must be periodically opened and cleared of filter cake. Larger processes require continuous filters. *Rotary-drum vacuum filters*, such as that shown in Figure 10.4, are the most widely-used filtration devices in the fermentation industry. A horizontal drum 0.5–3 m in diameter is covered with filter cloth and rotated slowly at 0.1–2 rpm. The cloth is partially immersed in an agitated reservoir containing material to be filtered. As a section of drum enters the liquid, a vacuum is applied from the interior of the drum. A cake forms on the face of the cloth while liquid is drawn through internal pipes to a collection tank. As the drum rotates out of the reservoir, the surface of the filter is sprayed with wash liquid which is drawn through the cloth and collected in a separate holding tank. After washing, the cake is dewatered by continued application of the vacuum. The vacuum is turned off as the drum reaches the discharge zone where the cake is removed by means of a scraper, knife or strings. Air pressure may be applied at this

Figure 10.4 Continuous rotary-drum vacuum filter. (From G.G. Brown, A.S. Foust, D.L. Katz, R. Schneidewind, R.R. White, W.P. Wood, G.M. Brown, L.E. Brownell, J.J. Martin, G.B. Williams, J.T. Banchero and J.L. York, 1950, *Unit Operations*, John Wiley, New York.)

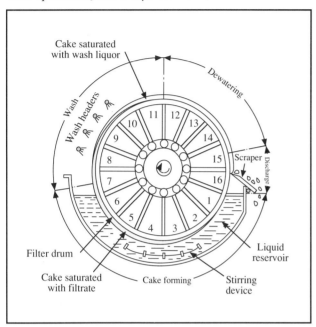

stage to help dislodge the filter cake from the cloth. After the cake is removed, the drum re-enters the reservoir for another filtration cycle.

10.1.3 Filtration Theory

Filtration theory is used to estimate rate of filtration. For a given pressure drop across the filter, rate of filtration is greatest just as filtering begins. This is because resistance to filtration is at a minimum when there are no deposited solids. Orientation of particles in the initial cake deposit is very important and can significantly influence the structure and permeability of the whole filter bed. Excessive pressure drop and high initial rates of filtration can result in plugging of the filter cloth and very high subsequent resistance to flow. Flow resistance due to the filter cloth can be considered constant if particles do not penetrate the material; resistance due to the cake increases with thickness.

Rate of filtration is usually measured as the rate at which liquid filtrate is collected. Filtration rate depends on the area of the filter cloth, the viscosity of the fluid, the pressure difference across the filter, and the resistance to filtration offered by the cloth and deposited filter cake. At any instant during filtration, rate of filtration is given by the equation:

$$\frac{1}{A} \frac{dV_f}{dt} = \frac{\Delta p}{\mu_f \left[\alpha \left(\frac{M_c}{A} \right) + r_m \right]}$$

(10.1)

where A is filter area, V_f is volume of filtrate, t is filtration time, Δp is pressure drop across the filter, μ_f is filtrate viscosity, M_c is the total mass of solids in the cake, α is the average *specific cake resistance* and r_m is the *filter medium resistance*. r_m includes the effect of the filter cloth and any particles wedged in it during the initial stages of filtration. α has dimensions LM^{-1}; r_m has dimensions L^{-1}. dV_f/dt is the filtrate flow rate or *volumetric rate of filtration*. Area A represents the capital cost of the filter; the bigger the area required to achieve a given filtration rate, the larger is the equipment and related investment. α is a measure of the resistance of the filter cake to flow; its value depends on the shape and size of the particles, the size of the interstitial spaces between them, and the mechanical stability of the cake. Resistance due to the filter medium is often negligible compared with cake resistance.

If the filter cake is incompressible, the specific cake resistance α does not vary with pressure drop across the filter. However, cakes from fermentation broths are seldom incompressible; as these cakes compress, filtration rates decline. For a compressible cake, α can be related empirically to Δp as follows:

$$\alpha = \alpha' (\Delta p)^s$$

(10.2)

where s is cake *compressibility* and α' is a constant dependent largely on the size and morphology of particles in the cake. The value of s is zero for rigid incompressible solids; for highly compressible material s is close to unity. α is also related to the average properties of particles in the cake as follows:

$$\alpha = \frac{K_v a^2 (1 - \varepsilon)}{\varepsilon^3 \rho_p}$$

(10.3)

where K_v is a factor depending on the shape of the particles, a is the specific surface area of the particles, ε is *porosity* of the cake and ρ_p is density of the particles. a and ε are defined as follows:

$$a = \frac{\text{surface area of a single particle}}{\text{volume of a single particle}} ;$$

(10.4)

$$\varepsilon = \frac{\text{total volume of the cake} - \text{volume of solids in the cake}}{\text{total volume of the cake}}$$

(10.5)

For compressible cakes, both ε and K_v depend on filtration pressure drop.

It is useful to consider methods for improving rate of filtration for a given batch of material. Various strategies can be deduced from the relationship between variables in Eq. (10.1).

(i) *Increase the filter area A.* When all other parameters remain constant, rate of filtration is improved if A is increased. However, this requires installation of larger filtration equipment and greater capital cost.

(ii) *Increase the filtration pressure drop Δp.* The problem with this approach for compressible cakes is that α increases with Δp as indicated in Eq. (10.2); higher α results in lower filtration rate. In practice, pressure drops are usually kept below 0.5 atm to minimise cake resistance. Improving filtration rate by increasing the pressure drop can only be achieved by reducing s, the compressibility of the cake. Addition of filter aid in the broth can reduce s to some extent.

(iii) *Reduce the cake mass M_c.* This is achieved in continuous rotary filtration by reducing the thickness of cake deposited per revolution of the drum, and ensuring that the scraper leaves minimal cake residue on the filter cloth.

(iv) *Reduce the liquid viscosity μ_f.* Material to be filtered is sometimes diluted if the starting viscosity is very high.

(v) *Reduce the specific cake resistance α.* From Eq. (10.3), possible methods of reducing α for compressible cakes are as follows:

(a) *Increase the porosity ε.* Cake porosity usually decreases as cells are filtered. Application of filter aid reduces this effect.

(b) *Reduce the shape factor of the particles K_v.* In the case of mycelial broths, it may be possible to change the morphology of the cells by manipulating fermentation conditions.

(c) *Reduce the specific surface area of the particles a.* Increasing the average size of the particles and minimising variation in particle size reduce the value of a. Changes in fermentation conditions and broth pretreatment are used to achieve these effects.

Integration of Eq. (10.1) allows us to calculate the time required to filter a given volume of material. Before carrying out the integration, let us substitute an expression for mass of solids in the cake as a function of filtrate volume:

$$M_c = c\, V_f$$

(10.6)

where c is the mass of solids deposited per volume of filtrate and is related to the concentration of solid in the material to be filtered. Substituting Eq. (10.6) into Eq. (10.1), the expression for rate of filtration becomes:

$$\frac{1}{A}\frac{dV_f}{dt} = \frac{\Delta p}{\mu_f\left[\alpha\left(\dfrac{cV_f}{A}\right)+r_m\right]}.$$

(10.7)

A filter can be operated in two different ways. If the pressure drop across the filter is kept constant, the filtration rate will become progressively smaller as resistance due to the cake increases. On the other hand, in constant-rate filtration the flow rate is maintained by gradually increasing the pressure drop. Filtrations are most commonly carried out at constant pressure. When this is the case, Eq. (10.7) can be integrated directly because V_f and t are the only variables; for a given filtration device and material to be filtered, each of the remaining parameters is constant.

It is convenient for integration to write Eq. (10.7) in its reciprocal form:

$$A\frac{dt}{dV_f} = \mu_f\alpha c\,\frac{V_f}{A\Delta p} + \frac{\mu_f r_m}{\Delta p}.$$

(10.8)

At the beginning of filtration $t=0$ and $V_f=0$; this is the initial condition for integration. Separating variables and placing constants outside the integral signs gives:

$$A\int dt = \left(\frac{\mu_f\alpha c}{A\,\Delta p}\right)\int V_f\, dV_f + \left(\frac{\mu_f r_m}{\Delta p}\right)\int dV_f.$$

(10.9)

Carrying out the integration:

$$A\,t = \left(\frac{\mu_f\alpha c}{2A\,\Delta p}\right)V_f^2 + \left(\frac{\mu_f r_m}{\Delta p}\right)V_f.$$

(10.10)

Thus, for constant-pressure filtration, Eq. (10.10) can be used to calculate either filtrate volume V_f or filtration time t provided all the constants are known. α and r_m for a particular filtration must be evaluated beforehand. For experimental determination of α and r_m, Eq. (10.10) is rearranged by dividing both sides of the equation by AV_f:

$$\frac{t}{V_f} = \left(\frac{\mu_f\alpha c}{2A^2\Delta p}\right)V_f + \left(\frac{\mu_f r_m}{A\Delta p}\right).$$

(10.11)

Eq. (10.11) can be written more simply as:

$$\frac{t}{V_f} = K_1 V_f + K_2 \qquad (10.12)$$

where:

$$K_1 = \frac{\mu_f \alpha c}{2A^2 \Delta p} \qquad (10.13)$$

and

$$K_2 = \frac{\mu_f r_m}{A\,\Delta p}. \qquad (10.14)$$

K_1 and K_2 are constant during constant-pressure filtration. Eq. (10.12) is therefore an equation of a straight line when t/V_f is plotted against V_f. The slope K_1 depends on the filtration pressure drop and properties of the cake; the intercept K_2 also depends on pressure drop but is independent of cake properties. α is calculated from the slope; r_m is determined from the intercept. Eq. (10.12) is valid for compressible cakes; however K_1 becomes a more complex function of Δp than is directly apparent from Eq. (10.13) because of the dependence of α on pressure.

Eq. (10.11) is the basic filtration equation for industrial-scale equipment such as the rotary-drum vacuum filter shown in Figure 10.4; rates of filtration during cake formation and washing determine the size of the filter required for the process. Simple modifications to Eq. (10.11) give equations which are directly applicable to rotary-drum filters [2].

Example 10.1 Filtration of mycelial broth

A 30-ml sample of broth from a penicillin fermentation is filtered in the laboratory on a 3 cm^2 filter at a pressure drop of 5 psi. The filtration time is 4.5 min. Previous studies have shown that filter cake of *Penicillium chrysogenum* is significantly compressible with $s = 0.5$. If 500 litres broth from a pilot-scale fermenter must be filtered in 1 hour, what size filter is required if the pressure drop is:

(a) 10 psi?
(b) 5 psi?

Resistance due to the filter medium is negligible.

Solution:
Properties of the filtrate and mycelial cake can be determined from results of the laboratory experiment. If r_m can be eliminated from Eq. (10.11):

$$\frac{t}{V_f} = \left(\frac{\mu_f \alpha c}{2A^2 \Delta p} \right) V_f.$$

Substituting α for a compressible cake from Eq. (10.2):

$$\frac{t}{V_f} = \left(\frac{\mu_f \alpha' (\Delta p)^{s-1} c}{2A^2} \right) V_f \qquad (1)$$

and rearranging:

$$\mu_f \alpha' c = \frac{2A^2 t}{(\Delta p)^{s-1} V_f^2}.$$

Substituting values:

$$\mu_f \alpha' c = \frac{2(3 \text{ cm}^2)^2 (4.5 \text{ min})}{(5 \text{ psi})^{0.5-1} (30 \text{ cm}^3)^2} = 0.201 \text{ cm}^{-2} (\text{psi})^{0.5} \text{ min.}$$

This value for $\mu_f \alpha' c$ is used to evaluate the area required for pilot-scale filtration. From (1):

$$A^2 = \left(\frac{\mu_f \alpha' c (\Delta p)^{s-1}}{2t} \right) V_f^2$$

therefore:

$$A = \left(\frac{\mu_f \alpha' c (\Delta p)^{s-1}}{2t} \right)^{1/2} V_f.$$

(a) Substituting values when $\Delta p = 10$ psi:

$$A = \left(\frac{(0.201 \text{ cm}^{-2} (\text{psi})^{0.5} \text{ min}) \cdot (10 \text{ psi})^{0.5-1}}{2 \,(1 \text{ h}) \cdot \left| \frac{60 \text{ min}}{1 \text{ h}} \right|} \right)^{1/2} (500 \text{ l}) \cdot \left| \frac{1000 \text{ cm}^3}{1 \text{ l}} \right|$$

$$= 1.15 \times 10^4 \text{ cm}^2 = 1.15 \text{ m}^2.$$

(b) When $\Delta p = 5$ psi:

$$A = \left(\frac{(0.201 \text{ cm}^{-2} (\text{psi})^{0.5} \text{ min}) \cdot (5 \text{ psi})^{0.5-1}}{2 \,(1 \text{ h}) \cdot \left| \frac{60 \text{ min}}{1 \text{ h}} \right|} \right)^{1/2} (500 \text{ l}) \cdot \left| \frac{1000 \text{ cm}^3}{1 \text{ l}} \right|$$

$$= 1.37 \times 10^4 \text{ cm}^2 = 1.37 \text{ m}^2.$$

Halving the pressure drop increases the area required by only 20% because at 5 psi the cake is less compressed and more porous than at 10 psi.

Example 10.1 underlines the importance of laboratory testing in design of filtration systems. Experiments are required to evaluate properties of the cake such as compressibility and specific resistance; these parameters cannot be calculated from theory. Experimental observations are also necessary to evaluate a wide range of other important filtration characteristics. These include filtrate clarity, ease of washing, dryness of the final cake, ease of cake removal, and effects of filter aids and broth pre-treatment. It is essential that any laboratory tests be conducted with the same materials as the large-scale process. Variables such as temperature, age of the broth, and presence of contaminants and cell debris have significant effects on filtration characteristics.

In general, fungal mycelia are filtered relatively easily because mycelial filter cake has sufficiently large porosity. Yeast and bacteria are much more difficult to handle because of their small size. Alternative filtration methods which eliminate the filter cake are becoming more accepted for bacterial and yeast separations. *Microfiltration* is achieved by developing large cross-flow fluid velocities across the filter surface while the velocity normal to the surface is relatively small. Build-up of filter cake and problems of high cake resistance are therefore prevented. Microfiltration is not covered in this text; further details are available elsewhere, e.g. [3, 4].

10.2 Centrifugation

Centrifugation is used to separate materials of different density when a force greater than gravity is desired. In

Figure 10.5 Separation of solids in a tubular-bowl centrifuge.

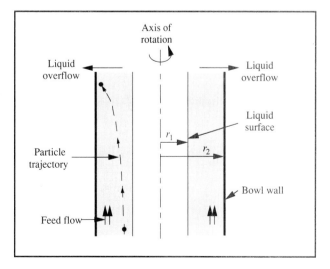

bioprocessing, centrifugation is used to remove cells from fermentation broth, to eliminate cell debris, to collect precipitates, and to prepare fermentation media such as in clarification of molasses or production of wort for brewing. Equipment for centrifugation is more expensive than for filtration; however centrifugation is often effective when the particles are very small and difficult to filter. Centrifugation of fermentation broth produces a thick, concentrated cell sludge or cream containing more liquid than filter cake.

Steam-sterilisable centrifuges are applied when either the cells or fermentation liquid is recycled to the fermenter or when product contamination must be prevented. Industrial centrifuges generate large amounts of heat due to friction; it is therefore necessary to have good ventilation and cooling. Aerosols created by fast-spinning centrifuges have been known to cause infections and allergic reactions in factory workers so that isolation cabinets are required for certain applications.

Centrifugation is most effective when the particles to be separated are large, the liquid viscosity is low, and the density difference between particles and fluid is great. It is also assisted by large centrifuge radius and high rotational speed. In centrifugation of biological solids such as cells, the particles are very small, the viscosity of the medium can be relatively high, and the particle density is very similar to the suspending fluid. These disadvantages are easily overcome in the laboratory with small centrifuges operated at high speed. However, problems arise in industrial centrifugation when large quantities of material must be treated. Centrifuge capacity cannot be increased by simply increasing the size of the equipment without limit; mechanical stress in centrifuges increases in proportion to (radius)^2 so that safe operating speeds are substantially lower in large equipment. The need for continuous throughput of material in industrial applications also restricts practical operating speeds. To overcome these difficulties, a

Figure 10.6 Disc-stack bowl centrifuge with continuous discharge of solids.

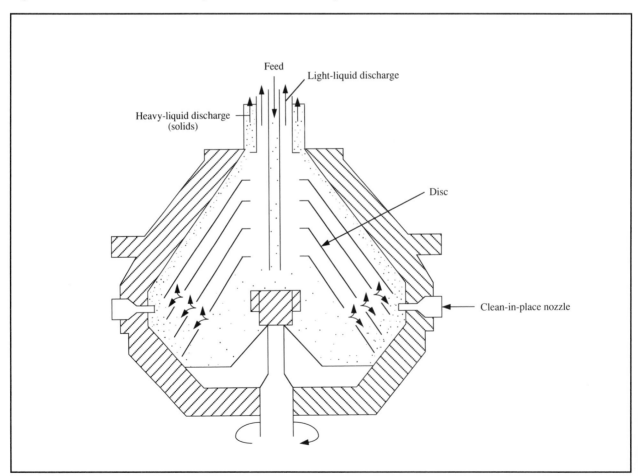

Figure 10.7 Mechanism of solids separation in a disc-stack bowl centrifuge. (From C.J. Geankoplis, 1983, *Transport Processes and Unit Operations*, 2nd edn, Allyn and Bacon, Boston.)

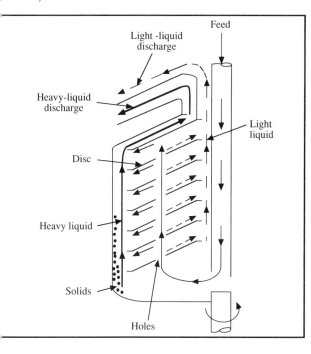

ange of centrifuges has been developed for bioprocessing. Types of centrifuge commonly used in industrial operations are described below.

10.2.1 Centrifuge Equipment

Centrifuge equipment is classified according to internal structure. The *tubular-bowl centrifuge* has the simplest configuration and is widely employed in the food and pharmaceutical industries. Feed enters under pressure through a nozzle at the bottom, is accelerated to rotor speed, and moves upwards through the cylindrical bowl. As the bowl rotates, particles travelling upward are spun out and collide with the walls of the bowl as illustrated schematically in Figure 10.5. Solids are removed from the liquid if they move with sufficient velocity to reach the wall of the bowl within the residence time of liquid in the machine. As the feed rate is increased the liquid layer moving up the wall of the centrifuge becomes thicker; this reduces performance of the centrifuge by increasing the distance a particle must travel to reach the wall. Liquid from the feed spills over a weir at the top of the bowl; solids which have collided with the walls are collected separately. When the thickness of sediment collecting in the bowl reaches the posi-

tion of the liquid-overflow weir, separation efficiency declines rapidly. This limits the capacity of the centrifuge. Tubular centrifuges are applied mainly for difficult separations requiring high centrifugal forces. Solids in tubular centrifuges are accelerated by forces between 13 000 and 16 000 times the force of gravity.

A type of narrow tubular-bowl centrifuge is the *ultracentrifuge*. This device is used for recovery of fine precipitates from high-density solutions, for breaking down emulsions, and for separation of colloidal particles such as ribosomes and mitochondria. It produces centrifugal forces 10^5–10^6 times the force of gravity. The bowl is usually air-driven and operated at low pressure or in an atmosphere of nitrogen to reduce generation of frictional heat. The main commercial application of ultracentrifuges has been in production of vaccines to separate viral particles from cell debris. Ultracentrifugation has also been tested for removal of very fine cell debris in isolation of enzymes, and for ribosome removal in purification of RNA polymerase. A typical ultracentrifuge operates discontinuously so its processing capacity is restricted by the need to empty the bowl manually. Continuous ultracentrifuges are available commercially; however safe operating speeds with these machines are not as high as in batch equipment.

An alternative to the tubular centrifuge is the *disc-stack bowl centrifuge*. Disc-stack centrifuges are common in bioprocessing. There are many types of disc centrifuge; the principal difference between them is the method used to discharge the accumulated solid. In simple disc centrifuges, solids must be removed periodically by hand. Continuous or intermittent discharge of solids is possible in a variety of disc centrifuges without reducing the bowl speed. Some centrifuges are equipped with peripheral nozzles for continuous solids removal; others have valves for intermittent discharge. Another method is to concentrate the solids in the periphery of the bowl and then discharge them at the top of the centrifuge using a paring device; the equipment configuration for this mode of operation is shown in Figure 10.6. A disadvantage of centrifuges with automatic discharge of solids is that the solids must remain sufficiently wet to flow through the machine. Extra nozzles may be provided for cleaning the bowl should blockages of the system occur.

Disc-stack centrifuges contain conical sheets of metal called discs which are stacked one on top of the other with clearances as small as 0.3 mm. The discs rotate with the bowl and their function is to split the liquid into thin layers. As shown in Figure 10.7, feed is released near the bottom of the centrifuge and travels upwards through matching holes in the discs. Between the discs, heavy components of the feed are thrown outward under the influence of centrifugal forces as lighter liquid is displaced towards the centre of the bowl. As they are flung out, the solids strike the undersides of the discs and slide down to the bottom edge of the bowl. At the same time, the lighter liquid flows in and over the upper surfaces of the discs to be discharged from the top of the bowl. Heavier liquid containing solids can be discharged either at the top of the centrifuge or through nozzles around the periphery of the bowl. Disc-stack centrifuges used in bioprocessing typically develop forces of 5000–15 000 times gravity. As a guide, the minimum density difference between solid and liquid for successful separation in a disc-stack centrifuge is approximately 0.01–0.03 kg m^{-3}; in practical operations at appropriate flow rates the minimum particle diameter separated is about 0.5 μm [5].

10.2.2 Centrifugation Theory

The particle velocity achieved in a particular centrifuge compared with the settling velocity which would occur under the influence of gravity characterises the effectiveness of centrifugation. The terminal velocity during gravity settling of a small spherical particle in dilute suspension is given by Stoke's law:

$$u_g = \frac{\rho_p - \rho_f}{18\,\mu}\,D_p^2\,g$$

(10.15)

where u_g is the sedimentation velocity under gravity, ρ_p is density of the particle, ρ_f is density of the liquid, μ is viscosity of the liquid, D_p is particle diameter and g is gravitational acceleration. In a centrifuge, the corresponding terminal velocity is:

$$u_c = \frac{\rho_p - \rho_f}{18\,\mu}\,D_p^2\,\omega^2\,r$$

(10.16)

where u_c is particle velocity in the centrifuge, ω is angular velocity of the bowl in units of rad s^{-1} and r is radius of the centrifuge drum. The ratio of velocity in the centrifuge to velocity under gravity is called the *centrifuge effect* or *g-number*, and is usually denoted Z. Therefore:

$$Z = \frac{\omega^2 r}{g}\,.$$

(10.17)

The force developed in a centrifuge is Z times the force of

gravity, and is often expressed as so many *g*-forces. Industrial centrifuges have Z factors from 300 to 16 000; for small laboratory centrifuges Z may be up to 500 000 [6].

Sedimentation occurs in a centrifuge as particles moving away from the centre of rotation collide with the walls of the centrifuge bowl. Increasing the velocity of motion will improve the rate of sedimentation. From Eq. (10.16), particle velocity in a given centrifuge can be increased by:

(i)   increasing the centrifuge speed ω;
(ii) increasing the particle diameter D_p;
(iii) increasing the density difference between particle and liquid, $\rho_p - \rho_f$; and
(iv) decreasing the viscosity of the suspending fluid, μ.

However, whether the particles reach the walls of the bowl also depends on the time of exposure to the centrifugal force. In batch centrifuges such as those used in the laboratory, centrifuge time is increased by running the equipment longer. In continuous-flow devices such as the disc-stack centrifuge, the residence time is increased by decreasing the feed flow rate.

Performance of centrifuges of different size can be compared using a parameter called the *sigma factor* Σ. Physically, Σ represents the cross-sectional area of a gravity settler with the same sedimentation characteristics as the centrifuge. For continuous centrifuges, Σ is related to the feed rate of material as follows:

$$\Sigma = \frac{Q}{2\,u_g}$$

(10.18)

where Q is the volumetric feed rate and u_g is the terminal velocity of the particles in a gravitational field. If two centrifuges perform with equal effectiveness:

$$\frac{Q_1}{\Sigma_1} = \frac{Q_2}{\Sigma_2}$$

(10.19)

where subscripts 1 and 2 denote the two centrifuges. Eq. (10.19) can be used to scale-up centrifuge equipment. Equations for evaluating Σ depend on the centrifuge design. For a disc-stack bowl centrifuge [7]:

$$\Sigma = \frac{2\pi\,\omega^2\,(N-1)}{3g\,\tan\theta}\,(r_2^3 - r_1^3)$$

(10.20)

where ω is angular velocity in rad s^{-1}, N is number of discs in the stack, r_2 is outer radius of the disc, r_1 is inner radius of the disc, g is gravitational acceleration, and θ is the half-cone angle of the disc. For a tubular-bowl centrifuge, the following equation is accurate to within 4% [8, 9]:

$$\Sigma = \frac{\pi \omega^2 b}{2g} (3r_2^2 + r_1^2)$$

(10.21)

where b is length of the bowl, r_1 is radius of the liquid surface and r_2 is radius of the inner wall of the bowl. Because r_1 and r_2 in a tubular-bowl centrifuge are about equal, Eq. (10.21) can be approximated as:

$$\Sigma = \frac{2\pi \omega^2 b r^2}{g}$$

(10.22)

where r is an average radius roughly equal to either r_1 or r_2.

The above equations for Σ are based on ideal operating conditions. Because different types of centrifuge deviate to varying degrees from ideal operation, Eq. (10.19) cannot generally be used to compare different centrifuge configurations. Performance of any centrifuge can deviate from theoretical prediction due to factors such as particle shape and size distribution, aggregation of particles, non-uniform flow distribution in the centrifuge and interaction between particles during sedimentation. Experimental tests must be performed to account for these factors.

Example 10.2 Cell recovery in a disc-stack centrifuge

A continuous disc-stack centrifuge is operated at 5000 rpm for separation of bakers' yeast. At a feed rate of 60 l min^{-1}, 50% of the cells are recovered. At constant centrifuge speed, solids recovery is inversely proportional to flow rate.

(a) What flow rate is required to achieve 90% cell recovery if the centrifuge speed is maintained at 5000 rpm?

(b) What operating speed is required to achieve 90% recovery at a feed rate of 60 l min^{-1}?

Solution:

(a) If solids recovery is inversely proportional to feed rate, the flow rate required is:

$$\frac{50\%}{90\%} (60 \, l \, min^{-1}) = 33.3 \, l \, min^{-1}.$$

(b) Eq. (10.19) relates operating characteristics of centrifuges achieving the same separation. From (a), 90% recovery is achieved at $Q_1 = 33.3$ l min^{-1} and $\omega_1 = 5000$ rpm. $Q_2 = 60$ l min^{-1}. From Eq. (10.19):

$$\frac{Q_1}{Q_2} = \frac{\Sigma_1}{\Sigma_2} = \frac{33.3 \, l \, min^{-1}}{60 \, l \, min^{-1}} = 0.56.$$

Because the same centrifuge is used and all the geometric parameters are the same, from Eq. (10.20):

$$\frac{\Sigma_1}{\Sigma_2} = \frac{\omega_1^2}{\omega_2^2} = 0.56.$$

Therefore:

$$\omega_2^2 = \frac{\omega_1^2}{0.56} = \frac{(5000 \, rpm)^2}{0.56} = 4.46 \times 10^7 \, rpm^2.$$

Taking the square root, $\omega_2 = 6680$ rpm.

10.3 Cell Disruption

Downstream processing of fermentation broths usually begins with separation of cells by filtration or centrifugation. The next step depends on location of the desired product. For substances such as ethanol, citric acid and antibiotics which are excreted from cells, product is recovered from the cell-free broth using unit operations such as those described later in this

Figure 10.8 Cell disruption in a high-pressure homogeniser.

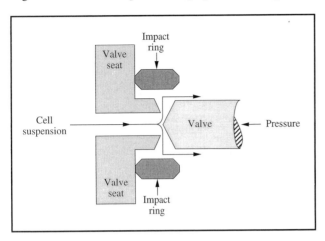

sufficient to completely disrupt many types of cell. A common apparatus for homogenisation of cells is the Manton–Gaulin homogeniser. As indicated in Figure 10.8, this high-pressure pump incorporates an adjustable valve with restricted orifice through which cells are forced at pressures up to 550 atm. The homogeniser is of general applicability for cell disruption although the homogenising valve can become blocked when used with highly filamentous organisms.

The following equation relates disruption of cells to operating conditions in the Manton–Gaulin homogeniser [10]:

$$\ln \left(\frac{R_{\mathrm{m}}}{R_{\mathrm{m}} - R} \right) = k N p^{\alpha}.$$

(10.23)

In Eq. (10.23), R_{m} is the maximum amount of protein available for release, R is the amount of protein released after N passes through the homogeniser, k is a temperature-dependent rate constant, and p is the operating pressure. The exponent α is a measure of the resistance of the cells to disruption. For *Saccharomyces cerevisiae* yeast, α has been determined as 2.9 [10]; however the exponent for a particular organism depends to some extent on growth conditions [11]. The strong dependence of protein release on pressure suggests that high-pressure operation is beneficial; complete disruption in a single pass may be possible if the pressure is sufficiently high. Reduction in the number of passes through the homogeniser is generally preferable because multiple passes produce fine cell debris which can cause problems in subsequent clarification steps.

chapter. Biomass separated from the liquid is discarded or sold as a by-product.

For products such as enzymes and recombinant proteins which remain in the biomass, cell disruption must be carried out to release the desired material. A variety of methods is available to disrupt cells. Mechanical options include grinding with abrasives, high-speed agitation, high-pressure pumping and ultrasound. Non-mechanical methods such as osmotic shock, freezing and thawing, enzymic digestion of cell walls, and treatment with solvents and detergents can also be applied.

A widely-used technique for cell disruption is high-pressure homogenisation. Shear forces generated in this treatment are

Figure 10.9 An ideal stage.

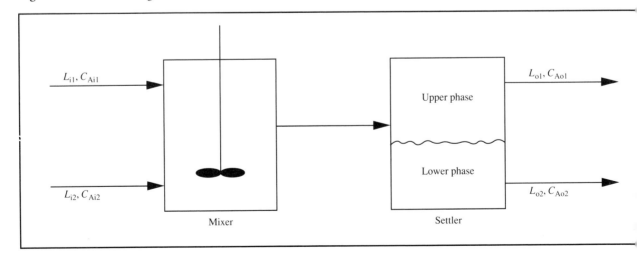

Release of protein is markedly dependent on temperature; protein recovery increases at elevated temperatures up to 50°C. However, cooling to 0–4°C is recommended during operation of homogenisers to minimise protein denaturation. Procedures for scale-up of homogenisers are not well developed. Methods which work well in the laboratory may give variable results when used at a larger scale.

10.4 The Ideal-Stage Concept

So far, we have considered only the initial steps of downstream processing: cell isolation and disruption. An important group of unit operations used for primary isolation of fermentation products relies on mass transfer to achieve separation of components between phases. Equipment for these separations sometimes consists of a series of stages in which the phases make intimate contact so that material can be transferred between them. Even if the equipment itself is not constructed in discrete stages, mass transfer can be considered to occur in stages. The effectiveness of each stage in accomplishing mass transfer depends on many factors, including equipment design, physical properties of the phases, and equilibrium relationships.

Consider the simple mixer-settler device of Figure 10.9 for extraction of component A from one liquid phase to another. The mass-transfer principles of liquid–liquid extraction have already been described in Chapter 9; unit operations for extraction are also discussed in the next section. Here, we will use the mixer-settler to explain the general concept of ideal stages. Two immiscible liquids enter the mixing vessel with volumetric flow rates L_{i1} and L_{i2}; these streams contain A at concentrations C_{Ai1} and C_{Ai2}, respectively. The two liquid phases are vigorously mixed together and A is transferred from one phase to the other. The mixture then passes to a settler where phase separation is allowed to occur under the influence of gravity. The flow rate of light liquid out of the settler is L_{o1}; the flow rate of heavy liquid is L_{o2}. Concentrations of A in these streams are C_{Ao1} and C_{Ao2}, respectively.

Operation of the mixer-settler relies on the liquids entering not being in equilibrium as far as concentration of A is concerned, i.e. C_{Ai1} and C_{Ai2} are not equilibrium concentrations. This means that there exists a driving force in the mixer for change of concentration as A is transferred from one liquid to the other. Depending on the ability of the mixer to promote mass transfer between the liquids and the effectiveness of the settler in allowing phase separation, when the liquids leave the system they will have been brought closer to equilibrium. If the mixer-settler were an *ideal stage* the two streams leaving it would be in equilibrium. No further mass transfer of A would

result from additional contact between the phases; C_{Ao1} and C_{Ao2} would be equilibrium concentrations and the device would be operating at maximum efficiency. In reality, stages are not ideal; the change in concentration achieved in a real stage is always less than in an ideal stage so that extra stages are necessary to achieve the desired separation. The relative performance of an actual stage compared with that of an ideal stage is expressed as the *stage efficiency*.

The concept of ideal stages is used in design calculations for several unit operations. The important elements in analysis of these operations are material balances, energy balances if applicable, and the equilibrium relationships between phases. Application of these principles is illustrated in the following sections.

10.5 Aqueous Two-Phase Liquid Extraction

Liquid extraction is used to isolate many pharmaceutical products from animal and plant sources. In liquid extraction of fermentation products, components dissolved in liquid are recovered by transfer into an appropriate solvent. Extraction of penicillin from aqueous broth using solvents such as butyl acetate, amyl acetate or methyl isobutyl ketone, and isolation of erythromycin using pentyl or amyl acetate are examples. Solvent extraction techniques are also applied for recovery of steroids, purification of vitamin B_{12} from microbial sources, and isolation of alkaloids such as morphine and codeine from raw plant material. The simplest equipment for liquid extraction is the separating funnel used for laboratory-scale product recovery. Liquids forming two distinct phases are shaken together in the separating funnel; solute in dilute solution in one solvent transfers to the other solvent to form a more concentrated solution. The two phases are then allowed to separate and the heavy phase is withdrawn from the bottom of the funnel. The phase containing the solute in concentrated form is processed further to purify the product. Whatever apparatus is used for extraction, it is important that contact between the liquid phases is maximised by vigorous mixing and turbulence to facilitate solute transfer.

Extraction with organic solvents is a major separation technique in bioprocessing, particularly for recovery of antibiotics. However, organic solvents are unsuitable for isolation of proteins and other sensitive biopolymers. Techniques are being developed for aqueous two-phase extraction of these molecules. Aqueous solvents which form two distinct phases provide favourable conditions for separation of proteins, cell fragments and organelles with protection of their biological activity. Two-phase aqueous systems are produced when particular polymers or a polymer and salt are dissolved together in

Table 10.2 Examples of aqueous two-phase systems

(From M.R. Kula, 1985, Liquid–liquid extraction of biopolymers. In: M. Moo-Young, Ed, Comprehensive Biotechnology, vol. 2, pp. 451–471, Pergamon Press, Oxford)

Component 1	Component 2
Polyethylene glycol	Dextran
	Polyvinyl alcohol
	Polyvinylpyrrolidone
	Ficoll
	Potassium phosphate
	Ammonium sulphate
	Magnesium sulphate
	Sodium sulphate
Polypropylene glycol	Polyvinyl alcohol
	Polyvinylpyrrolidone
	Dextran
	Methoxypolyethylene glycol
	Potassium phosphate
Ficoll	Dextran
Methylcellulose	Dextran
	Hydroxypropyldextran
	Polyvinylpyrrolidone

water above certain concentrations. The liquid partitions into two phases, each containing 85–99% water. Some components used to form aqueous two-phase systems are listed in Table 10.2. When added to these mixtures, biomolecules and cell fragments partition between the phases; by selecting appropriate conditions, cell fragments can be confined to one phase as the protein of interest partitions into the other phase. Aqueous two-phase separations are of special interest for extraction of enzymes and recombinant proteins from cell debris produced by cell disruption. After partitioning, product is removed from the extracting phase using other unit operations such as precipitation or crystallisation.

The extent of differential partitioning between phases depends on the equilibrium relationship for the system. The *partition coefficient K* is defined as:

$$K = \frac{C_{Au}}{C_{Al}}$$

(10.24)

where C_{Au} is the equilibrium concentration of component A in the upper phase and C_{Al} is the equilibrium concentration of A in the lower phase. If $K > 1$, component A favours the upper phase;

if $K < 1$, A is concentrated in the lower phase. In many aqueous systems K is constant over a wide range of concentrations provided the molecular properties of the phases are not changed. Partitioning is influenced by the size, electric charge and hydrophobicity of the particles or solute molecules; biospecific affinity for one of the polymers may also play a role in some systems. Surface free energy of the phase components and ionic composition of the liquids are of paramount importance in determining separation; K is related to both these parameters. Partitioning is also affected by other and sometimes interdependent factors so that it is impossible to predict partition coefficients from molecular properties. For single-stage extraction of enzymes, partition coefficients ≥ 3 are normally required [5].

Even when the partition coefficient is low, good *product recovery* or *yield* can be achieved by using a large volume of the phase preferred by the solute. Yield of A in the upper phase Y_u, is defined as:

$$Y_u = \frac{V_u C_{Au}}{V_0 C_{A0}} = \frac{V_u C_{Au}}{V_u C_{Au} + V_l C_{Al}}$$

(10.25)

where V_u is volume of the upper phase, V_l is volume of the lower phase, V_0 is the original volume of solution containing the product and C_{A0} is the original product concentration in that liquid. In the lower phase, yield Y_l is defined as:

$$Y_l = \frac{V_l C_{Al}}{V_0 C_{A0}} = \frac{V_l C_{Al}}{V_u C_{Au} + V_l C_{Al}}.$$

(10.26)

The maximum possible yield for an ideal extraction stage can be evaluated using Eqs (10.25) and (10.26) and the equilibrium partition coefficient, K. Dividing both numerator and denominator of Eq. (10.25) by C_{Au} and recognising that, at equilibrium, C_{Al}/C_{Au} is equal to $1/K$:

$$Y_u = \frac{V_u}{V_u + \dfrac{V_l}{K}}.$$

(10.27)

Similarly, dividing the numerator and denominator of Eq. (10.26) by C_{Al} gives:

$$Y_l = \frac{V_l}{V_u K + V_l}.$$

(10.28)

Example 10.3 Enzyme recovery using aqueous extraction

Aqueous two-phase extraction is used to recover α-amylase from solution. A polyethylene glycol–dextran mixture is added and the solution separates into two phases. The partition coefficient is 4.2. Calculate the maximum possible enzyme recovery when:

a) the volume ratio of upper to lower phases is 5.0; and
b) the volume ratio of upper to lower phases is 0.5.

Solution:

As the partition coefficient is greater than 1, enzyme prefers the upper phase. Yield at equilibrium is therefore calculated for the upper phase. Dividing both numerator and denominator of Eq. (10.27) by V_1 gives:

$$Y_u = \frac{\dfrac{V_u}{V_1}}{\dfrac{V_u}{V_1} + \dfrac{1}{K}}.$$

a) $\dfrac{V_u}{V_1} = 5.0$. Therefore:

$$Y_u = \frac{5.0}{5.0 + \dfrac{1}{4.2}} \times 100 = 95\%.$$

b) $\dfrac{V_u}{V_1} = 0.5$. Therefore:

$$Y_u = \frac{0.5}{0.5 + \dfrac{1}{4.2}} \times 100 = 68\%.$$

Increasing the relative volume of the extracting phase enhances recovery.

Another parameter used to characterise two-phase partitioning is the *concentration factor* or *purification factor*, δ_c, defined as the ratio of product concentration in the preferred phase to the initial product concentration:

$$\delta_c = \frac{C_{Al}}{C_{A0}} \quad \text{(when product partitions to the lower phase)}$$

$$(10.29)$$

$$\delta_c = \frac{C_{Au}}{C_{A0}} \quad \text{(when product partitions to the upper phase)}.$$

Aqueous extraction in polyethylene glycol–salt mixtures is an effective technique for separating proteins from cell debris. In this system, debris partitions to the lower phase while most of the target proteins are recovered from the upper phase. Extraction can be carried out in a single-stage operation such as that depicted in Figure 10.9 using a polymer mixture which provides a suitable partition coefficient. Equilibrium is approached in extraction operations but rarely reached; for industrial application it is important to consider the time taken for mass transfer and the ease of mechanical separation of the phases. The rate of approach to equilibrium depends on the surface area available for exchange between the phases; this is maximised by rapid mixing. Separation of the phases is sometimes a problem because of the low interfacial tension between aqueous phases; very rapid large-scale extractions can be achieved by combining mixed vessels with centrifugal separators. In many cases, recovery and concentration of product with yields exceeding 90% can be achieved using a single extraction step [5]. When single-stage extraction does not give sufficient recovery, repeated extractions can be carried out in a chain or cascade of contacting and separation units.

10.6 Adsorption

Adsorption is a surface phenomenon whereby components of a gas or liquid are concentrated on the surface of solid particles or at fluid interfaces. Adsorption is the result of electrostatic, van der Waals, reactive or other binding forces between individual atoms, ions or molecules. Four types of adsorption can be distinguished: exchange, physical, chemical and non-specific.

Adsorption serves the same function as extraction in isolating products from dilute fermentation liquors. Several different adsorption operations are used in bioprocessing, particularly for medical and pharmaceutical products. Ion-exchange adsorption is established practice for recovery of amino acids, proteins, antibiotics and vitamins. Adsorption onto activated charcoal is a method of long standing for purification of citric acid; adsorption of organic chemicals onto charcoal or porous polymeric adsorbents is common in wastewater treatment. Adsorption operations generally have higher selectivity but smaller capacity than extraction. Scale-up procedures for adsorption are less well defined than for extraction; therefore, more experimental data are required for design. Handling of solids in industrial processing is also somewhat more difficult compared with the liquids used in extraction. However, despite these disadvantages, adsorption is gaining increasing application primarily because of its suitability for protein isolation.

In adsorption operations, the substance being concentrated on the surface is called the *adsorbate*; the material to which the adsorbate binds is the *adsorbent*. The ideal adsorbent material has a high surface area per unit volume; this can be achieved if the solid contains a network of fine internal pores which provide an extremely large internal surface area. Carbons and synthetic resins based on styrene, divinylbenzene or acrylamide polymers are commonly used for adsorption of biological molecules. Commercially available adsorbents are porous granular or gel resins with void volumes of 30–50% and pore diameters generally less than 0.01 mm. As an example, the total surface area in particles of activated carbon ranges from 450 to 1800 m^2 g^{-1}. Not all of this area is necessarily available for adsorption; adsorbate molecules only have access to surfaces in pores of appropriate diameter.

Figure 10.10 Adsorption isotherms.

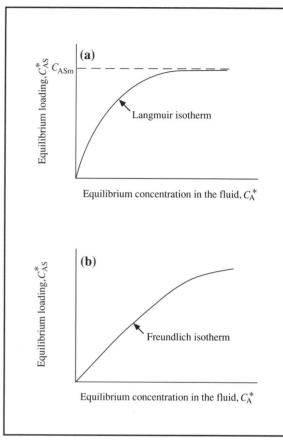

10.6.1 Adsorption Operations

A typical adsorption operation consists of the following stages: a *contacting* or adsorption step which loads solute onto the adsorptive resin, a *washing* step to remove residual unadsorbed material, *desorption* or *elution* of adsorbate with a suitable solvent, *washing* to remove residual eluant, and *regeneration* of the adsorption resin to its original condition. Because adsorbate is bound to the resin by physical or ionic forces, conditions used for desorption must overcome these forces. Desorption is normally accomplished by feeding a stream of different ionic strength or pH; elution with organic solvent or reaction of the sorbed material may be necessary in some applications. Eluant containing stripped solute in concentrated form is processed to recover the adsorbate; operations for final purification include spray drying, precipitation and crystallisation. After elution, the adsorbent undergoes a regenerative treatment to remove any impurities and regain its original adsorptive capacity. Despite regeneration, performance of the resin will decrease with use as complete removal of adsorbed material is impossible. Accordingly, after a few regenerations the adsorbate is replaced.

10.6.2 Equilibrium Relationships For Adsorption

Like extraction, analysis of adsorption depends somewhat on equilibrium relationships which determine the extent to which material can be adsorbed onto a particular surface. When an adsorbate and adsorbent are at equilibrium there is a defined distribution of solute between solid and fluid phases and no further net adsorption occurs. Adsorption equilibrium data are available as *adsorption isotherms*. For adsorbate A, an isotherm gives the concentration of A in the adsorbed phase versus the concentration in the unadsorbed phase at a given temperature. Adsorption isotherms are useful for selecting the most appropriate adsorbent; they also play a crucial role in predicting the performance of adsorption systems.

Several types of equilibrium isotherm have been developed to describe adsorption relationships. However, no single model is universally applicable; all involve assumptions which may or may not be valid in particular cases. One of the simplest adsorption isotherms that accurately describes certain practical systems is the *Langmuir isotherm* shown in Figure 10.10(a). The Langmuir isotherm can be expressed as follows:

$$C_{AS}^* = \frac{C_{ASm} K_A C_A^*}{1 + K_A C_A^*}.$$

(10.30)

In Eq. (10.30), C_{AS}^* is the equilibrium concentration or loading of A on the adsorbent in units of, e.g. kg solute kg^{-1} solid or kg solute m^{-3} solid. C_{ASm} is the maximum loading of adsorbate corresponding to complete monolayer coverage of all available adsorption sites, C_A^* is the equilibrium concentration of solute

in the fluid phase in units of, e.g. kg m^{-3}, and K_A is a constant. Because of the different units used for fluid- and solid-phase concentrations, K_A usually has units such as m^3 kg^{-1} solid.

Theoretically, Langmuir adsorption is applicable to systems where: (i) adsorbed molecules form no more than a monolayer on the surface; (ii) each site for adsorption is equivalent in terms of adsorption energy; and (iii) there are no interactions between adjacent adsorbed molecules. In many experimental systems at least one of these conditions is not met. For example, many commercial adsorbents possess highly irregular surfaces so that adsorption is favoured at particular points or 'strong sites' on the surface. Accordingly, each site is not equivalent. In addition, interactions between adsorbed molecules exist for almost all real adsorption systems. Recognition of these and other factors has led to application of other adsorption isotherms.

Of particular interest because of its widespread use in liquid–solid systems is the *Freundlich isotherm*, described by the relationship:

$$C_{AS}^* = K_F C_A^{*\,1/n}.$$

(10.31)

K_F and n are constants characteristic of the particular adsorption system; the dimensions of K_F depend on the dimensions of C_{AS}^* and C_A^* and the value of n. If adsorption is favourable n is > 1; if adsorption is unfavourable n is < 1. The form of the Freundlich isotherm is shown in Figure 10.10(b). Eq. (10.31) applies to adsorption of a wide variety of antibiotics, hormones and steroids.

There are many other forms of adsorption isotherm giving different C_{AS}^* – C_A^* curves [12]. Because the exact mechanisms of adsorption are not well understood, adsorption equilibrium data must be determined experimentally.

Example 10.4 Antibody recovery by adsorption

Cell-free fermentation liquor contains 8×10^{-5} mol l^{-1} immunoglobulin G. It is proposed to recover at least 90% of this antibody by adsorption on synthetic, non-polar resin. Experimental equilibrium data are correlated as follows:

$$C_{AS}^* = 5.5 \times 10^{-5} C_A^{*\,0.35}$$

where C_{AS}^* is mol solute adsorbed per cm^3 adsorbent and C_A^* is liquid-phase solute concentration in mol l^{-1}. What minimum quantity of resin is required to treat 2 m^3 fermentation liquor in a single-stage mixed tank?

Solution:
The quantity of resin required is minimum when equilibrium occurs. If 90% of the antibiotic is adsorbed, the residual concentration in the liquid is:

$$\frac{(100-90)\%}{100\%}\,(8\times10^{-5}\,\text{mol l}^{-1}) = 8\times10^{-6}\,\text{mol l}^{-1}.$$

Substituting this value for C_A^* in the isotherm expression gives the equilibrium loading of immunoglobulin:

$$C_{AS}^* = 5.5\times10^{-5}\,(8\times10^{-6})^{0.35} = 9.05\times10^{-7}\,\text{mol cm}^{-3}.$$

The amount of adsorbed antibody is:

$$\frac{90\%}{100\%}\,(8\times10^{-5}\,\text{mol l}^{-1})\,(2\,\text{m}^3)\cdot\left|\frac{1000\,\text{l}}{1\,\text{m}^3}\right| = 0.144\,\text{mol}.$$

Therefore, the mass of adsorbent needed is:

$$\frac{0.144\,\text{mol}}{9.05\times10^{-7}\,\text{mol cm}^{-3}} = 1.59\times10^5\,\text{cm}^3.$$

The minimum quantity of resin required is $1.6\times10^5\,\text{cm}^3$, or $0.16\,\text{m}^3$.

Figure 10.11 Movement of the adsorption zone and development of the breakthrough curve for a fixed-bed adsorber.

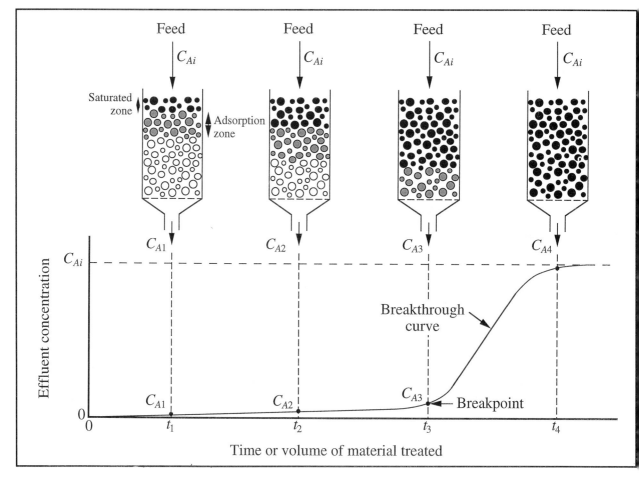

Figure 10.12 Relationship between the breakthrough curve, loss of solute in the effluent, and unused column capacity.

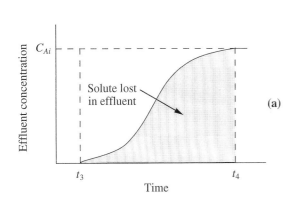

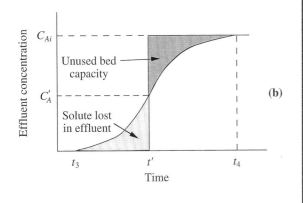

10.6.3 Performance Characteristics of Fixed-Bed Adsorbers

Various types of equipment have been developed for adsorption operations, including fixed beds, moving beds, fluidised beds and stirred-tank contactors. Of these, fixed-bed adsorbers are most commonly applied; the adsorption area available per unit volume is greater in fixed beds than in most other configurations. A fixed-bed adsorber is a vertical column or tube packed with adsorbent particles. Commercial adsorption operations are mostly performed as unsteady-state processes; liquid containing solute is passed through the bed and the loading or amount of product retained in the column increases with time.

Operation of a downflow fixed-bed adsorber is illustrated in Figure 10.11. Liquid solution containing adsorbate at concentration C_{Ai} is fed at the top of a column which is initially

free of adsorbate. At first, adsorbent resin at the top of the column takes up solute rapidly; solution passing through the column becomes depleted of solute and leaves the system with effluent concentration close to zero. As flow of solution continues, the region of the bed where most adsorption occurs, the *adsorption zone*, moves down the column as the top resin becomes saturated with solute in equilibrium with liquid concentration C_{Ai}. Movement of the adsorption zone usually occurs at a speed much lower than the velocity of fluid through the bed, and is called the *adsorption wave*. Eventually the lower edge of the adsorption zone reaches the bottom of the bed, the resin is almost completely saturated, and the concentration of solute in the effluent starts to rise appreciably; this is called the *breakpoint*. As the adsorption zone passes through the bottom of the bed, the resin can no longer adsorb solute and the effluent concentration rises to the inlet value, C_{Ai}. At this time the bed is completely saturated with adsorbate and must be regenerated. The curve in Figure 10.11 showing effluent concentration as a function of time or volume of material processed is known as the *breakthrough curve*.

The shape of the breakthrough curve greatly influences design and operation of fixed-bed adsorbers. Figure 10.12 shows the portion of the breakthrough curve between times t_3 and t_4 when solute appears in the column effluent. The amount of solute lost in the effluent is given by the area under the breakthrough curve. As indicated in Figure 10.12(a), if adsorption continues until the entire bed is saturated and the effluent concentration equals C_{Ai}, a considerable amount of solute is wasted. To avoid this, adsorption operations are usually stopped before the bed is completely saturated. As shown in Figure 10.12(b), if adsorption is halted at time t' when the effluent concentration is C'_A, only a small amount of solute is wasted compared with the process of Figure 10.12(a). The disadvantage is that some portion of the bed capacity is unused, as represented by the shaded area above the breakthrough curve. Because of the importance of the breakthrough curve in determining schedules of operation, much effort has been given to its prediction and to analysis of factors affecting it. This is discussed further in the next section.

10.6.4 Engineering Analysis of Fixed-Bed Adsorbers

In design of fixed-bed adsorbers, the quantity of resin and the time required for adsorption of a given quantity of solute must be estimated. Design procedures involve predicting the shape of the breakthrough curve and the time of appearance of the breakpoint. The form of the breakthrough curve is influenced by factors such as feed rate, concentration of solute in the feed, nature of the adsorption equilibrium and rate of adsorption.

Figure 10.13   Fixed-bed adsorber for mass-balance analysis.

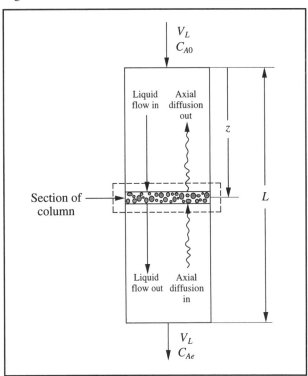

Performance of commercial adsorbers is usually controlled by adsorption rate. This in turn depends on mass-transfer processes within and outside the adsorbent particles. One approach to adsorber design is to conduct extensive pilot studies to examine the effects of major system variables. However, the duration and cost of these experimental studies can be minimised by prior mathematical analysis of the process. Because fixed-bed adsorption is an unsteady-state process and equations for adsorption isotherms are generally non-linear, the calculations involved in engineering analysis are relatively complex compared with many other unit operations. Non-homogeneous packing in adsorption beds and the difficulty of obtaining reproducible results in apparently identical beds add to these problems. It is beyond the scope of this book to consider design procedures in any depth as considerable effort and research is required to establish predictive models for adsorption systems. However, a simplified engineering analysis is presented below.

Let us consider the processes which cause changes in the liquid-phase concentration of adsorbate in a fixed-bed adsorber. The aim of this analysis is to derive an equation for effluent concentration as a function of time (the breakthrough curve). The technique used is the mass balance. Consider the column packed with adsorbent resin shown in Figure 10.13. Liquid containing solute A is fed at the top of the column and flows down

the bed. The total length of the bed is L. At distance z from the top is a section of column around which we can perform an unsteady-state mass balance. We will assume that this section is very thin so that z is approximately the same anywhere in the section. The system boundary is indicated in Figure 10.13 by dashed lines; four streams representing flow of material are shown to cross the boundary. The general mass-balance equation given in Chapter 4 can be applied to solute A:

$$
\begin{Bmatrix} \text{mass in} \\ \text{through} \\ \text{system} \\ \text{boundaries} \end{Bmatrix} - \begin{Bmatrix} \text{mass out} \\ \text{through} \\ \text{system} \\ \text{boundaries} \end{Bmatrix} + \begin{Bmatrix} \text{mass} \\ \text{generated} \\ \text{within} \\ \text{system} \end{Bmatrix} - \begin{Bmatrix} \text{mass} \\ \text{consumed} \\ \text{within} \\ \text{system} \end{Bmatrix} = \begin{Bmatrix} \text{mass} \\ \text{accumulated} \\ \text{within} \\ \text{system} \end{Bmatrix}.
$$

$$(4.1)$$

Let us consider each term of Eq. (4.1) to see how it applies to solute A in the designated section of the column. First, because we assume there are no chemical reactions taking place and A can be neither generated nor consumed, the third and fourth terms of Eq. (4.1) are zero. On the left-hand side, this leaves only the input and output terms. What are the mechanisms for input of component A to the section of column? A is brought into the section largely as a result of liquid flow down the column; this is indicated in Figure 10.13 by the solid arrow entering the section. Other mechanisms are related to local mixing and diffusion processes within the interstices or gaps between the resin particles. For example, some A may enter the section from the region just below it by countercurrent diffusion against the direction of flow; this is indicated in Figure 10.13 by the wiggly arrow entering the section from below. Let us now consider movement of A out of the system. The mechanisms for removal of A are the same as for entry: A is carried out of the section by liquid flow and by axial transfer along the length of the tube against the direction of flow. These processes are also indicated in Figure 10.13. The remaining term in Eq. (4.1) is accumulation of A. A will accumulate within the section due to adsorption onto the interior and exterior surfaces of the adsorbent particles. A may also accumulate in liquid trapped within the interstitial spaces or gaps between resin particles.

When appropriate mathematical expressions for rates of flow, axial dispersion and accumulation are substituted into Eq. (4.1), the following equation is obtained:

$$
\mathscr{D}_{Az}\frac{\partial^2 C_A}{\partial z^2} + u\frac{-\partial C_A}{\partial z} = \frac{\partial C_A}{\partial t} + \left(\frac{1-\varepsilon}{\varepsilon}\right)\frac{\partial C_{AS}}{\partial t}.
$$

| axial dispersion | flow | = accumulation in the interstices | accumulation by adsorption |

$$(10.32)$$

Eq. (10.32), C_A is the concentration of A in the liquid, z is d depth, t is time and C_{AS} is the average concentration of A the solid phase. $\mathscr{D}_{Az}$ is the *effective axial dispersion coefficient* r A in the column. In most packed beds $\mathscr{D}_{Az}$ is substantially eater than the molecular diffusion coefficient; the value of $_{Az}$ incorporates the effects of axial mixing in the column as e solid particles interrupt smooth liquid flow. ε is the *void iction* in the bed, defined as:

$$\varepsilon = \frac{V_T - V_s}{V_T}$$

(10.33)

here V_T is the total volume of the column and V_s is the volne of the resin particles. u is the interstitial liquid velocity, fined as:

$$u = \frac{F_L}{\varepsilon A_c}$$

(10.34)

here F_L is the volumetric liquid flow rate and A_c is the crossctional area of the column. In Eq. (10.32), $\partial/_{\partial t}$ denotes the rtial differential with respect to time, $\partial/_{\partial z}$ denotes the partial fferential with respect to distance, and $\partial^2/_{\partial z^2}$ denotes the cond partial differential with respect to distance. Although j. (10.32) looks complicated, it is useful to recognise the ysical meaning of its components. As indicated below each rm of the equation, rates of axial dispersion, flow, and accuulation in the liquid and solid phases are represented; cumulation is equal to the sum of the net rates of axial diffun and flow into the system.

There are four variables in Eq. (10.32): concentration of in the liquid, C_A; concentration of A in the solid, C_{AS}; disnce from the top of the column, z; and time, t. The other rameters can be considered constant. C_A and C_{AS} vary th time of operation and depth in the column. heoretically, with the aid of further information about the stem, Eq. (10.32) can be solved to provide an equation for e effluent concentration as a function of time: the breakrough curve. However, solution of Eq. (10.32) is generally ry difficult. To assist the analysis, simplifying assumptions e often made. For example, it is normally assumed that lute solutions are being processed; this results in nearly isoermal operation and eliminates the need for an companying energy balance for the system. In many cases e axial-diffusion term can be neglected; axial dispersion is nerally significant only at low flow rates. If the interstitial

fluid content of the bed is small compared with the total volume of feed, accumulation of A between the particles can also be neglected. With these simplifications, the first and third terms of Eq. (10.32) are eliminated and the design equation is reduced to:

$$u \frac{-\partial C_A}{\partial z} = \left(\frac{1 - \varepsilon}{\varepsilon} \right) \frac{\partial C_{AS}}{\partial t}.$$

(10.35)

To progress further with this analysis, information about $\partial C_{AS}/_{\partial t}$ is required. This term represents the rate of change of solid-phase adsorbate concentration and depends on the overall rate at which adsorption takes place. Overall rate of adsorption depends on two factors: the rate at which solute is transferred from liquid to solid by mass-transfer mechanisms, and the rate of the actual adsorption or attachment process. The mass-transfer pathway for adsorbate is analogous to that described in Section 9.5.2 for oxygen transfer. There are up to five steps which can pose significant resistance to adsorption as indicated in Figure 10.14. They are:

(i)   transfer from the bulk liquid to the liquid boundary layer surrounding the particle;
(ii)   diffusion through the relatively stagnant liquid film surrounding the particle;
(iii) transfer through the liquid in the pores of the particle to internal surfaces;
(iv)   the actual adsorption process; and
(v)   surface diffusion along the internal pore surfaces; i.e. migration of adsorbate molecules within the surface without prior desorption.

Normally only a small amount of adsorption occurs on the outer perimeter of the particle compared to within the pores; accordingly, external adsorption is not shown in Figure 10.14. Bulk transfer of solute is usually rapid because of mixing and convective flow of liquid passing over the solid; the effect of step (i) on overall adsorption rate can therefore be neglected. The adsorption step itself is sometimes very slow and can become the rate-limiting process; however in most cases adsorption occurs relatively quickly so that step (iv) is not rate-controlling. Step (ii) represents the major external resistance to mass transfer, while steps (iii) and (v) represent the major internal resistances. Any or all of these steps can control the overall rate of adsorption depending on the situation. Rate-controlling steps are usually identified experimentally using a small column with packing identical to the industrial-scale system; mass-transfer coefficients can then be measured under appropriate flow conditions.

Figure 10.14 Steps involved in adsorption of solute from liquid to porous adsorbent particle.

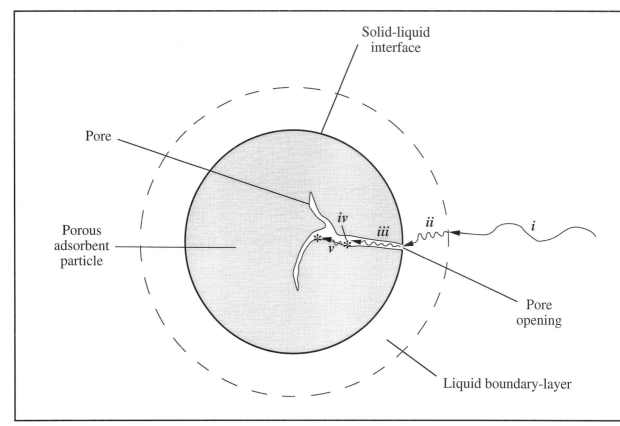

Unfortunately however, it is possible that the rate-controlling step changes as the process is scaled up, making rational design difficult.

Greatest simplification of Eq. (10.35) is obtained when the overall rate of mass transfer from liquid to internal surfaces is represented by a single equation. For example, by analogy with Eq. (9.34) or (9.35), we can write:

$$\frac{\partial C_{AS}}{\partial t} = \frac{K_L a}{1 - \varepsilon} (C_A - C_A^*)$$

$$(10.36)$$

where K_L is the overall mass-transfer coefficient, a is the surface area of the solid per unit volume, C_A is the liquid-phase concentration of A, and C_A^* is the liquid-phase concentration of A in equilibrium with C_{AS}. The value of K_L will depend on the properties of the liquid and the flow conditions; C_A^* can be related to C_{AS} by the equilibrium isotherm. In the end, after the differential equations are simplified as much as possible and then integrated with appropriate boundary conditions (usually with a computer), the result is a relationship between

effluent concentration (C_A at $z = L$) and time. The height the column required to achieve a certain recovery of solute c also be evaluated.

As mentioned already in this section, there are many unc tainties associated with design and scale up of adsorpti systems. The approach described here will give only approximate indication of design and operating paramete Other methods involving various simplifying assumptions c be employed [2]. The above analysis highlights the importa role played by mass transfer in practical adsorption operatio *Equilibrium is seldom achieved in commercial adsorption syster performance is controlled by the overall rate of adsorption.* Mc parameters determining the economics of adsorption are tho affecting mass transfer. Improvement in adsorber operatic can be achieved by enhancing rates of diffusion and reduci mass-transfer resistance.

10.7 Chromatography

Chromatography is a separation procedure for resolving m tures and isolating components. Many of the princip described in the previous section on adsorption apply also to chromatography. The basis of chromatography is *differential migration*, i.e. the selective retardation of solute molecules during passage through a bed of resin particles. A schematic description of chromatography is given in Figure 10.15; this diagram shows separation of three solutes from a mixture injected into a column. As solvent flows through the column, the solutes travel at different speeds depending on their relative affinities for the resin particles. As a result, they will be separated and appear for collection at the end of the column at different times. The pattern of solute peaks emerging from a chromatography column is called a *chromatogram*. The fluid carrying solutes through the column or used for elution is known as the *mobile phase*; the material which stays inside the column and effects the separation is called the *stationary phase*.

In *gas chromatography (GC)*, the mobile phase is a gas. Gas chromatography is used widely as an analytical tool for separating relatively volatile components such as alcohols, ketones, aldehydes and many other organic and inorganic compounds. However, of greater relevance to bioprocessing is *liquid chromatography*, which can take a variety of forms. Liquid chromatography finds application both as a laboratory method for sample analysis and as a preparative technique for large-scale purification of biomolecules. In recent years there have been rapid developments in the technology of liquid chromatography aimed at isolation of recombinant products from genetically engineered organisms. Chromatography is a high-resolution technique and therefore suitable for recovery of high-purity therapeutics and pharmaceuticals. Chromatographic methods available for purification of proteins, peptides, amino acids, nucleic acids, alkaloids, vitamins, steroids and many other biological materials include *adsorption chromatography, partition chromatography, ion-exchange chromatography, gel chromatography* and *affinity chromatography*. These methods differ in the principal mechanism by which molecules are retarded in the chromatography column.

(i)   *Adsorption chromatography.* Biological molecules have varying tendencies to adsorb onto polar adsorbents such as silica gel, alumina, diatomaceous earth and charcoal. Performance of the adsorbent relies strongly on the chemical composition of the surface, i.e. the types and concentrations of exposed atoms or groups. The order of elution of sample components depends primarily on molecule polarity. Because the mobile phase is in competition with solute for adsorption sites, solvent properties are also important. Polarity scales for solvents are available to aid mobile-phase selection [13].

(ii) *Partition chromatography.* Partition chromatography relies on the unequal distribution of solute between two immis-

Figure 10.15   Chromatographic separation of components in a mixture. Three different solutes are shown schematically as circles, squares and triangles. (From P.A. Belter, E.L. Cussler and W.-S. Hu, 1988, *Bioseparations: Downstream Processing For Biotechnology*, John Wiley, New York.)

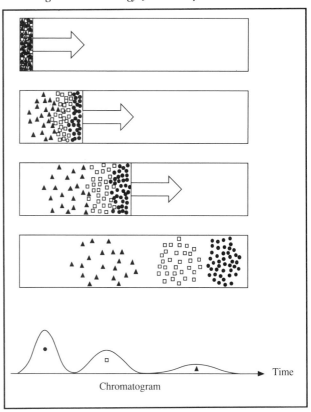

cible solvents. This is achieved by fixing one solvent (the stationary phase) to a support and passing the other solvent containing solute over it. The solvents make intimate contact allowing multiple extractions of solute to occur. Several methods are available to chemically bond the stationary solvent to supports such as silica [14]. When the stationary phase is more polar than the mobile phase, the technique is called *normal-phase chromatography*. When non-polar compounds are being separated it is usual to use a stationary phase which is less polar than the mobile phase; this is called *reverse-phase chromatography*. A common stationary phase for reverse-phase chromatography is hydrocarbon with 8 or 18 carbons bonded to silica gel; these materials are called C_8 and C_{18} packings, respectively. Solvent systems most frequently used are water–acetonitrile and water–methanol; aqueous buffers are also employed to suppress ionisation of sample components. Elution is generally in order of increasing solute hydrophobicity.

(iii) *Ion-exchange chromatography.* The basis of separation in this procedure is electrostatic attraction between the solute and dense clusters of charged groups on the column packing. Ion-exchange chromatography can give high resolution of macromolecules and is used commercially for fractionation of antibiotics and proteins. Column packings for low-molecular-weight compounds include silica, glass and polystyrene; carboxymethyl and diethylaminoethyl groups attached to cellulose, agarose or dextran provide suitable resins for protein chromatography. Solutes are eluted by changing the pH or ionic strength of the liquid phase; salt gradients are the most common way of eluting proteins from ion exchangers. Practical aspects of protein ion-exchange chromatography are described in greater detail elsewhere [15].

(iv) *Gel chromatography.* This technique is also known as *molecular-sieve chromatography, exclusion chromatography, gel filtration* and *gel-permeation chromatography.* Molecules in solution are separated in a column packed with gel particles of defined porosity. Gels most often used are cross-linked dextrans, agaroses and polyacrylamide gels. The speed with which components travel through the column depends on their effective molecular size. Large molecules are completely excluded from the gel matrix and move rapidly through the column to appear first in the chromatogram. Small molecules are able to penetrate the pores of the packing, traverse the column very slowly, and appear last in the chromatogram. Molecules of intermediate size enter the pores but spend less time there than the small solutes. Gel filtration can be used for separation of proteins and lipophilic compounds. Large-scale gel-filtration columns are operated with upward-flow elution.

(v) *Affinity chromatography.* This separation technique exploits the binding specificity of biomolecules. Enzymes, hormones, receptors, antibodies, antigens, binding proteins, lectins, nucleic acids, vitamins, whole cells and other components capable of specific and reversible binding are amenable to highly selective affinity purification. Column packing is prepared by linking a binding molecule called a *ligand* to an insoluble support; when sample is passed through the column, only solutes with appreciable affinity for the ligand are retained. The ligand must be attached to the support in such a way that its binding properties are not seriously affected; molecules called *spacer arms* are often used to set the ligand away from the support and make it more accessible to the solute. Many ready-made support–ligand preparations are available commercially and are suitable for a wide range of proteins. Conditions for elution depend on the specific binding complex formed: elution usually involves a change in pH, ionic strength or buffer composition. Enzyme proteins can be desorbed using a compound with higher affinity for the enzyme than the ligand, e.g. a substrate or substrate analogue. Affinity chromatography using antibody ligands is called *immuno-affinity chromatography.*

In this section we will consider principles of liquid chromatography for separation of biological molecules such as proteins and amino acids. Choice of stationary phase will depend to a large extent on the type of chromatography employed; however certain basic requirements must be met. For high capacity, the solid support must be porous with high internal surface area; it must also be insoluble and chemically stable during operation and cleaning. Ideally, the particles should exhibit high mechanical strength and show little or no non-specific binding. The low rigidity of many porous gels was initially a problem in industrial-scale chromatography; the weight of the packing material in large columns and the pressures developed during flow tended to compress the packing and impede operation. However, many macroporous gels and composite materials of high rigidity are now available for industrial use.

Two methods for carrying out chromatographic separations are high-performance liquid chromatography (HPLC) and fast protein liquid chromatography (FPLC). In principle, any of the types of chromatography described above can be executed using HPLC and FPLC techniques. Specialised equipment for HPLC and FPLC allows automated injection of sample, rapid flow of material through the column, collection of the separated fractions, and data analysis. Chromatographic separations traditionally performed under atmospheric pressure in vertical columns with manual sample feed and gravity elution are carried out faster and with better resolution using densely-packed columns and high flow rates in HPLC and FPLC systems. The differences between HPLC and FPLC lie in the flow rates and pressures used, the size of the packing material, and the resolution accomplished. In general, HPLC instruments are designed for small-scale, high-resolution analytical applications; FPLC is tailored for large-scale purification. In order to achieve the high resolutions characteristic of HPLC, stationary-phase particles 2–5 μm in diameter are commonly used. Because the particles are so small, HPLC systems are operated under high pressure (5–10 MPa) to achieve flow rates of 1–5 ml min^{-1}. FPLC instruments are not able to develop such high pressures (1–2 MPa), and are therefore operated with column packings of larger size. Resolution is poorer using FPLC compared with

Figure 10.16 Differential migration of two solutes A and B.

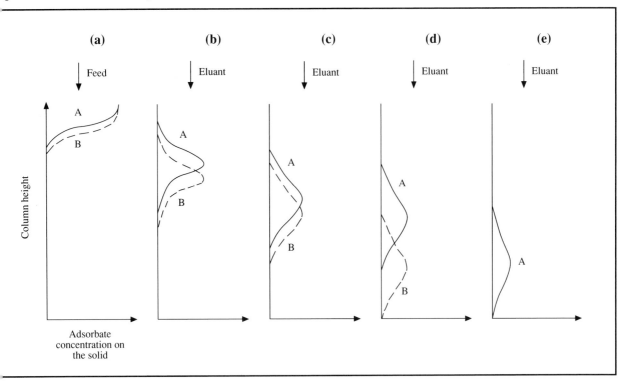

PLC; accordingly, it is common practice to collect only the central peak of the solute pulse emerging from the end of the column and to recycle or discard the leading and trailing edges. FPLC equipment is particularly suited to protein separations; many gels used for gel chromatography and affinity chromatography are compressible and cannot withstand the high pressures exerted in HPLC.

Chromatography is essentially a batch operation; however industrial chromatography systems can be monitored and controlled for easy automation. Cleaning the column in place is generally difficult. Depending on the nature of the impurities contained in the samples, rather harsh treatments with concentrated salt or dilute alkali solutions are required; these may affect swelling of the gel beads and, therefore, liquid flow in the column. Regeneration in place is necessary as re-packing of large columns can be laborious and time-consuming. Repeated use of chromatographic columns is essential because of their high cost.

0.7.1 Differential Migration

Differential migration provides the basis for chromatographic separation and is explained diagrammatically in Figure 10.15. A solution contains two solutes A and B which have different equilibrium affinities for a particular stationary phase. For the sake of brevity, let us say that the solutes are adsorbed onto the stationary phase although they may be adsorbed, bound or entrapped depending on the type of chromatography employed. Assume that A is adsorbed more strongly than B. If a small quantity of solution is passed through the column so that only a limited depth of packing is saturated, both solutes will be retained in the bed as shown in Figure 10.16(a).

A suitable eluant is now passed through the bed. As shown in Figures 10.16(b–e), both solutes will be alternately adsorbed and desorbed at lower positions in the column as flow of eluant continues. Because solute B is more easily desorbed than A, it moves forward more rapidly. Differences in migration velocities of solutes are related to differences in equilibrium distributions between stationary and mobile phases. In Figure 10.16(e), solute B has been separated from A and washed out of the system.

Several parameters are used to characterise differential migration. An important variable is the volume V_e of eluting solvent required to carry the solute through the column until it emerges at its maximum concentration. Each component separated by chromatography has a different elution volume. Another parameter commonly used to characterise elution is the *capacity factor*, k:

$$k = \frac{(V_e - V_o)}{V_o}$$

$$(10.37)$$

where V_o is the void volume, i.e. the volume of liquid in the column outside of the particles. For two solutes, the ratio of their capacity factors k_1 and k_2 is called the *selectivity* or *relative retention, δ*:

$$\delta = \frac{k_2}{k_1}.$$

$$(10.38)$$

Eqs (10.37) and (10.38) are normally applied to adsorption, partition, ion-exchange and affinity chromatography. In gel chromatography where separation is a function of effective molecular size, the elution volume is easily related to certain physical properties of the gel column. The total volume of a gel column is:

$$V_T = V_o + V_i + V_s$$

$$(10.39)$$

where V_T is total volume, V_o is void volume outside the particles, V_i is internal volume of liquid in the pores of the particles, and V_s is volume of the gel itself. The outer volume V_o can be determined by measuring the elution volume of a substance that is completely excluded from the stationary phase; a solute which does not penetrate the gel can be washed from the column using a volume of liquid equal to V_o. V_o is usually about one-third V_T. Solutes which are only partly excluded from the stationary phase elute with a volume described by the following equation:

$$V_e = V_o + K_p V_i$$

$$(10.40)$$

where K_p is the *gel partition coefficient*, defined as the fraction of internal volume available to the solute. For large molecules which do not penetrate the solid, $K_p = 0$. From Eq. (10.40):

$$K_p = \frac{(V_e - V_o)}{V_i}.$$

$$(10.41)$$

K_p is a convenient parameter for comparing separation results obtained with different gel-chromatography columns; it is independent of column size and packing density. However, experimental determination of K_p depends on knowledge of V_i, which is difficult to measure accurately. V_i is usually calculated using the equation:

$$V_i = a W_r$$

$$(10.42)$$

where a is mass of dry gel and W_r is the *water regain value*, defined as the volume of water taken up per mass of dry gel. The value for W_r is generally specified by the gel manufacturer. If, as is often the case, the gel is supplied already wet and swollen, the value of a is unknown and V_i is determined using the following equation:

$$V_i = \frac{W_r \rho_g}{(1 + W_r \rho_w)} (V_T - V_o)$$

$$(10.43)$$

where ρ_g is the density of wet gel and ρ_w is the density of water.

Example 10.5 Hormone separation using gel chromatography

A pilot-scale gel-chromatography column packed with Sephacryl resin is used to separate two hormones A and B. The column is 5 cm in diameter and 0.3 m high; the void volume is 1.9×10^{-4} m³. The water regain value of the gel is 3×10^{-3} m³ kg⁻¹ dry Sephacryl; the density of wet gel is 1.25×10^3 kg m⁻³. The partition coefficient for hormone A is 0.38; the partition coefficient for hormone B is 0.15. If the eluant flow rate is 0.7 l h⁻¹, what is the retention time for each hormone?

Solution:
The total column volume is:

$$V_T = \pi r^2 h = \pi (2.5 \times 10^{-2}\,\text{m})^2 (0.3\,\text{m}) = 5.89 \times 10^{-4}\,\text{m}^3.$$

$V_o = 1.9 \times 10^{-4}$ m³; $\rho_w = 1000$ kg m⁻³. From Eq. (10.43):

$$V_i = \frac{(3 \times 10^{-3}\,\text{m}^3\,\text{kg}^{-1})\,(1.25 \times 10^3\,\text{kg m}^{-3})}{1 + (3 \times 10^{-3}\,\text{m}^3\,\text{kg}^{-1})\,(1000\,\text{kg m}^{-3})} \; (5.89 \times 10^{-4}\,\text{m}^3 - 1.9 \times 10^{-4}\,\text{m}^3)$$

$$= 3.74 \times 10^{-4} \, \text{m}^3.$$

$K_{pA} = 0.38$; $K_{pB} = 0.15$. Therefore, from Eq. (10.40):

$$V_{eA} = 1.9 \times 10^{-4} \, \text{m}^3 + 0.38 \, (3.74 \times 10^{-4} \, \text{m}^3) = 3.32 \times 10^{-4} \, \text{m}^3$$

$$V_{eB} = 1.9 \times 10^{-4} \, \text{m}^3 + 0.15 \, (3.74 \times 10^{-4} \, \text{m}^3) = 2.46 \times 10^{-4} \, \text{m}^3 \, .$$

The times associated with these elution volumes are:

$$t_A = \frac{3.32 \times 10^{-4} \, \text{m}^3}{0.71 \, \text{h}^{-1} \cdot \left| \frac{1 \, \text{m}^3}{1000 \, \text{l}} \right| \cdot \left| \frac{1 \, \text{h}}{60 \, \text{min}} \right|} = 28 \, \text{min}.$$

$$t_B = \frac{2.46 \times 10^{-4} \, \text{m}^3}{0.71 \, \text{h}^{-1} \cdot \left| \frac{1 \, \text{m}^3}{1000 \, \text{l}} \right| \cdot \left| \frac{1 \, \text{h}}{60 \, \text{min}} \right|} = 21 \, \text{min}.$$

10.7.2 Zone Spreading

The effectiveness of chromatography depends not only on differential migration but on whether the elution bands for individual solutes remain compact and without overlap. Ideally, each solute should pass out of the column at a different instant in time. In practice, elution bands spread out somewhat so that each solute takes a finite period of time to pass across the end of the column. Zone spreading is not so important when migration rates vary widely because there is little chance that solute peaks will overlap. However if the molecules to be separated have similar structure, migration rates will also be similar and zone spreading must be carefully controlled.

As illustrated in Figure 10.15, typical chromatogram elution bands have a peak of high concentration at or about the centre of the pulse but are of finite width as the concentration trails off to zero before and after the peak. Spreading of the solute peak is caused by several factors represented schematically in Figure 10.17.

(i) *Axial diffusion.* As solute is carried through the column, molecular diffusion of solute will occur from regions of high concentration to regions of low concentration. Diffusion in the axial direction, i.e. along the length of the tube, is indicated in Figure 10.17(a) by broken arrows. Axial diffusion broadens the solute peak by transporting material upstream and downstream away from the region of greatest concentration.

(ii) *Eddy diffusion.* In columns packed with solid particles, actual flow paths of liquid through the bed can be highly variable. As indicated in Figure 10.17(a), some liquid will flow almost directly through the bed while other liquid will take longer and more tortuous paths through the gaps or interstices between the particles. Accordingly, some solute molecules carried in the fluid will move slower than the average rate of progress through the column while others will take shorter paths and move ahead of the average; the result is spreading of the solute band. Differential motion of material due to erratic local variations in flow velocity is known as *eddy diffusion*.

(iii) *Local non-equilibrium effects.* In most columns, lack of equilibrium is the most important factor affecting zone spreading, although perhaps the most difficult to understand. Consider the situation at position X indicated in Figure 10.17(a). A solute pulse is passing through the column; as shown in Figure 10.17(b) concentration within this pulse increases from the front edge to a maximum near the centre and then decreases to zero. As the solute pulse moves down the column, an initial gradual increase in solute concentration will be experienced at X. In response to this increase in mobile-phase solute concentration, solute will bind to the stationary phase and the stationary-phase concentration will start to increase towards an appropriate equilibrium value. Equilibrium is not established immediately however; it takes time for the solute to undergo the mass-transfer steps from liquid to solid as outlined in Section 10.6.4. Indeed, before equilibrium can be established, the mobile-phase concentration increases again as the centre of the solute pulse moves closer to X. Because concentration in the mobile

Figure 10.17 Zone spreading in a chromatography column.

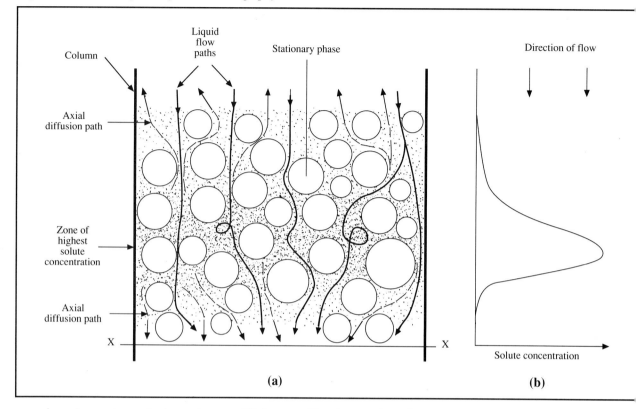

phase is continuously increasing, equilibrium at X remains always out of reach and the stationary-phase concentration lags behind equilibrium values. As a consequence, a higher concentration of solute remains in the liquid than if equilibrium were established, and the front edge of the solute pulse effectively moves ahead faster than the remainder of the pulse. As the peak of the solute pulse passes X, the mobile-phase concentration starts to decrease with time. In response, the solid phase must divest itself of solute to reach equilibrium with the lower liquid-phase concentrations. Again, because of delays due to mass transfer, equilibrium cannot be established with the continuously changing liquid concentration. As the solute pulse passes and the liquid concentration at X falls to zero, the solid phase still contains solute molecules that continue to be released into the liquid. Consequently, the rear of the solute pulse is effectively stretched out until the stationary phase reaches equilibrium with the liquid.

In general, conditions which improve mass transfer will increase the rate at which equilibrium is achieved between the phases and minimise zone spreading. For example, increasing the particle surface area per unit volume facilitates mass transfer and reduces non-equilibrium effects; surface area is usually increased by using smaller particles. On the other hand increasing the liquid flow rate will exacerbate non-equilibrium effects as the rate of adsorption fails to keep up with concentration changes in the mobile phase. Viscous solutions give rise to considerable zone broadening as a result of slower mass-transfer rates; zone broadening is also more pronounced if the solute molecules are large. Changes in temperature can affect zone broadening in several ways. Because viscosity is reduced at elevated temperatures, heating the column often decreases zone spreading. However, rates of axial diffusion increase at higher temperatures so that the overall effect depends on the system and temperature range tested.

10.7.3 Theoretical Plates in Chromatography

The concept of theoretical plates is often used to analyse zone broadening in chromatography. The idea is essentially the same as that described in Section 10.4 for an ideal equilibrium stage. The chromatography column is considered to be made up of a number of segments or plates of height H; the magnitude of H

Figure 10.18 Parameters for calculation of: (a) number of theoretical plates, and (b) resolution.

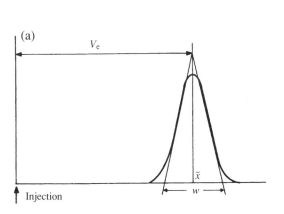

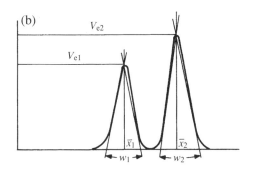

is of the same order as the diameter of the resin particles. Within each segment equilibrium is supposed to exist.

As in adsorption operations, equilibrium is not often achieved in chromatography so that the theoretical-plate concept does not accurately reflect conditions in the column. Nevertheless the idea of theoretical plates is applied extensively, mainly because it provides a parameter, the plate height H, which can be used to characterise zone spreading. Use of the plate height, which is also known as the *height equivalent to a theoretical plate* (*HETP*), is acceptable practice in chromatography design even though it is based on a poor model of column operation. HETP is a measure of zone broadening; in general, the lower the HETP value the narrower is the solute peak.

HETP depends on various processes which occur during elution of a chromatography sample. A popular and simple expression for HETP takes the form:

$$H = \frac{A}{u} + B u + C$$

(10.44)

where H is plate height, u is linear liquid velocity, and A, B and C are experimentally-determined kinetic constants. A, B and C include the effects of liquid–solid mass transfer, forward and backward axial dispersion, and non-ideal distribution of liquid around the packing. As outlined in Section 10.6.4, overall rates of solute adsorption and desorption in chromatography depend mainly on mass-transfer steps. Values of A, B and C are reduced by improving mass transfer between liquid and solid phases, resulting in a decrease in HETP and better column performance. Eq. (10.44) and other HETP models are discussed further in other references [16, 17].

HETP for a particular component is related to the elution volume and width of the solute peak as it appears on the chromatogram. If, as shown in Figure 10.18(a), the pulse has the standard symmetrical form of a normal distribution around a mean value $\bar{x}$, the *number of theoretical plates* can be calculated as follows:

$$N = 16 \left(\frac{V_e}{w} \right)^2$$

(10.45)

where N is number of theoretical plates, V_e is the distance on the chromatogram corresponding to the elution volume of the solute, and w is the base line width of the peak between lines drawn tangent to the inflection points of the curve. Eq. (10.45) applies if the sample is introduced into the column as a narrow pulse. Number of theoretical plates is related to HETP as follows:

$$N = \frac{L}{H}$$

(10.46)

where L is the length of the column. For a given column, the greater the number of theoretical plates the greater is the number of ideal equilibrium stages in the system and the more efficient is the separation. Values of H and N vary for a particular column depending on the component being separated.

10.7.4 Resolution

Resolution is a measure of zone overlap in chromatography and an indicator of column efficiency. For separation of two components, resolution is given by the following equation:

$$R_N = \frac{2(V_{e2} - V_{e1})}{(w_1 + w_2)} \tag{10.47}$$

where R_N is resolution, V_{e1} and V_{e2} are distances on the chromatogram corresponding to elution volumes for components 1 and 2, and w_1 and w_2 the baseline widths of the chromatogram peaks as shown in Figure 10.18(b). Column resolution is a dimensionless quantity; the greater the value of R_N the more separated are the two solute peaks. An R_N value of 1.5 corresponds to a baseline resolution of 99.8% or virtual complete separation; when $R_N = 1.0$ the two peaks overlap by about 2% of the total peak area.

Column resolution can be expressed in terms of HETP. Assuming w_1 and w_2 are approximately equal, Eq. (10.47) becomes:

$$R_N = \frac{(V_{e2} - V_{e1})}{w_2} . \tag{10.48}$$

Substituting for w_2 from Eq. (10.45):

$$R_N = \frac{1}{4} \sqrt{N} \left(\frac{V_{e2} - V_{e1}}{V_{e2}} \right). \tag{10.49}$$

The term

$$\frac{(V_{e2} - V_{e1})}{V_{e2}}$$

can be expressed in terms of k_2 and δ from Eqs (10.37) and (10.38), so that Eq. (10.49) becomes:

$$R_N = \frac{1}{4} \sqrt{N} \left(\frac{\delta - 1}{\delta} \right) \left(\frac{k_2}{k_2 + 1} \right). \tag{10.50}$$

Using the expression for N from Eq. (10.46), the equation for column resolution is:

$$R_N = \frac{1}{4} \sqrt{\frac{L}{H}} \left(\frac{\delta - 1}{\delta} \right) \left(\frac{k_2}{k_2 + 1} \right). \tag{10.51}$$

As is apparent from Eq. (10.51), peak resolution increases as a function of $\sqrt{L}$, where L is the column length. Resolution also increases with decreasing HETP; therefore, any enhancement of mass-transfer conditions reducing H will improve resolution. Derivation of Eq. (10.51) involves Eq. (10.45), which applies to chromatography systems where a relatively small quantity of sample is injected rapidly. Resolution is sensitive to increases in sample size; as the amount of sample increases, resolution declines. In laboratory analytical work it is common to use extremely small sample volumes, of the order of microlitres. However, depending on the type of chromatography used for production-scale purification, sample volumes 5–20% of the column volume and higher are used. Because resolution under these conditions is relatively poor, if the solute peak is collected for isolation of product the central portion of the peak is retained while the leading and trailing edges are recycled back to the feed.

10.7.5 Scaling-Up Chromatography

The aim in scale-up of chromatography is to retain the resolution and solute recovery achieved using a small-scale column while increasing the throughput of material. Strategies for scale up must take into account the dominance of mass-transfer effects in chromatography separations.

The easiest approach to scale-up is to simply increase the flow rate through the column. This gives unsatisfactory results; raising the liquid velocity increases zone spreading and produces high pressures in the column which compress the stationary phase and cause pumping difficulties. The pressure-drop problem can be alleviated by increasing the particle size; however this hinders the overall mass-transfer process and so decreases resolution. Increasing the column length can help regain any resolution lost by increasing the flow rate or particle diameter; however increasing L also has a strong effect in raising the pressure drop through the column.

The solution to scale-up is to keep the same column length, linear flow velocity and particle size as in the small column, but increase the column diameter. The larger capacity of the column is therefore due solely to its greater cross-sectional area. Sample volume and volumetric flow rate are increased in proportion to column volume. In this way, all the important parameters affecting the packing matrix, liquid flow, mass transfer and equilibrium conditions are kept constant; similar column performance can therefore be expected. Because liquid distribution in large-diameter packed columns tends to be poor, care must be taken to ensure liquid is fed evenly over the entire column cross-section.

In practice, variations in column properties and efficiency do occur with scale-up. As an example, compressible solids such as those used in gel chromatography get better support from the column wall in small columns than in large columns; as a result, lower linear flow rates must be used in large-scale systems. An advantage of using gels of high mechanical strength in laboratory systems is that they allow more direct scale-up to commercial operation. The elasticity and compressibility of gels used for fractionation of high-molecular-weight proteins preclude use of long columns in large-scale processes; bed heights in these systems are normally restricted to 0.6–1.0 m.

10.8 Summary of Chapter 10

At the end of Chapter 10 you should:

(i) know what is a *unit operation*;
(ii) be able to describe generally the steps of *downstream processing*;
(iii) understand the theory and practice of *filtration*;
(iv) understand the principles of *centrifugation*, including scale-up considerations;
(v) be familiar with methods used for *cell disruption*;
(vi) understand the concept of an *ideal stage*;
(vii) be able to analyse *aqueous two-phase extractions* in terms of the equilibrium *partition coefficient, product yield* and *concentration factor*;
(viii) understand the principles of *adsorption* operations and design of *fixed-bed adsorbers*;
(ix) know the different types of *chromatography* used for separation of biomolecules; and
(x) understand the concepts of *differential migration, zone spreading* and *resolution* in chromatography, know what operating conditions enhance chromatography performance, and be able to describe *scale-up procedures* for chromatography columns.

Problems

10.1 Bacterial filtration

A suspension of *Bacillus subtilis* cells is filtered under constant pressure for recovery of protease. A pilot-scale filter is used to measure filtration properties. The filter area is 0.25 m², the pressure drop is 360 mmHg, and the filtrate viscosity is 4.0 cP. The cell suspension deposits 22 g cake per litre of filtrate. The following data are measured.

Time (min)	2	3	6	10	15	20
Filtrate volume (l)	10.8	12.1	18.0	21.8	28.4	32.0

(a) Determine the specific cake resistance and filter medium resistance.
(b) What size filter is required to process 4000 l cell suspension in 30 min at a pressure drop of 360 mmHg?

10.2 Filtration of mycelial suspensions

Pelleted and filamentous forms of *Streptomyces griseus* are filtered separately using a small laboratory filter of area 1.8 cm². The mass of wet solids per ml of filtrate is 0.25 g ml^{-1} for the pelleted cells and 0.1 g ml^{-1} for the filamentous culture. Viscosity of the filtrate is 1.4 cP. Five filtration experiments at different pressures are carried out with each suspension. The results are as follows:

	Pressure drop (mmHg)				
	100	250	350	550	750
Filtrate volume (ml) *for pelleted* suspension			Time (s)		
10	22	12	9	7	5
15	52	26	20	14	12
20	90	49	36	28	22
25	144	75	60	43	34
30	200	110	88	63	51
35	285	149	119	84	70
40	368	193	154	110	90
45	452	240	195	140	113
50	–	301	238	175	141

	Pressure drop (mmHg)				
	100	250	350	550	750
Filtrate volume (ml) *for* filamentous suspension			Time (s)		
10	36	22	17	13	11
15	82	47	40	31	25
20	144	85	71	53	46
25	226	132	111	85	70
30	327	194	157	121	100
35	447	262	215	166	139
40	–	341	282	222	180
45	–	434	353	277	229
50	–	–	442	338	283

(a) Evaluate the specific cake resistance as a function of pressure for each culture.

(b) Determine the compressibility for each culture.

(c) A filter press with area 15 m^2 is used to process 20 m^3 filamentous *S. griseus* culture. If the filtration must be completed in one hour, what pressure drop is required?

10.3 Rotary-drum vacuum filtration

Continuous rotary vacuum filtration can be analysed by considering each revolution of the drum as a stationary batch filtration. Per revolution, each cm^2 of filter cloth is used to form cake only for the period of time it spends submerged in the liquid reservoir. A rotary-drum vacuum filter with drum diameter 1.5 m and filter width 1.2 m is used to filter starch from an aqueous slurry. The pressure drop is kept constant at 4.5 psi; the filter operates with 30% of the filter cloth submerged. Resistance due to the filter medium is negligible. Laboratory tests with a 5 cm^2 filter have shown that 500 ml slurry can be filtered in 23.5 min at a pressure drop of 12 psi; the starch cake was also found to be compressible with $s = 0.57$. Use the following steps to determine the drum speed required to produce 20 m^3 filtered liquid per hour.

(a) Evaluate $\mu_f \alpha' c$ from the laboratory test data.

(b) If N is the drum speed in revolutions per hour, what is the cycle time?

(c) From (b), for what period of time per revolution is each cm^2 of filter cloth used for cake formation?

(d) What volume of filtrate must be filtered per revolution to achieve the desired rate of 20 m^3 per hour?

(e) Apply Eq. (10.11) to a single revolution of the drum to evaluate N.

(f) The liquid level is raised so that the fraction of submerged filter area increases from 30% to 50%. What drum speed is required under these conditions?

10.4 Centrifugation of yeast

Yeast cells are to be separated from a fermentation broth. Assume that the cells are spherical with diameter 5 μm and density 1.06 g cm^{-3}. The viscosity of the culture broth is 1.36 × 10^{-3} N s m^{-2}. At the temperature of separation, the density of the suspending fluid is 0.997 g cm^{-3}. 500 litres broth must be treated every hour.

(a) Specify Σ for a suitably-sized disc-stack centrifuge.

(b) The small size and low density of microbial cells are disadvantageous in centrifugation. If instead of yeast, quartz particles of diameter 0.1 mm and specific gravity 2.0 are separated from the culture liquid, by how much is Σ reduced?

10.5 Centrifugation of food particles

Small food particles with diameter 10^{-2} mm and density 1.03 g cm^{-3} are suspended in liquid of density 1.00 g cm^{-3}. The viscosity of the liquid is 1.25 mPa s. A tubular-bowl centrifuge of length 70 cm and radius 11.5 cm is used to separate the particles. If the centrifuge is operated at 10 000 rpm, estimate the feed flow rate at which the food particles are just removed from the suspension.

10.6 Scale-up of disc-stack centrifuge

A pilot-scale disc-stack centrifuge is tested for recovery of bacteria. The centrifuge contains 25 discs with inner and outer diameters 2 cm and 10 cm, respectively. The half-cone angle is 35°. When operated at a speed of 3000 rpm with a feed rate of 3.5 litre min^{-1}, 70% of the cells are recovered. If a bigger centrifuge is to be used for industrial treatment of 80 litres min^{-1}, what operating speed is required to achieve the same sedimentation performance if the larger centrifuge contains 55 discs with outer diameter 15 cm, inner diameter 4.7 cm, and half-cone angle 45°?

10.7 Centrifugation of yeast and cell debris

A tubular-bowl centrifuge is used to concentrate a suspension of genetically-engineered yeast containing a new recombinant protein. At a speed of 12 000 rpm, the centrifuge treats 3 l broth min^{-1} with satisfactory results. It is proposed to use the same centrifuge to separate cell debris from homogenate produced by mechanical disruption of the yeast. If the average size of the debris is one-third that of the yeast and the viscosity of the homogenate is five times greater than the cell suspension, what flow rate can be handled if the centrifuge is operated at the same speed?

10.8 Cell disruption

Micrococcus bacteria are disrupted at 5°C in a Manton-Gaulin homogeniser operated at pressures between 200 and 550 kg$_f$ cm^{-2}. Data for protein release as a function of number of passes through the homogeniser are as follows:

	Pressure drop ($kg_f \, cm^{-2}$)				
	200	300	400	500	550
Number of passes			% protein release		
1	5.0	13.5	23.3	36.0	42.0
2	9.5	23.5	40.0	58.5	66.0
3	14.0	33.5	52.5	75.0	83.7
4	18.0	43.0	66.6	82.5	88.5
5	22.0	47.5	73.0	88.5	94.5
6	26.0	55.0	79.5	91.3	–

(a) How many passes are required to achieve 80% protein release at an operating pressure of 460 $kg_f \, cm^{-2}$?

(b) Estimate the pressure required to deliver 70% protein recovery in only two passes?

10.9 Enzyme purification using two-phase aqueous partitioning

Leucine dehydrogenase is recovered from a homogenate of disrupted *Bacillus cereus* cells using an aqueous two-phase polyethylene glycol–salt system. 150 litres of homogenate initially containing 3.2 units enzyme ml^{-1} are processed; a polyethylene glycol–salt mixture is added and two phases form. The enzyme partition coefficient is 3.5.

(a) What volume ratio of upper and lower phases must be chosen to achieve 80% recovery of enzyme in a single extraction step?

(b) If the volume of the lower phase is 100 litres, what is the concentration factor for 80% recovery?

10.10 Recovery of viral particles

Cells of the fall armyworm *Spodoptera frugiperda* are cultured in a fermenter to produce viral particles for insecticide. Viral particles are released into the culture broth after lysis of the host cells. The initial culture volume is 5 litres. An aqueous two-phase polymer solution of volume 2 litres is added to this liquid; the volume of the bottom phase is 1 litre. The virus partition coefficient is 10^{-2}.

(a) What is the yield of virus at equilibrium?

(b) Write a mass balance for viral particles in terms of concentrations and volumes of the phases, equating the amounts of virus present before and after addition of polymer solution.

(c) Derive an equation for the concentration factor in terms of liquid volumes and the partition coefficient only.

(d) Calculate the concentration factor for the viral extraction.

10.11 Gel chromatography scale-up

Gel chromatography is to be used for commercial-scale purification of a proteinaceous diphtheria toxoid from *Corynebacterium diphtheriae* supernatant. In the laboratory, a small column of 1.5 cm inner diameter and height 0.4 m is packed with 10 g dry Sephadex gel; the void volume is measured as 23 ml. A sample containing the toxoid and impurities is injected into the column. At a liquid flow rate of 14 ml min^{-1}, the elution volume for the toxoid is 29 ml; the elution volume for the principal impurity is 45 ml. A column of height 0.6 m and diameter 0.5 m is available for large-scale gel chromatography. The same type of packing is used; the void fraction and ratio of pore volume to total bed volume remain the same as in the bench-scale column. The liquid flow rate in the large column is scaled up in proportion to the column cross-sectional area; the flow patterns in both columns can be assumed identical. The water regain value for the packing is given by the manufacturer as 0.0035 $m^3 \, kg^{-1}$ dry gel.

(a) Which is the larger molecule, the diphtheria toxoid or the principal impurity?

(b) Determine the partition coefficients for the toxoid and impurity.

(c) Estimate the elution volumes in the commercial-scale column.

(d) What is the volumetric flow rate in the large column?

(e) Estimate the retention time of toxoid in the large column?

10.12 Protein separation using chromatography

Human insulin A and B from recombinant *Escherichia coli* are separated using pilot-scale affinity chromatography. Laboratory studies have shown that capacity factors for the proteins are 0.85 for the A-chain and 1.05 for the B-chain. The dependence of HETP on liquid velocity satisfies the following type of equation:

$$H = \frac{A}{u} + B u + C$$

where u is linear liquid velocity and A, B and C are constants. Values of A, B and C for the insulin system were found to be 2 $\times 10^{-9} \, m^2 \, s^{-1}$, 1.5 s and 5.7×10^{-5} m, respectively. Two columns with inner diameter 25 cm are available for the process; one is 1.0 m high, the other is 0.7 m high.

(a) Plot the relationship between H and u from $u = 0.1 \times 10^{-4} \, m \, s^{-1}$ to $u = 2 \times 10^{-4} \, m \, s^{-1}$.

(b) What is the minimum HETP? At what liquid velocity is the minimum HETP obtained?

(c) If the larger column is used with a liquid flow rate of 0.31 litres min^{-1}, will the two insulin chains be completely separated?

(d) If the smaller column is used, what is the maximum liquid flow rate that will give complete separation?

References

1. Dwyer, J.L. (1984) Scaling up bio-product separation with high performance liquid chromatography. *Bio/Technology* **2**, 957–964.

2. Belter, P.A., E.L. Cussler and W.-S. Hu (1988) *Bioseparations: Downstream Processing for Biotechnology*, John Wiley, New York.

3. Flaschel, E., C. Wandrey and M.-R. Kula (1983) Ultrafiltration for the separation of biocatalysts. *Adv. Biochem. Eng./Biotechnol.* **26**, 73–142.

4. Fane, A.G. and J.M. Radovich (1990) Membrane systems. In: J.A. Asenjo (Ed), *Separation Processes in Biotechnology*, pp. 209–262, Marcel Dekker, New York.

5. Kula, M.-R. (1985) Recovery operations. In: H.-J. Rehm and G. Reed (Eds), *Biotechnology*, vol. 2, pp. 725–760, VCH, Weinheim.

6. Hsu, H.-W. (1981) *Separations by Centrifugal Phenomena*, John Wiley, New York.

7. Ambler, C.M. (1952) The evaluation of centrifuge performance. *Chem. Eng. Prog.* **48**, 150–158.

8. Ambler, C.M. (1988) Centrifugation. In: P.A. Schweitzer (Ed), *Handbook of Separation Techniques for Chemical Engineers*, 2nd edn, pp. 4-59–4-88, McGraw-Hill, New York.

9. Perry, R.H., D.W. Green and J.O. Maloney (Eds) (1984) *Chemical Engineers' Handbook*, 6th edn, pp. 19-89–19-96, McGraw-Hill, New York.

10. Hetherington, P.J., M. Follows, P. Dunnill and M.D. Lilly (1971) Release of protein from bakers' yeast (*Saccharomyces cerevisiae*) by disruption in an industrial homogeniser. *Trans. IChE* **49**, 142–148.

11. Engler, C.R. and C.W. Robinson (1981) Effects of organism type and growth conditions on cell disruption by impingement. *Biotechnol. Lett.* **3**, 83–88.

12. Coulson, J.M. and J.F. Richardson (1991) *Chemical Engineering*, vol. 2, 4th edn, Chapters 17 and 18, Pergamon Press, Oxford.

13. Snyder, L.R. (1974) Classification of the solvent properties of common liquids. *J. Chromatog.* **92**, 223–230.

14. Johnson, E.L. and R. Stevenson (1978) *Basic Liquid Chromatography*, Varian Associates, Palo Alto.

15. Scopes, R.K. (1982) *Protein Purification*, Springer-Verlag, New York.

16. Giddings, J.C. (1965) *Dynamics of Chromatography*, Part I, Marcel Dekker, New York.

17. Heftmann, E. (Ed) (1967) *Chromatography*, 2nd edn, Reinhold, New York.

Suggestions for Further Reading

There is an extensive literature on downstream processing of biomolecules. The following is a small selection.

Downstream Processing (see also refs 2 and 5)

Asenjo, J.A. (1990) (Ed) *Separation Processes in Biotechnology*, Marcel Dekker, New York.

Atkinson, B. and F. Mavituna (1991) *Biochemical Engineering and Biotechnology Handbook*, 2nd edn, Chapters 16 and 17, Macmillan, Basingstoke.

van Brakel, J. and H.H. Kleizen (1990) Problems in downstream processing. In: M.A. Winkler (Ed), *Chemical Engineering Problems in Biotechnology*, pp. 95–165, Elsevier Applied Science, London.

Filtration (see also refs 2–4)

Coulson, J.M. and J.F. Richardson (1991) *Chemical Engineering*, vol. 2, 4th edn, Chapter 7, Pergamon Press, Oxford.

McCabe, W.L. and J.C. Smith (1976) *Unit Operations of Chemical Engineering*, 3rd edn, pp. 922–948, McGraw-Hill, Tokyo.

Nestaas, E. and D.I.C. Wang (1981) A new sensor, the 'filtration probe', for quantitative characterization of the penicillin fermentation. I. Mycelial morphology and culture activity. *Biotechnol. Bioeng.* **23**, 2803–2813.

Oolman, T. and T.-C. Liu (1991) Filtration properties of mycelial microbial broths. *Biotechnol. Prog.* **7**, 534–539.

Centrifugation (see also refs 2, 6 and 8)

Axelsson, H.A.C. (1985) Centrifugation. In: M. Moo-Young (Ed), *Comprehensive Biotechnology*, vol. 2, pp. 325–346, Pergamon Press, Oxford.

Coulson, J.M. and J.F. Richardson (1991) *Chemical Engineering*, vol. 2, 4th edn, Chapter 9, Pergamon Press, Oxford.

Cell Disruption (see also refs 10 and 11)

Chisti, Y. and M. Moo-Young (1986) Disruption of microbial cells for intracellular products. *Enzyme Microb. Technol.* **8**, 194–204.

Dunnill, P. and M.D. Lilly (1975) Protein extraction and recovery from microbial cells. In: S.R. Tannenbaum and D.I.C. Wang (Eds), *Single-Cell Protein II*, pp. 179–207, MIT Press, Cambridge, Massachusetts.

Engler, C.R. (1985) Disruption of microbial cells. In: M. Moo-Young (Ed), *Comprehensive Biotechnology*, vol. 2, pp. 305–324, Pergamon Press, Oxford.

Kula, M.-R. and H. Schütte (1987) Purification of proteins and the disruption of microbial cells. *Biotechnol. Prog.* **3**, 31–42.

Aqueous Two-Phase Liquid Extraction

Albertsson, P.-Å. (1971) *Partition of Cell Particles and Macromolecules*, 2nd edn, John Wiley, New York.

Diamond, A.D. and J.T. Hsu (1992) Aqueous two-phase systems for biomolecule separation. *Adv. Biochem. Eng./Biotechnol.* **47**, 89–135.

Kroner, K.H., H. Schütte, W. Stach and M.-R. Kula (1982) Scale-up of formate dehydrogenase by partition. *J. Chem. Tech. Biotechnol.* **32**, 130–137.

Kula, M.-R., K.H. Kroner and H. Hustedt (1982) Purification of enzymes by liquid–liquid extraction. *Adv. Biochem. Eng.* **24**, 73–118.

Kula, M.-R. (1985) Liquid–liquid extraction of biopolymers. In: M. Moo-Young (Ed), *Comprehensive Biotechnology*, vol. 2, pp. 451–471, Pergamon Press, Oxford.

Adsorption (see also refs 2 and 15)

Arnold, F.H., H.W. Blanch and C.R. Wilke (1985) Analysis of affinity separations. I. Predicting the performance of affinity adsorbers. *Chem. Eng. J.* **30**, B9–B23.

Hines, A.L. and R.N. Maddox (1985) *Mass Transfer: Fundamentals and Applications*, Chapter 14, Prentice-Hall, New Jersey.

Slejko, F.L. (1985) (Ed) *Adsorption Technology*, Marcel Dekker, New York.

Chromatography (see also refs 1, 16 and 17)

Chisti, Y. and M. Moo-Young (1990) Large scale protein separations: engineering aspects of chromatography. *Biotech. Adv.* **8**, 699–708.

Cooney, J.M. (1984) Chromatographic gel media for large scale protein purification. *Bio/Technology*, **2** 41–43, 46–51, 54–55.

Delaney, R.A.M. (1980) Industrial gel filtration of proteins. In: R.A. Grant (Ed), *Applied Protein Chemistry*, pp. 233–280, Applied Science, London.

Janson, J.-C. and P. Hedman (1982) Large-scale chromatography of proteins. *Adv. Biochem. Eng.* **25**, 43–99.

Ladisch, M.R. (1987) Separation by sorption. In: H.R. Bungay and G. Belfort (Eds), *Advanced Biochemical Engineering*, pp. 219–237, John Wiley, New York.

Robinson, P.J., M.A. Wheatley, J.-C. Janson, P. Dunnill and M.D. Lilly (1974) Pilot scale affinity chromatography: purification of β-galactosidase. *Biotechnol. Bioeng.* **16**, 1103–1112.

Part 4
Reactions
and Reactors

Homogeneous Reactions

The heart of a typical bioprocess is the reactor or fermenter. Flanked by unit operations which carry out physical changes for medium preparation and recovery of products, the reactor is where the major chemical and biochemical transformations occur. In many bioprocesses, characteristics of the reaction determine to a large extent the economic feasibility of the project.

Of most interest in biological systems are *catalytic* reactions. By definition, a catalyst is a substance which affects the rate of reaction without altering the reaction equilibrium or undergoing permanent change itself. Enzymes, enzyme complexes, cell organelles and whole cells perform catalytic roles; the latter may be viable or non-viable, growing or non-growing. Biocatalysts can be of microbial, plant or animal origin. Cell growth is an *autocatalytic reaction*: this means that the catalyst is a product of the reaction. The performance of catalytic reactions is characterised by variables such as the reaction rate and yield of product from substrate. These parameters must be taken into account when designing and operating reactors.

In engineering analysis of catalytic reactions, a distinction is made between *homogeneous* and *heterogeneous* reactions. A reaction is homogeneous if the temperature and all concentrations in the system are uniform. Most fermentations and enzyme reactions carried out in mixed vessels fall into this category. In contrast, heterogeneous reactions take place in the presence of concentration or temperature gradients. Analysis of heterogeneous reactions requires application of mass-transfer principles in conjunction with reaction theory. Heterogeneous reactions are treated in Chapter 12.

This chapter covers the basic aspects of reaction theory which allow us to quantify the extent and speed of homogeneous reactions and to identify important factors affecting reaction rate.

11.1 Basic Reaction Theory

Reaction theory has two fundamental parts: *reaction thermodynamics* and *reaction kinetics*. Reaction thermodynamics is concerned with *how far* the reaction can proceed; no matter how fast a reaction is, it cannot continue beyond the point of chemical equilibrium. On the other hand, reaction kinetics is concerned with the *rate* at which reactions proceed.

11.1.1 Reaction Thermodynamics

Consider a reversible reaction represented by the following equation:

$$A + b\,B \rightleftharpoons y\,Y + z\,Z. \tag{11.1}$$

A, B, Y and Z are chemical species; b, y and z are stoichiometric coefficients. If the components are left in a closed system for an infinite period of time, the reaction proceeds until *thermodynamic equilibrium* is reached. At equilibrium there is no net driving force for further change; the reaction has reached the limit of its capacity for chemical transformation in a closed system. Composition of the equilibrium mixture is determined exclusively by the thermodynamic properties of the reactants and products; it is independent of the way the reaction is executed. Equilibrium concentrations are related by the *equilibrium constant, K*. For the reaction of Eq. (11.1):

$$K = \frac{C_{Ye}{}^{y} C_{Ze}{}^{z}}{C_{Ae}\,C_{Be}{}^{b}} \tag{11.2}$$

where C_{Ae}, C_{Be}, C_{Ye} and C_{Ze} are equilibrium concentrations of A, B, Y and Z, respectively. The value of K depends on temperature as follows:

$$\ln K = \frac{-\Delta G^{\circ}_{rxn}}{RT} \tag{11.3}$$

where ΔG°_{rxn} is the *change in standard free energy* per mole of A reacted, R is the ideal gas constant and T is absolute temperature. Values of R are listed in Table 2.5 (p. 20). The superscript $^{\circ}$ in ΔG°_{rxn} indicates standard conditions. Usually, the standard condition for a substance is its most stable form at 1 atm pressure and 25°C; however, for biochemical reactions occurring in solution, other standard conditions may be used [1]. ΔG°_{rxn} is equal to the difference in *standard free energy of formation*, G°, between products and reactants:

$$\Delta G^\circ_{rxn} = y\,G^\circ_Y + z\,G^\circ_Z - G^\circ_A - b\,G^\circ_B.$$

(11.4)

Standard free energies of formation are available in handbooks such as those listed in Section 2.6.

Free energy G is related to enthalpy H, entropy S and absolute temperature T as follows:

$$\Delta G = \Delta H - T\Delta S.$$

(11.5)

Therefore, from Eq. (11.3):

$$\ln K = \frac{-\Delta H^\circ_{rxn}}{RT} + \frac{\Delta S^\circ_{rxn}}{R}.$$

(11.6)

Thus, for exothermic reactions with negative ΔH°_{rxn}, K decreases with increasing temperature. For endothermic reactions and positive ΔH°_{rxn}, K increases with temperature.

Example 11.1 Effect of temperature on glucose isomerisation

Glucose isomerase is used extensively in the USA for production of high-fructose syrup. The reaction is:

glucose $\rightleftharpoons$ fructose.

ΔH°_{rxn} for this reaction is 5.73 kJ gmol^{-1}; ΔS°_{rxn} is 0.0176 kJ gmol^{-1} K^{-1}.

(a) Calculate the equilibrium constants at 50°C and 75°C.
(b) A company aims to develop a sweeter mixture of sugars, i.e. one with a higher concentration of fructose. Considering equilibrium only, would it be more desirable to operate the reaction at 50°C or 75°C?

Solution:
(a) Convert temperatures to degrees Kelvin (K) using the formula of Eq. (2.24):

$T = 50°C = 323.15$ K
$T = 75°C = 348.15$ K.

From Table 2.5, $R = 8.3144$ J gmol^{-1} K^{-1} = 8.3144 × 10^{-3} kJ gmol^{-1} K^{-1}. Using Eq. (11.6)

$$\ln K\,(50°C) = \frac{-5.73\text{ kJ gmol}^{-1}}{(8.3144 \times 10^{-3}\text{ kJ gmol}^{-1}\text{K}^{-1})\,(323.15\text{K})} + \frac{0.0176\text{ kJ gmol}^{-1}\text{K}^{-1}}{8.3144 \times 10^{-3}\text{ kJ gmol}^{-1}\text{K}^{-1}}$$

$K\,(50°C) = 0.98.$

Similarly for $T = 75°C$:

$$\ln K\,(75°C) = \frac{-5.73\text{ kJ gmol}^{-1}}{(8.3144 \times 10^{-3}\text{ kJ gmol}^{-1}\text{K}^{-1})\,(348.15\text{K})} + \frac{0.0176\text{ kJ gmol}^{-1}\text{K}^{-1}}{8.3144 \times 10^{-3}\text{ kJ gmol}^{-1}\text{K}^{-1}}$$

$K\,(75°C) = 1.15.$

(b) As K increases, the fraction of fructose in the equilibrium mixture increases. Therefore, from an equilibrium point of view, it is more desirable to operate the reactor at 75°C. However, other factors such as enzyme deactivation at high temperatures should also be considered.

limited number of commercially-important enzyme conversions, such as glucose isomerisation and starch hydrolysis, are treated as reversible reactions. In these systems, the reaction mixture at equilibrium contains significant amounts of reactants as well as products. However, for many reactions ΔG°_{rxn} is negative and large in magnitude. As a result, K is also very large, the reaction favours the products rather than the reactants, and the reaction is regarded as *irreversible*. Most enzyme and cell reactions fall into this category. For example, the equilibrium constant for sucrose hydrolysis by invertase is about 10^4; for fermentation of glucose to ethanol and carbon dioxide, K is about 10^{30}. The equilibrium ratio of products to reactants is so overwhelmingly large for these reactions that they are considered to proceed to completion, i.e. the reaction stops only when the concentration of one of the reactants falls to zero. Equilibrium thermodynamics has therefore only limited application to enzyme and cell reactions. Moreover, the thermodynamic principles outlined in this section apply only to closed systems; true thermodynamic equilibrium does not exist in living cells which exchange matter with their surroundings. Metabolic processes in cells are in a dynamic state; products formed are constantly removed or broken down so that reactions are driven forward. Most reactions in biological systems proceed to completion in a finite period of time at a finite rate.

If we know that complete conversion will eventually take place, the most useful reaction parameter to know is the rate at which the transformation proceeds. Another important characteristic, especially for systems in which many different reactions take place at the same time, is the proportion of reactant that is converted to the desired products. These properties of reactions are discussed in the remainder of this chapter.

11.1.2 Reaction Yield

The extent to which reactants are converted to products is expressed as the reaction *yield*. Generally speaking, yield is the amount of product formed or accumulated per amount of reactant provided or consumed. Unfortunately, there is no strict definition of yield; several different yield parameters are applicable in different situations. The terms used to express yield in this text do not necessarily have universal acceptance and are defined here for our convenience. Be prepared for other books to use different definitions.

Consider the simple enzyme reaction:

$$\text{L-histidine} \rightarrow \text{urocanic acid} + NH_3$$

$$(11.7)$$

catalysed by histidase. According to the reaction stoichiometry, 1 gmol urocanic acid is produced for each gmol L-histidine consumed; the yield of urocanic acid from histidine is therefore 1 gmol gmol^{-1}. However, let us assume that the histidase used in this reaction is contaminated with another enzyme, histidine decarboxylase. Histidine decarboxylase catalyses the following reaction:

$$\text{L-histidine} \rightarrow \text{histamine} + CO_2.$$

$$(11.8)$$

If both enzymes are active, some L-histidine will react with histidase according to Eq. (11.7), while some will be decarboxylated according to Eq. (11.8). After addition of the enzymes to the substrate, analysis of the reaction mixture shows that 1 gmol urocanic acid and 1 gmol histamine are produced for every 2 gmol histidine consumed. The *observed* or *apparent yield* of urocanic acid from L-histidine is $\frac{1\ gmol}{2\ gmol} = 0.5$ gmol gmol^{-1}. The observed yield of 0.5 gmol gmol^{-1} is different from the *stoichiometric, true* or *theoretical yield* of 1 gmol gmol^{-1} calculated from reaction stoichiometry because the reactant was channelled in two separate reaction pathways. An analogous situation arises if product rather that reactant is consumed in other reactions; the observed yield of product would be lower than the theoretical yield. *When reactants or products are involved in additional reactions, the observed yield may be different from the theoretical yield.*

The above analysis leads to two useful definitions of yield for reaction systems:

$$\begin{pmatrix} \text{true, stoichiometric or} \\ \text{theoretical yield} \end{pmatrix} = \frac{\begin{pmatrix} \text{total mass or moles of} \\ \text{product formed} \end{pmatrix}}{\begin{pmatrix} \text{mass or moles of reactant used} \\ \text{to form that particular product} \end{pmatrix}}$$

$$(11.9)$$

and

$$\begin{pmatrix} \text{observed or} \\ \text{apparent yield} \end{pmatrix} = \frac{(\text{mass or moles of product present})}{\begin{pmatrix} \text{total mass or moles of reactant} \\ \text{consumed} \end{pmatrix}}.$$

$$(11.10)$$

There is a third type of yield applicable in certain situations. For reactions with incomplete conversion of reactant, it may be of interest to specify the amount of product formed per amount of reactant *provided to the reaction* rather than actually consumed. For example, consider the isomerisation reaction catalysed by glucose isomerase:

glucose $\rightleftharpoons$ fructose.

$$(11.11)$$

The reaction is carried out in a closed reactor with pure enzyme. At equilibrium the sugar mixture contains 55 mol% glucose and 45 mol% fructose. The *theoretical yield* of fructose from glucose is 1 gmol gmol^{-1} because, from stoichiometry, formation of 1 gmol fructose requires 1 gmol glucose. The *observed yield* would also be 1 gmol gmol^{-1} if the reaction occurs in isolation. However if the reaction is started with glucose present only, the equilibrium yield of fructose per gmol glucose added to the reactor is 0.45 gmol gmol^{-1}. This type of yield for incomplete reactions may be denoted *gross yield*.

Example 11.2 Incomplete enzyme reaction

An enzyme catalyses the reaction:

$$A \rightleftharpoons B.$$

At equilibrium, the reaction mixture contains 63 wt% A.

(a) What is the equilibrium constant?
(b) If the reaction starts with A only, what is the equilibrium yield of B from A?

Solution:
(a) From stoichiometry the molecular weights of A and B must be equal: therefore wt% = mol%. From Eq. (11.2):

$$K = \frac{C_{Be}}{C_{Ae}}.$$

Using a basis of 1 gmol l^{-1}, C_{Ae} is 0.63 gmol l^{-1} and C_{Be} is 0.37 gmol l^{-1}. The value of K therefore is $^{0.37}/_{0.63} = 0.59$.
(b) From stoichiometry, the true yield of B from A is 1 gmol gmol^{-1}. However the gross yield is $^{0.37}/_{1.0} = 0.37$ gmol gmol^{-1}.

11.1.3 Reaction Rate

Consider the general irreversible reaction:

$$a\,A + b\,B \ \rightarrow \ y\,Y + z\,Z.$$

$$(11.12)$$

The rate of this reaction can be represented by the rate of conversion of compound A; let us use the symbol R_A to denote the *rate of reaction with respect to* A. R_A has units of, for example kg s^{-1}.

How do we measure reaction rates? For a general reaction system, rate of reaction is related to rate of change of mass in the system by the unsteady-state mass-balance equation derived in Chapter 6:

$$\frac{dM}{dt} = \hat{M}_i - \hat{M}_o + R_G - R_C.$$

$$(6.5)$$

In Eq (6.5), M is mass, t is time, $\hat{M}_i$ is mass flow rate into the system, $\hat{M}_o$ is mass flow rate out of the system, R_G is mass rate of generation by reaction and R_C is mass rate of consumption by reaction. Let us apply Eq. (6.5) to compound A, assuming that the reaction of Eq. (11.12) is the only reaction taking place that involves A. Rate of consumption R_C is equal to R_A and $R_G = 0$. The mass-balance equation becomes:

$$\frac{dM_A}{dt} = \hat{M}_{Ai} - \hat{M}_{Ao} - R_A.$$

$$(11.13)$$

Therefore, rate of reaction R_A can be determined if we measure the rate of change in mass of A, $^{dM_A}/_{dt}$, and the rates of flow of A in and out of the system, $\hat{M}_{Ai}$ and $\hat{M}_{Ao}$. *In a closed system* where $\hat{M}_{Ai} = \hat{M}_{Ao} = 0$, Eq. (11.13) becomes:

$$R_A = \frac{-dM_A}{dt}$$

$$(11.14)$$

nd reaction rate is measured simply by monitoring the change n mass of A in the system. Most measurements of reaction rate re carried out in closed systems so that the data can be analysed according to Eq. (11.14). $\frac{dM_A}{dt}$ is negative when A is consumed by reaction; therefore the minus sign in Eq. (11.14) is necessary to make R_A a positive quantity. Rate of reaction is sometimes called *reaction velocity*. Reaction velocity can also be measured in terms of components B, Y or Z. In a closed system:

$$R_B = \frac{-dM_B}{dt} \qquad R_Y = \frac{dM_Y}{dt} \qquad R_Z = \frac{dM_Z}{dt}$$

(11.15)

where M_B, M_Y and M_Z are masses of B, Y and Z, respectively. When reporting reaction rate, the reactant being monitored should be specified. Because R_Y and R_Z are based on product accumulation, these reaction rates are called *production rates* or *productivity*.

Eqs (11.14) and (11.15) define the rate of reaction in a closed system. However, reaction rate can be expressed using different measurement bases. In bioprocess engineering there re three distinct ways of expressing reaction rate which can be applied in different situations.

(i) *Total rate.* Total reaction rate is defined in Eqs (11.14) and (11.15) and is expressed as either mass or moles per unit time. Total rate is useful for specifying the output of a particular reactor or manufacturing plant. Production rates for factories are often expressed as total rates; for example: 'The production rate is 100 000 tonnes per year'. If additional reactors are built so that the reaction volume in the plant is increased, then clearly the total reaction rate would increase. Similarly, if the amount of cells or enzyme used in each reactor were also increased, then the total production rate would be improved even further.

(ii) *Volumetric rate.* Because the total mass of reactant converted in a reaction mixture depends on the size of the system, it is often convenient to specify reaction rate as the rate per unit volume. Units of volumetric rate are, e.g. kg m^{-3} s^{-1}. Rate of reaction expressed on a volumetric basis is used to account for differences in volume between reaction systems. Therefore, if the reaction mixture in a closed system has volume V:

$$r_A = \frac{R_A}{V} = \frac{-1}{V} \frac{dM_A}{dt}$$

(11.16)

where r_A is the volumetric rate of reaction with respect to A. When V is constant, Eq. (11.16) can be written:

$$r_A = \frac{-dC_A}{dt}$$

(11.17)

where C_A is the concentration of A in units of, e.g. kg m^{-3}. Volumetric rates are particularly useful for comparing the performance of reactors of different size. A common objective in optimising reaction processes is to maximise volumetric productivity so that the desired total production rate can be achieved with reactors of minimum size and therefore minimum cost.

(iii) *Specific rate.* Biological reactions involve enzyme and cell catalysts. Because the total rate of conversion depends on the amount of catalyst present, it is sometimes useful to specify reaction rate as the rate per quantity of enzyme or cells involved in the reaction. In a closed system, specific reaction rate can be measured as follows:

$$r_A = -\left(\frac{1}{X} \text{ or } \frac{1}{E}\right) \frac{dM_A}{dt}$$

(11.18)

where r_A is the specific rate of reaction with respect to A, X is the quantity of cells, E is the quantity of enzyme and $\frac{dM_A}{dt}$ is the rate of change of mass of A in the system. As quantity of cells is usually expressed as mass, units of specific rate for a cell-catalysed reaction would be, e.g. kg (kg cells)$^{-1}$ s^{-1} or simply s^{-1}. On the other hand, the mass of a particular enzyme added to a reaction is rarely known; most commercial enzyme preparations contain several components in unknown and variable proportions depending on the batch obtained from the manufacturer. To overcome these difficulties, enzyme quantity is often expressed as *units of activity* measured under specified conditions. One unit of enzyme is usually taken as the amount which catalyses conversion of 1 μmol substrate per minute at the optimal temperature, pH and substrate concentration. Therefore, if E in Eq. (11.18) is expressed as units of enzyme activity, the specific rate of reaction under process conditions could be reported as, e.g. kg (unit enzyme)$^{-1}$ s^{-1}. In a closed system where the volume of reaction mixture remains constant, an alternative expression for specific reaction rate is:

$$r_A = -\left(\frac{1}{x} \text{ or } \frac{1}{e}\right) \frac{dC_A}{dt}$$

(11.19)

where x is cell concentration and e is enzyme concentration.

Volumetric and total rates are not a direct reflection of catalyst performance; this is represented by the specific rate. Specific rates are employed when comparing different cells or enzymes. Specific rate is the rate achieved per unit catalyst and, under usual circumstances, is not dependent on the size of the system or the amount of catalyst present. Some care is necessary when interpreting results for reaction rate. For example, if two fermentations are carried out with different cell lines and the volumetric rate of reaction is greater in the first fermentation than in the second, you should not jump to the conclusion that the cell line in the first experiment is 'better', or capable of greater metabolic activity. It could be that the faster volumetric rate is due to the first fermenter being operated at a higher cell density than the second, leading to measurement of a more rapid rate per unit volume. Different strains of organism should be compared in terms of specific reaction rates.

Total, volumetric and specific productivities are interrelated concepts in process design. For example, high total productivity could be achieved with a catalyst of low specific activity if the reactor is loaded with a high catalyst concentration. If this is not possible, the volumetric productivity will be relatively low and a larger reactor is required to achieve the desired total productivity. In this book, the symbol R_A will be used to denote total reaction rate with respect to component A; r_A represents either volumetric or specific rate.

11.1.4 Reaction Kinetics

As reactions proceed, the concentrations of reactants decrease. In general, rate of reaction depends on reactant concentration so that the specific rate of conversion decreases simultaneously. Reaction rate also varies with temperature; most reactions speed up considerably as the temperature rises. *Reaction kinetics* refers to the relationship between rate of reaction and conditions which affect reaction velocity, such as reactant concentration and temperature. These relationships are conveniently described using *kinetic expressions* or *kinetic equations*.

Consider again the general irreversible reaction of Eq. (11.12). Often but not always, the volumetric rate of this reaction can be expressed as a function of reactant concentrations using the following mathematical form:

$$r_A = k\, C_A^a\, C_B^b$$

(11.20)

where k is the *rate constant* or *rate coefficient* for the reaction. By definition, the rate constant is independent of the concentration of reacting species but is dependent on other variables that influence reaction rate, such as temperature. When the kinetic equation has the form of Eq. (11.20), the reaction is said to be of *order a* with respect to component A and order b with respect to B. The order of the overall reaction is $(a + b)$. It is not usually possible to predict the order of reactions from stoichiometry. The mechanism of single reactions and the functional form of the kinetic expression must be determined by experiment. The dimensions and units of k depend on the order of the reaction.

11.1.5 Effect of Temperature on Reaction Rate

Temperature has a significant kinetic effect on reactions. Variation of the rate constant k with temperature is described by the *Arrhenius equation*:

$$k = A\, e^{-E/RT}$$

(11.21)

where k is the rate constant, A is the *Arrhenius constant* or *frequency factor*, E is the *activation energy* for the reaction, R is the ideal gas constant, and T is absolute temperature. Values of A are listed in Table 2.5 (p. 20). According to the Arrhenius equation, as T increases, k increases exponentially. Taking the natural logarithm of both sides of Eq. (11.21):

$$\ln k = \ln A - \frac{E}{RT}.$$

(11.22)

Thus, a plot of $\ln k$ versus $1/T$ gives a straight line with slope $-E/R$. For many reactions the value of E is positive and large, indicating a rapid increase in reaction rate with temperature.

11.2 Calculation of Reaction Rates From Experimental Data

As outlined in Section 11.1.3, the volumetric rate of reaction in a closed system can be found by measuring the rate of change in the mass of reactant present, provided the reactant is involved in only one reaction. Most kinetic studies of biological reactions are carried out in closed systems with a constant volume of reaction mixture; therefore, Eq. (11.17) can be used to evaluate the volumetric reaction rate. The concentration of a particular reactant or product is measured as a function of time. For a reactant such as A in Eq. (11.12), the results will be similar to those shown in Figure 11.1(a); the concentration will decrease with time. The volumetric rate of reaction is equal to dC_A/dt, which can be evaluated as the slope of

Figure 11.1 (a) Change in reactant concentration with time during reaction. (b) Graphical differentiation of concentration data by drawing a tangent.

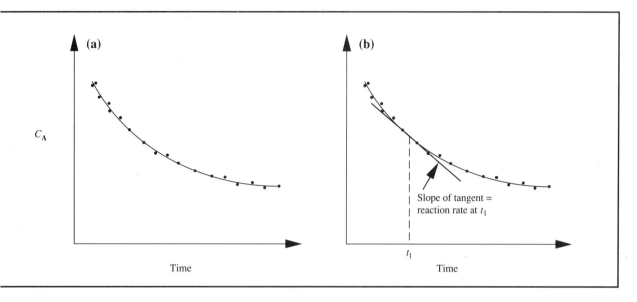

smooth curve drawn through the data points. The slope of the curve in Figure 11.1(a) changes with time; the reaction rate is greater at the beginning of the experiment than at the end.

One obvious way to determine reaction rate is to draw tangents to the curve of Figure 11.1(a) at various times and evaluate the slopes of the tangents; this is shown in Figure 11.1(b). If you have ever attempted this you will know that it can be extremely difficult, even though correct in principle. Drawing tangents to curves is a highly subjective procedure prone to great inaccuracy, even with special drawing devices designed for the purpose. The results depend strongly on the way the data are smoothed and the appearance of the curve at the points chosen. More reliable techniques are available for *graphical differentiation* of rate data. Graphical differentiation is valid only if the data can be presumed to differentiate smoothly.

11.2.1 Average Rate–Equal Area Method

This technique for determining rates is based on the *average rate–equal area construction*, and will be illustrated using data for oxygen uptake by immobilised cells. Results from measurement of oxygen concentration in a closed system as a function of time are listed in the first two columns of Table 11.1.

(i) Tabulate values of ΔC_A and Δt for each time interval as shown in Table 11.1. ΔC_A values are negative because C_A decreases over each interval.

Table 11.1 Graphical differentiation using the average rate–equal area construction

Time (t, min)	Oxygen concentration (C_A, ppm)	ΔC_A	Δt	$\Delta C_A/\Delta t$	dC_A/dt
0.0	8.00				−0.59
		−0.45	1.0	−0.45	
1.0	7.55				−0.38
		−0.33	1.0	−0.33	
2.0	7.22				−0.29
		−0.26	1.0	−0.26	
3.0	6.96				−0.23
		−0.20	1.0	−0.20	
4.0	6.76				−0.18
		−0.15	1.0	−0.15	
5.0	6.61				−0.14
		−0.12	1.0	−0.12	
6.0	6.49				−0.11
		−0.16	2.0	−0.08	
8.0	6.33				−0.06
		−0.08	2.0	−0.04	
10.0	6.25				−0.02

(ii) Calculate average oxygen uptake rates, $\Delta C_A/\Delta t$ for each time interval.

(iii) Plot $\Delta C_A/\Delta t$ on linear graph paper. Over each time interval a horizontal line is drawn to represent $\Delta C_A/\Delta t$ for that interval; this is shown in Figure 11.2.

(iv) Draw a smooth curve to cut the horizontal lines in such a manner that the shaded areas above and below the curve are equal for each time interval. The curve thus developed gives values of dC_A/dt for all points in time. Results for dC_A/dt at the times of sampling can be read from the curve and are tabulated in Table 11.1.

Figure 11.2 Graphical differentiation using the average rate–equal area construction.

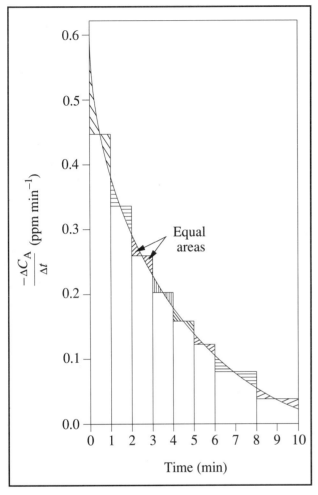

Figure 11.3 Average rate–equal area method for data with experimental error.

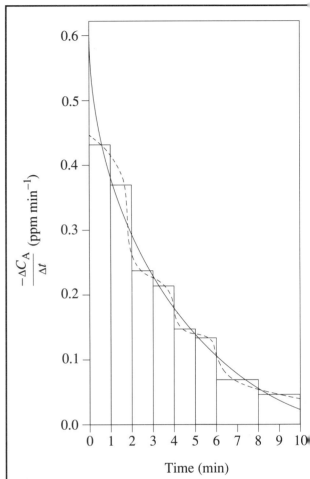

A disadvantage of the average rate–equal area method is that it is not easily applied if the data show scatter. If the concentration measurements are not very accurate, the horizontal lines representing $\Delta C_A/\Delta t$ may be located as shown in Figure 11.3. If we were to draw a curve equalising areas at each $\Delta C_A/\Delta t$ line, the rate curve would show complex behaviour oscillating up and down as indicated by the dashed line in Figure 11.3. Experience suggests that this is not a realistic representation of reaction rate. Because of the inaccuracies in measured data, we need several concentration measurements to define a change in rate. The data of Figure 11.3 are better represented using a smooth curve to equalise as far as possible the areas above and below adjacent groups of horizontal lines. For data showing even greater scatter, it may be necessary to average consecutive pairs of $\Delta C_A/\Delta t$ values to simplify graphical analysis.

A second graphical differentiation technique for evaluating dC_A/dt is described below.

11.2.2 Mid-Point Slope Method

In this method, the raw data are smoothed and values tabulated at intervals. The mid-point slope method is illustrated using the data of Table 11.1.

(i) Plot the raw data and smooth by hand. This is shown in Figure 11.4.

(ii) Mark off the smoothed curve at time intervals of ε. ε should be chosen so that the number of intervals is less than the number of data points measured; the less accurate the data the fewer should be the intervals. In this example, ε is taken as 1.0 min until $t = 6$ min; thereafter

Figure 11.4 Graphical differentiation using the mid-point slope method.

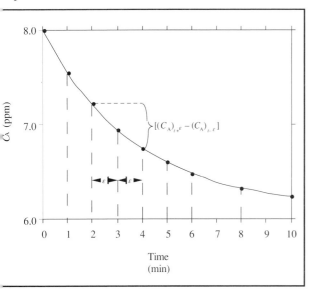

Table 11.2 Graphical differentiation using the mid-point slope method

Time (t, min)	Oxygen concentration (C_A, ppm)	ε	$[(C_A)_{t+\varepsilon} - (C_A)_{t-\varepsilon}]$	dC_A/dt
0.0	8.00	1.0	–	–
1.0	7.55	1.0	−0.78	−0.39
2.0	7.22	1.0	−0.59	−0.30
3.0	6.96	1.0	−0.46	−0.23
4.0	6.76	1.0	−0.35	−0.18
5.0	6.61	1.0	−0.27	−0.14
6.0	6.49	1.0	−0.22	−0.11
8.0	6.33	2.0	−0.24	−0.06
10.0	6.25	2.0	–	–

$= 2.0$ min. The intervals are marked in Figure 11.4 as dashed lines. Values of ε are entered in Table 11.2.

(iii) In the mid-point slope method, rates are calculated midway between two adjacent intervals of size ε. Therefore, the first rate determination is made for $t = 1$ min. Calculate the differences $[(C_A)_{t+\varepsilon} - (C_A)_{t-\varepsilon}]$ from Figure 11.4, where $(C_A)_{t+\varepsilon}$ denotes the concentration of A at time $t+\varepsilon$, and $(C_A)_{t-\varepsilon}$ denotes the concentration at time $t-\varepsilon$. A difference calculation is illustrated in Figure 11.4 for $t = 3$ min. Note that the concentrations are not taken from the list of original data but are read from the

smoothed curve. When $t = 6$ min, $\varepsilon = 1.0$; concentrations for the difference calculation are read from the curve at $t - \varepsilon = 5$ min and $t + \varepsilon = 7$ min. For the last rate determination at $t = 8$ min, $\varepsilon = 2.0$ and the concentrations are read from the curve at $t - \varepsilon = 6$ min and $t + \varepsilon = 10$ min.

(iv) The slope or rate is determined using the central-difference formula:

$$\frac{dC_A}{dt} = \frac{[(C_A)_{t+\varepsilon} - (C_A)_{t-\varepsilon}]}{2\varepsilon}.$$

(11.23)

These results are listed in Table 11.2. Values of dC_A/dt calculated using the two differentiation methods compare favourably. Application of both methods allows checking of the results.

11.3 General Reaction Kinetics For Biological Systems

The kinetics of many biological reactions are either zero-order, first-order or a combination of these called Michaelis–Menten kinetics. Kinetic expressions for biological systems are examined in this section.

11.3.1 Zero-Order Kinetics

If a reaction obeys zero-order kinetics, the reaction rate is independent of reactant concentration. The kinetic expression is:

$$r_A = k_0$$

(11.24)

where r_A is the volumetric rate of reaction with respect to A and k_0 is the *zero-order rate constant*. k_0 as defined in Eq. (11.24) is a volumetric rate constant with units of, e.g. kgmol m^{-3} s^{-1}. Because the volumetric rate of a catalytic reaction depends on the amount of catalyst present, when Eq. (11.24) is used to represent the rate of a cell or enzyme reaction, the value of k_0 includes the effect of catalyst concentration as well as the specific rate of reaction. We could write:

$$k_0 = k_0' e \quad \text{or} \quad k_0 = k_0'' x$$

(11.25)

where k_0' is the specific zero-order rate constant for enzyme reaction and e is the concentration of enzyme. Correspondingly, for cell reaction, k_0'' is the specific zero-order rate constant and x is cell concentration.

Let us assume we have collected concentration data for a particular reaction, and wish to determine the appropriate kinetic constant. If the reaction takes place in a closed, constant-volume system, rate of reaction can be evaluated directly as the rate of change in reactant concentration using the methods for graphical differentiation described in Section 11.2. From Eq. (11.24), if the reaction is zero-order the rate will be constant and equal to k_0 at all times during the reaction. Because the kinetic expression for zero-order reactions is relatively simple, rather than differentiate the concentration data it is easier to integrate Eq. (11.24) with $r_A = {-dC_A}/{dt}$ to obtain

an equation for C_A as a function of time. The experimenta data can then be checked directly against the integrated equa tion. Integrating Eq. (11.24) with initial condition $C_A = C_A$ at $t = 0$ gives:

$$C_A = \int -r_A \, dt = C_{A0} - k_0 t.$$

(11.26

Therefore, when the reaction is zero order, a plot of C_A versu time gives a straight line with slope $-k_0$. Application of Eq (11.26) is illustrated in Example 11.3.

Example 11.3 Kinetics of oxygen uptake

Serratia marcescens is cultured in minimal medium in a small stirred fermenter. Oxygen consumption is measured at a cell concentration of 22.7 g l^{-1} dry weight.

Time (min)	Oxygen concentration (mmol l^{-1})
0	0.25
2	0.23
5	0.21
8	0.20
10	0.18
12	0.16
15	0.15

(a) Determine the rate constant for oxygen uptake.
(b) If the cell concentration is reduced to 12 g l^{-1}, what is the value of the rate constant?

Solution:
(a) As indicated in Section 9.5.1, microbial oxygen consumption is a zero-order reaction over a wide range of oxygen concentrations above C_{crit}. To test if the measured data can be fitted using the zero-order model of Eq. (11.26), plot oxygen concentration as a function of time as shown in Figure 11E3.1.

Figure 11E3.1   Kinetic analysis of oxygen uptake.

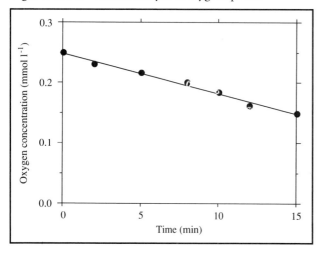

The zero-order model fits the data well. The slope is -6.7×10^{-3} mmol l^{-1} min^{-1}; therefore, $k_0 = 6.7 \times 10^{-3}$ mmol l^{-1} min^{-1}.

(b) For cells of the same age cultured under the same conditions, from Eq. (11.25), k_0 can be expected to be directly proportional to the number of cells present. Therefore, at a cell concentration of 12 g l^{-1}:

$$k_0 = \frac{12 \text{ g } l^{-1}}{22.7 \text{ g } l^{-1}} \; (6.7 \times 10^{-3} \text{ mmol } l^{-1} \text{ min}^{-1}) = 3.5 \times 10^{-3} \text{ mmol } l^{-1} \text{ min}^{-1}.$$

11.3.2 First-Order Kinetics

If a reaction obeys first-order kinetics, the relationship between reaction rate and reactant concentration is as follows:

$$r_A = k_1 C_A$$

(11.27)

where r_A is the volumetric rate of reaction and k_1 is the *first-order rate constant* with dimensions T^{-1}. Like the zero-order constant of the previous section, the value of k_1 depends on the catalyst concentration.

Let us assume we follow the progress of a particular reaction in a closed, constant-volume system by measuring the concentration of reactant A as a function of time. Under these conditions, $r_A = {-dC_A}/{dt}$. To determine whether the reaction follows first-order kinetics, we first integrate Eq. (11.27) with $r_A = {-dC_A}/{dt}$, and then check the measured concentration data against the resulting equation. Separating variables and integrating Eq. (11.27) with initial condition $C_A = C_{A0}$ at $t = 0$ gives:

$$C_A = C_{A0} e^{-k_1 t}.$$

(11.28)

Taking natural logarithms of both sides:

$$\ln C_A = \ln C_{A0} - k_1 t.$$

(11.29)

Therefore, for first-order reaction, a plot of $\ln C_A$ versus time gives a straight line with slope $-k_1$.

Example 11.4 Kinetics of gluconic acid production

Aspergillus niger is used to produce gluconic acid. Product synthesis is monitored in a fermenter; gluconic acid concentration is measured as a function of time for the first 39 h of culture.

Time (h)	Acid concentration (g l^{-1})
0	3.6
16	22
24	51
28	66
32	97
39	167

(a) Determine the rate constant.
(b) Estimate the product concentration after 20 h.

Solution:

(a) Test whether gluconic acid production can be modelled as a first-order reaction. If product concentration is measured rather than reactant concentration, in a closed reactor:

$$r_A = \frac{dC_A}{dt} = k_1 C_A$$

where A denotes gluconic acid. Integrating this equation and taking natural logarithms gives:

$$\ln C_A = \ln C_{A0} + k_1 t.$$

Therefore, a semi-log plot of gluconic acid concentration versus time will give a straight line with slope k_1. As shown in Figure 11E4.1, the first-order model fits the data well.

Figure 11E4.1   Kinetic analysis of gluconic acid production.

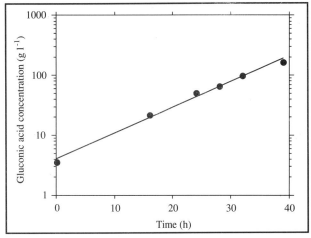

The slope and intercept are evaluated as described in Section 3.4.2; $k_1 = 0.10$ h^{-1}, $C_{A0} = 4.1$ g l^{-1}.
(b) The kinetic equation is:

$$C_A = 4.1 e^{0.10t}$$

where C_A has units g l^{-1} and t has units h. Therefore, at $t = 20$ h, $C_A = 30$ g l^{-1}.

11.3.3 Michaelis–Menten Kinetics

The kinetics of most enzyme reactions are reasonably well represented by the *Michaelis–Menten equation*:

$$r_A = \frac{v_{max} C_A}{K_m + C_A}$$

$$(11.30)$$

where r_A is the volumetric rate of reaction, C_A is the concentration of reactant A, v_{max} is the *maximum rate of reaction at infinite reactant concentration*, and K_m is the *Michaelis constant* for reactant A. v_{max} has the same dimensions as r_A; K_m has the same dimensions as C_A. Typical units for v_{max} are kgmol m^{-3} s^{-1}; typical units for K_m are kgmol m^{-3}. As defined in Eq. (11.30), v_{max} is a volumetric rate proportional to the amount of active enzyme present. The Michaelis constant K_m is equal to the reactant concentration at which $r_A = {}^{v_{max}}/_2$.

K_m values for some enzyme–substrate systems are listed in Table 11.3. K_m and other enzyme properties depend on the source of the enzyme.

If we adopt conventional symbols for biological reactions and call reactant A the *substrate*, Eq. (11.30) can be rewritten in the familiar form:

$$v = \frac{v_{max} s}{K_m + s}$$

$$(11.31)$$

where v is the volumetric rate of reaction and s is the substrate concentration. The biochemical basis of the Michaelis–Menten equation will not be covered here; discussion of enzyme reaction models and assumptions involved in derivation of Eq. (11.31) can be found in biochemistry texts [2, 3]. Suffice it to say here that the simplest reaction sequence which accounts for the kinetic properties of many enzymes is:

Table 11.3 Michaelis constants for some enzyme–substrate systems *(From B. Atkinson and F. Mavituna, 1991,* Biochemical Engineering and Biotechnology Handbook, *2nd edn, Macmillan, Basingstoke)*

Enzyme	Source	Substrate	K_m (mM)
Alcohol dehydrogenase	*Saccharomyces cerevisiae*	Ethanol	13.0
α-Amylase	*Bacillus stearothermophilus*	Starch	1.0
	Porcine pancreas	Starch	0.4
β-Amylase	Sweet potato	Amylose	0.07
Aspartase	*Bacillus cadaveris*	L-Aspartate	30.0
β-Galactosidase	*Escherichia coli*	Lactose	3.85
Glucose oxidase	*Aspergillus niger*	D-Glucose	33.0
	Penicillium notatum	D-Glucose	9.6
Histidase	*Pseudomonas fluorescens*	L-Histidine	8.9
Invertase	*Saccharomyces cerevisiae*	Sucrose	9.1
	Neurospora crassa	Sucrose	6.1
Lactate dehydrogenase	*Bacillus subtilis*	Lactate	30.0
Penicillinase	*Bacillus licheniformis*	Benzylpenicillin	0.049
Urease	Jack bean	Urea	10.5

$$E + S \underset{k_{-1}}{\overset{k_1}{\rightleftharpoons}} ES \overset{k_2}{\to} E + P \tag{11.32}$$

where E is enzyme, S is substrate and P is product. ES is the *enzyme–substrate complex*. Binding of substrate to the enzyme in the first step is considered reversible with forward reaction constant k_1 and reverse reaction constant k_{-1}. Decomposition of the enzyme–substrate complex to give the product is an irreversible reaction with rate constant k_2; k_2 is known as the *turnover number*. Analysis of this reaction sequence yields the relationship:

$$v_{max} = k_2\, e_a \tag{11.33}$$

where e_a is the concentration of active enzyme. As expected in catalytic reactions, enzyme E is recovered at the end of the reaction.

An essential feature of Michaelis–Menten kinetics is that the catalyst becomes saturated at high substrate concentrations. Figure 11.5 shows the form of Eq. (11.31); reaction rate v does not increase indefinitely with substrate concentration but approaches a limit, v_{max}. At high substrate concentrations $s \gg K_m$, K_m in the denominator of Eq. (11.31) is negligibly small compared with s so we can write:

$$v \approx \frac{v_{max}\, s}{s} \tag{11.34}$$

or

$$v \approx v_{max}. \tag{11.35}$$

Therefore, at high substrate concentrations, the reaction rate approaches a constant value independent of substrate concentration; in this concentration range, the reaction is essentially *zero order* with respect to substrate. On the other hand, at low substrate concentrations $s \ll K_m$, the value of s in the denominator of Eq. (11.31) is negligible compared with K_m, and Eq. (11.31) can be simplified to:

Figure 11.5   Michaelis–Menten plot.

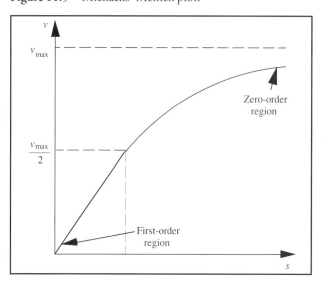

$$v \approx \frac{v_{max}}{K_m} s.$$

(11.36)

The ratio of constants v_{max}/K_m is, in effect, a first-order rate coefficient for the reaction. Therefore, at low substrate concentrations there is an approximate linear dependence of reaction rate on s; in this concentration range Michaelis–Menten reactions are essentially *first order* with respect to substrate.

The Michaelis–Menten equation is a satisfactory description of the kinetics of many industrial enzymes, although there are exceptions such as glucose isomerase and amyloglucosidase. More complex kinetic expressions must be applied if there are multiple substrates or inhibition effects [2–4]. Procedures for checking whether a particular reaction follows Michaelis-Menten kinetics and for evaluating v_{max} and K_m from experimental data are described in Section 11.4.

11.3.4 Effect of Conditions on Enzyme Reaction Rate

Rate of enzyme reaction is influenced by other conditions besides substrate concentration, such as temperature and pH. For enzymes with single rate-controlling steps, the effect of temperature is reasonably well described using the Arrhenius expression of Eq. (11.21) with v_{max} substituted for k. An example showing the relationship between temperature and rate of sucrose inversion by yeast invertase is given in Figure 11.6. Activation energies for enzyme reactions are of the order 40–80 kJ mol^{-1} [5]; as a rough guide, this means that a 10°C rise in temperature between 20°C and 30°C will increase the rate of reaction by a factor of 2–3.

Although an Arrhenius-type relationship between temperature and rate of reaction is observed for enzymes, the temperature range over which Eq. (11.21) is applicable is quite limited. Many proteins start to denature at 45–50°C; if the temperature is raised higher than this, thermal deactivation occurs and the reaction velocity quickly drops. Figure 11.7 illustrates how the Arrhenius relationship breaks down at high temperatures. In this experiment, the Arrhenius rate-law was obeyed between temperatures of about 0°C ($T = 273.15$ K; $1/T$ $= 3.66 \times 10^{-3}$ K^{-1}) and about 53°C ($T = 326.15$ K; $1/T = 3.07$ $\times 10^{-3}$ K^{-1}). With further increases in temperature the reaction rate declined rapidly due to thermal deactivation. Enzyme stability and rate of deactivation are important factors affecting overall catalytic performance in reactors. This topic is discussed further in the Section 11.5.

Figure 11.6 Arrhenius plot for inversion of sucrose by yeast invertase. (From I.W. Sizer, 1943, Effects of temperature on enzyme kinetics. *Adv. Enzymol.* **3**, 35–62.)

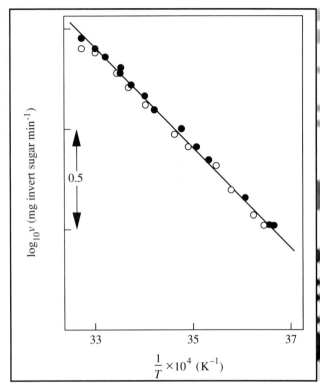

Figure 11.7 Arrhenius plot for catalase. The enzyme breaks down at high temperatures. (From I.W. Sizer, 1944, Temperature activation and inactivation of the crystalline catalase–hydrogen peroxide system. *J. Biol. Chem.* **154**, 461–473.)

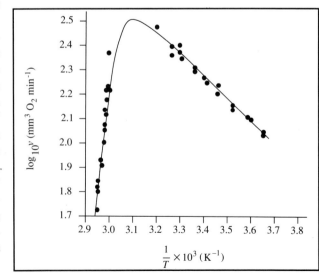

Figure 11.8   Effect of pH on enzyme activity. (From J.S. Fruton and S. Simmonds, 1958, *General Biochemistry*, 2nd edn, John Wiley, New York.)

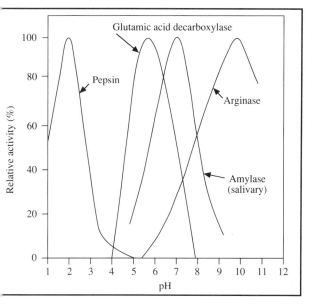

pH has a pronounced effect on enzyme kinetics, as illustrated in Figure 11.8. The reaction rate is maximum at some optimal pH and declines sharply if the pH is moved either side of the optimum value. Kinetic equations have been developed to describe the effect of pH on enzyme activity; however, the influence of pH is usually determined experimentally. Ionic strength and water activity also have considerable influence on rate of enzyme reaction but few correlations are available for prediction of these effects.

11.4 Determining Enzyme Kinetic Constants From Batch Data

To fully specify the kinetics of Michaelis–Menten reactions, two rate constants, v_{max} and K_m, must be evaluated. Estimating kinetic parameters for Michaelis–Menten reactions is not as straightforward as for zero- and first-order reactions. Several graphical methods are available; unfortunately some do not give accurate results.

The first step in kinetic analysis of enzyme reactions is to obtain data for rate of reaction v as a function of substrate concentration s. Rates of reaction can be determined from batch concentration data as described in Section 11.2. Typically, only *initial rate data* are used. This means that several batch experiments are carried out with different initial substrate concentrations; from each set of data the reaction rate is evaluated

at time zero. Initial rates and corresponding initial substrate concentrations are used as (v, s) pairs which can then be plotted in various ways for determination of v_{max} and K_m. Initial rate data are preferred for enzyme reactions because experimental conditions such as enzyme and substrate concentrations are known most accurately at the start of the reaction.

11.4.1 Michaelis–Menten Plot

This simple procedure involves plotting (v, s) values directly as shown in Figure 11.5. v_{max} and K_m can be estimated roughly from this graph; v_{max} is the rate as $s \to \infty$ and K_m is the value of s at $v = {}^{v_{max}}/_2$. The accuracy of this method is usually poor because of the difficulty of extrapolating to v_{max}.

11.4.2 Lineweaver–Burk Plot

This method uses a linearisation procedure to give a straight-line plot from which v_{max} and K_m can be determined. Inverting Eq. (11.31) gives:

$$\frac{1}{v} = \frac{K_m}{v_{max}\, s} + \frac{1}{v_{max}}$$

$$(11.37)$$

so that a plot of ${}^1/_v$ versus ${}^1/_s$ should give a straight line with slope ${}^{K_m}/_{v_{max}}$ and intercept ${}^1/_{v_{max}}$. This double-reciprocal plot is known as the *Lineweaver–Burk plot*, and is frequently found in the literature on enzyme kinetics. However, the linearisation process used in this method distorts the experimental error in v (see Section 3.3.4) so that these errors are amplified at low substrate concentrations. As a consequence, the Lineweaver–Burk plot often gives inaccurate results and is therefore not recommended [3].

11.4.3 Eadie–Hofstee Plot

If Eq. (11.37) is multiplied by

$$v\left(\frac{v_{max}}{K_m}\right)$$

and then rearranged, another linearised form of the Michaelis–Menten equation is obtained:

$$\frac{v}{s} = \frac{v_{max}}{K_m} - \frac{v}{K_m}.$$

$$(11.38)$$

According to Eq. (11.38), a plot of $v/_s$ versus v gives a straight line with slope $^{-1}/K_m$ and intercept v_{max}/K_m; this is called the *Eadie–Hofstee plot*. As with the Lineweaver–Burk plot, the Eadie–Hofstee linearisation distorts errors in the data so that the method has reduced accuracy.

11.4.4 Langmuir Plot

Multiplying Eq. (11.37) by s produces the linearised form of the Michaelis–Menten equation according to Langmuir:

$$\frac{s}{v} = \frac{K_m}{v_{max}} + \frac{s}{v_{max}}.$$

(11.39)

Therefore, a *Langmuir plot* of $^s/_v$ versus s should give a straight line with slope $^1/v_{max}$ and intercept K_m/v_{max}. Linearisation of data for the Langmuir plot minimises distortions in experimental error. Accordingly, its use for evaluation of v_{max} and K_m is recommended [6].

11.4.5 Direct Linear Plot

A different method for plotting enzyme kinetic data has been proposed by Eisenthal and Cornish-Bowden [7]. For each observation, reaction rate v is plotted on the vertical axis against s on the negative horizontal axis. This is shown in Figure 11.9 for four pairs of (v, s) data. A straight line is then drawn to join corresponding $(-s, v)$ points. In the absence of experimental error, lines for each $(-s, v)$ pair intersect at a unique point, (K_m, v_{max}). When real data containing errors are plotted, a family of intersection points is obtained. Each intersection gives one estimate of v_{max} and K_m; the median or middle v_{max} and K_m values are taken as the kinetic parameters for the reaction. This method is relatively insensitive to individual erroneous readings which may be far from the correct values. However a disadvantage of the procedure is that deviations from Michaelis–Menten behaviour are not easily detected. It is recommended therefore for enzymes which are known to obey Michaelis–Menten kinetics.

11.5 Kinetics of Enzyme Deactivation

Enzymes are protein molecules of complex configuration that can be destabilised by relatively weak forces. In the course of enzyme-catalysed reactions, enzyme deactivation occurs at a rate which is dependent on the structure of the enzyme and the reaction conditions. Environmental factors affecting enzyme stability include temperature, pH, ionic strength, mechanical forces and presence of denaturants such as solvents, detergents

Figure 11.9 Direct linear plot for determination of enzyme kinetic parameters. (From R. Eisenthal and A. Cornish-Bowden 1974, The direct linear plot: a new graphical procedure for estimating enzyme kinetic parameters. *Biochem. J.* **139**, 715–720.)

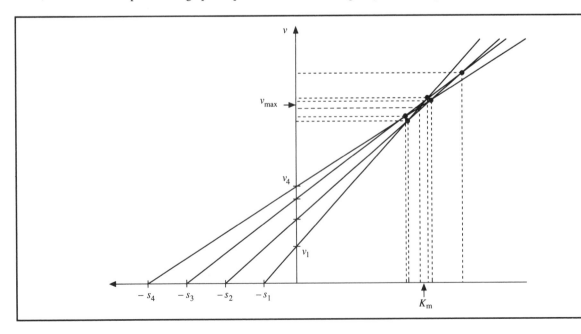

and heavy metals. Because the amount of active enzyme can decline considerably during reaction, in many applications the kinetics of enzyme deactivation are just as important as the kinetics of the reaction itself.

In the simplest model of enzyme deactivation, active enzyme E_a undergoes irreversible transformation to an inactive form E_i:

$$E_a \rightarrow E_i.$$

(11.40)

Rate of deactivation is generally considered to be first order in active enzyme concentration:

$$r_d = k_d \, e_a$$

(11.41)

where r_d is the volumetric rate of deactivation, e_a is the active enzyme concentration and k_d is the *deactivation rate constant*. In a closed system where enzyme deactivation is the only process affecting the concentration of active enzyme:

$$\frac{-de_a}{dt} = r_d = k_d \, e_a.$$

(11.42)

Integration of Eq. (11.42) gives an expression for active enzyme concentration as a function of time:

$$e_a = e_{a0} \, e^{-k_d t}$$

(11.43)

where e_{a0} is the concentration of active enzyme at time zero. According to Eq. (11.43), concentration of active enzyme decreases exponentially with time; the greatest rate of enzyme deactivation occurs when e_a is high.

As indicated in Eq. (11.33), the value of v_{max} for enzyme reaction depends on the amount of active enzyme present. Therefore, as e_a declines due to deactivation, v_{max} is also diminished. We can estimate the variation of v_{max} with time

by substituting into Eq. (11.33) the expression for e_a from Eq. (11.43):

$$v_{max} = k_2 e_{a0} e^{-k_d t} = v_{max0} e^{-k_d t}$$

(11.44)

where v_{max0} is the initial value of v_{max} before deactivation occurs.

Stability of enzymes is frequently reported in terms of *half-life*. Half-life is the time required for half the enzyme activity to be lost as a result of deactivation; after one half-life, the active enzyme concentration equals $e_{a0}/2$. Substituting $e_a = e_{a0}/2$ into Eq. (11.43), taking logarithms and rearranging yields the following expression:

$$t_h = \frac{\ln 2}{k_d}$$

(11.45)

where t_h is the enzyme half-life.

Rate of enzyme deactivation is strongly dependent on temperature. This dependency is generally well described using the Arrhenius equation:

$$k_d = A e^{-E_d/RT}$$

(11.46)

where A is the Arrhenius constant or frequency factor, E_d is the activation energy for enzyme deactivation, R is the ideal gas constant, and T is absolute temperature. According to Eq. (11.46), as T increases, rate of enzyme deactivation increases exponentially. Values of E_d are high, of the order 170–400 kJ gmol^{-1} for many enzymes [5]. Accordingly, a temperature rise of 10°C between 30°C and 40°C will increase the rate of enzyme deactivation by a factor between 10 and 150. The stimulatory effect of increasing temperature on rate of enzyme reaction has already been described in Section 11.3.4. However, as shown here, raising the temperature also reduces the amount of active enzyme present. It is clear that temperature has a critical effect on enzyme kinetics.

Example 11.5 Enzyme half-life

Amyloglucosidase from *Endomycopsis bispora* is immobilised in polyacrylamide gel. Activities of immobilised and soluble enzyme are compared at 80°C. Initial rate data measured at a fixed substrate concentration are listed below.

Time (min)	Enzyme activity (μmol ml^{-1} min^{-1})	
	Soluble enzyme	Immobilised enzyme
0	0.86	0.45
3	0.79	0.44
6	0.70	0.43
9	0.65	0.43
15	0.58	0.41
20	0.46	0.40
25	0.41	0.39
30	–	0.38
40	–	0.37

What is the half-life for each form of enzyme?

Solution:

From Eq. (11.31), at any fixed substrate concentration, the rate of enzyme reaction v is directly proportional to v_{max}. Therefore, k_d can be determined from Eq. (11.44) using enzyme activity v instead of v_{max}. Taking natural logarithms gives:

$$\ln v = \ln v_0 - k_d\, t$$

where v_0 is the initial enzyme activity before deactivation. So, if deactivation follows a first-order model, a semi-log plot of reaction rate versus time should give a straight line with slope $-k_d$. The data are plotted in Figure 11E5.1.

Figure 11E5.1 Kinetic analysis of enzyme deactivation.

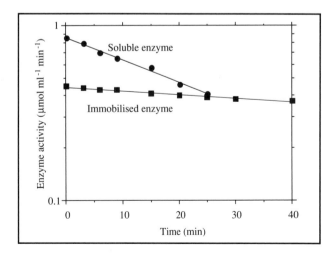

From the slopes, k_d for soluble enzyme is 0.03 min^{-1}; k_d for immobilised enzyme is 0.005 min^{-1}. Applying Eq. (11.45) for half-life:

$$t_h \text{ (soluble)} = \frac{\ln 2}{0.03 \text{ min}^{-1}} = 23 \text{ min}$$

$$t_h \text{ (immobilised)} = \frac{\ln 2}{0.005 \text{ min}^{-1}} = 139 \text{ min.}$$

Immobilisation significantly enhances stability of the enzyme.

11.6 Yields in Cell Culture

The basic concept of reaction yield was introduced in Section 11.1.2 for simple one-step reactions. When we consider processes such as cell growth, we are in effect lumping together many individual enzyme and chemical conversions. Despite this complexity, yield principles can be applied to cell metabolism to relate flow of substrate in metabolic pathways to formation of biomass and other products. Yields which are frequently reported and of particular importance are expressed using *yield coefficients* or *yield factors*. Several yield coefficients, such as yield of biomass from substrate, yield of biomass from oxygen, and yield of product from substrate, are in common use. Yield coefficients allow us to quantify the nutrient requirements and production characteristics of an organism.

Some metabolic yield coefficients: the biomass yield Y_{XS}, the product yield Y_{PS}, and the respiratory quotient RQ, were introduced in Chapter 4. Definition of yield coefficients can be generalised as follows:

$$Y_{FG} = \frac{-\Delta F}{\Delta G} \tag{11.47}$$

where Y_{FG} is the yield factor, F and G are substances involved in metabolism, ΔF is the mass or moles of F produced, and ΔG is the mass or moles of G consumed. The negative sign is required in Eq. (11.47) because ΔG for a consumed substance is negative in value; yield is calculated as a positive quantity. A list of frequently-used yield coefficients is given in Table 11.4. Note that in some cases, such as Y_{PX}, both substances represented by the yield coefficient are products of metabolism. Although the term 'yield' usually refers to the amount of product formed per amount of reactant, yields can also be used to relate other quantities. Some yield coefficients are based on quantities such as ATP formed or heat evolved during metabolism.

Table 11.4 Some metabolic yield coefficients

Symbol	Definition
Y_{XS}	Mass or moles of biomass produced per unit mass or mole of substrate consumed. (Moles of biomass can be calculated from the 'molecular formula' for biomass; see Section 4.6.1)
Y_{PS}	Mass or moles of product formed per unit mass or mole of substrate consumed
Y_{PX}	Mass or moles of product formed per unit mass or mole of biomass formed
Y_{XO}	Mass or moles of biomass formed per unit mass or mole of oxygen consumed
Y_{CS}	Mass or moles of carbon dioxide formed per unit mass or mole of substrate consumed
RQ	Moles of carbon dioxide formed per mole of oxygen consumed. This yield is called the *respiratory quotient.*
Y_{ATP}	Mass or moles of biomass formed per mole of ATP formed
Y_{kcal}	Mass or moles of biomass formed per kilocalorie of heat evolved during fermentation

11.6.1 Overall and Instantaneous Yields

A problem with application of Eq. (11.47) is that values of ΔF and ΔG depend on the time period over which they are measured. In batch culture, ΔF and ΔG can be calculated as the difference between initial and final values; this gives an *overall yield* representing some sort of average value for the entire culture period. Alternatively, ΔF and ΔG can be determined between two other points in time; this calculation might produce a different value of Y_{FG}. Yields can vary during culture, and it is sometimes necessary to evaluate the *instantaneous yield* at a particular point in time. For a closed, constant-volume reactor in which the reaction between F and G is the only reaction involving these components, if r_F and r_G are volumetric rates of production and consumption of F and G, respectively, instantaneous yield can be calculated as follows:

$$Y_{FG} = \lim_{\Delta G \to 0} \frac{-\Delta F}{\Delta G} = \frac{-dF}{dG} = \frac{-dF/_{dt}}{dG/_{dt}} = \frac{r_F}{r_G}.$$

(11.48)

For example, Y_{XS} at a particular instant in time is defined as:

$$Y_{XS} = \frac{r_X}{r_S} = \frac{\text{growth rate}}{\text{substrate consumption rate}}.$$

(11.49)

When yields for fermentation are reported, the time or time period to which they refer should be stated.

11.6.2 Theoretical and Observed Yields

As described in Section 11.1.2, it is necessary to distinguish between theoretical and observed yields. This is particularly important for cell metabolism because there are always many reactions occurring at the same time; theoretical and observed yields are therefore very likely to differ. Consider the example of biomass yield from substrate, Y_{XS}. If the total mass of substrate consumed is S_T, some proportion of S_T equal to S_G will be used for growth while the remainder, S_R, is channelled into other products and metabolic activities not related to growth. Therefore, the observed biomass yield based on total substrate consumption is:

$$Y'_{XS} = \frac{-\Delta X}{\Delta S_T} = \frac{-\Delta X}{\Delta S_G + \Delta S_R}$$

(11.50)

where ΔX is the amount of biomass produced and Y'_{XS} is the *observed biomass yield from substrate*. Values of observed biomass yields for several organisms and substrates are listed in Table 11.5. In comparison, the *true* or *theoretical biomass yield* from substrate is:

$$Y_{XS} = \frac{-\Delta X}{\Delta S_G}$$

(11.51)

as ΔS_G is the mass of substrate actually directed into biomass production. Because of the complexity of metabolism, ΔS_G is usually unknown and the observed yield is the only yield available. Theoretical yields are sometimes referred to as *maximum possible yields* because they represent the yield in the absence of competing reactions.

Table 11.5 Observed biomass yields for several microorganisms and substrates

(From S.J. Pirt, 1975, Principles of Microbe and Cell Cultivation, *Blackwell Scientific, Oxford)*

Microorganism	Substrate	Observed biomass yield Y'_{XS} (g g^{-1})
Aerobacter cloacae	Glucose	0.44
Penicillium chrysogenum	Glucose	0.43
Candida utilis	Glucose	0.51
	Acetic acid	0.36
	Ethanol	0.68
Candida intermedia	*n*-Alkanes (C$_{16}$–C$_{22}$)	0.81
Pseudomonas sp.	Methanol	0.41
Methylococcus sp.	Methane	1.01

Example 11.6 Yields in acetic acid production

The equation for aerobic production of acetic acid from ethanol is:

$$\underset{\text{(ethanol)}}{C_2H_5OH} + O_2 \to \underset{\text{(acetic acid)}}{CH_3CO_2H} + H_2O.$$

Acetobacter aceti bacteria are added to vigorously-aerated medium containing 10 g l^{-1} ethanol. After some time, the ethanol concentration is 2 g l^{-1} and 7.5 g l^{-1} acetic acid is produced. How does the overall yield of acetic acid from ethanol compare with the theoretical yield?

Solution:
Using a basis of 1 litre, the observed yield over the entire culture period is obtained from application of Eq. (11.10):

$$Y'_{PS} = \frac{7.5\ g}{(10-2)\ g} = 0.94\ g\ g^{-1}.$$

Theoretical yield is based on the mass of ethanol actually used for synthesis of acetic acid. From the stoichiometric equation:

$$Y'_{PS} = \frac{1\ gmol\ acetic\ acid}{1\ gmol\ ethanol} = \frac{60\ g}{46g} = 1.30\ g\ g^{-1}.$$

The observed yield is 72% theoretical.

11.7 Cell Growth Kinetics

The kinetics of cell growth are expressed using equations similar to those presented in Section 11.3. From a mathematical point of view there is little difference between the kinetic equations for enzymes and cells; after all, cell metabolism depends on the integrated action of a multitude of enzymes.

11.7.1 Batch Growth

Several phases of cell growth are observed in batch culture; a typical growth curve is shown in Figure 11.10. The different phases of growth are more readily distinguished when the natural logarithm of viable cell concentration is plotted against time; alternatively, a semi-log plot can be used. Rate of growth varies depending on the growth phase. During the lag phase immediately after inoculation, rate of growth is essentially zero. Cells use the lag phase to adapt to their new environment; new enzymes or structural components may be synthesised. Following the lag period, growth starts in the acceleration phase and continues through the growth and decline phases. If growth is exponential, the growth phase appears as a straight line on a semi-log plot. As nutrients in the culture medium become depleted or inhibitory products accumulate, growth slows down and the cells enter the decline phase. After this transition period, the stationary phase is reached during which no further growth occurs. Some cultures exhibit a death phase as the cells lose viability or are destroyed by lysis. Table 11.6 provides a summary of growth and metabolic activity during the phases of batch culture.

During the growth and decline phases, rate of cell growth is described by the equation:

$$r_X = \mu x \tag{11.52}$$

where r_X is the volumetric rate of biomass production with units of, for example, kg m^{-3} s^{-1}, x is viable cell concentration

Figure 11.10 Typical batch growth curve.

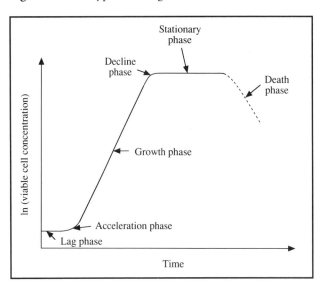

Table 11.6 Summary of batch cell growth

Phase	Description	Specific growth rate
Lag	Cells adapt to the new environment; no or very little growth	$\mu \approx 0$
Acceleration	Growth starts	$\mu < \mu_{max}$
Growth	Growth achieves its maximum rate	$\mu \approx \mu_{max}$
Decline	Growth slows due to nutrient exhaustion or build-up of inhibitory products	$\mu < \mu_{max}$
Stationary	Growth ceases	$\mu = 0$
Death	Cells lose viability and lyse	$\mu < 0$

with units of, for example, kg m^{-3}, and μ is the *specific growth rate*. Specific growth rate has dimensions T^{-1}. Eq. (11.52) has the same form as (11.27); cell growth is therefore considered a *first-order autocatalytic reaction*. In a closed system where growth is the only process affecting cell concentration, $r_X = \frac{dx}{dt}$, and integration of Eq. (11.52) gives an expression for x as a function of time. If μ is constant we can integrate directly with initial condition $x = x_0$ at $t = 0$ to give:

$$x = x_0 \, e^{\mu t}$$

(11.53)

where x_0 is the viable cell concentration at time zero. Eq. (11.53) represents *exponential growth*. Taking natural logarithms:

$$\ln x = \ln x_0 + \mu t.$$

(11.54)

According to Eq. (11.54), a plot of $\ln x$ versus time gives a straight line with slope μ. Because the relationship of Eq. (11.54) is strictly valid only if μ is unchanging, a plot of $\ln x$ versus t is often used to assess whether the specific growth rate is constant. As shown in Figure 11.10, μ is usually constant during the growth phase. It is always advisable to prepare a semi-log plot of cell concentration before identifying phases of growth. As shown in Figure 3.6, if cell concentrations are plotted on linear coordinates, growth often appears slow at the beginning of the culture. We might be tempted to conclude there was a lag phase of 1–2 hours for the culture represented in Figure 3.6(a). However, when the same data are plotted using logarithms as shown in Figure 3.6(b), it is clear that the culture did not experience a lag phase. Growth always appears much slower at the beginning of culture because the number of cells present is small.

Cell growth rates are often expressed in terms of the *doubling time* t_d. An expression for doubling time can be derived from Eq. (11.53). Starting with a cell concentration of x_0, the concentration at $t = t_d$ is $2x_0$. Substituting these values into Eq. (11.53):

$$2x_0 = x_0 \, e^{\mu t_d}$$

(11.55)

and cancelling x_0 gives:

$$2 = e^{\mu t_d}.$$

(11.56)

Taking the natural logarithm of both sides:

$$\ln 2 = \mu t_d$$

(11.57)

or

$$t_d = \frac{\ln 2}{\mu}.$$

(11.58)

11.7.2 Balanced Growth

In an environment favourable for growth, cells regulate their metabolism and adjust the rates of various internal reactions so that a condition of *balanced growth* occurs. During balanced growth, composition of the biomass remains constant. Balanced growth means that the cell is able to modulate the effect of external conditions and keep the cell composition steady despite changes in environmental conditions.

For biomass composition to remain constant during growth, the specific rate of production of each component in the culture must be equal to the cell specific growth-rate μ:

$$r_Z = \mu z$$

(11.59)

where Z is a cellular constituent such as protein, RNA, polysaccharide, etc., r_Z is the volumetric rate of production of Z, and z is the concentration of Z in the reactor volume. Therefore, during balanced growth the doubling time for each cell component must be equal to t_d for growth. Balanced growth cannot be achieved if environmental changes affect rate of growth. In most cultures, balanced growth occurs at the same time as exponential growth.

11.7.3 Effect of Substrate Concentration

During the growth and decline phases of batch culture, the specific growth rate of cells is dependent on the concentration of nutrients in the medium. Often, a single substrate exerts a dominant influence on rate of growth; this component is known as the *growth-rate-limiting substrate* or, more simply, the *growth-limiting substrate*. The growth-limiting substrate is often the carbon or nitrogen source, although in some cases it is oxygen or another oxidant such as nitrate. During balanced growth, the specific growth rate is related to the concentration of growth-limiting substrate by the *Monod equation*, a homologue of the Michaelis–Menten expression:

$$\mu = \frac{\mu_{max} s}{K_S + s}.$$

(11.60)

Figure 11.11   Relationship between specific growth rate and concentration of growth-limiting substrate in cell culture.

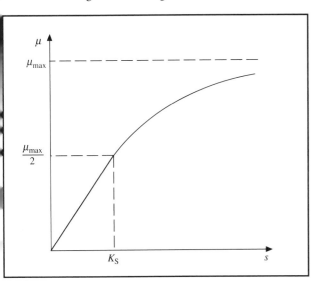

Table 11.7 K_S values for several organisms

(From S.J. Pirt, 1975, Principles of Microbe and Cell Cultivation, *Blackwell Scientific, Oxford; and D.I.C. Wang, C.L. Cooney, A.L. Demain, P. Dunnill, A.E. Humphrey and M.D. Lilly, 1979,* Fermentation and Enzyme Technology, *John Wiley, New York)*

Microorganism (genus)	Limiting substrate	K_S (mg l^{-1})
Saccharomyces	Glucose	25
Escherichia	Glucose	4.0
	Lactose	20
	Phosphate	1.6
Aspergillus	Glucose	5.0
Candida	Glycerol	4.5
	Oxygen	0.042–0.45
Pseudomonas	Methanol	0.7
	Methane	0.4
Klebsiella	Carbon dioxide	0.4
	Magnesium	0.56
	Potassium	0.39
	Sulphate	2.7
Hansenula	Methanol	120.0
	Ribose	3.0
Cryptococcus	Thiamine	1.4×10^{-7}

In Eq. (11.60), s is the concentration of growth-limiting substrate, μ_{max} is the *maximum specific growth rate*, and K_S is the *substrate constant*. μ_{max} has dimensions T^{-1}; K_S has the same dimensions as substrate concentration. The form of Eq. (11.60) is shown in Figure 11.11. μ_{max} and K_S are intrinsic parameters of the cell–substrate system; values of K_S for several organisms are listed in Table 11.7.

Typical values of K_S are very small, of the order of mg per litre for carbohydrate substrates and µg per litre for other compounds such as amino acids. The level of growth-limiting substrate in culture media is normally much greater than K_S. As a result, growth can be approximated using zero-order kinetics with growth rate independent of substrate concentration until s reaches very low values. By analogy with Michaelis–Menten kinetics described in Section 11.3.3, $\mu \approx \mu_{max}$ provided s is greater than about 10 K_S. Because K_S is usually very small compared with the starting substrate concentration, s remains > 10 K_S during most of the culture period. This explains why μ remains constant and equal to μ_{max} in batch culture until the medium is virtually exhausted of substrate. When s finally falls below 10 K_S, transition from growth to stationary phase can be very abrupt as the low level of residual substrate is rapidly consumed by the large number of cells present.

The Monod equation is by far the most frequently-used expression relating growth rate to substrate concentration. However, it is valid only for balanced growth and should not be applied when growth conditions are changing rapidly.

There are also other restrictions; for example, the Monod equation has been found to have limited applicability at extremely low substrate levels. When growth is inhibited by high substrate or product concentrations, extra terms can be added to the Monod equation to account for these effects. Several other kinetic expressions have been developed for cell growth; these provide better correlations with experimental data in certain situations [8–11].

11.8 Growth Kinetics With Plasmid Instability

A potential problem in culture of recombinant organisms is plasmid loss or inactivation. Plasmid instability occurs in individual cells which, by reproducing, can generate a large plasmid-free population in the reactor and reduce the overall rate of synthesis of plasmid-encoded products. Plasmid instability occurs as a result of DNA mutation or defective plasmid segregation. For segregational stability, the total number of plasmids present in the culture must double once per generation, and the plasmid copies must be equally distributed between mother and daughter cells.

A simple model has been developed for batch culture to describe changes in the fraction of plasmid-bearing cells as a function of time [12]. The important parameters in this model are the probability of plasmid loss per generation of cells, and the difference in the growth rates of plasmid-bearing and plasmid-free cells. Exponential growth of the host cells is assumed. If x^+ is the concentration of plasmid-carrying cells and x^- is the concentration of plasmid-free cells, the rates at which the two cell populations grow are:

$$r_{X^+} = (1 - p)\, \mu^+ x^+$$

(11.61)

and

$$r_{X^-} = p\, \mu^+ x^+ + \mu^- x^-$$

(11.62)

where r_{X^+} is the rate of growth of the plasmid-bearing population, r_{X^-} is the rate of growth of the plasmid-free population, p is the probability of plasmid loss per cell division ($p \leqslant 1$), μ^+ is the specific growth rate of plasmid-carrying cells, and μ^- is the specific growth rate of plasmid-free cells. The model assumes that all plasmid-containing cells are identical in growth rate and probability of plasmid loss; this is the same as assuming that all plasmid-containing cells have the same copy number. By comparing Eq. (11.61) with Eq. (11.52) we can see that the rate of growth of the plasmid-bearing population is reduced by $p\, \mu^+ x^+$. This is because some of the progeny of plasmid-bearing cells do not contain plasmid and do not join the plasmid-bearing population. On the other hand, growth of the plasmid-free population has two contributions as indicated in Eq. (11.62). Existing plasmid-free cells grow with specific growth rate μ^- as usual; in addition, this population is supplemented by generation of plasmid-free cells due to defective plasmid segregation by plasmid-carrying cells.

At any time, the fraction of cells in the culture with plasmid is:

$$F = \frac{x^+}{x^+ + x^-}.$$

(11.63)

In batch culture where rates of growth can be determined by monitoring cell concentration, $r_{X^+} = {}^{dx^+}/_{dt}$ and $r_{X^-} = {}^{dx^-}/_{dt}$. Therefore, Eqs (11.61) and (11.62) can be integrated simul-

taneously with initial condition $x^+ = x_0^+$ and $x^- = x_0^-$ at $t = 0$. After n generations of plasmid-containing cells:

$$F = \frac{1 - \alpha - p}{1 - \alpha - 2^{n(\alpha + p - 1)} p}$$

(11.64)

where

$$\alpha = \frac{\mu^-}{\mu^+}$$

(11.65)

and

$$n = \frac{\mu^+ t}{\ln 2}.$$

(11.66)

The value of F depends on α, the ratio of the specific growth rates of plasmid-free and plasmid-carrying cells. In the absence of selection pressure, presence of plasmid usually reduces the growth rate of organisms due to the additional metabolic requirements imposed by the plasmid DNA. Therefore α is usually > 1. In general, the difference between μ^+ and μ^- becomes more pronounced as the size of the

Table 11.8 Relative growth rates of plasmid-free and plasmid-carrying cells

(From T. Imanaka and S. Aiba, 1981, A perspective on the application of genetic engineering: stability of recombinant plasmid. Ann. N.Y. Acad. Sci. 369, 1–14)

Organism	Plasmid	$\alpha = \dfrac{\mu^-}{\mu^+}$
Escherichia coli C600	F′ *lac*	0.99–1.10
E. coli K12 EC1055	R1 *drd*-19	1.03–1.12
E. coli K12 IR713	TP120 (various)	1.50–2.31
E. coli JC7623	Col E1	1.29
	Col E1 derivative TnA insertion (various)	1.15–1.54
	Col E1 deletion mutant (various)	1.06–1.65
Pseudomonas aeruginosa PA01	TOL	2.00

Figure 11.12 Fraction of plasmid-carrying cells in batch culture after 25 generations. (From T. Imanaka and S. Aiba, 1981, A perspective on the application of genetic engineering: stability of recombinant plasmid. *Ann. N.Y. Acad. Sci.* **369**, 1–14.)

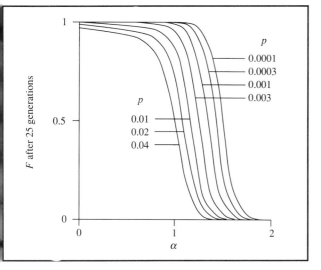

plasmid or copy number increases. Some values of α from the literature are listed in Table 11.8; typically $1.0 < \alpha < 2.0$. Under selection pressure α may equal zero; if the plasmid encodes biosynthetic enzymes for essential nutrients, loss of plasmid may result in $\mu^- = 0$. When this is the case, F remains close to 1 as the plasmid-free population cannot reproduce. F also depends on p, the probability of plasmid loss per generation, which can be as high as 0.1 if segregation occurs. When mutation or random insertions or deletions are the only cause of plasmid instability, p is usually much lower at about 10^{-6}. Plasmid fragmentation within a host cell can occur with higher frequency if the cloning vector is inherently unstable.

Batch culture of microorganisms usually requires 25 cell generations or more. Results for F after 25 generations have been calculated from Eq. (11.64) and are shown in Figure 11.12 as function of p and α. F deteriorates substantially as α increases from 1.0 to 2.0. Cultures with $p < 0.01$ and $\alpha < 1$ are relatively stable, with F after 25 generations remaining close to 1. Further application of Eq. (11.64) is illustrated in Example 11.7.

Example 11.7 Plasmid instability in batch culture

A plasmid-containing strain of *E. coli* is used to produce recombinant protein in a 250-litre fermenter. The probability of plasmid loss per generation is 0.005. The specific growth rate of plasmid-free cells is $1.4\ \text{h}^{-1}$; the specific growth-rate of plasmid-bearing cells is $1.2\ \text{h}^{-1}$. Estimate the fraction of plasmid-bearing cells after 18 h growth if the inoculum contains only cells with plasmid.

Solution:
The number of generations of plasmid-carrying cells in 18 h is calculated from Eq. (11.66):

$$n = \frac{(1.2\ \text{h}^{-1})\ 18\ \text{h}}{\ln 2} = 31.$$

Substituting this into Eq. (11.64) with $p = 0.005$ and $\alpha = {}^{1.4\ \text{h}^{-1}}/_{1.2\ \text{h}^{-1}} = 1.17$:

$$F = \frac{1 - 1.17 - 0.005}{1 - 1.17 - 2^{31(1.17 + 0.005 - 1)}0.005} = 0.45.$$

Therefore, after 18 h only 45% of the cells contain plasmid.

Alternative models for growth with plasmid instability have been developed [13]; some include equations for substrate utilisation and product formation [14]. A weakness in the simple model presented here is the assumption that all plasmid-containing cells are the same. In reality there are differences in copy number and therefore specific growth rate between plasmid-carrying cells; probability of plasmid loss also varies from cell to cell. More complex models that recognise the segregated nature of plasmid populations are available [15, 16].

11.9 Production Kinetics in Cell Culture

In this section we consider the kinetics of production of low-molecular-weight compounds, such as ethanol, amino acids, antibiotics and vitamins, which are excreted from cells in culture. As shown in Table 11.9, fermentation products can be classified according to the relationship between product synthesis and energy generation in the cell [10, 17]. Compounds in the first category are formed directly as end- or by-products of energy metabolism; these materials are synthesised in pathways which produce ATP. The second class of product is partly linked to energy generation but requires additional energy for synthesis. Formation of other products such as antibiotics involves reactions far removed from energy metabolism.

Table 11.9 Classification of low-molecular-weight fermentation products

Class of metabolite	Examples
Products directly associated with generation of energy in the cell	Ethanol, acetic acid, gluconic acid, acetone, butanol, lactic acid, other products of anaerobic fermentation
Products indirectly associated with energy generation	Amino acids and their products, citric acid, nucleotides
Products for which there is no clear direct or indirect coupling to energy generation	Penicillin, streptomycin, vitamins

Irrespective of the class of product, rate of product formation in cell culture can be expressed as a function of biomass concentration:

$$r_P = q_P x \tag{11.67}$$

where r_P is the volumetric rate of product formation with units of, for example, kg m^{-3} s^{-1}, x is biomass concentration, and q_P is the *specific rate of product formation* with dimensions T^{-1}. q_P can be evaluated at any time during fermentation as the ratio of production rate and biomass concentration; q_P is not necessarily constant during batch culture. Depending on whether the product is linked to energy metabolism or not, we can develop equations for q_P as a function of growth rate and other metabolic parameters.

11.9.1 Product Formation Directly Coupled With Energy Metabolism

For products formed in pathways which generate ATP, rate of production is related to cellular energy demand. Growth is usually the major energy-requiring function of cells; therefore, if production is coupled to energy metabolism, product will be formed whenever there is growth. However, ATP is also required for other activities called *maintenance*. Examples of maintenance functions include cell motility, turnover of cellular components and adjustment of membrane potential and internal pH. Maintenance activities are carried out by living cells even in the absence of growth. Products synthesised in energy pathways will be produced whenever maintenance functions are carried out because ATP is required. Kinetic expressions for product formation must account for growth-associated and maintenance-associated production, as in the following equation:

$$r_P = Y_{PX}\, r_X + m_P\, x. \tag{11.68}$$

In Eq. (11.68), r_X is the volumetric rate of biomass formation, Y_{PX} is the theoretical or true yield of product from biomass, m_P is the *specific rate of product formation due to maintenance*, and x is biomass concentration. m_P has dimensions T^{-1} and typical units kg product (kg biomass)$^{-1}$ s^{-1}. Eq. (11.68) states that rate of product formation depends partly on rate of growth but also partly on cell concentration. From Eq. (11.52), r_X is equal to μx; therefore:

$$r_P = (Y_{PX}\, \mu + m_P)x. \tag{11.69}$$

Comparison of Eqs (11.67) and (11.69) shows that, for products coupled to energy metabolism, q_P is equal to a combination of growth-associated and non-growth-associated terms:

$$q_P = Y_{PX}\, \mu + m_P. \tag{11.70}$$

11.9.2 Product Formation Indirectly Coupled With Energy Metabolism

When product is synthesised partly in metabolic pathways used for energy generation and partly in other pathways requiring energy, the relationship between product formation and growth can be complicated. We will not attempt to develop equations for q_P for this type of product. A generalised treatment of indirectly-coupled product formation is given by Roels and Kossen [10].

11.9.3 Product Formation Not Coupled With Energy Metabolism

Production not involving energy metabolism is difficult to relate to growth because growth and product synthesis are somewhat dissociated. However in some cases, rate of formation of non-growth-associated product is directly proportional to biomass concentration, so that production rate as defined in Eq. (11.67) can be applied with constant q_P. Sometimes q_P is a complex function of growth rate and must be expressed using empirical equations derived from experiment. An example is penicillin synthesis; equations for rate of penicillin production as a function of biomass concentration and specific growth rate have been derived by Heijnen *et al.* [18].

11.10 Kinetics of Substrate Uptake in Cell Culture

Cells consume substrate from the external environment and channel it into different metabolic pathways. Some substrate may be directed into growth and product synthesis; another fraction is used to generate energy for maintenance activities. Substrate requirements for maintenance vary considerably depending on the organism and culture conditions; a complete account of substrate uptake should include a maintenance component. The specific rate of substrate uptake for maintenance activities is known as the *maintenance coefficient*, m_S. The dimensions of m_S are T^{-1}; typical units are kg substrate (kg biomass)$^{-1}$ s^{-1}. Some examples of mainte-

nance coefficients for various microorganisms are listed in Table 11.10. Ionic strength greatly influences the value of m_S; significant amounts of energy are needed to maintain concentration gradients across cell membranes. The physiological significance of m_S has been the subject of much debate; there are indications that m_S for a particular organism may not be constant at all possible growth rates.

Rate of substrate uptake can be expressed as a function of biomass concentration by an equation analogous to Eq. (11.67):

$$r_S = q_S \, x$$

$$(11.71)$$

where r_S is the volumetric rate of substrate consumption with units of, for example, kg m^{-3} s^{-1}, q_S is the *specific rate of substrate uptake*, and x is biomass concentration. Like q_P, q_S has dimensions T^{-1}. In this section, we will develop equations for q_S as a function of growth rate and other relevant metabolic parameters.

11.10.1 Substrate Uptake in the Absence of Product Formation

In some cultures there is no extracellular product formation; for example, biomass itself is the product in manufacture of bakers' yeast and single-cell protein. In the absence of product formation, we assume that all substrate entering the cell is used for growth and maintenance functions. Rates of these cell activities are related as follows:

$$r_S = \frac{r_X}{Y_{XS}} + m_S x.$$

$$(11.72)$$

Table 11.10 Maintenance coefficients for several microorganisms with glucose as energy source

(From S.J. Pirt, 1975, Principles of Microbe and Cell Cultivation, *Blackwell Scientific, Oxford)*

Microorganism	Growth conditions	m_S (kg substrate (kg cells)$^{-1}$ h^{-1})
Saccharomyces cerevisiae	Anaerobic	0.036
	Anaerobic, 1.0 M NaCl	0.360
Azotobacter vinelandii	Nitrogen fixing, 0.2 atm dissolved-oxygen tension	1.5
	Nitrogen fixing, 0.02 atm dissolved-oxygen tension	0.15
Klebsiella aerogenes	Anaerobic, tryptophan-limited, 2 g l^{-1} NH$_4$Cl	2.88
	Anaerobic, tryptophan-limited, 4 g l^{-1} NH$_4$Cl	3.69
Lactobacillus casei		0.135
Aerobacter cloacae	Aerobic, glucose-limited	0.094
Penicillium chrysogenum	Aerobic	0.022

In Eq. (11.72), r_X is the volumetric rate of biomass production, Y_{XS} is the true yield of biomass from substrate, m_S is the maintenance coefficient, and x is biomass concentration. Eq. (11.72) states that rate of substrate uptake depends partly on the rate of growth but also varies with cell concentration. When r_X is expressed using Eq. (11.52), Eq. (11.72) becomes:

$$r_S = -\left(\frac{\mu}{Y_{XS}} + m_S\right) x.$$

(11.73)

If we now express μ as a function of substrate concentration using Eq. (11.60), Eq. (11.73) becomes:

$$r_S = -\left[\frac{\mu_{max}s}{Y_{XS}(K_S+s)} + m_S\right] x.$$

(11.74)

When s is zero, Eq. (11.74) predicts that substrate consumption will proceed at a rate equal to $m_S x$. Substrate uptake in the absence of substrate is impossible; this feature of Eq. (11.74) is therefore unrealistic. The problem arises because of implicit assumptions we have made about the nature of maintenance activities. It can be shown however that Eq. (11.74) is a realistic description of substrate uptake as long as there is external substrate available; when the substrate is exhausted maintenance energy is generally supplied by endogenous metabolism.

11.10.2 Substrate Uptake With Product Formation

Patterns of substrate flow in cells synthesising products depend on whether product formation is coupled to energy metabolism. When products are formed in energy-generating pathways, e.g. in anaerobic culture, product synthesis is an unavoidable consequence of cell growth and maintenance. Accordingly, as illustrated in Figure 11.13(a), there is no separate flow of substrate into the cell for product synthesis; product is formed from the substrate taken up to support growth and maintenance. Substrate consumed for maintenance does not contribute to growth; it therefore constitutes a separate substrate flow into the cell. In contrast, when production is not linked or only partly linked to energy metabolism, all or some of the substrate required for product synthesis is additional to, and separate from, that needed for growth and maintenance. Flow of substrate in this case is illustrated in Figure 11.13(b).

When products are directly linked to energy generation, equations for rate of substrate consumption do not include a

Figure 11.13 Substrate uptake with product formation: (a) production directly coupled to energy metabolism; (b) production not directly coupled to energy metabolism.

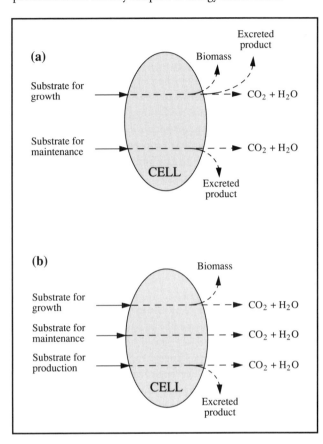

separate term for production; substrate requirements for product formation are already accounted for in terms for growth- and maintenance-associated substrate uptake. Accordingly, equations presented in the previous section for substrate consumption in the absence of product formation apply; rate of substrate uptake is related to growth and maintenance requirements by Eqs (11.73) and (11.74).

In cultures where product synthesis is only indirectly coupled to energy metabolism, rate of substrate consumption is a function of three factors: growth rate, rate of product formation and rate of substrate uptake for maintenance. These different cell functions can be related using yield and maintenance coefficients:

$$r_S = \frac{r_X}{Y_{XS}} + \frac{r_P}{Y_{PS}} + m_S x$$

(11.75)

where r_S is the volumetric rate of substrate consumption, r_X is the volumetric rate of biomass production, r_P is the volumetric rate of product formation, Y_{XS} is the true yield of biomass from substrate, Y_{PS} is the true yield of product from substrate, m_S is the maintenance coefficient, and x is biomass concentration. If we express r_X and r_P using Eqs (11.52) and (11.67):

$$r_S = \left(\frac{\mu}{Y_{XS}} + \frac{q_P}{Y_{PS}} + m_S \right) x. $$

$$(11.76)$$

11.11 Effect of Culture Conditions on Cell Kinetics

Temperature has a marked effect on metabolic rate. Temperature has a direct influence on reaction rate according to the Arrhenius law; it can also change the configuration of cell constituents, especially proteins and membrane components. In general, the effect of temperature on growth is similar to that already described in Section 11.3.4 for enzymes. There is an approximate two fold increase in specific growth rate for every 10°C rise in temperature, until structural breakdown of cell proteins and lipids starts to occur. Like other rate constants, the maintenance coefficient m_S has an Arrhenius-type temperature dependence [19]; this can have a significant kinetic effect on cultures where turnover of macromolecules is an important contribution to maintenance requirements. In contrast, temperature has only a minor effect on the biomass yield coefficient, Y_{XS} [19]. Other cellular responses to temperature are described elsewhere [1, 20, 21].

Growth rate depends on medium pH in much the same way as enzyme activity (Section 11.3.4); maximum growth rate is usually maintained over 1–2 pH units but declines with further variation. pH also affects the profile of product synthesis in anaerobic culture and can change maintenance-energy requirements [1, 20, 21].

11.12 Determining Cell Kinetic Parameters From Batch Data

In order to apply the equations presented in Sections 11.6–11.10 to real fermentations, we must know the kinetic and yield parameters for the system and have information about rates of growth, substrate uptake and product formation. Batch culture is the most frequently-applied method for investigating kinetic behaviour, but it is not always the best. Methods for determining reaction parameters from batch data are described below.

11.12.1 Rates of Growth, Product Formation and Substrate Uptake

Determining growth rates in cell culture requires measurement of cell concentration. Many different experimental procedures are applied for biomass estimation [20, 22]. Direct measurement can be made of cell number, dry or wet cell mass, packed cell volume or culture turbidity; alternatively, indirect estimates are obtained from measurements of product formation, heat evolution or cell composition. Cell viability is usually evaluated using plating or staining techniques. Each method for biomass estimation will give somewhat different results. For example, rate of growth determined using cell dry weight may differ from that obtained from cell number because dry weight in the culture can increase without a corresponding increase cell number.

Irrespective of how cell concentration is measured, the techniques described in Section 11.2 for graphical differentiation of concentration data are suitable for determining volumetric growth rates in batch culture. The results will depend to some extent on how the data are smoothed. For reasons described in Section 11.7.3, there is usually a relatively abrupt change in growth rate between growth and stationary phases; this feature requires extra care for accurate differentiation of batch growth curves. As discussed in Section 3.3.1, an advantage of hand-smoothing is that it allows us to judge the significance of individual points. Once the volumetric growth rate r_X is known, the specific growth rate μ is obtained by dividing r_X by the cell concentration.

For the growth phase of batch culture, an alternative method can be applied to calculate μ. Assuming growth can be represented by the first-order model of Eq. (11.52), the integrated relationship of Eq. (11.53) or (11.54) allows us to obtain μ directly. During the growth phase when μ is essentially constant, a plot of ln x versus time gives a straight line with slope μ. This is illustrated in Example 11.8.

Example 11.8 Hybridoma doubling time

A mouse-mouse hybridoma cell line is used to produce monoclonal antibody. Growth in batch culture is monitored with the following results.

Time (d)	Cell concentration (cells ml^{-1} $\times$ 10^{-6})
0.0	0.45
0.2	0.52
0.5	0.65
1.0	0.81
1.5	1.22
2.0	1.77
2.5	2.13
3.0	3.55
3.5	4.02
4.0	3.77
4.5	2.20

(a) Determine the specific growth rate during the growth phase.

(b) What is the culture doubling time?

Solution:

(a) The data are plotted as a semi-log graph in Figure 11E8.1.

Figure 11E8.1   Calculation of specific growth rate for hybridoma cells.

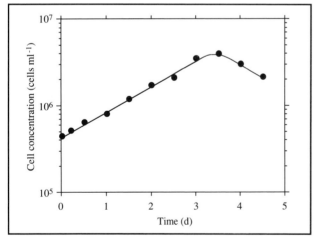

No lag phase is evident. As Eq. (11.54) applies only when μ is constant, i.e. during the exponential growth phase, we must determine which data points belong to the exponential growth phase. In Figure 11E8.1, the final three points appear to belong to the decline and death phases of the culture. Fitting a straight line to the remaining data gives a slope of 0.67. Therefore, $\mu = 0.67$ d^{-1}.

(b) From Eq. (11.58):

$$t_d = \frac{\ln 2}{0.67 \text{ d}^{-1}} = 1.0 \text{ d}.$$

This doubling time applies only during the growth phase of the culture.

Volumetric rates of substrate uptake and product formation, r_S and r_P, can be evaluated by graphical differentiation of substrate- and product-concentration data, respectively. Specific product-formation rate q_P and specific substrate-uptake rate q_S are obtained by dividing the respective volumetric rates by cell concentration.

11.12.2 μ_{max} and K_S

The Monod equation for specific growth rate, Eq. (11.60), is analogous mathematically to the Michaelis–Menten expression for enzyme kinetics. In principle therefore, the techniques described in Section 11.4 for determining v_{max} and K_m for enzyme reaction can be applied for evaluation of μ_{max} and K_S. However, because values of K_S in cell culture are usually very low, accurate determination of this parameter from batch data is difficult. Better estimation of K_S can be made using continuous culture of cells as discussed in Chapter 13. On the other hand, measurement of μ_{max} from batch data is relatively straightforward. As described in Section 11.7.3, if all nutrients are present in excess, the specific growth rate during exponential growth is equal to the maximum specific growth rate. Therefore, the specific growth rate calculated in Example 11.8 is equal to μ_{max}.

11.13 Effect of Maintenance on Yields

True yields such as Y_{XS}, Y_{PX} and Y_{PS} are often difficult to evaluate. Although true yields are essentially stoichiometric coefficients, the stoichiometry of biomass production and product formation is only known for relatively simple fermen-

tations. If the metabolic pathways are complex, stoichiometric calculations become too complicated. However, theoretical yields can be related to observed yields such as Y'_{XS}, Y'_{PX} and Y'_{PS}, which are more easily determined.

11.13.1 Observed Yields

Expressions for observed yield coefficients can be obtained by applying Eq. (11.48):

$$Y'_{XS} = \frac{-dX}{dS} = \frac{r_X}{r_S}$$

(11.77)

$$Y'_{PX} = \frac{dP}{dX} = \frac{r_P}{r_X}$$

(11.78)

and

$$Y'_{PS} = \frac{-dP}{dS} = \frac{r_P}{r_S}$$

(11.79)

where X, S and P are masses of cells, substrate and product, respectively, and r_X, r_S and r_P are observed rates evaluated from experimental data. Therefore, yield coefficients can be determined by plotting X, S or P against each other and evaluating the slope as illustrated in Figure 11.14. Alternatively, observed yield coefficients at a particular instant in time can be calculated

Figure 11.14 Evaluation of observed yields in batch culture from cell, substrate and product concentrations.

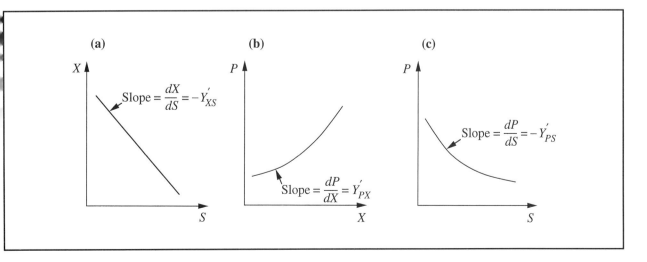

as the ratio of rates evaluated at that instant. Observed yields are not necessarily constant throughout batch culture; in some cases they exhibit significant dependence on environmental parameters such as substrate concentration and growth rate. Nevertheless, for many cultures, the observed biomass yield Y'_{XS} is approximately constant. Because of the errors in experimental data, considerable uncertainty is usually associated with measured yield coefficients.

11.13.2 Biomass Yield From Substrate

Equations for true biomass yield can be determined for systems without extracellular product formation or when product synthesis is directly coupled to energy metabolism. Substituting expressions for r_X and r_S from Eqs (11.52) and (11.73) into Eq. (11.77) gives:

$$Y'_{XS} = \frac{\mu}{\left(\dfrac{\mu}{Y_{XS}} + m_S\right)}.$$

(11.80)

Inverting Eq. (11.80) produces the expression:

$$\frac{1}{Y'_{XS}} = \frac{1}{Y_{XS}} + \frac{m_S}{\mu}.$$

(11.81)

Therefore, if Y_{XS} and m_S are relatively constant, a plot of $^1/Y'_{XS}$ versus $^1/\mu$ gives a straight line with slope m_S and intercept $^1/Y_{XS}$. Eq. (11.81) is not generally applied to batch growth data; under typical batch conditions, μ does not vary from μ_{max} for much of the culture period so it is difficult to plot Y'_{XS} as a function of specific growth rate. We will revisit Eq. (11.81) when we consider continuous cell culture in Chapter 13. As a rule of thumb, true biomass yield from glucose under aerobic conditions is around 0.5 g g^{-1}.

In processes such as production of bakers' yeast and single-cell protein where the required product is biomass, it is desirable to maximise the actual or observed yield of cells from substrate. The true yield Y_{XS} is limited by stoichiometric considerations. However, from Eq. (11.80), Y'_{XS} can be improved by decreasing the maintenance coefficient or increasing the growth rate. m_S may be reduced by lowering the temperature of fermentation, using a medium of lower ionic strength, or by applying a different organism or strain with lower maintenance-energy requirements. Assuming these changes do not reduce the growth rate, they can be employed to improve the biomass yield.

When the culture produces compounds not directly coupled with energy metabolism, Eqs (11.80) and (11.81) do not apply because a different expression for r_S must be used in Eq. (11.77). Determination of true yields and maintenance coefficients is more difficult in this case because of the number of terms involved.

11.13.3 Product Yield From Biomass

Observed yield of product from biomass Y'_{PX} is defined in Eq. (11.78). When product synthesis is directly coupled to energy metabolism, r_P is given by Eq. (11.69). Substituting this and Eq. (11.52) into Eq. (11.78) gives:

$$Y'_{PX} = Y_{PX} + \frac{m_P}{\mu}.$$

(11.82)

The extent of deviation of Y'_{PX} from Y_{PX} depends on the relative magnitudes of m_P and μ. To increase the observed yield of product for a particular process, m_P should be increased and μ decreased. Eq. (11.82) does not apply to products not directly coupled with energy metabolism; we do not have a general expression for r_P in terms of true yield coefficients for this class of product.

11.13.4 Product Yield From Substrate

Observed product yield from substrate Y'_{PS} is defined in Eq. (11.79). For products coupled to energy generation, expressions for r_P and r_S are available from Eqs (11.69) and (11.73). Therefore:

$$Y'_{PS} = \frac{Y_{PX}\mu + m_P}{\left(\dfrac{\mu}{Y_{XS}} + m_S\right)}.$$

(11.83)

In many anaerobic fermentations such as ethanol production, yield of product from substrate is a critical factor affecting process economics. At high Y'_{PS}, more ethanol is produced per mass of carbohydrate consumed so that the cost of production is reduced. Growth rate has a strong effect on Y'_{PS} for ethanol. Because Y'_{PS} is low when $\mu = \mu_{max}$, it is desirable to reduce the specific growth rate of the cells. Low growth rate can be obtained by depriving the cells of some essential nutrient, e.g. a nitrogen source, or by immobilising the cells to prevent growth. Increasing the rate of maintenance activity relative to growth

ill also enhance product yield. This can be done by using a
medium of high ionic strength, raising the temperature, or
selecting a mutant or different organism with high maintenance
requirements. Continuous culture provides more opportunity
for manipulating rates of growth than batch culture.

The effect of growth rate and maintenance on Y'_{PS} is difficult to determine for products not directly coupled with
energy metabolism unless information is available about the
effect of these parameters on q_p.

11.14 Kinetics of Cell Death

The kinetics of cell death is an important consideration in
design of sterilisation processes and in analysis of fermentations where substantial viability loss is expected. In a lethal
environment, cells in a population do not die all at once; deactivation of the culture occurs over a finite period of time
depending on the initial number of viable cells and the severity
of the conditions imposed. Loss of cell viability can be
described mathematically in much the same way as enzyme
deactivation; cell death is assumed to be a first-order process:

$$r_d = k_d N \tag{11.84}$$

where r_d is rate of cell death, N is number of viable cells, and k_d
is the *specific death constant*. Alternatively, rate of cell death can
be expressed using cell concentration rather than cell number:

$$r_d = k_d x \tag{11.85}$$

where k_d is the specific death constant based on cell concentration and x is the concentration of viable cells.

In a closed system with cell death the only process affecting
viable cell concentration, rate of cell death is equal to the rate
of decrease in cell number. Therefore, from Eq. (11.84):

$$r_d = \frac{-dN}{dt} = k_d N. \tag{11.86}$$

If k_d is constant, we can integrate Eq. (11.86) to derive an
expression for N as a function of time:

$$N = N_0 \, e^{-k_d t} \tag{11.87}$$

where N_0 is the number of viable cells at time zero. Taking natural logarithms of both sides of Eq. (11-87) gives:

$$\ln N = \ln N_0 - k_d \, t. \tag{11.88}$$

According to Eq. (11.88), if first-order death kinetics apply, a
plot of $\ln N$ versus t gives a straight line with slope $-k_d$.
Experimental measurements for many vegetative cells have
confirmed the relationship of Eq. (11.88); as an example, data
for thermal death of *Escherichia coli* at various temperatures are
shown in Figure 11.15. However, first-order death kinetics do
not always hold, particularly for bacterial spores immediately
after exposure to heat.

Like other kinetic constants, the value of the specific death
constant k_d depends on temperature. This effect can be
described using the Arrhenius relationship of Eq. (11.46).
Typical E_d values for thermal destruction of microorganisms
are high, of the order 250–290 kJ gmol^{-1} [23]. Therefore,
small increases in temperature have a significant effect on k_d
and rate of death.

Figure 11.15 Relationship between temperature and rate of
thermal death for vegetative *Escherichia coli* cells. (From S.
Aiba, A.E. Humphrey and N.F. Millis, 1965, *Biochemical
Engineering*, Academic Press, New York.)

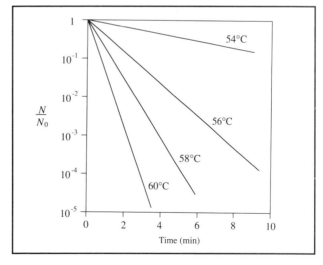

Example 11.9 Thermal death kinetics

The number of viable spores of a new strain of *Bacillus subtilis* is measured as a function of time at various temperatures.

Time (min)	Number of spores at:			
	$T = 85°C$	$T = 90°C$	$T = 110°C$	$T = 120°C$
0.0	2.40×10^9	2.40×10^9	2.40×10^9	2.40×10^9
0.5	2.39×10^9	2.38×10^9	1.08×10^9	2.05×10^7
1.0	2.37×10^9	2.30×10^9	4.80×10^8	1.75×10^5
1.5	–	2.29×10^9	2.20×10^8	1.30×10^3
2.0	2.33×10^9	2.21×10^9	9.85×10^7	–
3.0	2.32×10^9	2.17×10^9	2.01×10^7	–
4.0	2.28×10^9	2.12×10^9	4.41×10^6	–
6.0	2.20×10^9	1.95×10^9	1.62×10^5	–
8.0	2.19×10^9	1.87×10^9	6.88×10^3	–
9.0	2.16×10^9	1.79×10^9	–	–

(a) Determine the activation energy for thermal death of *B. subtilis* spores.
(b) What is the specific death constant at 100°C?
(c) Estimate the time required to kill 99% of spores in a sample at 100°C?

Solution:
(a) A semi-log plot of number of viable spores versus time is shown in Figure 11E9.1.

Figure 11E9.1   Thermal death of *Bacillus subtilis* spores.

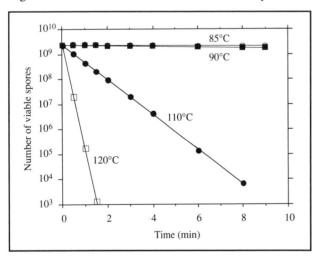

From Eq. (11.88), the slopes of the lines in Figure 11E9.1 are equal to $-k_d$ at the various temperatures. Fitting straight lines to the data gives the following results:

$k_d (85°C) = 0.012 \text{ min}^{-1}$
$k_d (90°C) = 0.032 \text{ min}^{-1}$
$k_d (110°C) = 1.60 \text{ min}^{-1}$
$k_d (120°C) = 9.61 \text{ min}^{-1}$.

The relationship between k_d and absolute temperature is given by Eq. (11.46). Therefore, a semi-log plot of k_d versus $1/T$ should yield a straight line with slope $= -E_d/R$ where T is absolute temperature. T is converted to degrees Kelvin using the formula of Eq. (2.24); $1/T$ values in units of K^{-1} are plotted in Figure 11E9.2.

Figure 11E9.2 Calculation of kinetic parameters for thermal death of spores.

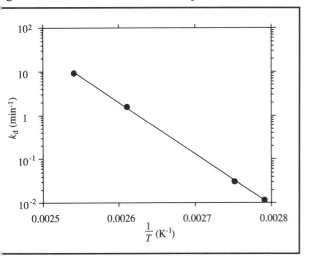

The slope is $-27\,030$ K. From Table 2.5, $R = 8.3144$ J K^{-1} gmol^{-1}. Therefore:

$$E_d = 27\,030 \text{ K } (8.3144 \text{ J K}^{-1} \text{ gmol}^{-1}) = 2.25 \times 10^5 \text{ J gmol}^{-1}$$
$$= 225 \text{ kJ gmol}^{-1}.$$

b) The equation to the line in Figure 11E9.2 is:

$$k_d = 6.52 \times 10^{30} \, e^{-27\,030/T}.$$

Therefore, at $T = 100°C = 373.15$ K, $k_d = 0.23$ min^{-1}.

c) From Eq. (11.88):

$$t = \frac{-(\ln N - \ln N_0)}{k_d}$$

or

$$t = \frac{-\ln\left(\dfrac{N}{N_0}\right)}{k_d}.$$

For N equal to 1% of N_0, $N/N_0 = 0.01$. At 100°C, $k_d = 0.23$ min^{-1} and the time required is:

$$t = \frac{-\ln(0.01)}{0.23 \text{ min}^{-1}} = 20 \text{ min}.$$

As contaminating organisms are being killed by heat sterilisation, nutrients in the medium may also be destroyed. The sensitivity of nutrient molecules to temperature is described by the Arrhenius equation of Eq. (11.46). Values of the activation energy E_d for thermal destruction of vitamins and amino acids are 84–92 kJ gmol^{-1}; for proteins E_d is about 165 kJ gmol^{-1} [23]. Because these values are somewhat lower than typical E_d values for microorganisms, raising the temperature has a greater effect on cell death than nutrient destruction. This means that sterilisation at higher temperatures for shorter periods of time has the advantage of killing cells with limited destruction of medium components.

11.15 Summary of Chapter 11

At the end of Chapter 11 you should:

(i) understand the difference between *reversible* and *irreversible reactions*, and the limitations of equilibrium thermodynamics in representing cell and enzyme reactions;

(ii) be able to calculate reaction rates from batch concentration data using *graphical differentiation*;

(iii) be familiar with kinetic relationships for *zero-order, first-order* and *Michaelis–Menten reactions*;

(iv) be able to determine enzyme kinetic parameters v_{max} and K_m from batch concentration data;

(v) be able to quantify the effect of temperature on rates of enzyme reaction and deactivation;

(vi) be able to calculate *yield coefficients* for cell culture;

(vii) know the basic relationships for cell growth kinetics and be able to evaluate growth, substrate uptake and production rates in batch culture;

(viii) be able to analyse growth in cultures with plasmid instability;

(ix) know how *maintenance activities* affect substrate utilisation in cells; and

(x) be able to describe the kinetics of *cell death*.

Problems

11.1 Reaction equilibrium

Calculate equilibrium constants for the following reactions under standard conditions:

(a) glutamine + $H_2O \rightarrow$ glutamate + NH_4^+
$\Delta G_{rxn}^o = -14.1$ kJ mol^{-1}

(b) malate $\rightarrow$ fumarate + H_2O
$\Delta G_{rxn}^o = 3.2$ kJ mol^{-1}.

Could either of these reactions be considered irreversible?

11.2 Equilibrium yield

The following reaction catalysed by phosphoglucomutase occurs during breakdown of glycogen:

glucose 1-phosphate $\rightleftharpoons$ glucose 6-phosphate.

A reaction is started by adding phosphoglucomutase to 0.04 gmol glucose 1-phosphate in 1 litre solution at 25°C. The reaction proceeds to equilibrium at which the concentration of glucose 1-phosphate is 0.002 M and the concentration of glucose 6-phosphate is 0.038 M.

(a) Calculate the equilibrium constant.
(b) What is the theoretical yield?
(c) What is the yield based on amount of reactant supplied?

11.3 Reaction rate

(a) The volume of a fermenter is doubled while keeping the cell concentration and other fermentation conditions the same.
 (i) How is the volumetric productivity affected?
 (ii) How is the specific productivity affected?
 (iii) How is the total productivity affected?

(b) If instead of (a) the cell concentration were doubled, what affect would this have on volumetric, specific and total productivities?

(c) A fermenter produces 100 kg lysine per day.
 (i) If the volumetric productivity is 0.8 g l^{-1} h^{-1}, what is the volume of the fermenter?
 (ii) The cell concentration is 20 g l^{-1} dry weight. Calculate the specific productivity.

11.4 Enzyme kinetics

Lactase, also known as β-galactosidase, catalyses the hydrolysis of lactose to produce glucose and galactose from milk and whey. Experiments are carried out to determine the kinetic parameters for the enzyme. Initial rate data are listed below.

Lactose concentration (mol l^{-1} × 10^2)	Initial reaction velocity (mol l^{-1} min^{-1} × 10^3)
2.50	1.94
2.27	1.91
1.84	1.85
1.35	1.80
1.25	1.78
0.730	1.46
0.460	1.17
0.204	0.779

Evaluate v_{max} and K_m.

11.5 Effect of temperature on hydrolysis of starch

α-Amylase from malt is used to hydrolyse starch. The dependence of initial reaction rate on temperature is determined experimentally. Results measured at fixed starch and enzyme concentrations are listed below.

Temperature (°C)	Rate of glucose production (mmol m^{-3} s^{-1})
20	0.31
30	0.66
40	1.20
50	6.33

(a) Determine the activation energy for this reaction.
(b) α-Amylase is used to break down starch in baby food. It is proposed to carry out the reaction at a relatively high temperature so that the viscosity is reduced. What is the reaction rate at 55°C compared with 25°C?
(c) Thermal deactivation of this enzyme is described by the equation:

$$k_d = 2.25 \times 10^{27} \, e^{-41\,630/RT}$$

where k_d is the deactivation rate constant in h^{-1}, R is the ideal gas constant in cal gmol^{-1} K^{-1}, and T is temperature in K. What is the half-life of the enzyme at 55°C compared with 25°C? Which of these two operating temperatures is more practical for processing baby food?

11.6 Enzyme reaction and deactivation

Lipase is being investigated as an additive to laundry detergent for removal of stains from fabric. The general reaction is:

fats → fatty acids + glycerol.

The Michaelis constant for pancreatic lipase is 5 mM. At 60°C, lipase is subject to deactivation with a half-life of 8 min. Fat hydrolysis is carried out in a well-mixed batch reactor which simulates a top-loading washing machine. The initial fat concentration is 45 gmol m^{-3}. At the beginning of the reaction the rate of hydrolysis is 0.07 mmol l^{-1} s^{-1}. How long does it take for the enzyme to hydrolyse 80% of the fat present?

11.7 Growth parameters for recombinant *E. coli*

Escherichia coli is being used for production of recombinant porcine growth hormone. The bacteria are grown aerobically in batch culture with glucose as growth-limiting substrate. Cell and substrate concentrations are measured as a function of culture time; the results are listed below.

Time (h)	Cell concentration, x (kg m^{-3})	Substrate concentration, s (kg m^{-3})
0.0	0.20	25.0
0.33	0.21	24.8
0.5	0.22	24.8
0.75	0.32	24.6
1.0	0.47	24.3
1.5	1.00	23.3
2.0	2.10	20.7
2.5	4.42	15.7
2.8	6.9	10.2
3.0	9.4	5.2
3.1	10.9	1.65
3.2	11.6	0.2
3.5	11.7	0.0
3.7	11.6	0.0

(a) Plot μ as a function of time.
(b) What is the value of μ_{max}?
(c) What is the observed biomass yield from substrate? Is Y'_{XS} constant?

11.8 Growth parameters for hairy roots

Hairy roots are produced by genetic transformation of plants using *Agrobacterium rhizogenes*. The following biomass and sugar concentrations were obtained during batch culture of *Atropa belladonna* hairy roots in a bubble-column fermenter.

Time (d)	Biomass concentration (g l^{-1} dry weight)	Sugar concentration (g l^{-1})
0	0.64	30.0
5	1.95	27.4
10	4.21	23.6
15	5.54	21.0
20	6.98	18.4
25	9.50	14.8
30	10.3	13.3
35	12.0	9.7
40	12.7	8.0
45	13.1	6.8
50	13.5	5.7
55	13.7	5.1

(a) Plot μ as a function of culture time. When is the growth rate maximum?

(b) Plot the specific rate of sugar uptake as a function of time.

(c) What is the observed biomass yield from substrate? Is Y'_{XS} constant?

11.9 Ethanol fermentation by yeast and bacteria

Ethanol is produced by anaerobic fermentation of glucose by *Saccharomyces cerevisiae*. For the particular strain of *S. cerevisiae* employed, the maintenance coefficient is 0.18 kg kg^{-1} h^{-1}, Y_{XS} is 0.11 kg kg^{-1}, Y_{PX} is 3.9 kg kg^{-1} and μ_{max} is 0.4 h^{-1}. It is decided to investigate the possibility of using *Zymomonas mobilis* bacteria instead of yeast for making ethanol. *Z. mobilis* is known to produce ethanol under anaerobic conditions using a different metabolic pathway to that employed by yeast. Typical values of Y_{XS} are lower than for yeast at about 0.06 kg kg^{-1}; on the other hand, the maintenance coefficient is higher at 2.2 kg kg^{-1} h^{-1}. Y_{PX} for *Z. mobilis* is 7.7 kg kg^{-1}; μ_{max} is 0.3 h^{-1}.

(a) From stoichiometry, what is the maximum theoretical yield of ethanol from glucose?

(b) Y'_{PS} is maximum and equal to the theoretical yield when there is zero growth and all substrate entering the cell is used for maintenance activities. If ethanol is the sole extracellular product of energy-yielding metabolism, calculate m_p for each organism.

(c) *S. cerevisiae* and *Z. mobilis* are cultured in batch fermenters. Predict the observed product yield from substrate for the two cultures.

(d) What is the efficiency of ethanol production by the two organisms? Efficiency is defined as the observed product yield from substrate divided by the maximum or theoretical product yield.

(e) How does the specific rate of ethanol production by *Z. mobilis* compare with that by *S. cerevisiae*?

(f) Using Eq. (11.70), compare the proportions of growth-associated and non-growth-associated ethanol production by *Z. mobilis* and *S. cerevisiae*. For which organism is non-growth-associated production more substantial?

(g) In order to achieve the same volumetric ethanol productivity from the two cultures, what yeast concentration is required compared with the concentration of bacteria?

(h) At zero growth, the efficiency of ethanol production is the same in both cultures. Under these conditions, if the same concentration of yeast and bacteria are employed, what size fermenter is required for the yeast compared with the bacteria in order to achieve the same total productivity?

(i) Predict the observed biomass yield from substrate for the two organisms. For which organism is biomass disposal less of a problem?

(j) Make a recommendation about which organism is better suited for industrial ethanol production, and give your reasons.

11.10 Plasmid loss during culture maintenance

A stock culture of plasmid-containing *Streptococcus cremoris* cells is maintained with regular sub-culturing for a period of 28 d. After this time, the fraction of plasmid-carrying cells is measured and found to be 0.66. The specific growth rate of plasmid-free cells at the storage temperature is 0.033 h^{-1}; the specific growth rate of plasmid-containing cells is 0.025 h^{-1}. If all the cells initially contained plasmid, estimate the probability per generation of plasmid loss.

11.11 Medium sterilisation

A steam steriliser is used to sterilise liquid medium for fermentation. The initial concentration of contaminating organisms is 10^8 per litre. For design purposes, the final acceptable level of contamination is usually taken to be 10^{-3}

ells; this corresponds to a risk that one batch in a thousand will remain contaminated even after the sterilisation process is complete. For how long should 1 m^3 medium be treated if the temperature is:

a) 80°C?
b) 121°C?
c) 140°C?

To be safe, assume that the contaminants present are spores of *Bacillus stearothermophilus*, one of the most heat-resistant microorganisms known. For these spores the activation energy for thermal death is 283 kJ gmol^{-1} and the Arrhenius constant is $10^{36.2}$ s^{-1} [24].

References

1. Atkinson, B. and F. Mavituna (1991) *Biochemical Engineering and Biotechnology Handbook*, 2nd edn, Macmillan, Basingstoke.

2. Stryer, L. (1981) *Biochemistry*, 2nd edn, W.H. Freeman, New York.

3. Cornish-Bowden, A. and C.W. Wharton (1988) *Enzyme Kinetics*, IRL Press, Oxford.

4. Dixon, M. and E.C. Webb (1964) *Enzymes*, 2nd edn, Longmans, London.

5. Sizer, I.W. (1943) Effects of temperature on enzyme kinetics. *Adv. Enzymol.* **3**, 35–62.

6. Moser, A. (1985) Rate equations for enzyme kinetics. In: H.-J. Rehm and G. Reed (Eds), *Biotechnology*, vol. 2, pp. 199–226, VCH, Weinheim.

7. Eisenthal, R. and A. Cornish-Bowden (1974) The direct linear plot: a new graphical procedure for estimating enzyme kinetic parameters. *Biochem. J.* **139**, 715–720.

8. Moser, A. (1985) Kinetics of batch fermentations. In: H.-J. Rehm and G. Reed (Eds), *Biotechnology*, vol. 2, pp. 243–283, VCH, Weinheim.

9. Bailey, J.E. and D.F. Ollis (1986) *Biochemical Engineering Fundamentals*, 2nd edn, Chapter 7, McGraw-Hill, New York.

10. Roels, J.A. and N.W.F. Kossen (1978) On the modelling of microbial metabolism. *Prog. Ind. Microbiol.* **14**, 95–203.

11. Shuler, M.L. and F. Kargi (1992) *Bioprocess Engineering*, Chapter 6, Prentice Hall, New Jersey.

12. Imanaka, T. and S. Aiba (1981) A perspective on the application of genetic engineering: stability of recombinant plasmid. *Ann. N.Y. Acad. Sci.* **369**, 1–14.

13. Cooper, N.S., M.E. Brown and C.A. Caulcott (1987) A mathematical model for analysing plasmid stability in micro-organisms. *J. Gen. Microbiol.* **133**, 1871–1880.

14. Ollis, D.F. and H.-T. Chang (1982) Batch fermentation kinetics with (unstable) recombinant cultures. *Biotechnol. Bioeng.* **24**, 2583–2586.

15. Bailey, J.E., M. Hjortso, S.B. Lee and F. Srienc (1983) Kinetics of product formation and plasmid segregation in recombinant microbial populations. *Ann. N.Y. Acad. Sci.* **413**, 71–87.

16. Wittrup, K.D. and J.E. Bailey (1988) A segregated model of recombinant multicopy plasmid propagation. *Biotechnol. Bioeng.* **31**, 304–310.

17. Stouthamer, A.H. and H.W. van Verseveld (1985) Stoichiometry of microbial growth. In: M. Moo-Young (Ed), *Comprehensive Biotechnology*, vol. 1, pp. 215–238, Pergamon Press, Oxford.

18. Heijnen, J.J., J.A. Roels and A.H. Stouthamer (1979) Application of balancing methods in modeling the penicillin fermentation. *Biotechnol. Bioeng.* **21**, 2175–2201.

19. Heijnen, J.J. and J.A. Roels (1981) A macroscopic model describing yield and maintenance relationships in aerobic fermentation processes. *Biotechnol. Bioeng.* **23**, 739–763.

20. Pirt, S.J. (1975) *Principles of Microbe and Cell Cultivation*, Blackwell Scientific, Oxford.

21. Forage, R.G., D.E.F. Harrison and D.E. Pitt (1985) Effect of environment on microbial activity. In: M. Moo-Young (Ed), *Comprehensive Biotechnology*, vol. 1, pp. 251–280, Pergamon Press, Oxford.

22. Wang, D.I.C., C.L. Cooney, A.L. Demain, P. Dunnill, A.E. Humphrey and M.D. Lilly (1979) *Fermentation and Enzyme Technology*, John Wiley, New York.

23. Cooney, C.L. (1985) Media sterilization. In: M. Moo-Young (Ed), *Comprehensive Biotechnology*, vol. 2, pp. 287–298, Pergamon Press, Oxford.

24. Deindoerfer, F.H. and A.E. Humphrey (1959) Analytical method for calculating heat sterilization times. *Appl. Microbiol.* **7**, 256–264.

Suggestions for Further Reading

Reaction Thermodynamics (see also ref. 2)

Lehninger, A.L. (1965) *Bioenergetics*, W.A. Benjamin, New York.

General Reaction Kinetics

Froment, G.F. and K.B. Bischoff (1979) *Chemical Reactor Analysis and Design*, Chapter 1, John Wiley, New York.

Holland, C.D. and R.G. Anthony (1979) *Fundamentals of*

Chemical Reaction Engineering, Chapter 1, Prentice-Hall, New Jersey.

Levenspiel, O. (1972) *Chemical Reaction Engineering*, 2nd edn, Chapters 1 and 2, John Wiley, New York.

Graphical Differentiation

Churchill, S.W. (1974) *The Interpretation and Use of Rate Data: The Rate Concept*, McGraw-Hill, New York.

Hougen, O.A., K.M. Watson and R.A. Ragatz (1962) *Chemical Process Principles*, Part I, 2nd edn, Chapter 1, John Wiley, New York.

Enzyme Kinetics and Deactivation (see also refs 2–6)

Hei, D.J. and D.S. Clark (1993) Estimation of melting curves from enzymatic activity–temperature profiles. *Biotechnol. Bioeng.* **42**, 1245–1251.

Laidler, K.J. and P.S. Bunting (1973) *The Chemical Kinetics of Enzyme Action*, 2nd edn, Clarendon, Oxford.

Lencki, R.W., J. Arul and R.J. Neufeld (1992) Effect of sub-unit dissociation, denaturation, aggregation, coagulation, and decomposition on enzyme inactivation kinetics. Parts I and II. *Biotechnol. Bioeng.* **40**, 1421–1434.

Lencki, R.W., A. Tecante and L. Choplin (1993) Effect of shear on the inactivation kinetics of the enzyme dextran-sucrase. *Biotechnol. Bioeng.* **42**, 1061–1067.

Cell Kinetics and Yield (see also refs 1, 8–11 and 17–22)

Roels, J.A. (1983) *Energetics and Kinetics in Biotechnology*, Elsevier Biomedical, Amsterdam.

Stouthamer, A.H. (1979) Energy production, growth, and product formation by microorganisms. In: O.K. Sebek and A.I. Laskin (Eds), *Genetics of Industrial Microorganisms*, American Society for Microbiology, Washington DC.

van't Riet, K. and J. Tramper (1991) *Basic Bioreactor Design*, Chapters 3 and 4, Marcel Dekker, New York.

Growth Kinetics With Plasmid Instability (see also refs 12–16)

Hjortso, M.A. and J.E. Bailey (1984) Plasmid stability in budding yeast populations: steady-state growth with selection pressure. *Biotechnol. Bioeng.* **26**, 528–536.

Sardonini, C.A. and D. DiBiasio (1987) A model for growth of *Saccharomyces cerevisiae* containing a recombinant plasmid in selective media. *Biotechnol. Bioeng.* **29**, 469–475.

Srienc, F., J.L. Campbell and J.E. Bailey (1986) Analysis of unstable recombinant *Saccharomyces cerevisiae* population growth in selective medium. *Biotechnol. Bioeng.* **18**, 996–1006.

Death Kinetics (see also refs 23 and 24)

Aiba, S., A.E. Humphrey and N.F. Millis (1965) *Biochemical Engineering*, Chapter 8, Academic Press, New York.

Richards, J.W. (1968) *Introduction to Industrial Sterilization*, Academic Press, London.

12

Heterogeneous Reactions

In the previous chapter, reaction rate was considered as a function of substrate concentration and temperature. Reaction systems were assumed to be homogeneous; local variations in concentration and rate of conversion were not examined. Yet, in many bioprocesses, concentrations of substrates and products differ from point to point in the reaction mixture. Concentration gradients arise in single-phase systems when mixing is poor; if different phases are present, local variations in composition can be considerable. As described in Chapter 9, concentration gradients occur within phase boundary layers around gas bubbles and solids. More severe gradients are found inside solid biocatalysts such as cell flocs, pellets, biofilms, and immobilised-cell and -enzyme beads.

Reactions occurring in the presence of significant concentration or temperature gradients are called *heterogeneous reactions*. Because biological reactions are not generally associated with large temperature gradients, we confine our attention in this chapter to concentration effects. When heterogeneous reactions occur in solid catalysts, not all reactive molecules are available for immediate conversion. Reaction takes place only after reactants are transported to the site of reaction. Thus, mass-transfer processes can have a considerable influence on the overall conversion rate.

Because rate of reaction is generally dependent on substrate concentration, when concentrations in the system vary, kinetic analysis becomes more complex. The principles of homogeneous reaction and the equations outlined in Chapter 11 remain valid for heterogeneous systems; however, the concentrations used in these equations must be those actually prevailing at the site of reaction. For solid biocatalysts, we must know the concentration of substrate at each point inside the solid in order to determine the local rate of conversion. In most cases these concentrations cannot be measured; fortunately, they can be estimated using diffusion-reaction theory.

In this chapter, methods are presented for analysing reactions affected by mass transfer. The mathematics required is more sophisticated than is applied elsewhere in this book; however, attention can be directed to the results of the analysis rather than to the mathematical derivations. The practical outcome of this chapter is simple criteria for assessing mass-transfer limitations which can be used directly in experimental design.

12.1 Heterogeneous Reactions in Bioprocessing

Reactions involving solid-phase catalysts are important in bioprocessing. Macroscopic flocs, clumps and pellets are produced naturally by certain bacteria and fungi; mycelial pellets are common in antibiotic fermentations. Some cells grow as biofilms on reactor walls; others form slimes such as in waste treatment processes. Plant cell suspensions invariably contain aggregates; microorganisms in soil crumbs play a crucial role in environmental bioremediation of land. Animal tissues are now being cultured on three-dimensional scaffolds for surgical transplantation and organ repair. More traditionally, many food fermentations involve microorganisms attached to solid particles. In all of these systems, rate of reaction depends on the rate of mass transfer outside or within the solid catalyst.

If cells or enzymes do not spontaneously form clumps or attach to solid surfaces, they can be induced to do so using *immobilisation* techniques. Many procedures are available for artificial immobilisation of cells and enzymes; the results of two commonly-used methods are illustrated in Figure 12.1. As shown in Figure 12.1(a), cells and enzymes can be immobilised by entrapment within gels such as alginate, agarose and carrageenan. Cells or enzymes are mixed with liquified gel before it is hardened or cross-linked and broken into small particles. The gel polymer must be porous and relatively soft to allow diffusion of reactants and products to and from the interior of the particle. As shown in Figure 12.1(b), an alternative to gel immobilisation is entrapment within porous solids

Figure 12.1 Immobilised biocatalysts: (a) cells entrapped in soft gel; (b) enzymes attached to the internal surfaces of a porous solid.

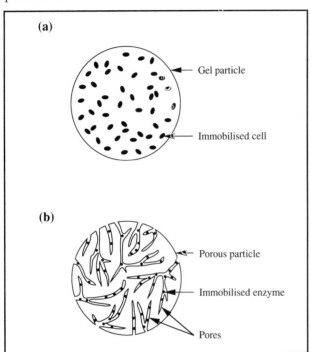

Figure 12.2 Typical substrate concentration profile for a spherical biocatalyst.

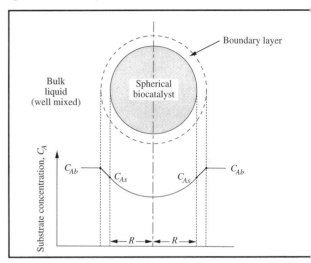

such as ceramics, porous glass and resin beads. Enzymes or cells migrate into the pores of these particles and attach to the internal surfaces; substrate must diffuse through the pores for reaction to occur. In both immobilisation methods, sites of reaction are distributed throughout the particle. Thus, a catalyst particle of higher activity can be formed by increasing the loading of cells or enzyme per volume of matrix.

Immobilised biocatalysts have many advantages in large-scale processing. One of the most important is continuous operation using the same catalytic material. For enzymes, an additional advantage is that immobilisation often enhances stability and increases the enzyme half-life. Further discussion of immobilisation methods and the rationale behind cell and enzyme immobilisation can be found in many articles and books; a selection of references is given at the end of this chapter.

In Chapter 11, enzymes and cells were considered as biological catalysts. In heterogeneous reactions involving a solid phase, the term 'catalyst' is also used to refer to the entire catalytically-active body, such as a particle or biofilm. Engineering analysis of heterogeneous reactions applies equally well to naturally occurring solid catalysts and artificially immobilised cells and enzymes.

12.2 Concentration Gradients and Reaction Rates in Solid Catalysts

Consider a spherical catalyst of radius R immersed in well mixed liquid containing substrate A. In the bulk liquid away from the particle the substrate concentration is uniform and equal to C_{Ab}. If the particle were inactive, after some time the concentration of substrate inside the solid would reach a constant value in equilibrium with C_{Ab}. However, when substrate is consumed by reaction, its concentration C_A decreases within the particle as shown in Figure 12.2. If immobilised cells or enzymes are distributed uniformly within the catalyst, the concentration profile is symmetrical with a minimum at the centre. Mass transfer of substrate to reaction sites in the particle is driven by the concentration difference between the bulk solution and particle interior.

In the bulk liquid, substrate is carried rapidly by convective currents. However, as substrate molecules approach the solid, they must be transported from the bulk liquid across the relatively stagnant boundary layer to the solid surface; this process is called *external mass transfer*. A concentration gradient develops across the boundary layer from C_{Ab} in the bulk liquid to C_{As} at the solid–liquid interface. If the particle were not porous and all enzyme or cells confined to its outer surface, external mass transfer would be the only transport process required. More often, reaction takes place inside the particle so that *internal mass transfer* through the solid is also required.

Although the form of the concentration gradient shown in Figure 12.2 is typical, other variations are possible. If mass transfer is much slower than reaction, it is possible that all substrate entering the particle will be consumed before reaching the

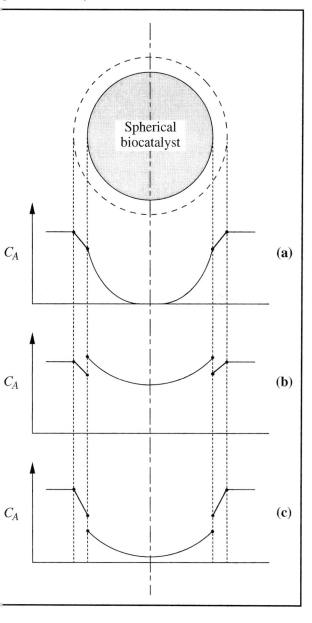

Figure 12.3 Variations in substrate concentration profile in spherical biocatalysts.

continuity of concentration at the solid–liquid interface shows that substrate distributes preferentially to the solid phase. Conversely, Figure 12.3(c) shows the concentrations when substrate is attracted more to the liquid than to the solid. The effect of mass transfer on intraparticle concentration can be magnified or diminished by substrate partitioning. Partitioning is important when the substrate and solid are charged or if strong hydrophobic interactions cause repulsion or attraction. Because most materials used for cell and enzyme immobilisation are very porous and contain a high percentage of water, partition effects can often be neglected. In our treatment of heterogeneous reaction, we will assume that partitioning is not significant.

12.2.1 True and Observed Reaction Rates

Because concentrations vary in solid catalysts, local rates of reaction also vary depending on position within the particle. Even for zero-order reactions, reaction rate changes with position if substrate is exhausted. Each cell or enzyme molecule responds to the substrate concentration at its location with a rate of reaction determined by the kinetic parameters of the catalyst. This local rate of reaction is known as the *true rate* or *intrinsic rate*. Like any reaction rate, intrinsic rates can be expressed using total, volumetric or specific bases as described in Section 11.1.3. The relationship between true reaction rate and local substrate concentration follows the principles outlined in Chapter 11 for homogeneous reactions.

True local reaction rates are difficult to measure in solid catalysts without altering the reaction conditions. It is possible, however, to measure the overall reaction rate for the entire catalyst. In a closed system, the rate of disappearance of substrate from the bulk liquid must equal the overall rate of conversion by reaction; in heterogeneous systems this is also called the *observed rate*. It is important to remember that the observed rate is not usually equal to the true activity of any cell or enzyme in the particle. Because intraparticle substrate levels are reduced inside solid catalysts, we expect the observed rate to be less than if the entire particle were exposed to the bulk liquid. The relationship between observed rate and bulk substrate concentration is not as simple as in homogeneous reactions. Kinetic equations for heterogeneous reactions also involve mass-transfer parameters.

True reaction rates depend on the kinetic parameters of the cells or enzyme. For example, rate of reaction by an immobilised enzyme obeying Michaelis–Menten kinetics depends on the values of v_{max} and K_m for the enzyme in its immobilised state. These parameters are sometimes called *true kinetic parameters* or *intrinsic kinetic parameters*. Because kinetic parameters

centre. In this case, the concentration falls to zero within the solid as illustrated in Figure 12.3(a). Cells or enzyme near the centre are starved of substrate and the core of the particle becomes inactive. In the examples of Figures 12.3(b) and 12.3(c), the *partition coefficient* for the substrate is not equal to unity. This means that, at equilibrium and in the absence of reaction, the concentration of substrate in the solid is naturally higher or lower than in the liquid. In Figure 12.3(b), the discan be altered during immobilisation as a result of cell or enzyme damage, configurational change and steric hindrance, values measured before immobilisation may not apply. Unfortunately, true kinetic parameters for immobilised biocatalysts can be difficult to determine because measured reaction rates incorporate mass transfer effects. The problem of evaluating true kinetic parameters is discussed further in Section 12.9.

12.2.2 Interaction Between Mass Transfer and Reaction

Rates of reaction and substrate mass transfer are not independent in heterogeneous systems. Rate of mass transfer depends on the concentration gradient established in the system; this in turn depends on the rate of substrate depletion by reaction. On the other hand, rate of reaction depends on the availability of substrate; this of course depends on the rate of mass transfer. One of the objectives in analysing heterogeneous reactions is to determine the relative influences of mass transfer and reaction on observed reaction rates. One can conceive, for example, that if a reaction proceeds slowly even in the presence of adequate substrate, it is likely that mass transfer will be rapid enough to meet the reaction demand. In this case, the observed rate would be determined more directly by the reaction process than mass transfer. Conversely, if the reaction tends to be very rapid, it is likely that mass transfer will be too slow to supply substrate at the rate required. The observed rate would then reflect strongly the rate of mass transfer. As will be shown in the remainder of this chapter, there are mathematical criteria for assessing the extent to which mass transfer influences the observed reaction rate. Reactions which are significantly affected are called *mass-transfer limited* or *diffusion-limited* reactions. It is also possible to distinguish the relative influence of internal and external mass transfer. Improvement of mass transfer and the elimination of mass-transfer limitations are desired objectives in heterogeneous catalysis. Once the effect and location of major mass-transfer resistances are identified, it is then possible to devise strategies for their elimination.

12.3 Internal Mass Transfer and Reaction

Let us now concentrate on the processes occurring within a solid biocatalyst; external mass transfer will be examined later in the chapter. The exact equations and procedures used in this analysis depend on the geometry of the system and the reaction kinetics. First, let us consider the case of cells or enzymes immobilised in a spherical particle.

Figure 12.4   Shell mass balance on a spherical particle.

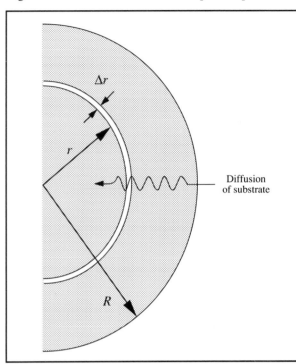

12.3.1 Steady-State Shell Mass Balance

Mathematical analysis of heterogeneous reactions involves technique called the *shell mass balance*. In this section, we will perform a shell mass balance on a spherical catalyst particle o radius R. Imagine a thin spherical shell of thickness Δr located at radius r from the centre, as shown in Figure 12.4. It may be helpful to think of this shell as the thin wall of a ping-pong ball encased inside and concentric with a larger cricket ball o radius R. Substrate diffusing into the sphere must cross the shell to reach the centre.

A mass balance of substrate is performed around the shell by considering the processes of mass transfer and reaction occurring at radius r. The system considered for the mass balance is the shell only; the remainder of the sphere is ignored for the moment. Substrate diffuses into the shell at radius $(r + \Delta r)$ and leaves at radius r; within the shell, immobilised cells o enzyme consume substrate by reaction. Flow of mass through the shell can be analysed using the general mass-balance equation derived in Chapter 4:

$$\begin{Bmatrix} \text{mass in} \\ \text{through} \\ \text{system} \\ \text{boundaries} \end{Bmatrix} - \begin{Bmatrix} \text{mass out} \\ \text{through} \\ \text{system} \\ \text{boundaries} \end{Bmatrix} + \begin{Bmatrix} \text{mass} \\ \text{generated} \\ \text{within} \\ \text{system} \end{Bmatrix} - \begin{Bmatrix} \text{mass} \\ \text{consumed} \\ \text{within} \\ \text{system} \end{Bmatrix} = \begin{Bmatrix} \text{mass} \\ \text{accumulated} \\ \text{within} \\ \text{system} \end{Bmatrix}.$$

(4.1

Before application of Eq. (4.1), certain assumptions must be made so that each term in the equation can be expressed mathematically [1].

i) *The particle is isothermal.* Kinetic parameters for enzyme and cell reactions are strong functions of temperature. If temperature in the particle varies, different values of the kinetic parameters must be applied. However, as temperature gradients generated by immobilised cells and enzymes are generally negligible, assuming constant temperature throughout the particle is reasonable and greatly simplifies the mathematical analysis.

ii) *Mass transfer occurs by diffusion only.* We will assume that the particle is impermeable to flow, so that convection within the pores is negligible. This assumption is valid for many solid-phase biocatalysts. However, some anomalies have been reported [2, 3]; depending on pore size, pressure gradients can induce convection of liquid through the particle and significantly enhance nutrient supply. When convective transport occurs, the analysis of mass transfer and reaction presented in this chapter must be modified [4–6].

iii) *Diffusion can be described using Fick's law with constant effective diffusivity.* We will assume that diffusive transport through the particle is governed by Fick's law (Section 9.1.1). Interaction of substrate with other concentration gradients and phenomena affecting transport of charged species are ignored. Fick's law will be applied using the *effective diffusivity* of substrate in the solid, $\mathscr{D}_{Ae}$. The value of $\mathscr{D}_{Ae}$ is a complex function of the molecular-diffusion characteristics of the substrate, the tortuousness of the diffusion path within the solid, and the fraction of the particle volume available for diffusion. We will assume that $\mathscr{D}_{Ae}$ is constant and independent of substrate concentration in the particle; this means that $\mathscr{D}_{Ae}$ does not change with position.

iv) *The particle is homogeneous.* Immobilised enzymes or cells are assumed to be distributed uniformly within the particle. Properties of the immobilisation matrix should also be uniform.

v) *The substrate partition coefficient is unity.* This assumption is valid for most substrates and particles, and ensures there is no discontinuity of concentration at the solid–liquid interface.

vi) *The particle is at steady state.* This assumption is usually valid if there is no change in activity of the catalyst, for example, due to enzyme deactivation, cell growth or differentiation. It is not valid when the system exhibits rapid transients such as when cells quickly consume and store substrates for subsequent metabolism.

vii) *Substrate concentration varies with a single spatial variable.* For the sphere of Figure 12.4, we will assume that concentration varies only in the radial direction, and that substrate diffuses radially through the particle from the external surface towards the centre.

Eq. (4.1) is applied according to these assumptions. Substrate is transported into and out of the shell by diffusion; therefore, the first and second terms are expressed using Fick's law with constant effective diffusivity. The third term is zero as no substrate is generated. Substrate is consumed by reaction inside the shell at a rate equal to the volumetric rate of reaction r_A multiplied by the volume of the shell. According to assumption (vi) listed above, the system is at steady state. Thus, its composition and mass must be unchanging, substrate cannot accumulate in the shell, and the right-hand side of Eq. (4.1) is zero. After substituting the appropriate expressions and applying calculus to reduce the dimensions of the shell to an infinitesimal thickness, the result of the shell mass balance is a second-order differential equation for substrate concentration as a function of radius in the particle.

For a shell mass-balance on substrate A, the terms of Eq. (4.1) are expressed as follows:

Rate of input by diffusion: $\left. \left(\mathscr{D}_{Ae} \dfrac{dC_A}{dr} \, 4\pi r^2 \right) \right|_{r+\Delta r}$

Rate of output by diffusion: $\left. \left(\mathscr{D}_{Ae} \dfrac{dC_A}{dr} \, 4\pi r^2 \right) \right|_{r}$

Rate of generation: 0

Rate of consumption by reaction: $r_A 4\pi r^2 \Delta r$

Rate of accumulation at steady state: 0.

$\mathscr{D}_{Ae}$ is the effective diffusivity of substrate A, C_A is the concentration of A in the particle, r is distance measured radially from the centre, Δr is thickness of the shell, and r_A is the rate of reaction per unit volume particle. Each of the above terms has dimensions MT^{-1} or NT^{-1} with units of, for example, $kg\ h^{-1}$ or $gmol\ s^{-1}$. The first two terms are derived from Fick's law of Eq. (9.1); the area of the spherical shell available for diffusion is $4\pi r^2$. The term

$$\left. \left(\mathscr{D}_{Ae} \dfrac{dC_A}{dr} \, 4\pi r^2 \right) \right|_{r+\Delta r}$$

means $\left(\mathscr{D}_{\mathrm{Ae}} \dfrac{\mathrm{d}C_{\mathrm{A}}}{\mathrm{d}r} \, 4\pi \, r^2 \right)$ evaluated at radius $(r+\Delta r)$;

$$\left. \left(\mathscr{D}_{\mathrm{Ae}} \frac{\mathrm{d}C_{\mathrm{A}}}{\mathrm{d}r} \, 4\pi \, r^2 \right) \right|_r$$

means $\left(\mathscr{D}_{\mathrm{Ae}} \dfrac{\mathrm{d}C_{\mathrm{A}}}{\mathrm{d}r} \, 4\pi \, r^2 \right)$ evaluated at r.

The shell volume is $4\pi \, r^2 \, \Delta r$.

From Eq. (4.1) we obtain the following steady-state mass-balance equation:

$$\left. \left(\mathscr{D}_{\mathrm{Ae}} \frac{\mathrm{d}C_{\mathrm{A}}}{\mathrm{d}r} \, 4\pi \, r^2 \right) \right|_{r+\Delta r} - \left. \left(\mathscr{D}_{\mathrm{Ae}} \frac{\mathrm{d}C_{\mathrm{A}}}{\mathrm{d}r} \, 4\pi \, r^2 \right) \right|_r$$
$$- \, r_{\mathrm{A}} \, 4\pi \, r^2 \, \Delta r = 0. \tag{12.1}$$

Dividing each term by $4\pi \, \Delta r$ gives:

$$\frac{\left. \left(\mathscr{D}_{\mathrm{Ae}} \dfrac{\mathrm{d}C_{\mathrm{A}}}{\mathrm{d}r} \, r^2 \right) \right|_{r+\Delta r} - \left. \left(\mathscr{D}_{\mathrm{Ae}} \dfrac{\mathrm{d}C_{\mathrm{A}}}{\mathrm{d}r} \, r^2 \right) \right|_r}{\Delta r} - r_{\mathrm{A}} \, r^2 = 0. \tag{12.2}$$

Eq. (12.2) can be written in the form:

$$\frac{\Delta \left(\mathscr{D}_{\mathrm{Ae}} \dfrac{\mathrm{d}C_{\mathrm{A}}}{\mathrm{d}r} \, r^2 \right)}{\Delta r} - r_{\mathrm{A}} \, r^2 = 0 \tag{12.3}$$

where

$$\Delta \left(\mathscr{D}_{\mathrm{Ae}} \frac{\mathrm{d}C_{\mathrm{A}}}{\mathrm{d}r} \, r^2 \right) \text{ means the change in } \left(\mathscr{D}_{\mathrm{Ae}} \frac{\mathrm{d}C_{\mathrm{A}}}{\mathrm{d}r} \, r^2 \right)$$

across Δr.

Eq. (12.3) is valid for a spherical shell of thickness Δr. To develop an equation which applies to any *point* in the sphere, we must shrink Δr to zero. As Δr appears only in the first term of Eq. (12.3), taking the limit of Eq. (12.3) as $\Delta r \to 0$ gives:

$$\lim_{\Delta r \to 0} \frac{\Delta \left(\mathscr{D}_{\mathrm{Ae}} \dfrac{\mathrm{d}C_{\mathrm{A}}}{\mathrm{d}r} \, r^2 \right)}{\Delta r} - r_{\mathrm{A}} \, r^2 = 0. \tag{12.4}$$

Invoking the definition of the derivative from Section D.2 of the Appendix, Eq. (12.4) is identical to the second-order differential equation:

$$\frac{\mathrm{d}}{\mathrm{d}r} \left(\mathscr{D}_{\mathrm{Ae}} \frac{\mathrm{d}C_{\mathrm{A}}}{\mathrm{d}r} \, r^2 \right) - r_{\mathrm{A}} \, r^2 = 0. \tag{12.5}$$

According to assumption (iii), $\mathscr{D}_{\mathrm{Ae}}$ is independent of r and can be moved outside the differential:

$$\mathscr{D}_{\mathrm{Ae}} \frac{\mathrm{d}}{\mathrm{d}r} \left(\frac{\mathrm{d}C_{\mathrm{A}}}{\mathrm{d}r} \, r^2 \right) - r_{\mathrm{A}} \, r^2 = 0. \tag{12.6}$$

In Eq. (12.6) we have a differential equation representing diffusion and reaction in a spherical biocatalyst. That Eq. (12.6) is a second-order differential equation becomes clear if the first term is written in its expanded form:

$$\mathscr{D}_{\mathrm{Ae}} \left(\frac{\mathrm{d}^2 C_{\mathrm{A}}}{\mathrm{d}r^2} \, r^2 + 2r \, \frac{\mathrm{d}C_{\mathrm{A}}}{\mathrm{d}r} \right) - r_{\mathrm{A}} \, r^2 = 0. \tag{12.7}$$

Eq. (12.7) can be solved by integration to yield an expression for the concentration profile in the particle: C_{A} as a function of r. However, we cannot integrate Eq. (12.7) as it stands because the reaction rate r_{A} is in most cases a function of C_{A}. Let us consider solutions of Eq. (12.7) with r_{A} representing first-order, zero-order and Michaelis–Menten kinetics.

12.3.2 Concentration Profile: First-Order Kinetics and Spherical Geometry

For first-order kinetics, Eq. (12.7) becomes:

$$\mathscr{D}_{\mathrm{Ae}} \left(\frac{\mathrm{d}^2 C_{\mathrm{A}}}{\mathrm{d}r} \, r^2 + 2r \, \frac{\mathrm{d}C_{\mathrm{A}}}{\mathrm{d}r} \right) - k_1 C_{\mathrm{A}} \, r^2 = 0 \tag{12.8}$$

where k_1 is the intrinsic first-order rate constant with dimensions T^{-1}. For biocatalytic reactions, k_1 depends on the ensity of cells or enzyme in the particle. According to assumptions (i), (iii) and (iv) in Section 12.3.1, k_1 and $\mathscr{D}_{Ae}$ for a given article can be considered constant. Accordingly, as the only ariables in Eq. (12.8) are C_A and r, the equation is ready for ntegration. Because Eq. (12.8) is a second-order differential quation we need two boundary conditions. These are:

$$C_A = C_{As} \qquad \text{at } r = R$$
(12.9)

$$\frac{dC_A}{dr} = 0 \qquad \text{at } r = 0$$
(12.10)

where C_{As} is the concentration of substrate at the outer surface f the particle. For the present we will assume C_{As} is known or an be measured. Eq. (12.10) is called the *symmetry condition*. s indicated in Figures 12.2 and 12.3, the substrate concentration profile is symmetrical about the centre of the sphere.

Therefore, the substrate concentration is minimum with slope $^{dC_A/}dr = 0$ at $r = 0$. Integration of Eq. (12.8) with boundary conditions Eqs (12.9) and (12.10) gives the following expression for substrate concentration as a function of radius [7]:

$$C_A = C_{As} \; \frac{R}{r} \; \frac{\sinh\left(r\sqrt{k_1/\mathscr{D}_{Ae}}\right)}{\sinh\left(R\sqrt{k_1/\mathscr{D}_{Ae}}\right)}.$$
(12.11)

In Eq. (12.11), sinh is the abbreviation for *hyperbolic sine*; $\sinh x$ is defined as:

$$\sinh x = \frac{e^x - e^{-x}}{2}.$$
(12.12)

Eq. (12.11) may appear complex, but contains simple exponential terms relating C_A and r, $\mathscr{D}_{Ae}$ representing rate of mass transfer, and k_1 representing rate of reaction.

Example 12.1 Concentration profile for immobilised enzyme

Enzyme is immobilised in 8 mm diameter agarose beads at a concentration of 0.018 kg protein m^{-3} gel. Ten beads are mmersed in a well-mixed solution containing 3.2×10^{-3} kg m^{-3} substrate. The effective diffusivity of substrate in agarose gel is .1 × 10^{-9} m^2 s^{-1}. Kinetics of the enzyme can be approximated as first order with specific rate constant 3.11×10^5 s^{-1} per kg rotein. Mass transfer effects outside the particles are negligible. Plot the steady-state substrate concentration profile as a function of particle radius.

Solution:

$R = 4 \times 10^{-3}$ m; $\mathscr{D}_{Ae} = 2.1 \times 10^{-9}$ m^2 s^{-1}. In the absence of external mass-transfer effects, $C_{As} = 3.2 \times 10^{-3}$ kg m^{-3}.

$$\text{Volume per bead} = \frac{4}{3}\pi R^3 = \frac{4}{3}\pi (4 \times 10^{-3}\,\text{m})^3 = 2.68 \times 10^{-7}\,\text{m}^3.$$

Therefore, 10 beads have volume 2.68×10^{-6} m^3. The amount of enzyme present is:

$$2.68 \times 10^{-6}\,\text{m}^3\,(0.018\,\text{kg m}^{-3}) = 4.83 \times 10^{-8}\,\text{kg}.$$

Therefore:

$$k_1 = 3.11 \times 10^5\,\text{s}^{-1}\,\text{kg}^{-1}\,(4.83 \times 10^{-8}\,\text{kg}) = 0.015\,\text{s}^{-1}$$

and:

$$R\sqrt{\frac{k_1}{\mathscr{D}_{Ae}}} = 10.693.$$

The denominator of Eq. (12.11) is:

$$\sinh\left(R\sqrt{k_1/\mathscr{D}_{\mathrm{Ae}}}\right) = \frac{e^{10.693} - e^{-10.693}}{2} = 2.202 \times 10^4.$$

C_A is calculated as a function of r from Eq. (12.11) and plotted in Figure 12E1.1. Substrate concentration drops rapidly inside the particle to reach virtually zero 2 mm from the centre.

Figure 12E1.1 Substrate concentration profile in an immobilised-enzyme bead.

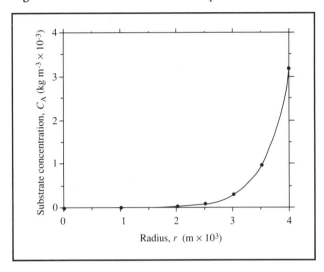

12.3.3 Concentration Profile: Zero-Order Kinetics and Spherical Geometry

From Eq. (12.7), the differential equation for zero-order kinetics is:

$$\mathscr{D}_{\mathrm{Ae}}\left(\frac{d^2 C_A}{dr^2}\, r^2 + 2\,r\, \frac{dC_A}{dr}\right) - k_0\, r^2 = 0$$

(12.13)

where k_0 is the intrinsic zero-order rate constant with units of, for example, gmol s^{-1} m^{-3} particle. Like k_1 for first-order reactions, k_0 varies with cell or enzyme density in the catalyst.

Zero-order reactions are unique in that, provided substrate is present, reaction rate is independent of substrate concentration. In solving Eq. (12.13) we must account for the possibility that substrate becomes depleted within the particle. As illustrated in Figure 12.5, if we assume this occurs at some radius R_0, the rate of reaction for $0 < r \leqslant R_0$ is zero. Everywhere else inside the particle, i.e., $r > R_0$, the volumetric reaction rate is constant and equal to k_0 irrespective of substrate concentration. For this situation the boundary conditions are:

$$C_A = C_{\mathrm{As}} \qquad \text{at } r = R$$

(12.9)

$$\frac{dC_A}{dr} = 0 \qquad \text{at } r = R_0.$$

(12.14)

Solution of Eq. (12.13) with these boundary conditions gives the following expression for C_A as a function of r [7]:

$$C_A = C_{\mathrm{As}} + \frac{k_0 R^2}{6\mathscr{D}_{\mathrm{Ae}}}\left(\frac{r^2}{R^2} - 1 + \frac{2R_0^3}{rR^2} - \frac{2R_0^3}{R^3}\right).$$

(12.15)

Eq. (12.15) is difficult to apply in practice because R_0 is generally not known. However, the equation can be simplified if C_A remains > 0 everywhere so that R_0 no longer exists. Substituting $R_0 = 0$ into Eq. (12.15) gives:

$$C_A = C_{\mathrm{As}} + \frac{k_0}{6\mathscr{D}_{\mathrm{Ae}}}\,(r^2 - R^2).$$

(12.16)

In bioprocess applications, it is important that the core of catalyst particles does not become starved of substrate. The likelihood of this happening increases with size of the particle. For zero-order reactions we can calculate the maximum particle radius for which C_A remains > 0. In such a particle, substrate is depleted just at the centre point. Therefore, calculating R from Eq. (12.16) with $C_A = r = 0$:

$$R_{max} = \sqrt{\frac{6 \mathscr{D}_{Ae} C_{As}}{k_0}}$$

(12.17)

where R_{max} is the maximum particle radius for $C_A > 0$.

Figure 12.5 Concentration and reaction zones in a spherical particle with zero-order reaction. Substrate is depleted at radius R_0.

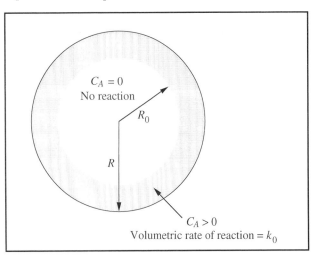

Example 12.2 Maximum particle size for zero-order reaction

Non-viable yeast cells are immobilised in alginate beads. The beads are stirred in glucose medium under anaerobic conditions. The effective diffusivity of glucose in the beads depends on cell density according to the relationship:

$$\mathscr{D}_{Ae} = 6.33 - 7.17 y_C$$

where $\mathscr{D}_{Ae}$ is effective diffusivity $\times 10^{10}$ m^2 s^{-1} and y_C is the weight fraction of yeast in the gel. Rate of glucose uptake can be assumed to be zero order; the rate constant at a yeast density in alginate of 15 wt% is 0.5 g l^{-1} min^{-1}. For maximum reaction rate, the concentration of glucose inside the particles should remain above zero.

(a) Plot the maximum allowable particle size as a function of bulk glucose concentration between 5 g l^{-1} and 60 g l^{-1}.
(b) For 30 g l^{-1} glucose, plot R_{max} as a function of cell loading between 10 and 45 wt%.

Solution:

(a) At $y_C = 0.15$, $\mathscr{D}_{Ae} = 5.25 \times 10^{-10}$ m^2 s^{-1}. Converting k_0 to units of kg, m and s:

$$k_0 = 0.5 \text{ g l}^{-1} \text{ min}^{-1} \cdot \left| \frac{1 \text{ kg}}{1000 \text{ g}} \right| \cdot \left| \frac{1000 \text{ l}}{1 \text{ m}^3} \right| \cdot \left| \frac{1 \text{ min}}{60 \text{ s}} \right|$$

$$= 8.33 \times 10^{-3} \text{ kg m}^{-3} \text{ s}^{-1}.$$

Assume C_{As} is equal to the bulk glucose concentration; C_{As} in g l^{-1} is the same as kg m^{-3}. R_{max} is calculated from Eq. (12.17).

C_{As} (kg m^{-3})	R_{max} (m)
5	1.38×10^{-3}
15	2.38×10^{-3}
25	3.07×10^{-3}
45	4.13×10^{-3}
60	4.76×10^{-3}

These results are plotted in Figure 12E2.1. At low external glucose concentrations, particles are restricted to small radii. Th driving force for diffusion increases with C_{As} so that larger particles may be used.

Figure 12E2.1 Maximum particle radius as a function of external substrate concentration.

Figure 12E2.2 Maximum particle radius as a function of cell density.

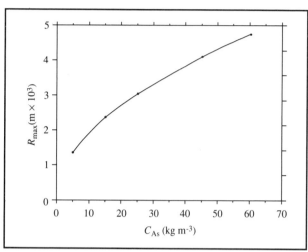

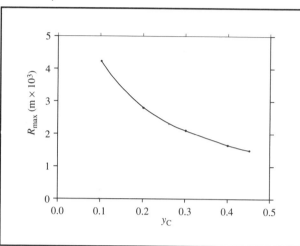

(b) $C_{As} = 30$ kg m^{-3}. As y_C varies, values of $\mathscr{D}_{Ae}$ and k_0 are affected. Changes in $\mathscr{D}_{Ae}$ can be calculated from the equation provided. We assume k_0 is directly proportional to cell density as described in Eq. (11.25), i.e. there is no steric hindrance or interaction between cells as y_C increases. Results as a function of y_C are listed below.

y_C	$\mathscr{D}_{Ae}$ (m^2 s^{-1})	k_0 (kg m^{-3} s^{-1})	R_{max} (m)
0.1	5.61×10^{-10}	5.55×10^{-3}	4.27×10^{-3}
0.2	4.90×10^{-10}	1.11×10^{-2}	2.82×10^{-3}
0.3	4.18×10^{-10}	1.67×10^{-2}	2.12×10^{-3}
0.4	3.46×10^{-10}	2.22×10^{-2}	1.67×10^{-3}
0.45	3.10×10^{-10}	2.50×10^{-2}	1.50×10^{-3}

The results are plotted in Figure 12E2.2. As y_C increases, $\mathscr{D}_{Ae}$ declines and k_0 increases. Lower $\mathscr{D}_{Ae}$ reduces the rate of diffusion into the particles; higher k_0 increases the demand for substrate. Therefore, increasing the cell density exacerbates mass-transfer restrictions. To ensure adequate supply of substrate under these conditions, the particle size must be reduced.

12.3.4 Concentration Profile: Michaelis–Menten Kinetics and Spherical Geometry

If reaction in the particle follows Michaelis–Menten kinetics, r_A takes the form of Eq. (11.30). Eq. (12.7) becomes:

$$\mathscr{D}_{Ae} \left(\frac{d^2 C_A}{dr^2} r^2 + 2r \frac{dC_A}{dr} \right) - \frac{v_{max} C_A}{K_m + C_A} r^2 = 0$$

(12.18)

where v_{max} and K_m are intrinsic kinetic parameters for the reaction. v_{max} has units of, for example, kg s^{-1} m^{-3} particle; its value depends on the concentration of cells or enzyme in the particle.

Owing to the non-linearity of the Michaelis–Menten expression, simple analytical integration of Eq. (12.18) is not possible. However, results for C_A as a function of r can be obtained using numerical methods, usually by computer. Because Michaelis–Menten kinetics lie somewhere between zero- and first-order kinetics (see Section 11.3.3), explicit

Figure 12.6 Measured and calculated oxygen concentrations in a spherical agarose bead containing immobilised enzyme. Particle diameter = 4 mm; C_{Ab} = 0.2 mol m^{-3}. Enzyme loadings are: 0.0025 kg m^{-3} gel (■); 0.005 kg m^{-3} gel (□); 0.0125 kg m^{-3} gel (△); and 0.025 kg m^{-3} gel (●). Measured concentrations are shown using symbols; calculated profiles are shown as lines. (From C.M. Hooijmans, S.G.M. Geraats and K.Ch.A.M. Luyben, 1990, Use of an oxygen microsensor for the determination of intrinsic kinetic parameters of an immobilised oxygen reducing enzyme. *Biotechnol. Bioeng.* **35**, 1078–1087.)

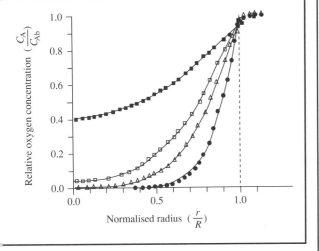

Figure 12.7 Substrate concentration profile in an infinite flat plate without boundary-layer effects.

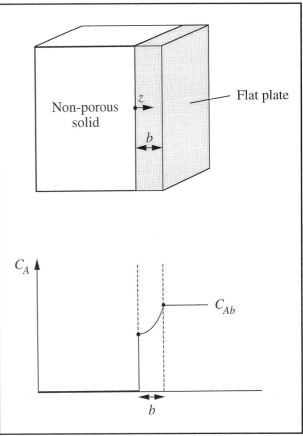

solutions found in Sections 12.3.2 and 12.3.3 can be used to estimate the extreme limits for Michaelis–Menten reactions.

Concentration profiles calculated from the equations presented in this section have been verified experimentally in several studies. Using special microelectrodes with tip diameters of the order 1 μm, it is possible to measure concentrations of oxygen and ions inside soft solids and cell slimes. As an example, oxygen concentrations measured in immobilised-enzyme beads are shown in Figure 12.6. The experimental data are very close to the calculated concentration profiles. Similar results have been found in other systems [8–10].

12.3.5 Concentration Profiles in Other Geometries

Our attention so far has been focussed on spherical catalysts. However, equations similar to Eq. (12.7) can be obtained from shell mass balances on other geometries. Of all other shapes, the one of most interest in bioprocessing is the flat plate. A typical substrate concentration profile for this geometry without external boundary-layer effects is illustrated in

Figure 12.7. Equations for flat-plate geometry are used to analyse reactions in cell films attached to inert solids; the biofilm constitutes the flat plate. Even if the surface supporting the biofilm is curved rather than flat, if the film thickness b is very small compared with the radius of curvature, equations for flat-plate geometry are applicable. To simplify mathematical treatment and keep the problem one-dimensional (as required by assumption (vii) of Section 12.3.1), the flat plate is assumed to have infinite length. In practice, this assumption is reasonable if its length is much greater than its thickness. If not, it must be assumed that the ends of the plate are sealed to eliminate axial concentration gradients.

Another catalyst shape of some relevance to bioprocessing is the hollow cylinder; this is useful in analysis of hollow-fibre membrane reactors. However, because of its relatively limited application, we will not consider this geometry further.

Concentration profiles for spherical and flat-plate geometries and first- and zero-order kinetics are summarised in Table

12.1. Boundary conditions for the flat plate similar to Eqs (12.9) and (12.10) are as follows:

$$C_A = C_{As} \qquad \text{at } z = b$$

$$(12.19)$$

$$\frac{dC_A}{dz} = 0 \qquad \text{at } z = 0$$

$$(12.20)$$

where C_{As} is the concentration of A at the solid–liquid interface, z is distance measured from the inner surface of the plate and b is the plate thickness.

12.3.6 Prediction of Observed Reaction Rate

Equations for intracatalyst substrate concentration such as those in Table 12.1 allow us to predict overall rates of reaction. Let us consider the situation for spherical particles and first-order, zero-order and Michaelis–Menten kinetics. Analogous equations can be derived for other geometries.

(i) *First-order kinetics.* Rate of reaction at any point in the sphere depends on the first-order kinetic constant k_1 and the concentration of substrate at that point. The overall rate for the entire particle is equal to the sum of all such rates at every location in the solid. This sum is mathematically equivalent to integrating the expression $k_1 C_A$ over the entire particle volume, taking into account the variation of C_A with radius expressed in Eq. (12.11). The result is an equation for the observed reaction rate $r_{A,obs}$ in a single particle:

$$r_{A,obs} = 4\pi R \mathscr{D}_{Ae} C_{As} \left[R\sqrt{k_1/\mathscr{D}_{Ae}} \coth \left(R\sqrt{k_1/\mathscr{D}_{Ae}} \right) - 1 \right]$$

$$(12.21)$$

where coth is the abbreviation for *hyperbolic cotangent* defined by:

$$\coth x = \frac{e^x + e^{-x}}{e^x - e^{-x}}.$$

$$(12.22)$$

(ii) *Zero-order kinetics.* As long as substrate is present, zero-order reactions occur at a fixed rate independent of substrate concentration. Therefore, if $C_A > 0$ everywhere in the particle, the overall rate of reaction is equal to the zero-order rate constant k_0 multiplied by the particle volume:

Table 12.1 Steady-state concentration profiles

First-order reaction: $r_A = k_1 C_A$

Sphere [a] $\quad C_A = C_{As} \, \dfrac{R}{r} \, \dfrac{\sinh\left(r\sqrt{k_1/\mathscr{D}_{Ae}}\right)}{\sinh\left(R\sqrt{k_1/\mathscr{D}_{Ae}}\right)}$

Flat plate [b] $\quad C_A = C_{As} \, \dfrac{\cosh\left(z\sqrt{k_1/\mathscr{D}_{Ae}}\right)}{\cosh\left(b\sqrt{k_1/\mathscr{D}_{Ae}}\right)}$

Zero-order reaction: $r_A = k_0$

Sphere [c] $\quad C_A = C_{As} + \dfrac{k_0}{6\mathscr{D}_{Ae}} \, (r^2 - R^2)$

Flat plate [c] $\quad C_A = C_{As} + \dfrac{k_0}{2\mathscr{D}_{Ae}} \, (z^2 - b^2)$

---

[a] Sinh is the abbreviation of hyperbolic sine. Sinh x is defined as:

$$\sinh x = \frac{e^x - e^{-x}}{2}.$$

[b] Cosh is the abbreviation of hyperbolic cosine. Cosh x is defined as:

$$\cosh x = \frac{e^x + e^{-x}}{2}.$$

[c] For $C_A > 0$ everywhere within the catalyst.

$$r_{A,obs} = \frac{4}{3} \pi R^3 k_0.$$

$$(12.23)$$

However, if C_A falls to zero at some radius R_0, the inner volume $\frac{4}{3} \pi R_0^3$ is inactive. In this case, the rate of reaction per particle is equal to k_0 multiplied by the active particle volume:

$$r_{A,obs} = \left(\frac{4}{3} \pi R^3 - \frac{4}{3} \pi R_0^3 \right) k_0 = \frac{4}{3} \pi (R^3 - R_0^3) \, k_0.$$

$$(12.24)$$

(iii) *Michaelis–Menten kinetics.* Observed rates for Michaelis–Menten reactions cannot be expressed explicitly because we do not have an equation for C_A as a function of radius. $r_{A,obs}$ can be evaluated, however, using numerical methods.

2.4 The Thiele Modulus and Effectiveness Factor

Charts based on the equations of the previous section allow us to determine $r_{A,obs}$ relative to r_{As}^*, the reaction rate that would occur if all cells or enzyme were exposed to the external substrate concentration. Differences between $r_{A,obs}$ and r_{As}^* show immediately the extent to which reaction is affected by internal mass transfer. Comparison of these rates requires application of theory as described in the following sections.

2.4.1 First-Order Kinetics

If a catalyst particle is unaffected by mass transfer, the concentration of substrate inside the particle is constant and equal to the surface concentration, C_{As}. Thus, the rate of first-order reaction without internal mass-transfer effects is equal to $k_1 C_{As}$ multiplied by the particle volume:

$$r_{As}^* = \frac{4}{3} \pi R^3 k_1 C_{As}.$$

$$(12.25)$$

The extent to which $r_{A,obs}$ is different from r_{As}^* is expressed by means of the *internal effectiveness factor* η_i:

$$\eta_i = \frac{r_{A,obs}}{r_{As}^*} = \frac{\text{(observed rate)}}{\left(\begin{array}{c}\text{rate that would occur if } C_A = C_{As} \\ \text{everywhere in the particle}\end{array}\right)}.$$

$$(12.26)$$

In the absence of mass-transfer limitations, $r_{A,obs} = r_{As}^*$ and $\eta_i = 1$; when mass-transfer effects reduce $r_{A,obs}$, $\eta_i < 1$. For calculation of η_i, $r_{A,obs}$ and r_{As}^* should have the same units, for example, kg s^{-1} m^{-3}, gmol s^{-1} per particle, etc. We can substitute expressions for $r_{A,obs}$ and r_{As}^* from Eqs (12.21) and (12.25) into Eq. (12.26) to derive an expression for η_{i1}, the internal effectiveness factor for first-order reaction:

$$\eta_{i1} = \frac{3 \mathscr{D}_{Ae}}{R^2 k_1} \left[R \sqrt{k_1/\mathscr{D}_{Ae}} \coth \left(R \sqrt{k_1/\mathscr{D}_{Ae}} \right) - 1 \right].$$

$$(12.27)$$

Thus, the internal effectiveness factor for first-order reaction depends on only three parameters: R, k_1 and $\mathscr{D}_{Ae}$. These parameters are usually grouped together to form a dimensionless variable called the *Thiele modulus*. There are several definitions of the Thiele modulus in the literature; as it was formulated originally [11], application of the modulus was cumbersome because a separate definition was required for different reaction kinetics and catalyst geometries. Generalised moduli which apply to any catalyst shape and reaction kinetics have since been proposed [12–14]. The *generalised Thiele modulus ϕ* is defined as:

$$\phi = \frac{V_p}{S_x} \frac{r_A|_{C_{As}}}{\sqrt{2}} \left(\int_{C_{A,eq}}^{C_{As}} \mathscr{D}_{Ae} \, r_A \, dC_A \right)^{-1/2}$$

$$(12.28)$$

where V_p is catalyst volume, S_x is external surface area, C_{As} is substrate concentration at the surface of the catalyst, r_A is reaction rate, $r_A|_{C_{As}}$ is the reaction rate when $C_A = C_{As}$, $\mathscr{D}_{Ae}$ is effective diffusivity of substrate and $C_{A,eq}$ is the equilibrium substrate concentration. As explained in Section 11.1.1, fermentations and many enzyme reactions are irreversible so that $C_{A,eq}$ is zero for most biological applications. From geometry, $V_p/S_x = R/3$ for spheres and b for flat plates. Expressions determined from Eq. (12.28) for first-order, zero-order and Michaelis–Menten kinetics are listed in Table 12.2 as ϕ_1, ϕ_0 and ϕ_m, respectively. ϕ represents a dimensionless combination of the important parameters affecting mass transfer and reaction in heterogeneous systems: catalyst size (R or b), effective diffusivity ($\mathscr{D}_{Ae}$), surface concentration (C_{As}), and intrinsic rate parameters (k_0, k_1 or v_{max} and K_m). Only the Thiele modulus for first-order reactions does not depend on substrate concentration.

When parameters R, k_1 and $\mathscr{D}_{Ae}$ in Eq. (12.27) are grouped together as ϕ_1, the result is:

$$\eta_{i1} = \frac{1}{3\phi_1^2} (3\phi_1 \coth 3\phi_1 - 1)$$

$$(12.29)$$

where coth is defined by Eq. (12.22). Eq. (12.29) applies to spherical geometry and first-order reaction; an analogous equation for flat plates is listed in Table 12.3. Plots of η_{i1} versus ϕ_1 for sphere, cylinder and flat-plate catalysts are shown in Figure 12.8. The curves coincide exactly for $\phi_1 \to 0$ and $\phi_1 \to \infty$, and fall within 10–15% for the remainder of the

Table 12.2 Generalised Thiele moduli

First-order reaction: $r_A = k_1 C_A$

$$\phi_1 = \frac{V_p}{S_x} \sqrt{\frac{k_1}{\mathscr{D}_{Ae}}}$$

Sphere $\quad \phi_1 = \frac{R}{3} \sqrt{\frac{k_1}{\mathscr{D}_{Ae}}}$

Flat plate $\quad \phi_1 = b \sqrt{\frac{k_1}{\mathscr{D}_{Ae}}}$

Zero-order reaction: $r_A = k_0$

$$\phi_0 = \frac{1}{\sqrt{2}} \frac{V_p}{S_x} \sqrt{\frac{k_0}{\mathscr{D}_{Ae} C_{As}}}$$

Sphere $\quad \phi_0 = \frac{R}{3\sqrt{2}} \sqrt{\frac{k_0}{\mathscr{D}_{Ae} C_{As}}}$

Flat plate $\quad \phi_0 = \frac{b}{\sqrt{2}} \sqrt{\frac{k_0}{\mathscr{D}_{Ae} C_{As}}}$

Michaelis–Menten reaction: $r_A = \dfrac{v_{max} C_A}{K_m + C_A}$

$$\phi_m = \frac{1}{\sqrt{2}} \frac{V_p}{S_x} \sqrt{\frac{v_{max}}{\mathscr{D}_{Ae} C_{As}} \left(\frac{1}{1+\beta}\right) \left[1 + \beta \ln\left(\frac{\beta}{1+\beta}\right)\right]^{-1/2}}$$

$$\beta = \frac{K_m}{C_{As}}$$

Sphere $\quad \phi_m = \dfrac{R}{3\sqrt{2}} \sqrt{\dfrac{v_{max}}{\mathscr{D}_{Ae} C_{As}} \left(\dfrac{1}{1+\beta}\right) \left[1 + \beta \ln\left(\dfrac{\beta}{1+\beta}\right)\right]^{-1/2}}$

Flat plate $\quad \phi_m = \dfrac{b}{\sqrt{2}} \sqrt{\dfrac{v_{max}}{\mathscr{D}_{Ae} C_{As}} \left(\dfrac{1}{1+\beta}\right) \left[1 + \beta \ln\left(\dfrac{\beta}{1+\beta}\right)\right]^{-1/2}}$

range. Figure 12.8 can be used to evaluate η_{i1} for any catalyst shape provided ϕ_1 is calculated using Eq. (12.28). Because of the errors involved in estimating the parameters defining ϕ_1, it has been suggested that effectiveness factor curves be viewed as diffuse bands rather than precise functions [15].

Thus, if the first-order Thiele modulus is known, we can use Figure 12.8 to find the internal effectiveness factor and Eq. (12.25) and (12.26) to predict the overall reaction rate for the catalyst. At low values of $\phi_1 < 0.3$, $\eta_{i1} \approx 1$ and the rate of reac-tion is not adversely affected by internal mass transfer. However as ϕ_1 increases above 0.3, η_{i1} falls as mass-transfer limitations come into play. Therefore, the value of the Thiel

Table 12.3 Effectiveness factors (ϕ for each geometry and kinetic order is defined in Table 12.2)

First-order reaction: $r_A = k_1 \, C_A$

Sphere[a] $\eta_{i1} = \dfrac{1}{3\phi_1^2} \, (3\phi_1 \coth 3\phi_1 - 1)$

Flat plate [b] $\eta_{i1} = \dfrac{\tanh \phi_1}{\phi_1}$

Zero-order reaction: $r_A = k_0$

Sphere [c] $\eta_{i0} = 1$ for $0 < \phi_0 \leqslant 0.577$

Sphere [c] $\eta_{i0} = 1 - \left[\dfrac{1}{2} + \cos \left(\dfrac{\Psi + 4\pi}{3} \right) \right]^3$

       where $\Psi = \cos^{-1} \left(\dfrac{2}{3\phi_0^2} - 1 \right)$ $\Bigg\}$ for $\phi_0 > 0.577$

Flat plate $\eta_{i0} = 1$ for $0 < \phi_0 \leqslant 1$

       $\eta_{i0} = \dfrac{1}{\phi_0}$ for $\phi_0 > 1$

[a] Coth is the abbreviation of hyperbolic cotangent. Coth x is defined as:

$$\coth x = \frac{e^x + e^{-x}}{e^x - e^{-x}}.$$

[b] Tanh is the abbreviation of hyperbolic tangent. Tanh x is defined as:

$$\tanh x = \frac{e^x - e^{-x}}{e^x + e^{-x}}.$$

[c] Cos is the abbreviation of cosine. The notation $\cos^{-1} x$ (or arccos x) denotes any angle whose cosine is x. Angles used to determine cos and $\cos^{-1}$ are in radians.

modulus indicates immediately whether the rate of reaction is diminished due to diffusional effects, or whether the catalyst is performing at its maximum rate at the prevailing surface concentration. For strong diffusion limitations at $\phi_1 > 10$, η_{i1} for all geometries can be estimated as:

$$\eta_{i1} \approx \frac{1}{\phi_1}.$$

$$(12.30)$$

12.4.2 Zero-Order Kinetics

When substrate is present throughout the catalyst, evaluation of the zero-order internal effectiveness factor η_{i0} is straightforward. Under these conditions, the reaction proceeds at the same rate that would occur if $C_A = C_{As}$ throughout in the particle. Therefore, from Eq. (12.26), $\eta_{i0} = 1$ and:

$$r_{A,obs} = r_{As}^* = \frac{4}{3}\pi R^3 k_0.$$

$$(12.31)$$

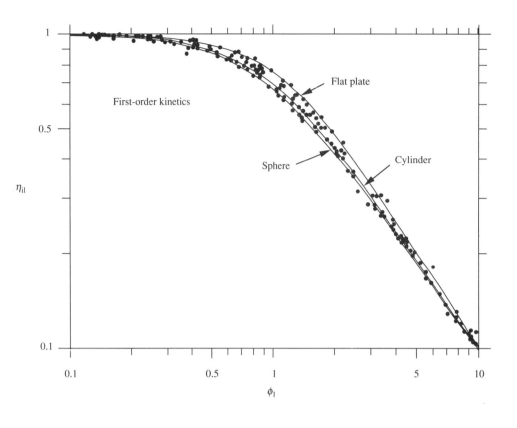

Figure 12.8 Internal effectiveness factor η_{i1} as a function of the generalised Thiele modulus ϕ_1 for first-order kinetics and spherical, cylindrical and flat-plate geometries. The dots represent calculations on finite or hollow cylinders and parallelepipeds. (From R. Aris, 1975, *The Mathematical Theory of Diffusion and Reaction in Permeable Catalysts*, vol. 1, Oxford University Press, London.)

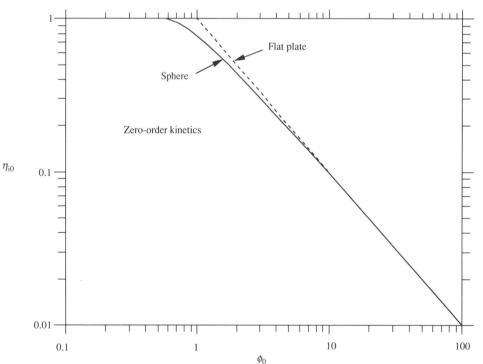

Figure 12.9 Internal effectiveness factor η_{i0} as a function of the generalised Thiele modulus ϕ_0 for zero-order kinetics and spherical and flat-plate geometries.

If C_A falls to zero within the pellet, the effectiveness factor must be evaluated differently. In this case, $r_{A,obs}$ is given by Eq. (12.24) and the internal effectiveness factor is:

$$\eta_{i0} = \frac{\frac{4}{3}\pi(R^3 - R_0^3)k_0}{\frac{4}{3}\pi R^3 k_0} = 1 - \left(\frac{R_0}{R}\right)^3 .$$

(12.32)

According to the above analysis, to evaluate η_{i0} for zero-order kinetics, first we must know whether or not substrate is depleted in the catalyst, then, if it is, the value of R_0. Usually this information is unavailable because we cannot easily measure intraparticle concentrations. Fortunately, further mathematical analysis [7] overcomes this problem by representing the system in terms of measurable properties such as R, $\mathscr{D}_{Ae}$, C_{As} and k_0 rather than R_0. These parameters define the Thiele modulus for zero-order reaction, ϕ_0. The results are summarised in Table 12.3 and Figure 12.9. C_A remains > 0 and $\eta_{i0} = 1$ for $0 < \phi_0 \le 0.577$; for $\phi_0 > 0.577$, η_{i0} declines as more and more of the particle becomes inactive. Effectiveness factors for flat-plate systems are also shown in Table 12.3 and Figure 12.9; in flat films, $\phi_0 = 1$ represents the threshold condition for substrate depletion. η_{i0} curves for spherical and flat-plate geometries coincide exactly at small and large values of ϕ_0.

12.4.3 Michaelis–Menten Kinetics

For a spherical catalyst, the rate of Michaelis–Menten reaction in the absence of internal mass-transfer effects is:

$$r_{As}^* = \frac{4}{3}\pi R^3 \left(\frac{v_{max}C_{As}}{K_m + C_{As}}\right) .$$

(12.33)

Our analysis cannot proceed further, however, because we do not have an equation for $r_{A,obs}$. Accordingly, we cannot develop an analytical expression for η_{im} as a function of ϕ_m. Diffusion–reaction equations for Michaelis–Menten kinetics are generally solved by numerical computation. As an example, the results for flat-plate geometry are shown in Figure 12.10 as a function of β, which is equal to K_m/C_{As}.

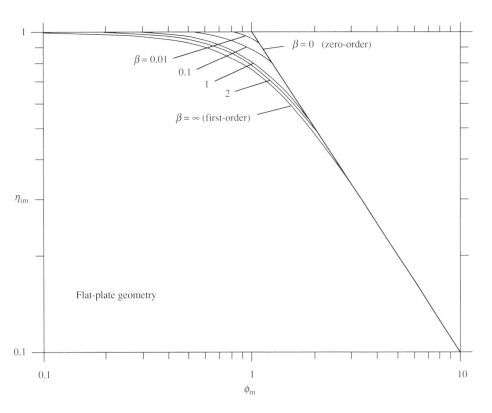

Figure 12.10 Internal effectiveness factor η_{im} as a function of the generalised Thiele modulus ϕ_m and parameter β for Michaelis–Menten kinetics and flat-plate geometry. $\beta = K_m/C_{As}$. (From R. Aris, 1975, *The Mathematical Theory of Diffusion and Reaction in Permeable Catalysts*, vol. 1, Oxford University Press, London.)

We can obtain approximate values for η_{im} by considering the zero- and first-order asymptotes of the Michaelis–Menten equation. As indicated in Figure 12.10, curves for η_{im} fall between the lines for zero- and first-order reactions. The exact position depends on the value of β. As $\beta \rightarrow \infty$, Michaelis–Menten kinetics can be approximated as first order and the internal effectiveness factor evaluated from Figure 12.8 with $k_1 = v_{max}/K_m$. When $\beta \rightarrow \infty$, zero-order reaction and Figure 12.9 apply with $k_0 = v_{max}$. Effectiveness factors for β between zero and infinity must be evaluated using numerical methods. Use of the generalised Thiele modulus as defined in Table 12.2 eliminates almost all variation in the internal effectiveness factor with changing β, except in the vicinity of $\phi_m = 1$. As in Figures 12.8 and 12.9, the generalised modulus also brings together effectiveness-factor curves for all shapes of catalyst at the two asymptotes $\phi_m \rightarrow 0$ and $\phi_m \rightarrow \infty$. Therefore, Figure 12.10 is valid for spherical catalysts if ϕ_m is much less or much greater than 1. It should be noted however that variation between geometries in the intermediate region

around $\phi_m = 1$ can be significant [16].

If values of ϕ_m and β are such that Michaelis–Menten kinetics cannot be approximated by either zero- or first-order equations, η_{im} can be estimated using an equation proposed by Moo-Young and Kobayashi [17]:

$$\eta_{im} = \frac{\eta_{i0} + \beta \eta_{i1}}{1 + \beta}$$

(12.34)

where $\beta = K_m/C_{As}$. η_{i0} is the zero-order internal effectiveness factor obtained using values of ϕ_0 evaluated with $k_0 = v_{max}$; η_{i1} is the first-order effectiveness factor obtained using ϕ_1 calculated with $k_1 = v_{max}/K_m$. For flat-plate geometry, the largest deviation of Eq. (12.34) from exact values of η_{im} is 0.089; this occurs at $\phi_m = 1$ and $\beta = 0.2$. For spherical geometry, the greatest deviations occur around $\phi_m = 1.7$ and $\beta = 0.3$; the maximum error in this region is 0.09. Further details can be found in the original paper [17].

Example 12.3 Reaction rates for free and immobilised enzyme

Invertase is immobilised in ion-exchange resin of average diameter 1 mm. The amount of enzyme in the beads is measured by protein assay as 0.05 kg m^{-3}. 20 cm^3 beads are packed into a small column reactor; 75 ml sucrose solution at a concentration of 16 mM is pumped rapidly through the bed. In another reactor an identical quantity of free enzyme is mixed into the same volume of sucrose solution. Assume the kinetic parameters for free and immobilised enzyme are equal: K_m is 8.8 mM and the turnover number is 2.4×10^{-3} gmol glucose (g enzyme)$^{-1}$ s^{-1}. The effective diffusivity of sucrose in the ion-exchange resin is 2×10^{-6} cm^2 s^{-1}.

(a) What is the rate of reaction by free enzyme?
(b) What is the rate of reaction by immobilised enzyme?

Solution:
The invertase reaction is:

$$\underset{\text{sucrose}}{C_{12}H_{22}O_{11}} + H_2O \rightarrow \underset{\text{glucose}}{C_6H_{12}O_6} + \underset{\text{fructose}}{C_6H_{12}O_6}.$$

Convert the data provided to units of gmol, m and s.

$$K_m = \frac{8.8 \times 10^{-3} \text{ gmol}}{\text{litre}} \cdot \left| \frac{1000 \text{ litres}}{1 \text{ m}^3} \right| = 8.8 \text{ gmol m}^{-3}$$

$$\mathscr{D}_{Ae} = 2 \times 10^{-6} \text{ cm}^2 \text{ s}^{-1} \cdot \left| \frac{1 \text{ m}}{100 \text{ cm}} \right|^2 = 2 \times 10^{-10} \text{ m}^2 \text{ s}^{-1}$$

$$R = \frac{1 \text{ mm}}{2} \cdot \left| \frac{1 \text{ m}}{10^3 \text{ mm}} \right| = 5 \times 10^{-4} \text{ m}.$$

If flow through the reactor is rapid, we can assume C_{As} is equal to the bulk sucrose concentration C_{Ab}:

$$C_{As} = C_{Ab} = 16\,\text{mM} = \frac{16 \times 10^{-3}\,\text{gmol}}{\text{litre}} \cdot \left| \frac{1000\,\text{litres}}{1\,\text{m}^3} \right| = 16\,\text{gmol m}^{-3}.$$

Also:

$$\text{Mass of enzyme} = 20\,\text{cm}^3 \cdot \frac{0.05\,\text{kg}}{\text{m}^3} \cdot \left| \frac{1\,\text{m}}{100\,\text{cm}} \right|^3 = 10^{-6}\,\text{kg}.$$

(a) In the free-enzyme reactor:

$$\text{Enzyme concentration} = \frac{10^{-6}\,\text{kg}}{75\,\text{cm}^3} \cdot \left| \frac{100\,\text{cm}}{1\,\text{m}} \right|^3 = 1.33 \times 10^{-2}\,\text{kg m}^{-3}.$$

Production of 1 gmol glucose requires consumption of 1 gmol sucrose; therefore $k_2 = 2.4 \times 10^{-3}$ gmol sucrose (g enzyme)$^{-1}$ s^{-1}. From Eq. (11.33), v_{max} is obtained by multiplying the turnover number by the concentration of active enzyme. Assuming all enzyme present is active:

$$v_{max} = \left(\frac{2.4 \times 10^{-3}\,\text{gmol}}{\text{g s}} \right) (1.33 \times 10^{-2}\,\text{kg m}^{-3}) \cdot \left| \frac{1000\,\text{g}}{1\,\text{kg}} \right|$$

$$= 3.19 \times 10^{-2}\,\text{gmol m}^{-3}\,\text{s}^{-1}.$$

Free-enzyme reaction takes place at uniform sucrose concentration, C_{Ab}. The volumetric rate of reaction is given by the Michaelis–Menten equation:

$$v = \frac{v_{max}\,C_{Ab}}{K_m + C_{Ab}} = \frac{(3.19 \times 10^{-2}\,\text{gmol m}^{-3}\,\text{s}^{-1})\,(16\,\text{gmol m}^{-3})}{8.8\,\text{gmol m}^{-3} + 16\,\text{gmol m}^{-3}}$$

$$= 2.06 \times 10^{-2}\,\text{gmol m}^{-3}\,\text{s}^{-1}.$$

The total rate of reaction is v multiplied by the liquid volume:

$$\text{Rate of reaction} = (2.06 \times 10^{-2}\,\text{gmol m}^{-3}\,\text{s}^{-1})\,(75\,\text{cm}^3) \cdot \left| \frac{1\,\text{m}}{100\,\text{cm}} \right|^3$$

$$= 1.55 \times 10^{-6}\,\text{gmol s}^{-1}.$$

(b) For heterogeneous reactions, v_{max} is expressed on a catalyst-volume basis. Therefore:

$$v_{max} = \left(\frac{2.4 \times 10^{-3}\,\text{gmol}}{\text{g s}} \right) (0.05\,\text{kg m}^{-3}) \cdot \left| \frac{1000\,\text{g}}{1\,\text{kg}} \right|$$

$$= 0.12\,\text{gmol s}^{-1}\,\text{m}^{-3}\,\text{particle}.$$

To determine the effect of mass transfer we must calculate η_{im}. The method used depends on the values of β and ϕ_m:

$$\beta = \frac{K_m}{C_{As}} = \frac{8.8\,\text{gmol m}^{-3}}{16\,\text{gmol m}^{-3}} = 0.55.$$

From Table 12.2:

$$\phi_m = \frac{R}{3\sqrt{2}} \sqrt{\frac{v_{max}}{\mathscr{D}_{Ae}\,C_{As}} \left(\frac{1}{1+\beta} \right) \left[1 + \beta \ln \left(\frac{\beta}{1+\beta} \right) \right]^{-1/2}}$$

$$\phi_{\mathrm{m}} = \frac{5 \times 10^{-4}\,\mathrm{m}}{3\sqrt{2}} \sqrt{\frac{0.12\,\mathrm{gmol\,m^{-3}\,s^{-1}}}{(2 \times 10^{-10}\,\mathrm{m^2\,s^{-1}})\,(16\,\mathrm{gmol\,m^{-3}})}} \left(\frac{1}{1+0.55}\right) \left[1 + 0.55\ln\left(\frac{0.55}{1+0.55}\right)\right]^{-1/2}$$

$$= 0.71.$$

Because both β and ϕ_{m} have intermediate values, Figure 12.10 cannot be applied for spherical geometry. Instead, we must use Eq. (12.34). From Table 12.2:

$$\phi_0 = \frac{R}{3\sqrt{2}} \sqrt{\frac{k_0}{\mathscr{D}_{\mathrm{Ae}}\,C_{\mathrm{As}}}}$$

$$= \frac{R}{3\sqrt{2}} \sqrt{\frac{v_{\mathrm{max}}}{\mathscr{D}_{\mathrm{Ae}}\,C_{\mathrm{As}}}}$$

$$= \frac{5 \times 10^{-4}\,\mathrm{m}}{3\sqrt{2}} \sqrt{\frac{0.12\,\mathrm{gmol\,m^{-3}\,s^{-1}}}{(2 \times 10^{-10}\,\mathrm{m^2\,s^{-1}})\,(16\,\mathrm{gmol\,m^{-3}})}}$$

$$= 0.72.$$

From Figure 12.9 or Table 12.3, $\eta_{i0} = 0.93$. Similarly:

$$\phi_1 = \frac{R}{3} \sqrt{\frac{k_1}{\mathscr{D}_{\mathrm{Ae}}}}$$

$$= \frac{R}{3} \sqrt{\frac{v_{\mathrm{max}}}{K_{\mathrm{m}}\mathscr{D}_{\mathrm{Ae}}}}$$

$$= \frac{5 \times 10^{-4}\,\mathrm{m}}{3} \sqrt{\frac{0.12\,\mathrm{gmol\,m^{-3}\,s^{-1}}}{(8.8\,\mathrm{gmol\,m^{-3}})\,(2 \times 10^{-10}\,\mathrm{m^2\,s^{-1}})}}$$

$$= 1.4.$$

From Figure 12.8 or Table 12.3, $\eta_{i1} = 0.54$. Substituting these results into Eq. (12.34):

$$\eta_{\mathrm{im}} = \frac{0.93 + 0.55\,(0.54)}{1 + 0.55} = 0.79.$$

The rate of immobilised-enzyme reaction without diffusional limitations is the same as that for free enzyme: $1.55 \times 10^{-6}\,\mathrm{gmol\,s^{-1}}$. Rate of reaction for the immobilised enzyme is 79% that of free enzyme even though the amount of enzyme present and external substrate concentration are the same:

$$\text{Observed rate} = 0.79\,(1.55 \times 10^{-6}\,\mathrm{gmol\,s^{-1}}) = 1.22 \times 10^{-6}\,\mathrm{gmol\,s^{-1}}.$$

12.4.4 The Observable Thiele Modulus

Diffusion-reaction theory as presented in the previous sections allows us to quantify the effect of mass transfer on rate of reaction. However, a drawback to the methods outlined so far is that they are useful only if we know the true kinetic parameters for the reaction: k_0, k_1 or v_{max} and K_{m}. In many cases these values are not known and, as discussed in Section 12.9, can be difficult to evaluate for biological systems. A way to circumvent this problem is to apply the *observable Thiele modulus* Φ, sometimes called *Weisz's modulus* [18], which is defined as:

$$\Phi = \left(\frac{V_{\mathrm{P}}}{S_{\mathrm{x}}}\right)^2 \frac{r_{\mathrm{A,obs}}}{\mathscr{D}_{\mathrm{Ae}}\,C_{\mathrm{As}}}$$

$$(12.35)$$

where V_p is catalyst volume, S_x is external surface area, $r_{A,obs}$ is the observed reaction rate *per unit volume of catalyst*, $\mathscr{D}_{Ae}$ is effective diffusivity of substrate, and C_{As} is the substrate concentration at the external surface. Expressions for Φ for spheres and flat plates are listed in Table 12.4. Evaluation of the observable Thiele modulus does not rely on prior knowledge of kinetic parameters; Φ is defined in terms of the measured reaction rate, $r_{A,obs}$.

For the observable Thiele modulus to be useful, we need to relate Φ to the internal effectiveness factor η_i. Some mathematical consideration of the equations already presented in this chapter yields the following relationships for first-order, zero-order and Michaelis–Menten kinetics:

First order
$$\Phi = \phi_1^2 \eta_{i1} \tag{12.36}$$

Zero order
$$\Phi = 2\phi_0^2 \eta_{i0} \tag{12.37}$$

Michaelis–Menten
$$\Phi = 2\phi_m^2 \eta_{im}(1+\beta)\left[1 + \beta\ln\left(\frac{\beta}{1+\beta}\right)\right] \tag{12.38}$$

Table 12.4 Observable Thiele moduli

Sphere	$\Phi = \left(\dfrac{R}{3}\right)^2 \dfrac{r_{A,obs}}{\mathscr{D}_{Ae}\,C_{As}}$
Flat plate	$\Phi = b^2\,\dfrac{r_{A,obs}}{\mathscr{D}_{Ae}\,C_{As}}$

Eqs (12.36)–(12.38) apply to all catalyst geometries and allow us to develop plots of Φ versus η_i from relationships between ϕ and η_i developed in the previous sections and represented in Figures 12.8–12.10. Curves for spherical catalysts and first-order, zero-order and Michaelis–Menten kinetics are given in Figure 12.11; results for flat-plate geometry are shown in Figure 12.12. All curves for β between zero and infinity are bracketed by the first- and zero-order lines. At each value of β, curves for all geometries coincide in the asymptotic regions $\Phi \to 0$ and $\Phi \to \infty$; at intermediate values of Φ the variation between effectiveness factors for different geometries can be significant. For $\Phi > 10$:

First-order kinetics
$$\eta_{i1} \approx \frac{1}{\Phi} \tag{12.39}$$

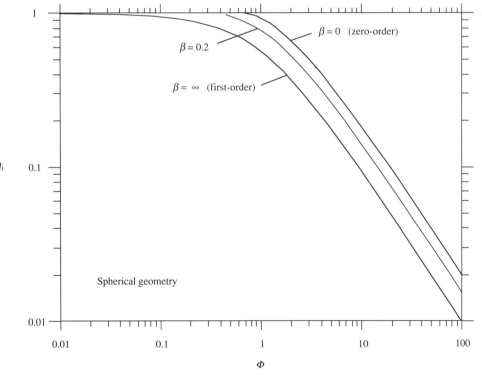

Figure 12.11 Internal effectiveness factor η_i as a function of the observable Thiele modulus Φ for spherical geometry and first-order, zero-order and Michaelis–Menten kinetics. $\beta = K_m/C_{As}$. (From W.H. Pitcher, 1975, Design and operation of immobilized enzyme reactors. In: R.A. Messing, Ed, *Immobilized Enzymes For Industrial Reactors*, pp. 151–199, Academic Press, New York.)

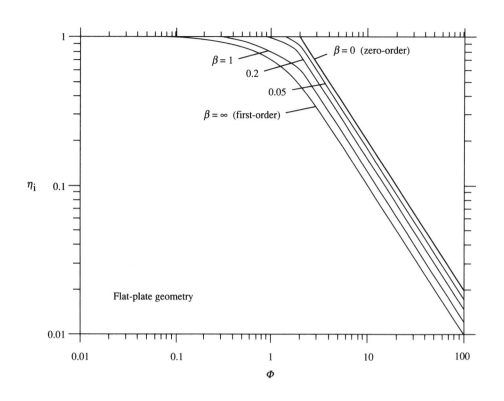

Figure 12.12
Internal effectiveness factor η_i as a function of the observable Thiele modulus Φ for flat-plate geometry and first-order, zero-order and Michaelis–Menten kinetics. $\beta = {}^{K}\!m/c_{As}$. (From W.H. Pitcher, 1975, Design and operation of immobilized enzyme reactors. In: R.A. Messing, Ed, *Immobilized Enzymes For Industrial Reactors*, pp. 151–199, Academic Press, New York.)

Zero-order kinetics $\qquad \eta_{i0} \approx \dfrac{2}{\Phi}.$

$$(12.40)$$

Although Φ is an observable modulus independent of kinetic parameters such as v_{max} and K_m, use of Figures 12.11 and 12.12 for Michaelis–Menten reactions requires knowledge of K_m for evaluation of β. This makes application of the observable Thiele modulus difficult for Michaelis–Menten kinetics. However, we know that effectiveness factors for Michaelis–Menten reactions lie between the first- and zero-order curves of Figures 12.11 and 12.12; therefore, we can always estimate the upper and lower bounds of η_{im}.

12.4.5 Weisz's Criteria

The following general observations can be made from Figures 12.11 and 12.12.

If $\Phi < 0.3$, $\eta_i \approx 1$ and internal mass-transfer limitations are insignificant.

If $\Phi > 3$, η_i is substantially < 1 and internal mass-transfer limitations are significant.

The above statements are known as *Weisz's criteria*, and are valid for all geometries and reaction kinetics. For Φ in the intermediate range $0.3 < \Phi < 3$, closer analysis is required to determine the influence of mass transfer on reaction rate.

Example 12.4 Internal oxygen transfer to immobilised cells

Baby hamster kidney cells are immobilised in alginate beads. The average particle diameter is 5 mm. Rate of oxygen consumption at a bulk concentration of 8×10^{-3} kg O_2 m^{-3} is 8.4×10^{-5} kg s^{-1} m^{-3} catalyst. The effective diffusivity of oxygen in the beads is 1.88×10^{-9} m^2 s^{-1}. Assume that the oxygen concentration at the surface of the catalyst is equal to the bulk concentration, and that oxygen uptake follows zero-order kinetics.

(a) Are internal mass-transfer effects significant?
(b) What reaction rate would be observed if diffusional resistance were eliminated?

Solution:

(a) To assess internal mass transfer, calculate the observable Thiele modulus. From Table 12.4 for spherical geometry:

$$\Phi = \left(\frac{R}{3}\right)^2 \frac{r_{A,obs}}{\mathscr{D}_{Ae} C_{As}}.$$

With

$$R = \frac{5 \times 10^{-3}\,\text{m}}{2} = 2.5 \times 10^{-3}\,\text{m}$$

then

$$\Phi = \left(\frac{2.5 \times 10^{-3}\,\text{m}}{3}\right)^2 \frac{8.4 \times 10^{-5}\,\text{kg s}^{-1}\,\text{m}^{-3}}{(1.88 \times 10^{-9}\,\text{m}^2\,\text{s}^{-1})\,(8 \times 10^{-3}\,\text{kg m}^{-3})} = 3.88.$$

From Weisz's criteria, internal mass-transfer effects are significant.

(b)   For spherical catalysts and zero-order reaction, from Figure 12.11, at $\Phi = 3.9$, $\eta_{i0} = 0.4$. From Eq. (12.26), without diffusional restrictions the reaction rate would be:

$$r_{As}^* = \frac{r_{A,obs}}{\eta_i} = \frac{8.4 \times 10^{-5}\,\text{kg m}^{-3}\,\text{s}^{-1}}{0.4} = 2.1 \times 10^{-4}\,\text{kg s}^{-1}\,\text{m}^{-3}\ \text{catalyst}.$$

12.4.6 Minimum Intracatalyst Substrate Concentration

It is sometimes of interest to know the minimum concentration $C_{A,min}$ inside solid catalysts. We can use this information to check, for example, that the concentration does not fall below some critical value for cell metabolism. $C_{A,min}$ is easily estimated for zero-order reactions. If Φ is such that $\eta_i < 1$, $C_{A,min}$ is zero because $\eta_i = 1$ if $C_A > 0$ throughout the particle. For $\eta_i = 1$, simple manipulation of the equations already presented in this chapter allow us to estimate $C_{A,min}$. The results are summarised in Table 12.5.

12.5 External Mass Transfer

Many equations in Sections 12.3 and 12.4 contain the term C_{As}, the concentration of substrate A at the external surface of the catalyst. This term made its way into the analysis in the boundary conditions used for solution of the shell mass balance. It was assumed that C_{As} is a known quantity. However, because surface concentrations are very difficult to measure accurately, we must find ways to estimate C_{As} using theoretical principles.

Reduction in substrate concentration from C_{Ab} in the bulk liquid to C_{As} at the catalyst surface occurs across the boundary layer surrounding the solid. In the absence of the boundary

Table 12.5   Minimum intracatalyst substrate concentration with zero-order kinetics (Φ for each geometry is defined in Table 12.4)

Sphere		
	$C_{A,min} = 0$	for $\Phi \geqslant 0.667$
	$C_{A,min} = C_{As}\left(1 - \frac{3}{2}\,\Phi\right)$	for $\Phi < 0.667$
Flat plate		
	$C_{A,min} = 0$	for $\Phi \geqslant 2$
	$C_{A,min} = C_{As}\left(1 - \frac{1}{2}\,\Phi\right)$	for $\Phi < 2$

layer, $C_{As} = C_{Ab}$, which is easily measured. When the boundary layer is present, C_{As} takes some value less than C_{Ab}. Rate of mass transfer across a liquid boundary layer is represented by the following equation (Section 9.4):

$$N_A = k_S\, a\,(C_{Ab} - C_{As}) \qquad (12.41)$$

where N_A is the rate of mass transfer, k_S is the *liquid-phase mass-transfer coefficient* with dimensions LT^{-1}, and a is the external surface area of the catalyst. If N_A is expressed per volume of catalyst with units of, for example, kgmol s^{-1} m^{-3}, to be consistent, a must also be expressed on a catalyst-volume basis with units of, for example, m^2 m^{-3} or m^{-1}. Using the previous notation of this chapter, a in Eq. (12.41) is equal to S_x/V_p for the catalyst. At steady state, the rate of substrate transfer across the boundary layer must be equal to the rate of consumption by the catalyst, $r_{A,obs}$. Therefore:

$$r_{A,obs} = k_S\, \frac{S_x}{V_p}\,(C_{Ab} - C_{As}) \qquad (12.42)$$

where $r_{A,obs}$ is the rate per volume of catalyst. Rearranging gives:

$$\frac{C_{As}}{C_{Ab}} = 1 - \frac{V_p}{S_x}\,\frac{r_{A,obs}}{k_S\,C_{Ab}}. \qquad (12.43)$$

Eq. (12.43) can be used to evaluate C_{As} before applying equations in the previous sections to calculate internal substrate concentrations and effectiveness factors. The magnitude of external mass-transfer effects can be gauged from Eq. (12.43). $C_{As}/C_{Ab} \approx 1$ indicates no external mass-transfer limitations; the substrate concentration at the surface is approximately equal to that in the bulk. On the other hand, $C_{As}/C_{Ab} \ll 1$ indicates a very steep concentration gradient in the boundary layer and severe external mass-transfer effects. We can define from Eq. (12.43) an *observable modulus for external mass transfer*, Ω:

$$\Omega = \frac{V_p}{S_x}\,\frac{r_{A,obs}}{k_S\,C_{Ab}}. \qquad (12.44)$$

Expressions for Ω for different catalyst geometries are listed in Table 12.6. Criteria for assessing external mass transfer are as follows:

If $\Omega \ll 1$, $C_{As} \approx C_{Ab}$ and external mass-transfer effects are insignificant.

Table 12.6 Observable moduli for external mass transfer

Sphere	$\Omega = \dfrac{R}{3}\,\dfrac{r_{A,obs}}{k_S\,C_{Ab}}$
Flat plat	$\Omega = b\,\dfrac{r_{A,obs}}{k_S\,C_{Ab}}$

Table 12.7 External effectiveness factors

First-order reaction: $r_A = k_1\,C_A$

$$\eta_{e1} = \frac{C_{As}}{C_{Ab}}$$

Zero-order reaction: $r_A = k_0$

$$\eta_{e0} = 1$$

Michaelis–Menten reaction: $r_A = \dfrac{v_{max}\,C_A}{K_m + C_A}$

$$\eta_{em} = \frac{C_{As}\,(K_m + C_{Ab})}{C_{Ab}\,(K_m + C_{As})}$$

Otherwise, $C_{As} < C_{Ab}$ and external mass-transfer effects are significant.

For reaction affected by both internal and external mass-transfer restrictions, we can define a *total effectiveness factor* η_T:

$$\eta_T = \frac{r_{A,obs}}{r^*_{Ab}} = \frac{\text{(observed rate)}}{\left(\begin{array}{c}\text{rate that would occur if } C_A = C_{Ab} \\ \text{everywhere in the particle}\end{array}\right)}. \qquad (12.45)$$

η_T can be related to the internal effectiveness factor η_i. Eq. (12.45) may be written as:

$$\eta_T = \left(\frac{r_{A,obs}}{r^*_{As}}\right)\left(\frac{r^*_{As}}{r^*_{Ab}}\right) = \eta_i\,\eta_e \qquad (12.46)$$

where η_e is the *external effectiveness factor* and η_i is defined in Eq. (12.26). Therefore, η_e has the following meaning:

$$\eta_e = \frac{r^*_{As}}{r^*_{Ab}} = \frac{\left(\begin{array}{c}\text{rate that would occur if } C_A = C_{As} \\ \text{everywhere in the particle}\end{array}\right)}{\left(\begin{array}{c}\text{rate that would occur if } C_A = C_{Ab} \\ \text{everywhere in the particle}\end{array}\right)}. \qquad (12.47)$$

Expressions for η_e for first-order, zero-order and Michaelis–Menten kinetics are listed in Table 12.7. For zero-order reactions, as long as C_{As} and $C_{Ab} > 0$, $\eta_{e0} = 1$. However, $\eta_{e0} = 1$ does not imply that an external boundary layer does not exist; even if there is a reduction in concentration across the boundary layer, because r^*_{As} and r^*_{Ab} are independent of C_A, $\eta_{e0} = 1$. Furthermore, $\eta_{e0} = 1$ does not imply that eliminating the external boundary layer could not improve the reaction rate. Removing the boundary layer would increase the value of C_{As}, thus establishing a greater driving force for internal mass-transfer and reducing the likelihood of C_A falling to zero inside the particle.

Example 12.5 Effect of mass transfer on bacterial denitrification

Denitrifying bacteria are immobilised in gel beads and used in a stirred reactor for removal of nitrate from groundwater. At a nitrate concentration of 3 g m^{-3}, the conversion rate is 0.011 g s^{-1} m^{-3} catalyst. The effective diffusivity of nitrate in the gel is 1.5×10^{-9} m^2 s^{-1}, the beads are 6 mm in diameter, and the liquid–solid mass-transfer coefficient is 10^{-5} m s^{-1}. K_m for the immobilised bacteria is approximately 25 g m^{-3}.

(a) Does external mass transfer influence the reaction rate?
(b) Are internal mass-transfer effects significant?
(c) By how much would the reaction rate be improved if both internal and external mass-transfer resistances were eliminated?

Solution:

(a)
$$R = \frac{6 \times 10^{-3}\,\text{m}}{2} = 3 \times 10^{-3}\,\text{m}.$$

The effect of external mass transfer is found by calculating Ω. From Table 12.6:

$$\Omega = \frac{R}{3}\frac{r_{A,obs}}{k_S\,C_{Ab}} = \frac{3 \times 10^{-3}\,\text{m}}{3}\frac{0.011\,\text{g s}^{-1}\,\text{m}^{-3}}{(10^{-5}\,\text{m s}^{-1})\,(3\,\text{g m}^{-3})} = 0.37.$$

As this value of Ω is relatively large, external mass transfer influences the reaction rate. From Eq. (12.43):

$$\frac{C_{As}}{C_{Ab}} = 1 - \Omega = 0.63$$

$$C_{As} = 0.63\,C_{Ab} = 0.63\,(3\,\text{g m}^{-3}) = 1.9\,\text{g m}^{-3}.$$

(b) Calculate the observable Thiele modulus with $C_{As} = 1.9$ g m^{-3} using the equation for spheres from Table 12.4:

$$\Phi = \left(\frac{R}{3}\right)^2 \frac{r_{A,obs}}{\mathscr{D}_{Ae}\,C_{As}}$$

$$\Phi = \left(\frac{3 \times 10^{-3}\,\text{m}}{3}\right)^2 \frac{0.011\,\text{g s}^{-1}\,\text{m}^{-3}}{(1.5 \times 10^{-9}\,\text{m}^2\,\text{s}^{-1})\,(1.9\,\text{g m}^{-3})} = 3.9.$$

From Weisz's criteria, internal mass-transfer effects are significant.

(c) Because the nitrate concentration is much smaller than K_m, we can assume the reaction is first order. From Table 12.7:

$$\eta_{e1} = \frac{C_{As}}{C_{Ab}} = 0.63.$$

Reaction rate in the absence of mass-transfer effects is r^*_{Ab}, which is related to $r_{A,obs}$ by Eq. (12.45). Therefore, we can calculate r^*_{Ab} if we know η_{T1}. From Figure 12.11, at $\Phi = 3.9$ $\eta_{i1} = 0.21$. As $\eta_{e1} = 0.63$, from Eq. (12.46):

$$\eta_{T1} = \eta_{i1}\,\eta_{e1} = 0.21\,(0.63) = 0.13.$$

Therefore, from Eq. (12.45):

$$r_{Ab}^* = \frac{r_{A,obs}}{\eta_{T1}} = \frac{0.011 \, \text{g s}^{-1} \, \text{m}^{-3}}{0.13} = 0.085 \, \text{g s}^{-1} \, \text{m}^{-3}.$$

The reaction rate would be increased by a factor of about 7.7.

12.6 Liquid–Solid Mass-Transfer Correlations

The mass-transfer coefficient k_S must be known before we can account for external mass-transfer effects. k_S depends on reactor hydrodynamics and liquid properties such as viscosity, density and diffusivity. It is difficult to determine k_S accurately, especially for particles which are neutrally buoyant. However, values can be estimated using correlations from the literature; these are usually accurate under the conditions specified to within 10–20%. Selected correlations are presented below. Further details can be found in chemical engineering texts [19, 20].

Correlations for k_S are expressed using the following dimensionless groups:

$$Re_p = (\text{particle}) \text{ Reynolds number} = \frac{D_p u_{pL} \rho_L}{\mu_L}$$

(12.48)

$$Sc = \text{Schmidt number} = \frac{\mu_L}{\rho_L \mathscr{D}_{AL}}$$

(12.49)

$$Sh = \text{Sherwood number} = \frac{k_S D_p}{\mathscr{D}_{AL}}$$

(12.50)

and

$$Gr = \text{Grashof number} = \frac{g D_p^3 \rho_L (\rho_p - \rho_L)}{\mu_L^2}$$

(12.51)

where D_p is particle diameter, u_{pL} is linear velocity of the particle relative to the bulk liquid, ρ_L is liquid density, μ_L is liquid viscosity, $\mathscr{D}_{AL}$ is the molecular diffusivity of component A in the liquid, k_S is the liquid–solid mass-transfer coefficient, g is gravitational acceleration, and ρ_p is particle density. The Sherwood number contains the mass-transfer coefficient and represents the ratio of overall and diffusive mass-transfer rates through the boundary layer. The Schmidt number represents the ratio of momentum diffusivity and mass diffusivity, and is made up of

physical properties of the system. At constant temperature, pressure and composition, Sc is constant for Newtonian fluids. The Grashof number represents the ratio of gravitational forces to viscous forces, and is important when the particles are neutrally buoyant. The form of the correlation used to evaluate Sh and therefore k_S depends on the configuration of the mass-transfer system, the flow conditions and other factors.

12.6.1 Free-Moving Spherical Particles

The equations presented here apply to solid particles suspended in stirred vessels. Rate of mass transfer depends on the velocity of the solid relative to the liquid; this is known as the *slip velocity*. Slip velocity in suspensions is difficult to measure and must therefore be estimated before calculating k_S. The following equations allow evaluation of the particle Reynolds number which incorporates the slip velocity, u_{pL} [7]:

For $Gr < 36$ $\qquad Re_p = {}^{Gr}/_{18}$

(12.52)

For $36 < Gr < 8 \times 10^4$ $\qquad Re_p = 0.153 \, Gr^{0.71}$

(12.53)

For $8 \times 10^4 < Gr < 3 \times 10^9$ $\qquad Re_p = 1.74 \, Gr^{0.5}.$

(12.54)

Once Re_p is known, Sh can be determined using equations such as [7, 19–22]:

For $Re_p Sc < 10^4$ $\qquad Sh = \sqrt{4 + 1.21 \, (Re_p Sc)^{0.67}}$

(12.55)

For $Re_p < 10^3$ $\qquad Sh = 2 + 0.6 \, Re_p^{0.5} Sc^{0.33}.$

(12.56)

12.6.2 Spherical Particles in a Packed Bed

k_S in a packed bed depends on the liquid velocity around the particles. For the range $10 < Re_p < 10^4$, Sherwood number in packed beds has been correlated by the equation [23]:

$$Sh = 0.95 \, Re_p^{0.5} Sc^{0.33}.$$

(12.57)

2.7 Experimental Aspects

Applying diffusion-reaction theory to real biocatalysts requires prior measurement of several parameters. This section considers some experimental aspects of heterogeneous reactions.

2.7.1 Observed Reaction Rate

Calculation of Weisz's modulus Φ and the external modulus Ω requires knowledge of $r_{A,obs}$, the observed rate of reaction per volume of catalyst. This information can be obtained from batch concentration data using the methods described in Section 11.2.

When substrate levels change relatively slowly during reaction, batch rate data can be obtained by removing samples of the reaction mixture at various times and analysing for substrate concentration. However if substrate is consumed rapidly or if oxygen uptake rates are required, continuous *in situ* monitoring is necessary. The equipment shown in Figure 12.13 is configured for measurement of oxygen consumption by immobilised cells or enzymes. Catalyst particles are packed into a column and liquid medium recirculated through the packed bed. Oxygen is sparged into a stirred vessel connected to the column; dissolved-oxygen tension is measured using an electrode. The catalyst is allowed to reach steady state at a particular oxygen tension; the gas supply is then switched off and the initial rate of oxygen uptake recorded. Observed rates must be measured at steady state so that the results do not reflect rapidly-changing transient conditions. Unless sufficient time is allowed for the system to establish steady state, the experimental results will be unreliable.

The system of Figure 12.13 is assumed to be well mixed so that the oxygen tension measured by the electrode is the same everywhere in the bulk liquid. Actually, oxygen concentrations entering and leaving the packed bed will be different because oxygen is consumed by the catalyst. However, if only 1–2% of the oxygen entering the column is consumed during each pass through the bed, from a practical point of view the system can be regarded as perfectly mixed. For this to occur, the recirculation flow rate must be sufficiently high [24]. In oxygen uptake experiments, temperature control is also very important as the solubility and mass-transfer properties of oxygen are temperature dependent (see Sections 9.6.4 and 9.8.2).

Observed reaction rates for catalytic reactions depend on the cell or enzyme loading per volume of catalyst. Therefore, experimental measurements apply only to the cell or enzyme density tested. Increasing the amount of cells or enzyme in the catalyst will increase $r_{A,obs}$ and the likelihood of mass-transfer restrictions.

Figure 12.13 Batch recirculation reactor for measuring oxygen uptake by immobilised cells or enzymes.

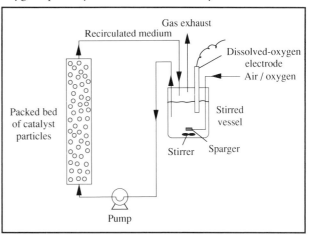

12.7.2 Effective Diffusivity

The value of the effective diffusivity $\mathscr{D}_{Ae}$ reflects the ease with which compound A diffuses within the catalyst matrix, and depends strongly on the pore structure of the solid. Effective diffusivities are normally lower than corresponding molecular diffusivities in water because porous solids offer more resistance to diffusion. Some experimental $\mathscr{D}_{Ae}$ values are given in Table 12.8.

As cells can pose a significant barrier to diffusion, $\mathscr{D}_{Ae}$ for immobilised-cell preparations must be determined with the biomass present. During the measurement, external mass-transfer limitations must be overcome with high liquid flow rates around the catalyst. If experimental values of $\mathscr{D}_{Ae}$ are higher than the molecular diffusion coefficient, this may indicate the presence of convective mass transfer in the catalyst [3–6]. Techniques for measuring effective diffusivity are described in several papers [28–33]; however, accurate measurement of $\mathscr{D}_{Ae}$ is difficult in most systems.

12.8 Minimising Mass-Transfer Effects

To improve overall rates of reaction in bioprocesses, mass-transfer restrictions must be minimised or eliminated. In this section we consider practical ways of achieving this objective based on the equations presented in Tables 12.4 and 12.6.

12.8.1 Internal Mass Transfer

Internal mass-transfer effects are eliminated when the internal effectiveness factor is equal to 1; η_i approaches unity as the observable Thiele modulus Φ is reduced. From Table 12.4, Φ is decreased by:

Table 12.8 Effective diffusivity values

Substance	Catalyst	Temperature (°C)	$\mathscr{D}_{Ae}$ (m² s⁻¹)	$\dfrac{\mathscr{D}_{Ae}}{\mathscr{D}_{Aw}}$*	Ref.
Oxygen	Agar (2% w/v) containing *Candida lipolytica* cells	30°C	1.94×10^{-9}	0.70	[25]
	Microbial aggregates from domestic waste-treatment plant	20°C	1.37×10^{-9}	0.62	[26]
	Trickling-filter slime	25°C	0.82×10^{-9}	–	[27]
Glucose	Microbial aggregates from domestic waste-treatment plant	20°C	0.25×10^{-9}	0.37	[26]
	Glass-fibre discs containing *Saccharomyces uvarum* cells	30°C	0.30×10^{-9}	0.43	[3]
	Calcium alginate (3 wt%)	30°C	0.62×10^{-9}	0.87	[28]
	Calcium alginate (2.4–2.8 wt%) containing 50 wt% bakers' yeast	30°C	0.26×10^{-9}	0.37	[28]
Sucrose	Calcium alginate (2% w/v)	25°C	0.48×10^{-9}	0.86	[29]
	Calcium alginate (2% w/v) containing 12.5% (v/v) *Catharanthus roseus* cells	25°C	0.14×10^{-9}	0.25	[29]
L-tryptophan	Calcium alginate (2%)	30°C	0.67×10^{-9}	1.0	[30]
	κ-Carrageenan (4%)	30°C	0.58×10^{-9}	0.88	[31]
Lactate	Agar (1%) containing 1% Ehrlich ascites tumour cells	37°C	1.4×10^{-9}	0.97	[32]
	Agar (1%) containing 6% Ehrlich ascites tumour cells	37°C	0.7×10^{-9}	0.48	[32]
Ethanol	κ-Carrageenan (4%)	30°C	1.01×10^{-9}	0.92	[31]
	Calcium alginate (1.4–3.8 wt%)	30°C	1.25×10^{-9}	0.92	[28]
	Calcium alginate (2.4–2.8 wt%) containing 50 wt% bakers' yeast	30°C	0.45×10^{-9}	0.33	[28]
Nitrate	Compressed film of nitrifying organisms	–	1.4×10^{-9}	0.90	[26]
Ammonia	Compressed film of nitrifying organisms	–	1.3×10^{-9}	0.80	[26]

* $\mathscr{D}_{Aw}$ is the molecular diffusivity in water at the temperature of measurement.

) reducing the observed reaction rate $r_{A,obs}$;

i) reducing the size of the catalyst;

ii) increasing the effective diffusivity $\mathscr{D}_{Ae}$; and

v) increasing the surface concentration of substrate C_{As}.

ll these changes impact directly on the effectiveness of mass ansfer in supplying substrate to the particle.

Paradoxically, reducing the reaction rate $r_{A,obs}$ improves ie effectiveness of mass transfer aimed at increasing the reacon rate. When the catalyst is very active with a high demand or substrate, mass transfer is likely to be slow relative to reacon so that steep concentration gradients are produced. Iowever, limiting the reaction rate by operating at sub-opti- ium conditions or using an organism or enzyme with low itrinsic activity does not achieve the overall goal of higher onversion rates. Because $r_{A,obs}$ is the reaction rate per volume f catalyst, another way of reducing $r_{A,obs}$ is to reduce the cell r enzyme loading in the solid. This reduces the demand for ibstrate per particle so that mass transfer has a better chance f supplying it at a sufficient rate. Therefore, if the same mass f cells or enzyme is distributed between more particles, the te of conversion will increase. However, using more particles iay mean that a larger reactor is required.

Because Φ is proportional to the square of catalyst size (R^2 or spheres or b^2 for flat plates), reducing the catalyst size has a iore dramatic effect on Φ than changes in any other variable. is therefore a good way to improve the reaction rate. In prin- ple, mass-transfer limitations can be completely overcome if ie particle size is decreased sufficiently. However, it is often xtremely difficult in practice to reduce particle dimensions to iis extent [34]. Even if mass-transfer effects were eliminated, peration of large-scale reactors with tiny, highly compressible el particles raises new problems. Some degree of internal iass-transfer restriction must usually be tolerated.

2.8.2 External Mass Transfer

xternal mass-transfer effects decrease as the observable mod- lus Ω is reduced. From the equations in Table 12.6, this can e achieved by:

) reducing the observed reaction rate $r_{A,obs}$;

i) reducing the size of the catalyst;

ii) increasing the mass-transfer coefficient k_S; and

v) increasing the bulk concentration of substrate C_{Ab}.

Decreasing the catalyst size and increasing the mass-transfer oefficient reduce the thickness of the boundary layer and facili- te external mass transfer. k_S is increased most readily by raising ie liquid velocity outside the catalyst. k_S is a function of other vstem properties such as liquid viscosity, density, and substrate

Figure 12.14 Spinning-basket reactor for minimising external mass-transfer effects: (a) reactor configuration; (b) detail of spinning basket. (From D.G. Tajbl, J.B. Simons and J.J. Carberry, 1966, Heterogeneous catalysis in a continuous stirred tank reactor. *Ind. Eng. Chem. Fund.* 5, 171–175.)

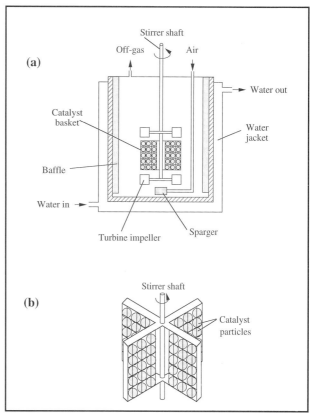

diffusivity; changes in these variables can also reduce boundary-layer effects to some extent. External mass transfer is more rapid at high bulk substrate concentrations; the higher the concentration, the greater is the driving force for mass transfer across the boundary layer. By reducing the demand for substrate transfer across the boundary layer, decreasing $r_{A,obs}$ as described in the previous section also reduces external limitations.

In large-scale reactors, external mass-transfer problems may be unavoidable if sufficiently high liquid velocities cannot be achieved. However, when evaluating biocatalyst kinetics in the laboratory, it is advisable to eliminate fluid boundary layers to simplify analysis of the data. Several laboratory reactor configurations allow almost complete elimination of interparticle and interphase concentration gradients [35]. Recycle reactors such as that shown in Figure 12.13 have been employed extensively for study of immobilised cell and enzyme reactions; operation with high liquid velocity through the bed reduces

Figure 12.15 Stirred laboratory reactor for minimising external mass-transfer effects. (From K. Sato and K. Toda, 1983, Oxygen uptake rate of immobilized growing *Candida lipolytica. J. Ferment. Technol.* **61**, 239–245.)

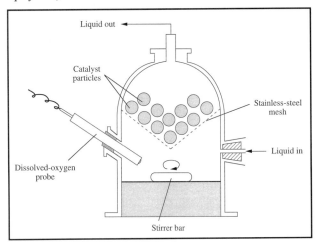

Figure 12.16 Relationship between observed reaction rate and external liquid velocity.

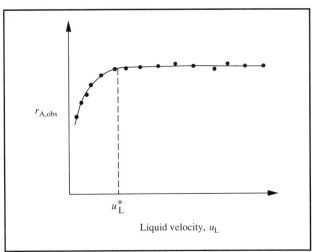

boundary-layer effects. Another suitable configuration is the *spinning-basket reactor* [36, 37] illustrated in Figure 12.14. By rotating the baskets at high speed, high relative velocity is achieved between the particles and the surrounding fluid. Slip velocities obtained using this apparatus are significantly greater than with freely-suspended particles. Another laboratory design aimed at increasing the slip velocity is the stirred vessel shown in Figure 12.15. In this vessel, catalyst particles are held relatively stationary in a wire-mesh cage while liquid is agitated at high speed.

Elimination of external mass-transfer effects can be verified by calculating the observable parameter Ω as described in Section 12.5. However, if the mass-transfer coefficient is not known accurately, an experimental test may be used instead [24]. Consider again the apparatus of Figure 12.13. If boundary layers around the particles affect the reaction, increased liquid velocity through the bed will improve conversion rates by reducing the film thickness and bringing C_{As} closer to C_{Ab}. As illustrated in Figure 12.16, at sufficiently high liquid velocity, external mass-transfer effects can be removed using this procedure; further increases in pump speed do not change the overall reaction rate. Therefore, if we can identify a liquid velocity u_L^* at which reaction rate becomes independent of liquid velocity, operation at $u_L > u_L^*$ will ensure that $\eta_e = 1$. For stirred reactors, a similar relationship holds between $r_{A,obs}$ and agitation speed.

12.9 Evaluating True Kinetic Parameters

The intrinsic kinetics of zero-order, first-order and Michaelis–Menten reactions are represented by the parame-

ters k_0, k_1, v_{max} and K_m. In general, it cannot be assumed that the values of these parameters will be the same before and after cell or enzyme immobilisation; significant changes can be wrought during the immobilisation process. As an example, Figure 12.17 shows Lineweaver–Burk plots for free and immobilised β-galactosidase enzyme. According to Eq. (11.37), the slopes and intercepts of the lines in Figure 12.17 indicate values of K_m/v_{max} and $1/v_{max}$, respectively. The steeper slopes and higher intercepts obtained for the immobilised enzyme indicate that immobilisation reduces v_{max}; this is a commonly observed result. The value of K_m can also be affected [9].

As described in Sections 11.3 and 11.4, kinetic parameters for homogeneous reactions can be determined directly from experimental rate data. However, evaluating the true kinetic parameters of immobilised cells and enzymes is somewhat more difficult. The observed rate of reaction is not the true rate at all points in the catalyst; mass-transfer processes effectively 'mask' true kinetic behaviour. For example, v_{max} and K_m for immobilised catalysts cannot be estimated using the classical plots described in Section 11.4; under the influence of mass transfer, these plots no longer give straight lines over the entire range of substrate concentration [38, 39].

As illustrated in Figure 12.17 for β-galactosidase, Lineweaver–Burk plots for immobilised enzymes are non-linear; the shape and slope of the curves depend on the magnitude of the Thiele modulus. Deviation from linearity in such plots is easily obscured by the scatter in real experimental data and the distortion of errors due to the Lineweaver–Burk linearisation. At low substrate concentrations, i.e. large value

Figure 12.17 Lineweaver–Burk plots for free and immobilised β-galactosidase. Enzyme concentrations within the gel are: 0.10 mg ml^{-1} (●); 0.17 mg ml^{-1} (□); and 0.50 mg ml^{-1} (■). (Data from P.S. Bunting and K.J. Laidler, 1972, Kinetic studies on solid-supported β-galactosidase. *Biochemistry* **11**, 4477–4483.)

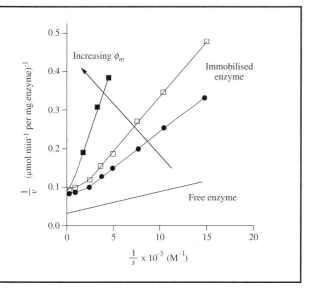

Several methods have been proposed for determining v_{max} and K_m in heterogeneous catalysts [38–41]. The most straightforward approach is experimental; it involves reducing the particle size and catalyst loading and increasing the external liquid velocity to eliminate all mass-transfer resistances. Measured rate data can then be analysed for kinetic parameters as if the reaction were homogeneous. However, because it is usually very difficult or impossible to completely remove intraparticle effects, procedures involving a series of experiments coupled with theoretical analysis have also been proposed. In these methods, rate data are collected at high and low substrate concentrations using different particle sizes. At high substrate levels, it is assumed that the reaction is zero order with $\eta_i = 1$; at low concentrations, first-order kinetics are assumed. These assumptions simplify the analysis, but may not always be valid.

When adequate computing facilities are available, true values of v_{max} and K_m can be extracted from diffusion-limited data using iterative calculations based on numerical integration and non-linear regression [7, 9]. Many iteration loops may be required before convergence to the final parameter values.

12.10 General Comments on Heterogeneous Reactions in Bioprocessing

Before concluding this chapter, some general observations and rules of thumb for heterogeneous reactions are outlined.

(i) *Importance of oxygen mass-transfer limitations.* In aerobic reactions, mass transfer of oxygen is more likely to limit reaction rate than mass transfer of most other substrates. The reason is the poor solubility of oxygen in aqueous solutions. Whereas the sugar concentration in a typical fermentation broth is around 20–50 kg m^{-3}, oxygen concentration under 1 atm air pressure is limited to about 8×10^{-3} kg m^{-3}. Therefore, because C_{As} is so low for oxygen, the observable Thiele modulus Φ can be several orders of magnitude greater than for other substrates. In anaerobic systems, the rate-limiting substrate is more difficult to identify.

(ii) *Relationship between internal effectiveness factor and concentration gradient.* Depending on the reaction kinetics, the severity of intraparticle concentration gradients can be inferred from the value of the internal effectiveness factor. For first-order kinetics, reaction rate is directly proportional to substrate concentration; therefore $\eta_{i1} = 1$ implies that concentration gradients do not exist in the catalyst. Conversely, if $\eta_{i1} < 1$ we can conclude that concentration

of $1/_s$, the Lineweaver–Burk plot appears linear as first-order kinetics are approached. Even if we mistakenly interpret one of the immobilised-enzyme curves in Figure 12.17 as a straight line and evaluate apparent values of v_{max} and K_m, we can check whether or not we have found the true kinetic parameters by changing the particle radius and substrate concentration over a wide range of values. True kinetic parameters do not vary with these conditions which affect mass transfer into the catalyst; if the slope and/or intercept change with Thiele modulus, it is soon evident that the system is subject to diffusional limitations. Thus, apparent linearity of Lineweaver–Burk or similar plots is insufficient demonstration of the absence of mass-transfer restrictions. Studies have shown that the effect of diffusion is more pronounced in Eadie–Hofstee plots than Lineweaver–Burk or Langmuir plots [39]; however, all three can be approximated by straight lines over certain intervals.

Although the Lineweaver–Burk plot for immobilised β-galactosidase is non-linear, we should not conclude that immobilised enzymes fail to obey Michaelis–Menten kinetics. The kinetic form of reactions is generally maintained upon immobilisation of cells and enzymes [9]; non-linearity of the Lineweaver–Burk plot is due to the effect of mass transfer on $r_{A,obs}$.

gradients are present. In contrast, for zero-order reactions $\eta_{i0} = 1$ does not imply that gradients are absent; as long as $C_A > 0$, the reaction rate is unaffected. Concentration gradients can be so steep that C_A is reduced to almost zero within the catalyst, but η_{i0} remains equal to 1. On the other hand, $\eta_{i0} < 1$ implies that the concentration gradient is very severe and that some fraction of the particle volume is starved of substrate.

(iii) *Relative importance of internal and external mass-transfer limitations.* For porous catalysts, it has been demonstrated with realistic values of mass transfer and diffusion parameters that external mass-transfer limitations do not exist unless internal limitations are also present [42]. Concentration differences between the bulk liquid and external catalyst surface are never observed without larger internal gradients developing within the particle. On the other hand, if internal limitations are known to be present, external limitations may or may not be important depending on conditions. Significant external mass-transfer effects can occur when reaction does not take place inside the catalyst, for example, if cells or enzymes are attached only to the exterior surface.

(iv) *Operation of catalytic reactors.* Certain solid-phase properties are desirable for operation of immobilised-cell and -enzyme reactors. For example, in packed-bed reactors, large, rigid and uniformly-shaped particles promote well-distributed and stable liquid flow. Solids in packed columns should also have sufficient mechanical strength to withstand their own weight. These requirements are in direct conflict with those needed for rapid intraparticle mass transfer; diffusion is facilitated in particles that are small, soft and porous. Because blockages and large pressure drops through the bed must be avoided, mass-transfer rates are usually compromised. In stirred reactors, soft, porous gels are readily destroyed at the agitation speeds needed to eliminate external boundary-layer effects.

(v) *Product effects.* Products formed by reaction inside catalysts must diffuse out under the influence of a concentration gradient. The concentration profile for product is the reverse of that for substrate; concentration is highest at the centre of the catalyst and lowest in the bulk liquid. If product inhibition affects cell or enzyme activity, high intraparticle concentrations may inhibit progress of the reaction. Immobilised enzymes which produce or consume H^+ ions are often affected; because enzyme reactions are very sensitive to pH, small local variations due to slow diffusion of ions can have a significant influence on reaction rate [43].

12.11 Summary of Chapter 12

At the end of Chapter 12 you should:

(i) know what heterogeneous reactions are and when they occur in bioprocessing;

(ii) understand the difference between *observed* and *true reaction rates*;

(iii) know how concentration gradients arise in solid-phase catalysts;

(iv) understand the concept of the *effectiveness factor*;

(v) be able to apply the *Thiele modulus* and *observable Thiele modulus* to determine the effect of internal mass transfer on reaction rate;

(vi) be able to quantify external mass-transfer effects from measured data;

(vii) know how to minimise internal and external mass-transfer restrictions; and

(viii) understand that it is generally difficult to determine true kinetic parameters for heterogeneous biological reactions.

Problems

12.1 Diffusion and reaction in a waste treatment lagoon

Industrial wastewater is often treated in large shallow lagoons. Consider such a lagoon covering land of area A. Microorganisms form a sludge layer of thickness L at the bottom of the lagoon; this sludge remains essentially undisturbed by movement of the liquid. As indicated in Figure 12P1.1, distance from the bottom of the lagoon is measured by coordinate z. Assume that microorganisms are distributed uniformly in the sludge.

Figure 12P1.1 Lagoon for wastewater treatment.

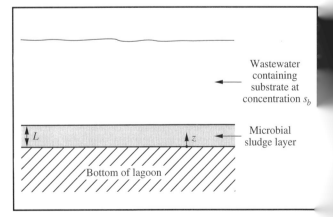

At steady state, wastewater is fed into the lagoon so that the bulk concentration of digestible substrate remains constant at s_b. Cells consume substrate diffusing into the sludge layer, thereby establishing a concentration gradient across thickness L.

(a) Set up a shell mass balance on substrate by considering a thin slice of sludge of thickness Δz perpendicular to the direction of diffusion. The rate of microbial reaction per unit volume sludge is:

$$r_S = k_1 s$$

where s is the concentration of substrate in the sludge layer (gmol cm^{-3}) and k_1 is the first-order rate constant (s^{-1}). The effective diffusivity of substrate in sludge is $\mathscr{D}_{Se}$. Obtain a differential equation relating s and z.

Hint: Area A is constant for flat-plate geometry and can be cancelled from all terms of the mass-balance equation.

(b) External mass-transfer effects at the liquid–sludge interface are negligible. What are the boundary conditions for this problem?

(c) The differential equation obtained in (a) is solved by making the substitution:

$$s = N e^{pz}$$

where N and p are constants.

(i) Substitute this expression for s into the differential equation derived in (a) to obtain an equation for p. (Remember that $\sqrt{p^2} = p$ or $-p$.)

(ii) Because there are two possible values of p, let:

$$s = N e^{pz} + M e^{-pz}.$$

Apply the boundary condition at $z = 0$ to this expression and obtain a relationship between N and M.

(iii) Use the boundary condition at $z = L$ to find N and M explicitly. Obtain an expression for s as a function of z.

(iv) Use the definition of $\cosh x$:

$$\cosh x = \frac{e^x + e^{-x}}{2}$$

to prove that:

$$\frac{s}{s_b} = \frac{\cosh\left(z\sqrt{k_1/\mathscr{D}_{Se}}\right)}{\cosh\left(L\sqrt{k_1/\mathscr{D}_{Se}}\right)}.$$

(d) At steady state, the rate of substrate consumption must be equal to the rate at which substrate enters the sludge. As substrate enters the sludge by diffusion, the overall rate of reaction can be evaluated using Fick's law:

$$r_{A,obs} = \mathscr{D}_{Se} A \left.\frac{ds}{dz}\right|_{z=L}$$

where

$$\left.\frac{ds}{dz}\right|_{z=L} \text{ means } \frac{ds}{dz} \text{ evaluated at } z = L.$$

Use the equation for s from (c) to derive an equation for $r_{A,obs}$.

Hint: The derivative of $\cosh ax = a \sinh ax$ where a is a constant and:

$$\sinh x = \frac{e^x - e^{-x}}{2}.$$

e) Show from the result of (d) that the internal effectiveness factor is given by the expression:

$$\eta_{il} = \frac{\tanh \phi_1}{\phi_1}$$

where

$$\phi_1 = L \sqrt{\frac{k_1}{\mathscr{D}_{Se}}}$$

and:

$$\tanh x = \frac{\sinh x}{\cosh x}.$$

(f) Plot the concentration profiles through a sludge layer of thickness 2 cm for the following sets of conditions:

	(1)	(2)	(3)
k_1 (s^{-1})	4.7×10^{-8}	2.0×10^{-7}	1.5×10^{-4}
$\mathscr{D}_{Se}$ (cm^2 s^{-1})	7.5×10^{-7}	2.0×10^{-7}	6.0×10^{-6}

Take s_b to be 10^{-5} gmol cm^{-3}. Label the profiles with corresponding values of ϕ_1 and η_{il}. Comment on the general relationship between ϕ_1, the shape of the concentration profile, and the value of η_{il}.

12.2 Oxygen profile in immobilised-enzyme catalyst

L-Lactate 2-monooxygenase from *Mycobacterium smegmatis* is immobilised in spherical agarose beads. The enzyme catalyses the reaction:

$$C_3H_6O_3 + O_2 \rightarrow C_2H_4O_2 + CO_2 + H_2O.$$
(lactic acid)   (acetic acid)

Beads 4 mm in diameter are immersed in a well-mixed solution containing 0.5 mM oxygen. A high lactic acid concentration is provided so that oxygen is the rate-limiting substrate. The effective diffusivity of oxygen in agarose is 2.1×10^{-9} m^2 s^{-1}. K_m for the immobilised enzyme is 0.015 mM; v_{max} is 0.12 mol s^{-1} per kg enzyme. The beads contain 0.012 kg enzyme m^{-3} gel. External mass-transfer effects are negligible.

(a) Plot the oxygen concentration profile inside the beads.
(b) What fraction of the catalyst volume is active?
(c) Determine the largest bead size that allows the maximum conversion rate?

12.3 Effect of oxygen transfer on recombinant cells

Recombinant *E. coli* cells contain a plasmid derived from pBR322 incorporating genes for the enzymes β-lactamase and catechol 2,3-dioxygenase from *Pseudomonas putida*. To produce the desired enzymes the organism requires aerobic conditions. The cells are immobilised in spherical beads of carrageenan gel. The effective diffusivity of oxygen is 1.4×10^{-9} m^2 s^{-1}. Uptake of oxygen is zero-order with intrinsic rate constant 10^{-3} mol s^{-1} m^{-3} particle. The concentration of oxygen at the surface of the catalyst is 8×10^{-3} kg m^{-3}. Cell growth is negligible.

(a) What is the maximum particle diameter for aerobic conditions throughout the catalyst?
(b) For particles half the diameter calculated in (a), what is the minimum oxygen concentration in the beads?
(c) The density of cells in the gel is reduced by a factor of five. If specific activity is independent of cell loading, what is the maximum particle size for aerobic conditions?

12.4 Ammonia oxidation by immobilised cells

Thiosphaera pantotropha is being investigated for aerobic oxidation of ammonia to nitrite in wastewater treatment. The organism is immobilised in spherical agarose particles of diameter 3 mm. The effective diffusivity of oxygen in the particles is 1.9×10^{-9} m^2 s^{-1}. The immobilised cells are placed in a flow chamber for measurement of oxygen uptake rate. Using published correlations, the liquid–solid mass-transfer coefficient for oxygen is calculated as 6×10^{-5} m s^{-1}. When the bulk oxygen concentration is 6×10^{-3} kg m^{-3}, the observed rate of oxygen consumption is 2.2×10^{-5} kg s^{-1} m^{-3} catalyst.

(a) What effect does external mass-transfer have on respiration rate?
(b) What is the effectiveness factor?
(c) For optimal activity of *T. pantotropha*, oxygen levels must be kept above the critical level, 1.2×10^{-3} kg m^{-3}. Is this condition satisfied?

12.5 Microcarrier culture and external mass transfer

Mammalian cells form a monolayer on the surface of microcarrier beads of diameter 120 μm and density 1.2×10^3 kg m^{-3}. The culture is maintained in spinner flasks in serum-free medium of viscosity 10^{-3} N s m^{-2} and density 10^3 kg m^{-3}. The diffusivity of oxygen in the medium is 2.3×10^{-9} m^2 s^{-1}. The observed rate of oxygen uptake is 0.015 mol s^{-1} m^{-3} at a bulk oxygen concentration of 0.2 mol m^{-3}. What is the effect of external mass transfer on reaction rate?

12.6 Immobilised-enzyme reaction kinetics

Invertase catalyses the reaction:

$$C_{12}H_{22}O_{11} + H_2O \rightarrow C_6H_{12}O_6 + C_6H_{12}O_6.$$
(sucrose)   (glucose)   (fructose)

Invertase from *Aspergillus oryzae* is immobilised in porous resin particles of diameter 1.6 mm at a density of 0.1 μmol enzyme g^{-1}. The effective diffusivity of sucrose in the resin is 1.3×10^{-11} m^2 s^{-1}. The resin is placed in a spinning-basket reactor operated so that external mass-transfer effects are eliminated. At a sucrose concentration of 0.85 kg m^{-3}, the observed rate of conversion is 1.25×10^{-3} kg s^{-1} m^{-3} resin. K_m for the immobilised enzyme is 3.5 kg m^{-3}.

(a) Calculate the effectiveness factor.
(b) Determine the true first-order reaction constant for immobilised invertase.
(c) Assume that specific enzyme activity is not affected by steric hindrance or conformational changes as enzyme loading increases. This means that k_1 should be directly proportional to enzyme concentration in the resin. Plot changes in effectiveness factor and reaction rate as a function of enzyme loading from 0.01 μmol g^{-1} to

2.0 μmol g^{-1}. Comment on the relative benefit of increasing the concentration of enzyme in the resin.

12.7 Mass-transfer effects in plant cell culture

Suspended *Catharanthus roseus* cells form spherical clumps approximately 1.5 mm in diameter. Oxygen uptake is measured using the apparatus of Figure 12.13; medium is recirculated with a superficial liquid velocity of 0.83 cm s^{-1}. At a bulk concentration of 8 mg l^{-1}, oxygen is consumed at a rate of 0.28 mg per g wet weight per hour. Assume that the density and viscosity of the medium are similar to water, the specific gravity of wet cells is 1, and oxygen uptake is zero order. The effective diffusivity of oxygen in the clumps is 9 × 10^{-6} cm^2 s^{-1}, or half that in the medium.

(a) Does external mass transfer affect the oxygen-uptake rate?
(b) To what extent does internal mass transfer affect oxygen uptake?
(c) Roughly, what would you expect the profile of oxygen concentration to be within the aggregates?

12.8 Respiration in mycelial pellets

Aspergillus niger cells are observed to form aggregates of average diameter 5 mm. The effective diffusivity of oxygen in the aggregates is 1.75 × 10^{-9} m^2 s^{-1}. In a fixed-bed reactor, the oxygen-consumption rate at a bulk oxygen concentration of 8 × 10^{-3} kg m^{-3} is 8.7 × 10^{-5} kg s^{-1} m^{-3} biomass. The liquid–solid mass-transfer coefficient is 3.8 × 10^{-5} m s^{-1}.

(a) Is oxygen uptake affected by external mass transfer?
(b) What is the external effectiveness factor?
(c) What reaction rate would be observed if both internal and external mass-transfer resistances were eliminated?
(d) If only external mass-transfer effects were removed, what would be the reaction rate?

References

1. Karel, S.F., S.B. Libicki and C.R. Robertson (1985) The immobilization of whole cells: engineering principles. *Chem. Eng. Sci.* **40**, 1321–1354.

2. Wittler, R., H. Baumgartl, D.W. Lübbers and K. Schügerl (1986) Investigations of oxygen transfer into *Penicillium chrysogenum* pellets by microprobe measurements. *Biotechnol. Bioeng.* **28**, 1024–1036.

3. Bringi, V. and B.E. Dale (1990) Experimental and theoretical evidence for convective nutrient transport in an immobilized cell support. *Biotechnol. Prog.* **6**, 205–209.

4. Nir, A. and L.M. Pismen (1977) Simultaneous intraparticle forced convection, diffusion and reaction in a porous catalyst. *Chem. Eng. Sci.* **32**, 35–41.

5. Rodrigues, A.E., J.M. Orfao and A. Zoulalian (1984) Intraparticle convection, diffusion and zero order reaction in porous catalysts. *Chem. Eng. Commun.* **27**, 327–337.

6. Stephanopoulos, G. and K. Tsiveriotis (1989) The effect of intraparticle convection on nutrient transport in porous biological pellets. *Chem. Eng. Sci.* **44**, 2031–2039.

7. van't Riet, K. and J. Tramper (1991) *Basic Bioreactor Design*, Marcel Dekker, New York.

8. Hooijmans, C.M., S.G.M. Geraats, E.W.J. van Neil, L.A. Robertson, J.J. Heijnen and K.Ch.A.M. Luyben (1990) Determination of growth and coupled nitrification/denitrification by immobilized *Thiosphaera pantotropha* using measurement and modeling of oxygen profiles. *Biotechnol. Bioeng.* **36**, 931–939.

9. Hooijmans, C.M., S.G.M. Geraats and K.Ch.A.M. Luyben (1990) Use of an oxygen microsensor for the determination of intrinsic kinetic parameters of an immobilized oxygen reducing enzyme. *Biotechnol. Bioeng.* **35**, 1078–1087.

10. de Beer, D. and J.C. van den Heuvel (1988) Gradients in immobilized biological systems. *Anal. Chim. Acta* **213**, 259–265.

11. Thiele, E.W. (1939) Relation between catalytic activity and size of particle. *Ind. Eng. Chem.* **31**, 916–920.

12. Aris, R. (1965) A normalization for the Thiele modulus. *Ind. Eng. Chem. Fund.* **4**, 227–229.

13. Bischoff, K.B. (1965) Effectiveness factors for general reaction rate forms. *AIChE J.* **11**, 351–355.

14. Froment, G.F. and K.B. Bischoff (1979) *Chemical Reactor Analysis and Design*, Chapter 3, John Wiley, New York.

15. Aris, R. (1965) *Introduction to the Analysis of Chemical Reactors*, Prentice-Hall, New Jersey.

16. Aris, R. (1975) *The Mathematical Theory of Diffusion and Reaction in Permeable Catalysts*, vol. 1, Oxford University Press, London.

17. Moo-Young, M. and T. Kobayashi (1972) Effectiveness factors for immobilized-enzyme reactions. *Can. J. Chem. Eng.* **50**, 162–167.

18. Weisz, P.B. (1973) Diffusion and chemical transformation: an interdisciplinary excursion. *Science* **179**, 433–440.

19. Sherwood, T.K., R.L. Pigford and C.R. Wilke (1975) *Mass Transfer*, Chapter 6, McGraw-Hill, New York.

20. McCabe, W.L. and J.C. Smith (1976) *Unit Operations of Chemical Engineering*, 3rd edn, Chapter 22, McGraw-Hill, Tokyo.

21. Brian, P.L.T. and H.B. Hales (1969) Effects of transpira-

tion and changing diameter on heat and mass transfer to spheres. *AIChE J.* **15**, 419–425.

22. Ranz, W.E. and W.R. Marshall (1952) Evaporation from drops. Parts I and II. *Chem. Eng. Prog.* **48**, 141–146, 173–180.

23. Moo-Young, M. and H.W. Blanch (1981) Design of biochemical reactors: mass transfer criteria for simple and complex systems. *Adv. Biochem. Eng.* **19**, 1–69.

24. Ford, J.R., A.H. Lambert, W. Cohen and R.P. Chambers (1972) Recirculation reactor system for kinetic studies of immobilized enzymes. *Biotechnol. Bioeng. Symp.* **3**, 267–284.

25. Sato, K. and K. Toda (1983) Oxygen uptake rate of immobilized growing *Candida lipolytica*. *J. Ferment. Technol.* **61**, 239–245.

26. Matson, J.V. and W.G. Characklis (1976) Diffusion into microbial aggregates. *Water Res.* **10**, 877–885.

27. Chen, Y.S. and H.R. Bungay (1981) Microelectrode studies of oxygen transfer in trickling filter slimes. *Biotechnol. Bioeng.* **23**, 781–792.

28. Axelsson, A. and B. Persson (1988) Determination of effective diffusion coefficients in calcium alginate gel plates with varying yeast cell content. *Appl. Biochem. Biotechnol.* **18**, 231–250.

29. Pu, H.T. and R.Y.K. Yang (1988) Diffusion of sucrose and yohimbine in calcium alginate gel beads with or without entrapped plant cells. *Biotechnol. Bioeng.* **32**, 891–896.

30. Tanaka, H., M. Matsumura and I.A. Veliky (1984) Diffusion characteristics of substrates in Ca-alginate gel beads. *Biotechnol. Bioeng.* **26**, 53–58.

31. Scott, C.D., C.A. Woodward and J.E. Thompson (1989) Solute diffusion in biocatalyst gel beads containing biocatalysis and other additives. *Enzyme Microb. Technol.* **11**, 258–263.

32. Chresand, T.J., B.E. Dale, S.L. Hanson and R.J. Gillies (1988) A stirred bath technique for diffusivity measurements in cell matrices. *Biotechnol. Bioeng.* **32**, 1029–1036.

33. Omar, S.H. (1993) Oxygen diffusion through gels employed for immobilization: Parts 1 and 2. *Appl. Microbiol. Biotechnol.* **40**, 1–6, 173–181.

34. Rovito, B.J. and J.R. Kittrell (1973) Film and pore diffusion studies with immobilized glucose oxidase. *Biotechnol. Bioeng.* **15**, 143–161.

35. Shah, Y.T. (1979) *Gas–Liquid–Solid Reactor Design*, McGraw-Hill, New York.

36. Carberry, J.J. (1964) Designing laboratory catalytic reactors. *Ind. Eng. Chem.* **56**, 39–46.

37. Tajbl, D.G., J.B. Simons and J.J. Carberry (1966) Heterogeneous catalysis in a continuous stirred tank reactor. *Ind. Eng. Chem. Fund.* **5**, 171–175.

38. Hamilton, B.K., C.R. Gardner and C.K. Colton (1974) Effect of diffusional limitations on Lineweaver–Burk plots for immobilized enzymes. *AIChE J.* **20**, 503–510.

39. Engasser, J.-M. and C. Horvath (1973) Effect of internal diffusion in heterogeneous enzyme systems: evaluation of true kinetic parameters and substrate diffusivity. *J. Theor. Biol.* **42**, 137–155.

40. Clark, D.S. and J.E. Bailey (1983) Structure–function relationships in immobilized chymotrypsin catalysis. *Biotechnol. Bioeng.* **25**, 1027–1047.

41. Lee, G.K., R.A. Lesch and P.J. Reilly (1981) Estimation of intrinsic kinetic constants for pore diffusion-limited immobilized enzyme reactions. *Biotechnol. Bioeng.* **23**, 487–497.

42. Petersen, E.E. (1965) *Chemical Reaction Analysis*, Prentice-Hall, New Jersey.

43. Stewart, P.S. and C.R. Robertson (1988) Product inhibition of immobilized *Escherichia coli* arising from mass transfer limitation. *App. Environ. Microbiol.* **54**, 2464–2471.

Suggestions for Further Reading

Immobilised Cells and Enzymes

de Bont, J.A.M., J. Visser, B. Mattiasson and J. Tramper (Eds) (1990) *Physiology of Immobilized Cells*, Elsevier, Amsterdam.

Katchalski-Katzir, E. (1993) Immobilized enzymes – learning from past successes and failures. *Trends in Biotechnol.* **11**, 471–478.

Klein, J. and K.-D. Vorlop (1985) Immobilization techniques – cells. In: M. Moo-Young (Ed), *Comprehensive Biotechnology*, vol. 2, pp. 203–224, Pergamon Press, Oxford.

Messing, R.A. (Ed) (1975) *Immobilized Enzymes for Industrial Reactors*, Academic Press, New York.

Messing, R.A. (1985) Immobilization techniques – enzymes. In: M. Moo-Young (Ed), *Comprehensive Biotechnology*, vol. 2, pp. 191–201, Pergamon Press, Oxford.

Phillips, C.R. and Y.C. Poon (1988) *Immobilization of Cells*, Springer-Verlag, Berlin.

Engineering Analysis of Mass Transfer and Reaction (see also refs 1, 7, 14 and 38–41)

Engasser, J.-M. and C. Horvath (1976) Diffusion and kinetics with immobilized enzymes. *Appl. Biochem. Biotechnol.* **1**, 127–220.

Satterfield, C.N. (1970) *Mass Transfer in Heterogeneous Catalysis*, MIT Press, Cambridge, Massachusetts.

13

Reactor Engineering

The reactor is the heart of any fermentation or enzyme conversion process. Design of bioreactors is a complex task, relying on scientific and engineering principles and many rules of thumb. Specifying aspects of the reactor and its operation involves several critical decisions.

i) *Reactor configuration.* For example, should the reactor be a stirred tank or an air-driven vessel without mechanical agitation?

ii) *Reactor size.* What size reactor is required to achieve the desired rate of production?

iii) *Processing conditions inside the reactor.* What reaction conditions such as temperature, pH and dissolved-oxygen tension should be maintained in the vessel, and how will these parameters be controlled? How will contamination be avoided?

iv) *Mode of operation.* Will the reactor be operated batch-wise or as a continuous-flow process? Should substrate be fed intermittently? Should the reactor be operated alone or in series with others?

Decisions made in reactor design have a significant impact on overall process performance, yet there are no simple or standard design procedures available which specify all aspects of the vessel and its operation. Reactor engineering brings together much of the material already covered in Chapters 5–12 of this book. Knowledge of reaction kinetics is essential for understanding how biological reactors work. Other areas of bioprocess engineering such as mass and energy balances, mixing, mass transfer and heat transfer are also required.

13.1 Reactor Engineering in Perspective

Before starting to design a reactor, some objectives have to be defined. Simple aims like 'Produce 1 g of monoclonal antibody per day', or 'Produce 10 000 tonnes of amino acid per year', provide the starting point. Other objectives are also relevant; in industrial processes the product should be made at the lowest possible cost to maximise the company's commercial advantage. In some cases, economic objectives are overridden by safety concerns, the need for high-product purity or regula-

tory considerations. The final reactor design will be a reflection of all these process requirements and, in most cases, represents a compromise solution to conflicting demands.

In this section, we will consider the various contributions to bioprocessing costs for different types of product, and examine the importance of reactor engineering in improving overall process performance. As shown in Figure 13.1, the value of products made by bioprocessing covers a wide range. Typically, products with the highest value are those from

Figure 13.1 Range of value of fermentation products. (From P.N. Royce, 1993, A discussion of recent developments in fermentation monitoring and control from a practical perspective. *Crit. Rev. Biotechnol.* **13**, 117–149.)

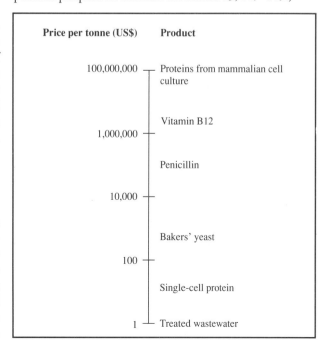

Figure 13.2 Contributions to total production cost in bioprocessing.

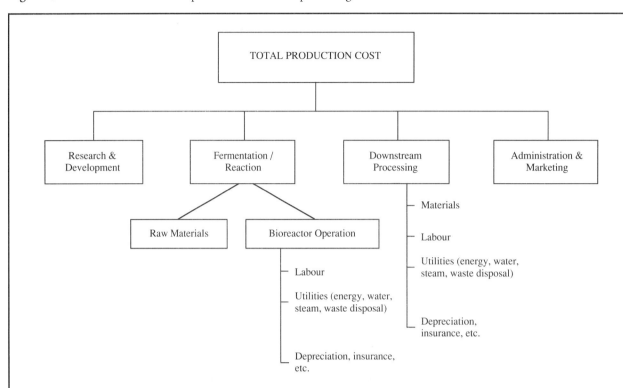

mammalian cell culture, such as therapeutic proteins and monoclonal antibodies. At the opposite end of the scale is treatment of waste, where the overriding objective is minimal financial outlay for the desired level of purity. To reduce the cost of any bioprocess, it is first necessary to identify which aspect of it is cost-determining. Break-down of production costs varies from process to process; however, a general scheme is shown in Figure 13.2. The following components are important: (i) research and development; (ii) the fermentation or reaction step; (iii) downstream processing; and (iv) administration and marketing. In most bioprocesses, the cost of administration and marketing is relatively small. Products for which the cost of reaction dominates include biomass such as bakers' yeast and single-cell protein, catabolic metabolites such as ethanol and lactic acid, and bioconversion products such as high-fructose corn syrup and 6-aminopenicillanic acid. Intracellular products such as proteins have high downstream-processing costs compared with reaction; other examples in this category are antibiotics, vitamins and amino acids. For new, high-value biotechnology products such as recombinant proteins and antibodies, actual processing costs are only a small part of the total because of the enormous

investment required for research and development and regulatory approval. Getting the product into the marketplace quickly is the most important cost-saving measure in these cases; any savings made by improving the efficiency of the reactor are generally trivial in comparison. However, for the majority of fermentation products outside this high-value category, bioprocessing costs make a significant contribution to the final price.

If the reaction step dominates the cost structure, this may be because of the high cost of the raw materials required or the high cost of reactor operation. The relative contributions of these factors depends on the process. As an example, to produce high-value antibiotics, the cost of 100 m^3 media is US$25 000–100 000 [1]. In contrast, the cost of energy, i.e electricity, to operate a 100 m^3 stirred-tank fermenter including agitation, air compression and cooling water for a 6-d antibiotic fermentation is about US$8000 [1, 2]. Clearly then, energy costs for reactor operation are much less important than raw-material costs for this fermentation process. For high-value, low-yield products such as antibiotics, vitamins, enzymes and pigments, media represents 60–90% of the fermentation costs [1]. For low-cost, high-yield metabolites such

Figure 13.3   Strategies for bioreactor design as a function of the cost-determining factors in the process.

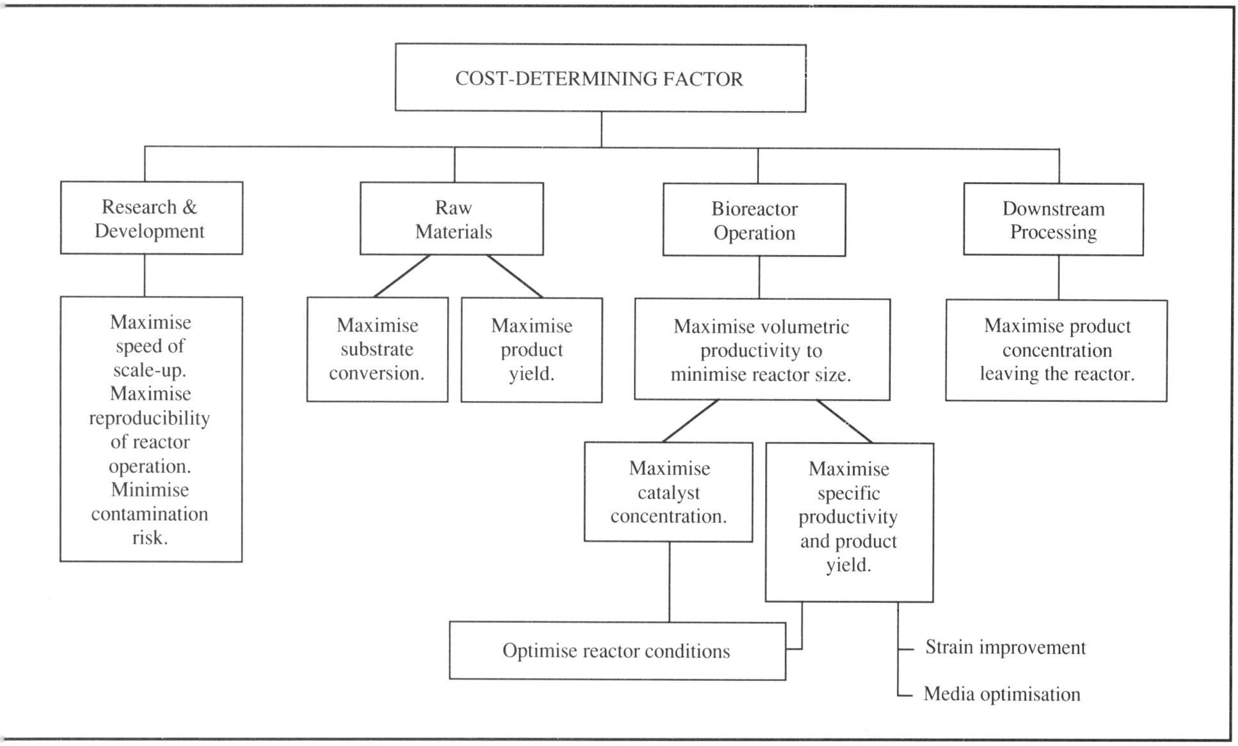

ethanol, citric acid, biomass and lactic acid, raw material costs range from 40% of fermentation costs for citric acid to about 70% for ethanol produced from molasses [1, 3]. The remainder of the operating cost of bioreactors consists mainly of labour and utilities costs.

As indicated in Figure 13.3, identifying the cost structure of bioprocesses assists in defining the objectives for reactor design. Even if the reaction itself is not cost-determining, aspects of reactor design may still be important. If the cost of research and development is dominating, design of the reactor is directed towards the need for rapid scale-up; this is more important than maximising conversion or minimising operating costs. For new biotechnology products intended for therapeutic use, regulatory guidelines require that the entire production scheme be validated and process control guaranteed for consistent quality and safety; reproducibility of reactor operation is therefore critical. When the cost of raw materials is significant, maximising substrate conversion and product yield in the reactor have high priority. If downstream processing is expensive, the reactor is designed and operated to maximise the product concentration leaving the vessel; this avoids the expense of recovering product from dilute solutions. When reaction costs are significant, the reactor should

be as small as possible to reduce both operating and capital costs. To achieve the desired total production rate using a small vessel, the volumetric productivity of the reactor must be sufficiently high (see Section 11.1.3).

As indicated in Figure 13.3, volumetric productivity depends on the concentration of catalyst and its specific rate of production. To achieve high volumetric rates, the reactor must therefore allow maximum catalyst activity at the highest practical catalyst concentration. For tightly packed cells or cell organelles, the physical limit on concentration is of the order 200 kg dry weight m^{-3}; for enzymes in solution, the maximum concentration depends on the solubility of enzyme in the reaction mixture. The extent to which these limiting concentrations can be approached depends on the functioning of the reactor. For example, if mixing or mass transfer is inadequate, oxygen or nutrient starvation will occur and the maximum cell density achieved will be low. Alternatively, if shear levels in the reactor are too high, cells will be disrupted and enzymes inactivated so that the effective concentration of catalyst is reduced.

Maximum specific productivity is obtained when the catalyst is capable of high levels of production and conditions in the reactor allow the best possible catalytic function. For simple metabolites such as ethanol, butanol and acetic acid which are linked to energy production in the cell, the maximum theoretical yield is limited by the thermodynamic and stoichiometric principles outlined in Section 4.6. Accordingly, there is little scope for increasing production titres of these materials; reduced production costs and commercial advantage rely mostly on improvements in reactor operation which allow the system to achieve close to the maximum theoretical yield. In contrast, it is not unusual for strain improvement and media optimisation programmes to improve yields of antibiotics and enzymes by over 100-fold, particularly in the early stages of process development. Therefore, for these products, identification of high-producing strains and optimal environmental conditions is initially more rewarding than improving the reactor design and operation.

13.2 Bioreactor Configurations

The cylindrical tank, either stirred or unstirred, is the most common reactor in bioprocessing. Yet, a vast array of fermenter configurations is in use in different bioprocess industries. Novel bioreactors are constantly being developed for special applications and new forms of biocatalyst such as plant and animal tissue and immobilised cells and enzymes.

Much of the challenge in reactor design lies in the provision of adequate mixing and aeration for the large proportion of fermentations requiring oxygen; reactors for anaerobic culture are usually very simple in construction without sparging or agitation. In the following discussion of bioreactor configurations, aerobic operation will be assumed.

13.2.1 Stirred Tank

A conventional stirred, aerated bioreactor is shown schematically in Figure 13.4. Mixing and bubble dispersion are achieved by mechanical agitation; this requires a relatively high input of energy per unit volume. Baffles are used in stirred reactors to reduce vortexing. A wide variety of impeller sizes and shapes is available to produce different flow patterns inside the vessel; in tall fermenters, installation of multiple impellers improves mixing. The mixing and mass-transfer functions of stirred reactors are described in detail in Chapters 7 and 9.

Typically, only 70–80% of the volume of stirred reactors is filled with liquid; this allows adequate headspace for disengagement of droplets from the exhaust gas and to accommodate any foam which may develop. If foaming is a problem, a supplementary impeller called a *foam breaker* may

Figure 13.4   Typical stirred-tank fermenter for aerobic culture.

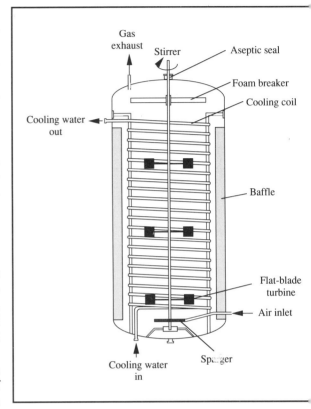

be installed as shown in Figure 13.4. Alternatively, chemic antifoam agents are added to the broth; because antifoar reduce the rate of oxygen transfer (see Section 9.6.3), mecha ical foam dispersal is generally preferred.

The aspect ratio of stirred vessels, i.e. the ratio of height diameter, can be varied over a wide range. The least expensi shape to build has an aspect ratio of about 1; this shape has t smallest surface area and therefore requires the least material construct for a given volume. However, when aeration required, the aspect ratio is usually increased. This provides longer contact times between the rising bubbles and liquid a produces a greater hydrostatic pressure at the bottom of t vessel.

As shown in Figure 13.4, temperature control and he transfer in stirred vessels can be accomplished using interr cooling coils. Alternative cooling equipment for bioreactors illustrated in Figure 8.1 (p. 165). The relative advantages a disadvantages of different heat-exchange systems are discuss in Section 8.1.1.

Stirred fermenters are used for free- and immobilise

Figure 13.5 Bubble-column bioreactor.

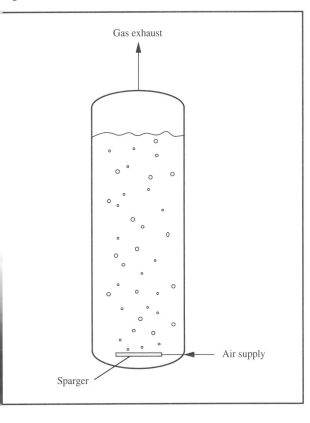

Gas exhaust

Air supply

Sparger

enzyme reactions, and for culture of suspended and immobilised cells. Care is required with particulate catalysts which may be damaged or destroyed by the impeller at high speeds. As discussed in Section 7.14, high levels of shear can also damage sensitive cells, particularly in plant and animal cell culture.

13.2.2 Bubble Column

Alternatives to the stirred reactor include vessels with no mechanical agitation. In bubble-column reactors, aeration and mixing are achieved by gas sparging; this requires less energy than mechanical stirring. Bubble columns are applied industrially for production of bakers' yeast, beer and vinegar, and for treatment of wastewater.

Bubble columns are structurally very simple. As shown in Figure 13.5, they are generally cylindrical vessels with height greater than twice the diameter. Other than a sparger for entry of compressed air, bubble columns typically have no internal structures. A height-to-diameter ratio of about 3:1 is common in bakers' yeast production; for other applications, towers with height-to-diameter ratios of 6:1 have been used. Perforated

horizontal plates are sometimes installed in tall bubble columns to break up and redistribute coalesced bubbles. Advantages of bubble columns include low capital cost, lack of moving parts, and satisfactory heat- and mass-transfer performance. As in stirred vessels, foaming can be a problem requiring mechanical dispersal or addition of antifoam to the medium.

Bubble-column hydrodynamics and mass-transfer characteristics depend entirely on the behaviour of the bubbles released from the sparger. Different flow regimes occur depending on the gas flow rate, sparger design, column diameter and medium properties such as viscosity. *Homogeneous flow* occurs only at low gas flow rates and when bubbles leaving the sparger are evenly distributed across the column cross-section. In homogeneous flow, all bubbles rise with the same upward velocity and there is no backmixing of the gas phase. Liquid mixing in this flow regime is also limited, arising solely from entrainment in the wakes of the bubbles. Under normal operating conditions at higher gas velocities, large chaotic circulatory flow cells develop and *heterogeneous flow* occurs as illustrated in Figure 13.6. In this regime, bubbles and liquid tend to rise up the centre of the column while a corresponding downflow of liquid occurs near the walls. Liquid circulation entrains bubbles so that some backmixing of gas occurs.

Liquid mixing time in bubble columns depends on the flow regime. For heterogeneous flow, the following equation has been proposed [4] for the upward liquid velocity at the centre of the column for $0.1 < D < 7.5$ m and $0 < u_G < 0.4$ m s^{-1}:

$$u_L = 0.9(\, g\, D\, u_G)^{0.33}$$

$$(13.1)$$

where u_L is linear liquid velocity, g is gravitational acceleration, D is column diameter, and u_G is gas superficial velocity. u_G is equal to the volumetric gas flow rate at atmospheric pressure divided by the reactor cross-sectional area. From this equation, an expression for the mixing time t_m (see Section 7.9.4) can be obtained [5]:

$$t_m = 11 \frac{H}{D} \, (g\, u_G D^{-2})^{-0.33}$$

$$(13.2)$$

where H is the height of the bubble column.

As discussed in Section 9.6.1, values for gas–liquid mass-transfer coefficients in reactors depend largely on bubble diameter and gas hold-up. In bubble columns containing non-viscous liquids, these variables depend solely on the gas flow rate. However, as exact bubble sizes and liquid circulation

Figure 13.6 Heterogeneous flow in a bubble column.

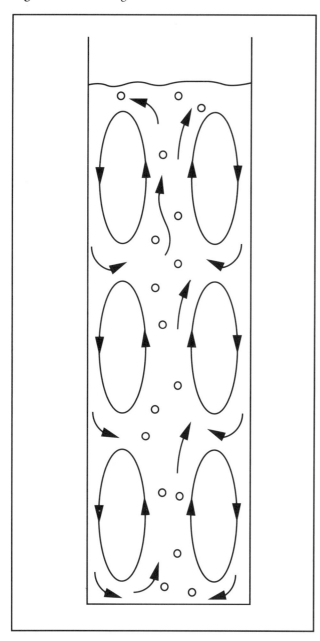

patterns are impossible to predict in bubble columns, accurate estimation of the mass-transfer coefficient is difficult. The following correlation has been proposed for non-viscous media in heterogeneous flow [4, 5]:

$$k_L a \approx 0.32 u_G^{0.7}$$

(13.3)

where $k_L a$ is the combined volumetric mass-transfer coefficient and u_G is the gas superficial velocity. Eq. (13.3) is valid for bubbles with mean diameter about 6 mm, 0.08 m < D < 11.6 m, 0.3 m < H < 21 m, and 0 < u_G < 0.3 m s^{-1}. If smaller bubbles are produced at the sparger and the medium is non-coalescing, $k_L a$ will be larger than the value calculated using Eq. (13.3), especially at low values of u_G less than about 10^{-2} m s^{-1} [4].

13.2.3 Airlift Reactor

As in bubble columns, mixing in airlift reactors is accomplished without mechanical agitation. Airlift reactors are often chosen for culture of plant and animal cells and immobilised catalysts because shear levels are significantly lower than in stirred vessels.

Several types of airlift reactor are in use. Their distinguishing feature compared with the bubble column is that patterns of liquid flow are more defined owing to the physical separation of up-flowing and down-flowing streams. As shown in Figure 13.7, gas is sparged into only part of the vessel cross-section called the *riser*. Gas hold-up and decreased fluid density cause liquid in the riser to move upwards. Gas disengages at the top of the vessel leaving heavier bubble-free liquid to recirculate through the *downcomer*. Liquid circulates in airlift reactors as a result of the density difference between riser and downcomer.

Figure 13.7 illustrates the most common airlift configurations. In the *internal-loop vessels* of Figures 13.7(a) and 13.7(b), the riser and downcomer are separated by an internal baffle or *draft tube*; air may be sparged into either the draft tube or the annulus. In the *external-loop* or *outer-loop* airlift of Figure 13.7(c), separate vertical tubes are connected by short horizontal sections at the top and bottom. Because the riser and downcomer are further apart in external-loop vessels, gas disengagement is more effective than in internal-loop devices. Fewer bubbles are carried into the downcomer, the density difference between fluids in the riser and downcomer is greater, and circulation of liquid in the vessel is faster. Accordingly, mixing is usually better in external-loop than internal-loop reactors.

Airlift reactors generally provide better mixing than bubble columns except at low liquid velocities when circulatory flow-patterns similar to those shown in Figure 13.6 develop. The airlift configuration confers a degree of stability to liquid flow compared with bubble columns; therefore, higher gas flow rates can be used without incurring operating problems such as slug flow or spray formation. Several empirical correlations have been developed for liquid velocity, circulation time and

Figure 13.7 Airlift reactor configurations.

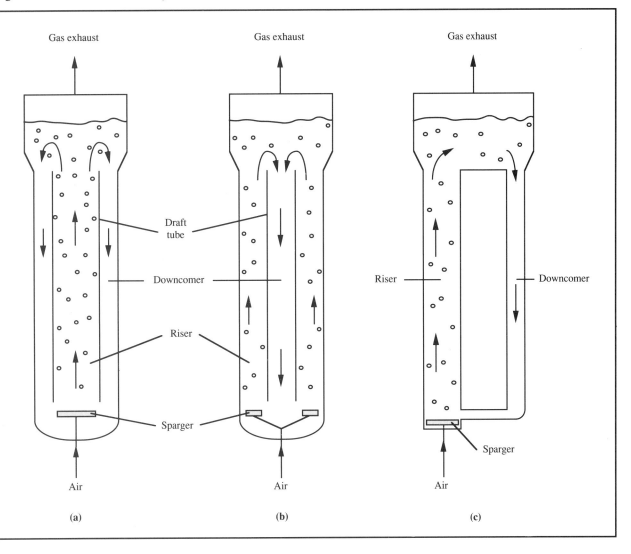

Gas exhaust   Gas exhaust   Gas exhaust

Draft tube

Downcomer

Riser

Sparger

Air

(a)   (b)   (c)

Riser   Downcomer

Sparger

Air

mixing time in airlift reactors; however there is considerable discrepancy between the results [6]. Equations derived from hydrodynamic models are also available [6, 7]; these are usually relatively complex and, because liquid velocity and gas hold-up are not independent, require iterative numerical solution.

Gas hold-up and gas–liquid mass-transfer rates in internal-loop airlifts are similar to those in bubble columns [6]. However, in external-loop devices, near-complete gas disengagement increases the liquid velocity and decreases the air hold-up [8, 9] so that mass-transfer rates at identical gas velocities are lower than in bubble columns [6]. Therefore, by comparison with Eq. (13.3) for bubble columns, for external-loop airlifts:

$$k_L a < 0.32\, u_G^{0.7}.$$

(13.4)

Several other empirical mass-transfer correlations have been developed for Newtonian and non-Newtonian fluids in airlift reactors [6].

Performance of airlift devices is influenced significantly by the details of vessel construction [6, 10, 11]. For example, in internal-loop airlifts, changing the distance between the lower edge of the draft tube and the base of the reactor alters the pressure drop in this region and affects liquid velocity and gas hold-up. The depth of draft-tube submersion from the top of the liquid also influences mixing and mass-transfer characteristics.

Airlift reactors have been applied in production of single-cell protein from methanol and gas oil; they are also used for plant and animal cell culture and in municipal and industrial waste treatment. Large airlift reactors with capacities of thousands of cubic metres have been constructed. Tall internal-loop airlifts built underground are known as *deep-shaft reactors*; very high hydrostatic pressure at the bottom of these vessels considerably improves gas–liquid mass-transfer. The height of airlift reactors is typically about 10 times the diameter; for deep-shaft systems the height-to-diameter ratio may be increased up to 100.

13.2.4 Stirred and Air-Driven Reactors: Comparison of Operating Characteristics

For low-viscosity fluids, adequate mixing and mass transfer can be achieved in stirred tanks, bubble columns and airlift vessels. When a large fermenter (50–500 m^3) is required for low-viscosity culture, a bubble column is an attractive choice because it is simple and cheap to install and operate. Mechanically-agitated reactors are impractical at volumes greater than about 500 m^3 as the power required to achieve adequate mixing becomes extremely high (see Section 7.11).

If the culture has high viscosity, sufficient mixing and mass transfer cannot be provided by air-driven reactors. Stirred vessels are more suitable for viscous liquids because greater power can be input by mechanical agitation. Nevertheless, mass-transfer rates decline rapidly in stirred vessels at viscosities greater than 50–100 cP [5].

Heat transfer can be an important consideration in the choice between air-driven and stirred reactors. Mechanical agitation generates much more heat than sparging of compressed gas. When the heat of reaction is high, such as in production of single-cell protein from methanol, removal of frictional stirrer heat can be a problem so that air-driven reactors may be preferred.

Stirred-tank and air-driven vessels account for the vast majority of bioreactor configurations used for aerobic culture. However, other reactor configurations may be used in particular processes.

13.2.5 Packed Bed

Packed-bed reactors are used with immobilised or particulate biocatalysts. The reactor consists of a tube, usually vertical, packed with catalyst particles. Medium can be fed either at the top or bottom of the column and forms a continuous liquid phase between the particles. Damage due to particle attrition is minimal in packed beds compared with stirred reactors. Packed-bed reactors have been used commercially with

Figure 13.8 Packed-bed reactor with medium recycle.

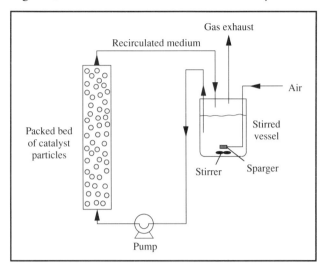

immobilised cells and enzymes for production of aspartate and fumarate, conversion of penicillin to 6-aminopenicillanic acid, and resolution of amino acid isomers.

Mass transfer between the liquid medium and solid catalyst is facilitated at high liquid flow rates through the bed; to achieve this, packed beds are often operated with liquid recycle as shown in Figure 13.8. The catalyst is prevented from leaving the column by screens at the liquid exit. The particles should be relatively incompressible and able to withstand their own weight in the column without deforming and occluding liquid flow. Recirculating medium must also be clean and free of debris to avoid clogging the bed. Aeration is generally accomplished in a separate vessel; if air is sparged directly into the bed, bubble coalescence produces gas pockets and flow *channelling* or maldistribution. Packed beds are unsuitable for processes which produce large quantities of carbon dioxide or other gases which can become trapped in the packing.

13.2.6 Fluidised Bed

When packed beds are operated in upflow mode with catalyst beads of appropriate size and density, the bed expands at high liquid flow rates due to upward motion of the particles. This is the basis for operation of fluidised-bed reactors as illustrated in Figure 13.9. Because particles in fluidised beds are in constant motion, channelling and clogging of the bed are avoided and air can be introduced directly into the column. Fluidised-bed reactors are used in waste treatment with sand or similar material supporting mixed microbial populations. They are also used with flocculating organisms in brewing and for production of vinegar.

Figure 13.9   Fluidised-bed reactor.

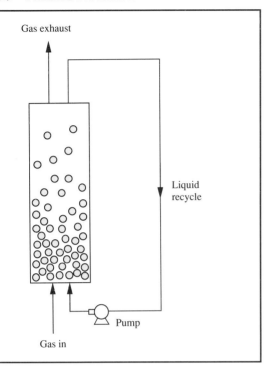

Figure 13.10   Trickle-bed reactor.

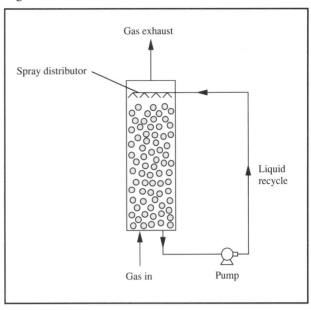

13.2.7 Trickle Bed

The trickle-bed reactor is another variation of the packed bed. As illustrated in Figure 13.10, liquid is sprayed onto the top of the packing and trickles down through the bed in small rivulets. Air may be introduced at the base; because the liquid phase is not continuous throughout the column, air and other gases move with relative ease around the packing. Trickle-bed reactors are used widely for aerobic wastewater treatment.

13.3 Practical Considerations For Bioreactor Construction

Industrial bioreactors for sterile operation are usually designed as steel pressure vessels capable of withstanding full vacuum up to about 3 atm positive pressure at 150–180°C. A hole is provided on large vessels to allow workers entry into the tank for cleaning and maintenance; on smaller vessels the top is removable. Flat headplates are commonly used with laboratory-scale fermenters; for larger vessels a domed construction is less expensive. Large fermenters are equipped with a lighted vertical sight-glass for inspecting the contents of the reactor. Nozzles for medium, antifoam, acid and alkali addition, air-exhaust pipes, pressure gauge, and a rupture disc for emergency pressure release, are normally located on the head-plate. Side ports for pH, temperature and dissolved-oxygen sensors are a minimum requirement; a steam-sterilisable sample outlet should also be provided. The vessel must be fully draining via a harvest nozzle located at the lowest point of the reactor. If the vessel is mechanically agitated, either a top- or bottom-entering stirrer is installed.

13.3.1 Aseptic Operation

Most fermentations outside of the food and beverage industry are carried out using pure cultures and aseptic conditions. Keeping the reactor free of unwanted organisms is especially important for slow-growing cultures which can be quickly over-run by contamination. Fermenters must be capable of operating aseptically for a number of days, sometimes months.

Typically, 3–5% of fermentations in an industrial plant are lost due to failure of sterilisation procedures. However, the frequency and causes of contamination vary considerably from process to process. For example, the nature of the product in antibiotic fermentations affords some protection from contamination; fewer than 2% of production-scale antibiotic fermentations are lost through contamination by microorganisms or phage [12]. In contrast, a contamination rate of 17% has been reported for industrial-scale production of β-interferon from human fibroblasts cultured in 50-litre bioreactors [13].

Figure 13.11   Pinch valve.

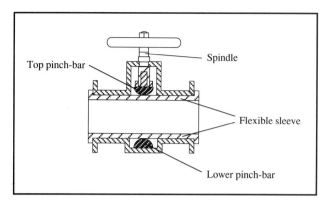

Industrial fermenters are designed for *in situ* steam sterilisation under pressure. The vessel should have a minimum number of internal structures, ports, nozzles, connections and other attachments to ensure that steam reaches all parts of the equipment. For effective sterilisation, all air in the vessel and pipe connections must be displaced by steam. The reactor should be free of crevices and stagnant areas where liquid or solids can accumulate; polished welded joints are used in preference to other coupling methods. Small cracks or gaps in joints and fine fissures in welds are a haven for microbial contaminants and are avoided in fermenter construction whenever possible. After sterilisation, all nutrient medium and air entering the fermenter must be sterile. As soon as flow of steam into the fermenter is stopped, sterile air is introduced to maintain a slight positive pressure in the vessel and discourage

entry of air-borne contaminants. Filters preventing passage of microorganisms are fitted to exhaust-gas lines; this serves to contain the culture inside the fermenter and insures against contamination should there be a drop in operating pressure.

Flow of liquids to and from the fermenter is controlled using valves. Because valves are a potential entry point for contaminants, their construction must be suitable for aseptic operation. Common designs such as simple gate and globe valves have a tendency to leak around the valve stem and accumulate broth solids in the closing mechanism. Although used in the fermentation industry, they are unsuitable if a high level of sterility is required. Pinch and diaphragm valves such as those shown in Figures 13.11 and 13.12 are recommended for fermenter construction. These designs make use of flexible sleeves or diaphragms so that the closing mechanism is isolated from the contents of the pipe and there are no dead spaces in the valve structure. Rubber or neoprene capable of withstanding repeated sterilisation cycles is used to fashion the valve closure; the main drawback is that these components must be checked regularly for wear to avoid valve failure. To minimise costs, ball and plug valves are also used in fermenter construction.

With stirred reactors, another potential entry point for contamination is where the stirrer shaft enters the vessel. The gap between the rotating stirrer shaft and the fermenter body must be sealed; if the fermenter is operated for long periods, wear at the seal opens the way for air-borne contaminants. Several types of stirrer seal have been developed to prevent contamination. On large fermenters, mechanical seals are commonly used [14].

Figure 13.12   Weir-type diaphragm valve in (a) closed and (b) open positions.

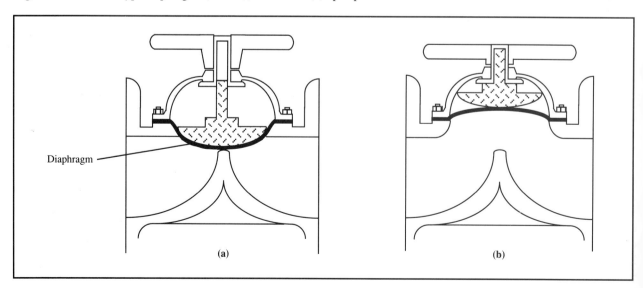

Figure 13.13 Pipe and valve connections for aseptic transfer of inoculum to a large-scale fermenter. (From A. Parker, 1958, Sterilization of equipment, air and media. In: R. Steel, Ed, *Biochemical Engineering*, pp. 97–121, Heywood, London.)

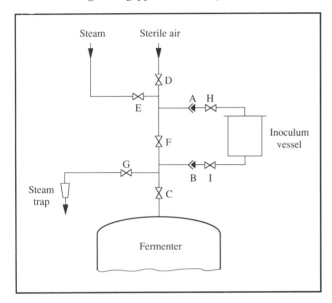

transfer method is to pressurise the inoculum vessel using sterile air; culture is then effectively blown into the larger fermenter. An example of the pipe and valve connections required for this type of transfer is shown in Figure 13.13. The fermenter and its piping and the inoculum tank and its piping including valves H and I are sterilised separately before culture is added to the inoculum tank. With valves H and I closed, the small vessel is joined to the fermenter at connections A and B. Because these connectors were open prior to being joined, they must be sterilised before the inoculum tank is opened. With valves D, H, I and C closed and A and B slightly open, steam flows through E, F and G and bleeds slowly from A and B. After about 20 minutes' steam sterilisation, valves E and G and connectors A and B are closed; the route from inoculum tank to fermenter is now sterile. Valves D and C are opened for flow of sterile air into the fermenter to cool the line under positive pressure. Valve F is then closed, valves H and I are opened and sterile air is used to force the contents of the inoculum tank into the fermenter. The line between the vessels is emptied of most residual liquid by blowing through with sterile air. Valves D, C, H and I are then closed to isolate both the fermenter and the empty inoculum tank which can now be disconnected at A and B.

Sampling ports are fitted to fermenters to allow removal of broth for analysis. An arrangement for sampling which preserves aseptic operation is shown in Figure 13.14. Initially, valves A and D are closed; valves B and C are open to maintain a steam barrier between the reactor and the outside environment. Valve C is then closed, valve B partially closed and valve D partially opened to allow steam and condensate to bleed from the sampling port D. For sampling, A is opened briefly to cool the pipe and carry away any condensate that would dilute the sample; this broth is discarded. Valve B is then closed and a sample collected through D. When sampling is complete, valve A is closed and B opened for re-sterilisation of the sample line; this prevents any contaminants which entered while D was open from travelling up to the fermenter. Valve D is then closed and valve C re-opened.

13.3.3 Materials of Construction

Fermenters are constructed from materials that can withstand repeated steam sterilisation and cleaning cycles. Materials contacting the fermentation medium and broth should also be non-reactive and non-absorptive. Glass is used to construct fermenters up to about 30 litres capacity. The advantages of glass are that it is smooth, non-toxic, corrosion-proof and transparent for easy inspection of the vessel contents. Because entry ports are required for medium, inoculum, air and instruments such as pH

In these devices, one part of the assembly is stationary while the other rotates on the shaft; the precision-machined surfaces of the two components are pressed together by springs or expanding bellows and cooled and lubricated with water. Mechanical seals with running surfaces of silicon carbide paired with tungsten carbide are often specified for fermenter application. Stirrer seals are especially critical if the reactor is designed with a bottom-entering stirrer; double mechanical seals may be installed to prevent fluid leakage. On smaller vessels, magnetic drives can be used to couple the stirrer shaft with the motor; with these devices, the shaft does not pierce the fermenter body. A magnet in a housing on the outside of the fermenter is driven by the stirrer motor; inside, another magnet is attached to the end of the stirrer shaft and held in place by bearings. Sufficient power can be transmitted using magnetic drives to agitate vessels up to at least 800 litres in size [15]. However, the suitability of magnetic drives for viscous broths, especially when high oxygen-transfer rates are required, is limited.

13.3.2 Fermenter Inoculation and Sampling

Consideration must be given in design of fermenters for aseptic inoculation and sample removal. Inocula for large-scale fermentations are transferred from smaller reactors; to prevent contamination during this operation, both vessels are maintained under positive air pressure. The simplest aseptic

Figure 13.14 Pipe and valve connections for a simple fermenter sampling port.

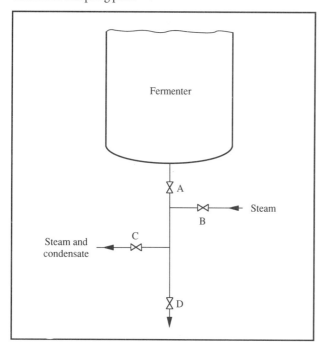

and temperature sensors, glass fermenters are usually equipped with stainless steel headplates containing many screw fittings.

Most pilot- and large-scale fermenters are made of corrosion-resistant stainless steel, although mild steel with stainless steel cladding has also been used. Cheaper grades of stainless steel may be used for the jacket and other surfaces isolated from the broth. Copper and copper-containing materials must be avoided in all parts of the fermenter contacting the culture because of its toxic effect on cells. Interior steel surfaces are polished to a bright 'mirror' finish to facilitate cleaning and sterilisation of the reactor; welds on the interior of the vessel are ground flush before polishing. Electropolishing is preferred over mechanical polishing which leaves tiny ridges and grooves in the metal to accumulate dirt and microorganisms.

13.3.4 Sparger Design

The sparger, impeller and baffles determine the effectiveness of mixing and oxygen mass transfer in stirred bioreactors. Design of impellers and baffles is discussed in Section 7.9. Three types of sparger are commonly used in bioreactors: porous, orifice and nozzle. *Porous spargers* of sintered metal, glass or ceramic are used mainly in small-scale applications; gas throughput is limited because the sparger poses a high resistance to flow. Cells growing through the fine holes and

blocking the sparger can also be a problem. *Orifice spargers*, also known as *perforated pipes*, are constructed by making small holes in piping which is then fashioned into a ring or cross and placed at the base of the reactor; individual holes must be large enough to minimise blockages. Orifice spargers have been used to a limited extent for production of yeast and single-cell protein and in waste treatment. *Nozzle spargers* are used in many agitated fermenters from laboratory to production scale. These spargers consist of a single open or partially-closed pipe providing a stream of air bubbles; advantages compared with other sparger designs include low resistance to gas flow and small risk of blockage. Other sparger designs have also been developed. In *two-phase ejector–injectors*, gas and liquid are pumped concurrently through a nozzle to produce tiny bubbles; in *combined sparger–agitator* designs for smaller fermenters, a hollow stirrer-shaft is used for delivery of air. Irrespective of sparger design, provision should be made for in-place cleaning of the interior of the pipe.

13.3.5 Evaporation Control

Aerobic cultures are continuously sparged with air; however, most components of air are inert and leave directly through the exhaust gas line. If air entering the fermenter is dry, water is continually stripped from the medium and leaves the system as vapour. Over a period of days, evaporative water loss can be significant. This problem is more pronounced in air-driven reactors because the gas flow rates required for good mixing and mass transfer are generally higher than in stirred reactors.

To combat evaporation problems, air sparged into fermenters may be pre-humidified by bubbling through columns of water outside the fermenter; humid air entering the fermenter has less capacity for evaporation than dry air. Fermenters are also equipped with water-cooled condensers to return to the broth any vapours carried by the exit gas. Evaporation can be a particular problem when products or substrates are more volatile than water. For example, *Acetobacter* species are used to produce acetic acid from ethanol in a highly aerobic process requiring large quantities of air. It has been reported for stirred-tank reactors operated at air flow rates between 0.5 and 1.0 vvm (vvm means volume of gas per volume of liquid per minute) that from a starting alcohol concentration of 5%, 30–50% of the substrate is lost within 48 hours due to evaporation [16].

13.4 Monitoring and Control of Bioreactors

The environment inside bioreactors should allow optimal catalytic activity. Parameters such as temperature, pH, dissolved-oxygen concentration, medium flow rate, stirrer speed and sparging rate have a significant effect on the outcome of fermentation and enzyme reactions. To provide the desired environment, system properties must be monitored and control action taken to rectify any deviations from the desired values. Fermentation monitoring and control is an active area of research aimed at improving the performance of bioprocesses and achieving uniform and reliable fermenter operation.

Various levels of process control exist in the fermentation industry. Manual control is the simplest, requiring a human operator to manipulate devices such as pumps, motors and valves. Automatic feedback control is used to maintain parameters at specified values. With increasing application of computers in the fermentation industry, there is also scope for implementing advanced control and optimisation strategies based on fermentation models.

13.4.1 Fermentation Monitoring

Any attempt to understand or control the state of a fermentation depends on knowledge of critical variables which affect the process. These parameters can be grouped into three categories: physical, chemical and biological. Examples of process variables in each group are given in Table 13.1. Many of the physical measurements listed are well established in the fermentation industry; others are currently being developed or are the focus of research into new instrumentation.

Despite the importance of fermentation monitoring, industrially-reliable instruments and sensors capable of rapid, accurate and direct measurements are not available for many process variables. For effective control of fermentations based on measured data, the time taken to complete the measure-

ment should be compatible with the rate of change of the variable being monitored. For example, in a typical fermentation, the time scale for change in pH and dissolved-oxygen tension is several minutes, while the time scale for change in culture fluorescence is less than 1 second. For other variables such as biomass concentration, an hour or more may pass before measurable changes occur. The frequency and speed of measurement must be consistent with these time scales. Ideally, measurements should be made *in situ* and on-line, i.e. in or near the reactor during operation, so that the result is available for timely control action. Many important variables such as biomass concentration and broth composition cannot currently be measured on-line because of the lack of appropriate instruments. Instead, samples must be removed from the reactor and taken to the laboratory for off-line analysis. Because fermentation conditions can change during laboratory analysis, control action based on the measurement is not as effective. Off-line measurements are used in industrial fermentations for analysis of biomass, carbohydrate, protein, phosphate and lipid concentrations, enzyme activity and broth rheology. Samples are usually taken manually every 4–8 h; the results are available 2–24 h later.

Examples of measurements which can be made on-line in industry are temperature, pressure, pH, dissolved-oxygen tension, flow rate, stirrer speed, power consumption, foam level, broth weight and gas composition. Instruments for taking these measurements are relatively commonplace; detailed descriptions can be found elsewhere [17–19]. The availability of an on-line measurement or its use in the laboratory does not necessarily mean it is applied in commercial-scale processing. Owing to the cost of installation and the financial consequences of instrument failure during fermentation, measurement devices used in industry must meet stringent performance

Table 13.1   Parameters measured or controlled in bioreactors

Physical	*Chemical*	*Biological*
Temperature	pH	Biomass concentration
Pressure	Dissolved O_2	Enzyme concentration
Reactor weight	Dissolved CO_2	Biomass composition
Liquid level	Redox potential	(such as DNA, RNA,
Foam level	Exit gas composition	protein, ATP/ADP/AMP,
Agitator speed	Conductivity	NAD/NADH levels)
Power consumption	Broth composition	Viability
Gas flow rate	(substrate, product, ion	Morphology
Medium flow rate	concentrations, etc.)	
Culture viscosity		
Gas hold-up		

Figure 13.15 Operating principle for biosensors. (From E.A.H. Hall, 1991, *Biosensors*, Prentice-Hall, New Jersey.)

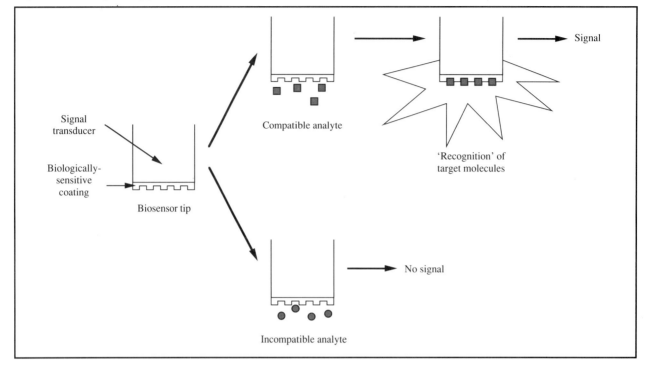

criteria. These include: accuracy to within 1–2% of full scale, reliable operation for at least 80% of the time, low maintenance needs, steam sterilisability, simple and fast calibration, and maximum drift of less than 1–2% of full scale [1]. Sterilisable pH probes have been proven reliable if properly grounded and are used widely in industrial fermentations; their low failure rate is assisted by replacement every four or five fermentations. However, a variety of problems is associated in the industrial environment with electrodes for dissolved carbon dioxide, redox potential and specific ions; these instruments are therefore confined mostly to laboratory applications. Drift and interference from air bubbles are well-known limitations with dissolved-oxygen probes in industrial fermenters; deposits and microbial fouling of the membrane also reduce measurement reliability. Even if electrodes are steam sterilisable, probe life can be significantly reduced by repeated sterilisation cycles.

Development of new on-line measurements for chemical and biological variables is a challenging area of bioprocess engineering research. One of the areas targeted is *biosensors* [20, 21]. There are many different biosensor designs; however, in general, biosensors incorporate a biological sensing element in close proximity or integrated with a signal transducer. As illustrated schematically in Figure 13.15, the binding specificity of particular biological molecules immobilised on the probe is used to 'sense' or recognise target species in the fermentation broth; ultimately, an electrical signal is generated after the interaction between sensing molecules and analyte is detected by the transducer. Biosensors which use immobilised cells to recognise broth components are also being developed. Biosensors with enzyme, antibody and cell sensing-elements have been tested for a wide range of substrates and products including glucose, sucrose, ethanol, acetic acid, ammonia, penicillin, urea, riboflavin, cholesterol and amino acids. However, although several biosensors are available commercially, they are not used for routine monitoring of large-scale fermentations because of intractable problems with robustness, sterilisation, limited life-span and long-term stability. At this stage, because of the sensitivity of their biological components, biosensors appear to be more suitable for medical diagnostics and environmental analyses than *in situ* bioprocess applications. Nevertheless, biosensors may be used for rapid analysis of broth samples removed from fermenters.

Another approach to on-line chemical and biological measurements involves use of automatic sampling devices linked to analytical equipment for high-performance liquid chromatography (HPLC), image analysis, NMR, flow cytometry or fluorometry. For these techniques to work, methods must be available for on-line aseptic sampling and analysis without blockage or interference from gas bubbles and cell solids.

Development in this area has centred around *flow injection analysis* (*FIA*), a sample-handling technique for removing cell-free medium from the fermenter and delivering a pulse of analyte to *ex situ* measurement devices. To date, routine on-line measurement of the most fundamental fermentation parameter, biomass concentration, has not been achieved. Several procedures based on culture turbidity, light scattering, fluorescence, calorimetry, piezoelectric membranes, radio-frequency dielectric permittivity and acoustics have been developed; however, they have not proved sufficiently accurate and reliable under industrial conditions. Air bubbles and background particles readily interfere with optical readings, while the relationship between biomass density and other culture properties such as fluorescence is affected by pH, dissolved-oxygen tension and substrate levels [19]. Other difficulties have been encountered in development of automatic sampling devices; the most important of these is the high risk of contamination and blockage. Further details about on-line measurement techniques can be found elsewhere [22, 23].

In a large fermentation factory, thousands of different vari-ables may be monitored at any given time. The traditional device for recording on-line process information is the chart recorder; however, with increasing application of computers in the fermentation industry, digital data logging is widespread. Because of the enormous quantity of information obtained from continuous monitoring of bioprocesses, an increasing problem in industry is the management and effective utilisation of gigabytes of data.

13.4.2 Measurement Analysis

Any attempt to analyse or apply the results of fermentation monitoring must consider the errors and spurious or transient results incorporated into the data. Noise and variability are particular problems with certain fermentation measurements; for example, probes used for pH and dissolved-oxygen measurements are exposed to rapid fluctuations and heterogeneities in the broth so that noise can seriously affect the accuracy of point readings. Figure 13.16 shows typical results from on-line measurement of dilution rate and carbon dioxide

Figure 13.16　On-line measurements of dilution rate and off-gas carbon dioxide during an industrial mycelial fermentation. (From G.A. Montague, A.J. Morris and A.C. Ward, 1989, Fermentation monitoring and control: a perspective. *Biotechnol. Genet. Eng. Rev.* 7, 147–188.)

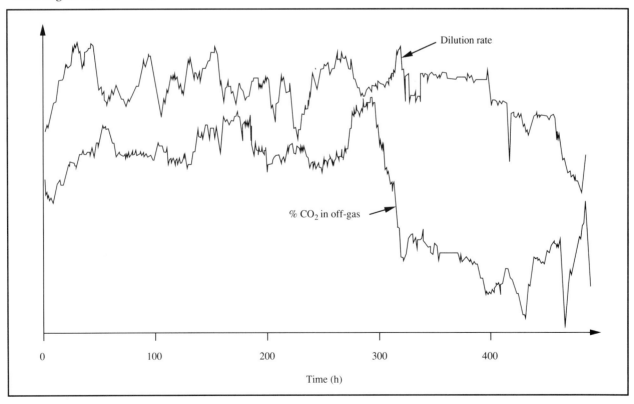

evolution in a production-scale fermenter; in many cases, *signal conditioning* or *smoothing* must be carried out to reduce the noise in these data before they can be applied for process control or modelling. Most modern data acquisition and logging systems contain facilities for signal conditioning. Unwanted pseudo-random noise can be filtered out using analogue filter circuits or by averaging values over successive measurements. Alternatively, unfiltered signals can be digitised and filtering algorithms applied using computer software. Measurement drift cannot be corrected using electronic circuitry; instruments must be periodically recalibrated during long fermentations to avoid loss of accuracy.

Raw data are sometimes used to calculate *derived variables* which characterise the performance of the fermenter. The most common derived variables are oxygen-uptake rate, rate of carbon dioxide evolution, the respiratory quotient *RQ*, and the mass-transfer coefficient $k_L a$. As oxygen-uptake rate is usually calculated from the difference between two quantities of similar magnitude (gas inlet and outlet oxygen levels; see Section 9.10.1), noise in this variable can be significant [24] and affect the quality of other dependent variables such as $k_L a$ and *RQ*.

13.4.3 Fault Analysis

Faults in reactor operation affect 15–20% of fermentations [25]. Fermentation measurements can be used to detect faults; for example, signals from a flow sensor could be used to detect blockages in a pipe and trigger an alarm in the factory control room. Normally however, the sensors themselves are the most likely components to fail; rates of failure of some fermentation instruments are listed in Table 13.2. Failure of a sensor might

be detected as an unexpected change or rate of change in its signal, or a change in the noise characteristics. Several approaches can be used to reduce the impact of faults on large-scale fermentations; these include comparison of current measurements with historical values, cross-checking between independent measurements, using multiple and back-up sensors, and building hardware redundancy into computer systems and power supplies.

13.4.4 Process Modelling

In modern approaches to fermentation control, a reasonably accurate mathematical model of the reaction and reactor environment is required. Using process models, we can progress beyond environmental control of bioreactors into the realm of direct biological control. Development of fermentation models is aided by information from measurements taken during process operation.

Models are mathematical relationships between variables. Traditionally, models are based on a combination of 'theoretical' relationships which provide the structure of the model, and experimental observations which provide the numerical values of coefficients. For biological processes, specifying the model structure can be difficult because of the complexity of cellular processes and the large number of environmental factors which affect cell culture. Usually, bioprocess models are much-simplified approximate representations deduced from observation rather than from theoretical laws of science. As an example, a frequently-used mathematical model for batch fermentation consists of the Monod equation for growth and an expression for rate of substrate consumption as a function of biomass concentration:

Table 13.2   Reliability of fermentation equipment

(From S.W. Carleysmith, 1987, Monitoring of bioprocessing, In: J.R. Leigh, Ed, Modelling and Control of Fermentation Processes, pp. 97–117, Peter Peregrinus, London; and P.N. Royce, 1993, A discussion of recent developments in fermentation monitoring and control from a practical perspective, Crit. Rev. Biotechnol. 13, 117–149)

Equipment	Reliability	Mean time between failures (weeks)
Temperature probe	–	150–200
pH probe	98%	9–48
Dissolved-oxygen probe	50–80%	9–20
Mass spectrometer	–	10–50
Paramagnetic O_2 analyser	–	24
Infrared CO_2 analyser	–	52
Computer system	–	4

$$\frac{dx}{dt} = \mu x = \frac{\mu_{max}\, s\, x}{K_S + s} \tag{13.5}$$

$$\frac{-ds}{dt} = \frac{\mu x}{Y_{XS}} + m_S x. \tag{13.6}$$

This model represents a combination of Eqs (11.52), (11.60) and (11.72). The form of the equations was determined from experimental observation of a large number of different culture systems. In principle, once values of the parameters μ_{max}, K_S, Y_{XS} and m_S have been determined, we can use the model to estimate cell concentration x and substrate concentration s as a function of time. A common problem with fermentation models is that the model parameters can be difficult to measure or change with time.

We have already encountered several model equations in this book; examples include the kinetic, yield and maintenance relationships introduced in Chapter 11 and the stoichiometric equations of Chapter 4 relating masses of substrates, biomass and products during reaction. Other models were derived in Chapter 12 for heterogeneous reactions; dependence of culture parameters such as cell concentration on physical conditions in the bioreactor are represented in simple form by the equations of Sections 8.6.1 and 9.5.2. Process models vary in form but have the unifying feature that they predict outputs from a set of inputs. When models are used for fermentation control, they are usually based on mass- and energy-balance equations for the system.

Development of a comprehensive model covering all key aspects of a particular bioprocess and able to predict the effects of a wide range of culture variables is a demanding exercise. Accurate models applicable to a range of process conditions are rare. As many aspects of fermentation are poorly understood, it difficult to devise mathematical models covering these areas. For example, the response of cells to spatial variations in dissolved-oxygen and substrate levels in fermenters, or the effect of impeller design on microbial growth and productivity, is not generally incorporated into models because the subject has been inadequately studied. Evidence that all important fermentation variables have not yet been identified is the significant batch-to-batch variation that occurs in the fermentation industry.

13.4.5 State Estimation

As described in Section 13.4.1, it is not possible to measure on-line all the key variables or states of a fermentation process.

Often, considerable delays are involved in off-line measurement of important variables such as biomass, substrate and product concentrations. Such delays make effective control of the reactor difficult if control action is dependent on the value of these parameters, but must be undertaken more quickly than off-line analysis allows. One approach to this problem is to use available on-line measurements in conjunction with mathematical models of the process to estimate unknown variables. The computer programs and numerical procedures developed to achieve this are called *software sensors, estimators* or *observers*. The *Kalman filter* is a well-known type of observer applicable to linear process equations; non-linear systems can be treated using the *extended Kalman filter* [26]. The success of Kalman filters and other observers depends largely on the accuracy and robustness of the process model used.

Several techniques for state estimation are available. As a simple example, on-line measurements of carbon dioxide in fermenter off-gas can be applied with an appropriate process model to estimate biomass concentration during penicillin fermentation [27]. As shown in Figure 13.17, the results were satisfactory for 100-m³ fed-batch fermentations over a period of more than 8 days. Direct state estimation can achieve reasonable results as long as the process model remains valid and the estimator is able to reduce measurement noise. However, if

Figure 13.17 Measured and estimated biomass concentrations in a large-scale fed-batch penicillin fermentation. (From G.A. Montague, A.J. Morris and J.R. Bush, 1988, Considerations in control scheme development for fermentation process control. *IEEE Contr. Sys. Mag.* **8**, April, 44–48.)

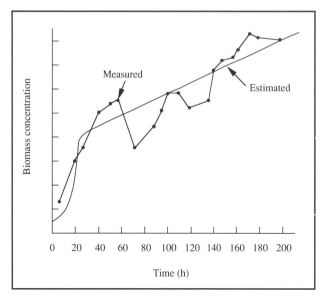

Figure 13.18 Components of a feedback control loop.

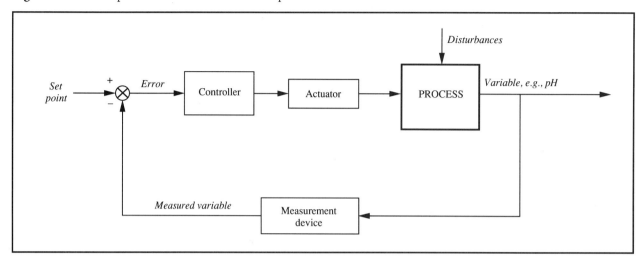

major fluctuations occur in operating conditions or if cell properties and model parameters change with time, the model may become inadequate and estimation accuracy will decline. *Adaptive estimators* are used to adjust faulty model parameters as the fermentation proceeds to alleviate problems caused by error in the equations. Off-line measurements can also be incorporated into the estimation procedure as they become available to improve the accuracy and reliability of prediction. Another technique involves 'generic' software sensors which use generally-structured models rather than model equations specific to the process of interest. Process characteristics are then 'learnt' or incorporated into the model structure as information becomes available from on- and off-line measurements. The primary advantage of generic sensors is that the software-development time is significantly reduced [28].

Mass-balancing techniques are another approach to state estimation. As described in Chapter 4, biomass can be estimated from stoichiometry and other process measurements using elemental balances. This method is suitable for fermentation of defined media, but is difficult to apply with complex medium ingredients such as molasses, casein hydrolysate, soybean meal and corn steep liquor which have undefined elemental composition. Another disadvantage is that the accuracy of biomass estimation often depends on measurement of substrate or product concentration which must be done off-line.

13.4.6 Feedback Control

Let us assume we wish to maintain the pH in a bioreactor at a constant value against a variety of disturbances, for example,

from metabolic activity. One of the simplest control schemes is a conventional *feedback control loop*, the basic elements of which are shown in Figure 13.18. A measurement device senses the value of the pH and sends the signal to a controller. At the controller, the measured value is compared with the desired value known as the *set point*. The deviation between measured and desired values is the *error*, which is used by the controller to determine what action must be taken to correct the process. The controller may be a person who monitors the process measurements and decides what to do; more often the controller is an automatic electronic, pneumatic or computer device. The controller produces a signal which is transmitted to the actuator, which executes the control action. In a typical system for pH control, an electrode would serve as the measurement device and a pump connected to a reservoir of acid or alkali as the actuator. Simple *on–off control* is generally sufficient for pH; if the measured pH falls below the set point, the controller switches on the pump which adds alkali to the fermenter. When enough alkali is added and the pH returned to the set value, the pump is switched off. Small deviations from the set point are usually tolerated in on–off control to avoid rapid switching and problems due to measurement delay.

On–off control is used when the actuator is an on–off device, such as a single-speed pump. If function of the actuator covers a continuous range, such as a variable-speed pump for supply of cooling water or a valve determining the rate of air flow, it is common to use *proportional-integral-derivative* or *PID control*. With PID control, the control action is determined in proportion to the error, the integral of the error and the derivative of the error with respect to time. The relative weightings given to these functions determine the response of he controller and the overall 'strength' of the control action. PID control is used to determine, for example, whether the pump speed for cooling water in the coils of a fermenter should be increased by a small or large amount when a certain increase in reactor temperature is detected above the set point. Proper adjustment of PID controllers usually provides excellent regulation of the measured variable. However, poorly tuned controllers can destabilise a process and cause continuous or accentuated fluctuations in culture conditions. Considerations for adjusting PID controllers are covered in texts on process control, for example [29].

Close fermentation control requires simultaneous monitoring and adjustment of many parameters. Instead of individual controllers for each function, it is becoming commonplace to use a single computer or microprocessor for several feedback control loops. The computer logs measurements from a range of sensors in a time sequence and generates electronic signals which may be used directly or indirectly to drive various actuators. Application of computers requires digitisation of signals from the sensor; after digital-to-analogue conversion of the output, the computer or microprocessor can provide the same PID functions as a conventional analogue controller. If computers are used to drive the actuator devices, the system is said to be under *direct digital control* (DDC).

13.4.7 Indirect Metabolic Control

Maintaining particular values of temperature or pH is a rather indirect approach to bioreactor control; the wider objective is to optimise performance of the catalyst and maximise production of the desired product. Certain derived variables such as the oxygen-uptake rate and respiratory quotient can be calculated from on-line fermentation measurements; these variables reflect to some extent the biological state of the culture. It can be advantageous to base control actions on deviation of these metabolic variables from desired values rather than on environmental conditions.

Indirect metabolic control is often used in fed-batch culture of bakers' yeast. Due to the Crabtree effect, yeast metabolism can switch from respiratory to fermentative mode depending on prevailing glucose and dissolved-oxygen concentrations. Maximum biomass yield from substrate is achieved at relatively low glucose concentrations in the presence of adequate oxygen; fermentative metabolism occurs if the glucose concentration rises above a certain level even though oxygen may be present. Fermentative metabolism should be avoided for biomass production because the yield of cells is reduced as ethanol and carbon dioxide are formed as end-products. The respiratory quotient RQ is a convenient indicator of metabolic state in this process; RQ values above about 1 indicate ethanol formation. In industrial bakers' yeast production, the feed rate of glucose to the culture is controlled to maintain RQ values within the desired range.

13.4.8 Programmed Control

Because of the inherent time-varying character of batch and fed-batch fermentations, maintaining a constant environment or constant values of metabolic variables is not always the optimal control strategy. Depending on the process, changes in variables such as pH and temperature at critical times can improve production rate and yield. Varying the rate of the feed is important in fed-batch bakers' yeast fermentations to minimise the Crabtree effect and maximise biomass production. Feed rate is also manipulated in *E. coli* fermentations to reduce by-product synthesis. In secondary-metabolite fermentations, specific growth rate should be high at the start of the culture but, at high cell densities, different conditions are required to slow growth and stimulate product formation. Similar strategies are needed to optimise protein synthesis from recombinant organisms. Expression of recombinant product is usually avoided at the start of the culture because cell growth is adversely affected; however, later in the batch an inducer is added to switch on protein synthesis.

For many bioprocesses, a particular time sequence of pH, temperature, dissolved-oxygen tension, feed rate and other variables is required to develop the culture in such a way that productivity is maximised. A control strategy that can accommodate wide-ranging changes in fermentation variables is *programmed control*, also known as *batch fermenter scheduling*. In programmed control, the control policy consists of a schedule of control functions to be implemented at various times during the process. This type of control requires a detailed understanding of the requirements of the process at various stages and a reasonably complete and accurate mathematical model of the system.

13.4.9 Application of Artificial Intelligence in Bioprocess Control

Several different approaches to bioprocess control have been described in the preceding sections. However, fermentation systems are by nature multivariable with non-linear and time-varying properties, and conventional control strategies may not be totally satisfactory. In the industrial environment, additional problems are caused by unexpected disturbances which affect process operation. Most bioprocesses cannot be described exactly by a mathematical model; it is also difficult to identify beforehand the optimum values of metabolic or environmental parameters and controller response functions. The flexibility of computer software is a significant advantage in complex situations; researchers continue to make progress in developing optimal, adaptive and self-tuning algorithms for bioprocess control. However, although these techniques offer much for improving process performance, the usual mathematical approach may not be the best way to solve bioprocess control problems.

A relatively recent development in control engineering is application of artificial intelligence techniques, especially *knowledge-based expert systems*. The term 'expert system' usually refers to computer software that processes linguistically-formulated 'knowledge' about a particular subject; this knowledge is represented by simple rules. The most important step in building a useful expert system is extraction of heuristic or subjective rules of thumb from experimental data and human experts. Use can be made of the ever-increasing quantity of measured fermentation data from industrial processes to synthesise the wide range of rules required for expert systems. The knowledge available is encoded using a computer as IF/THEN type rules, e.g. IF the cell density is high, THEN dilution with sterile water is recommended, or IF the dissolved-oxygen concentration increases quickly AND the rate of carbon dioxide evolution decreases quickly AND the sugar concentration drops AND the pH increases, THEN broth harvest is advised. Information about the progress of highly productive fermentations is also stored in the knowledge base. As measured data become available from fermentations in progress, pattern recognition techniques are applied to assess the results and, in conjunction with the knowledge base and its rules, handle a variety of operating problems or disturbances. The expert system can also 'learn' new information about process behaviour by upgrading its knowledge base. To maximise the potential of expert systems for intelligent supervisory control of fermentation processes, large and representative data bases of microbiological information and engineering knowledge must be established for use in rule formulation and interpretation of process phenomena. Expert systems can also be used for fault diagnosis, estimating unmeasurable fermentation properties, reconciling contradictory data and computer-aided modelling of metabolism [30–33].

Another area of artificial intelligence with applications in fermentation control is the theory of *neural networks*. Neural networks are particularly suited for extracting useful information from complex and uncertain data such as fermentation measurements, and for formulating generalisations from previous experience. They offer the ability to learn complex, non-linear relationships between variables and may therefore be useful in development of process models and for estimating unknown fermentation parameters. Neural network technology is based on an analogy with the brain in that information is stored in the form of connected computational elements or weights (synapses) between artificial neurons. The most commonly-used neural structure is the feed-forward network in which neurons are arranged in layers; incoming signals at the input layer are fed forward through the network connections to the output layer. The topology of the network provides it with powerful data processing capabilities. To solve a problem, a network structure is chosen and examples of the knowledge to be acquired are shown to the network which adjusts the synaptic strength of its neural connections so that, in effect, the knowledge is integrated within the structure. When the real data set is presented to the system, the network is able to predict outcomes based on the learning set. For example, information about feed flow rate and substrate and biomass concentrations as a function of time can be used to develop a neural network for analysis of transient behaviour in continuous fermentation which is able to predict future changes in substrate and cell concentrations [34–36]. Although a considerable amount of research remains to be done in this area, neural networks are a promising tool for modelling, estimating and predicting fermentation characteristics.

13.5 Ideal Reactor Operation

So far in this chapter we have considered the configuration of bioreactors and aspects of their construction and control. Another important factor affecting reactor performance is mode of operation. There are three principal modes of bioreactor operation: batch, fed-batch and continuous. Choice of operating strategy has a significant effect on substrate conversion, product concentration, susceptibility to contamination and process reliability.

Characteristics such as final substrate, product and biomass concentrations and the time required for conversion can be determined for different reactor operating schemes using mass balances. For a general reaction system, we can relate rates of change of component masses in the system to the rate of reaction using Eq. (6.5):

$$\frac{dM}{dt} = \hat{M}_i - \hat{M}_o + R_G - R_C$$

$$(6.5)$$

where M is mass of component A in the vessel, t is time, $\hat{M}_i$ is the mass flow rate of A entering the reactor, $\hat{M}_o$ is the mass flow rate of A leaving, R_G is the mass rate of generation of A by reaction, and R_C is the mass rate of consumption of A by reaction. In this section we consider application of Eq. (6.5) to batch, ed-batch and continuous reactors for enzyme conversion and ermentation.

3.5.1 Batch Operation of a Mixed Reactor

Batch processes operate in closed systems; substrate is added at the beginning of the process and products removed only at the end. Aerobic reactions are not batch operations in the strictest sense; the low solubility of oxygen in aqueous media means that it must be supplied continuously while carbon dioxide and other off-gases are removed. However, reactors with either input nor output of liquid are classified as batch. If there are no leaks or evaporation from the vessel, the liquid volume in batch reactors can be considered constant.

Most commercial bioreactors are mixed vessels operated in batch. The classic mixed reactor is the stirred tank; however mixed reactors can also be of bubble column, airlift or other configuration as long as concentrations of substrate, product and catalyst inside the vessel are uniform. The cost of running batch reactor depends on the time taken to achieve the desired product concentration or level of substrate conversion; operating costs are reduced if the reaction is completed quickly. It is therefore useful to be able to predict the time required for batch reactions.

3.5.1.1 *Enzyme reaction*

Let us apply Eq. (6.5) to the limiting substrate in a batch enzyme reactor such as that shown in Figure 13.19. $\hat{M}_i = \hat{M}_o$ = 0 because there is no substrate flow into or out of the vessel; mass of substrate in the reactor, M, is equal to the substrate concentration s multiplied by the liquid volume V. As substrate is not generated in the reaction, $R_G = 0$. Rate of substrate consumption R_C is equal to the volumetric rate of reaction v multiplied by V; v is given by Eq. (11.31). Therefore, the mass balance from Eq. (6.5) is:

$$\frac{d(sV)}{dt} = \frac{-v_{max}\,s}{K_m + s}\,V \tag{13.7}$$

where v_{max} is the maximum rate of enzyme reaction and K_m is the Michaelis constant. Because V is constant in batch reactors, we can take it outside of the differential and cancel through the equation to give:

$$\frac{ds}{dt} = \frac{-v_{max}\,s}{K_m + s} \tag{13.8}$$

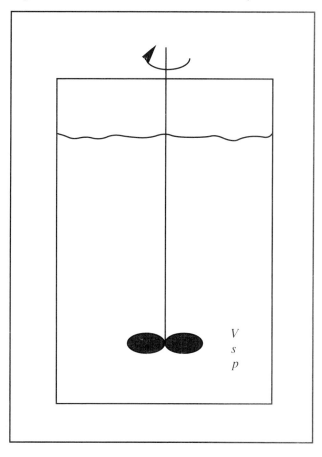

Figure 13.19   Flowsheet for a stirred batch enzyme-reactor.

Integration of this differential equation provides an expression for batch reaction time. Assuming v_{max} and K_m are constant during the reaction, separating variables:

$$-\int dt = \int \frac{K_m + s}{v_{max}\,s}\,ds \tag{13.9}$$

and integrating with initial condition $s = s_0$ at $t = 0$ gives:

$$t_b = \frac{K_m}{v_{max}}\,\ln\frac{s_0}{s_f} + \frac{s_0 - s_f}{v_{max}} \tag{13.10}$$

where t_b is the batch reaction time required to reduce the substrate concentration from s_0 to s_f. The batch time required to produce a certain concentration of product can be determined from Eq. (13.10) and stoichiometric relationships.

Example 13.1 Time course for batch enzyme conversion

An enzyme is used to produce a compound used in manufacture of sunscreen lotion. v_{max} for the enzyme is 2.5 mmol m^{-3} s^{-1} K_m is 8.9 mM. The initial concentration of substrate is 12 mM. Plot the time required for batch reaction as a function of substrate conversion.

Solution:

s_0 = 12 mM. Converting units of v_{max} to mM h^{-1}:

$$v_{max} = 2.5 \, \text{mmol m}^{-3} \, \text{s}^{-1} \cdot \left| \frac{3600 \, \text{s}}{1 \, \text{h}} \right| \cdot \left| \frac{1 \, \text{m}^3}{1000 \, \text{l}} \right|$$

$$= 9 \, \text{mmol l}^{-1} \, \text{h}^{-1}$$
$$= 9 \, \text{mM h}^{-1}.$$

Results from application of Eq. (13.10) are tabulated below and plotted in Figure 13E1.1.

Substrate conversion (%)	s_f (mM)	t_b (h)
0	12.0	0.00
10	10.8	0.24
20	9.6	0.49
40	7.2	1.04
50	6.0	1.35
60	4.8	1.71
80	2.4	2.66
90	1.2	3.48
95	0.60	4.23
99	0.12	5.87

Figure 13E1.1 Batch reaction time as a function of substrate conversion for a mixed enzyme reactor.

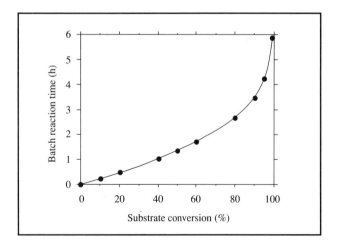

At high substrate conversions, the time required to achieve an incremental increase in conversion is greater than at low conversions. Accordingly, the benefit gained from conversions greater than 80–90% must be weighed against the significantly greater reaction time and reactor operating costs involved.

As discussed in Section 11.5, enzymes are subject to deactivation. Thus, the concentration of active enzyme in the reactor, and therefore the value of v_{max}, may change during reaction. When deactivation is significant, variation of v_{max} with time can be expressed using Eq. (11.44) so that Eq. (13.8) becomes:

$$\frac{ds}{dt} = \frac{-v_{max0}e^{-k_d t}s}{K_m + s}$$

(13.11)

where v_{max0} is the value of v_{max} before deactivation occurs and k_d is the first-order deactivation rate constant. Separating variables gives:

$$-\int e^{-k_d t}\,dt = \int \frac{K_m + s}{v_{max0}\,s}\,ds.$$

(13.12)

Integrating Eq. (13.12) with initial condition $s = s_0$ at $t = 0$ gives:

$$t_b = \frac{-1}{k_d}\ \ln\left[1 - k_d\left(\frac{K_m}{v_{max0}}\ \ln\frac{s_0}{s_f} + \frac{s_0 - s_f}{v_{max0}}\right)\right]$$

(13.13)

where t_b is the batch reaction time and s_f is the final substrate concentration.

Example 13.2 Batch reaction time with enzyme deactivation

The enzyme of Example 13.1 deactivates with half-life 4.4 h. Compare with Figure 13E1.1 the batch reaction time required to achieve 90% substrate conversion.

Solution:

$s_0 = 12$ mM; $v_{max0} = 9$ mM h^{-1}; $K_m = 8.9$ mM. $s_f = (0.1\ s_0) = 1.2$ mM. The deactivation rate constant is calculated from the half-life t_h using Eq. (11.45):

$$k_d = \frac{\ln 2}{t_h} = \frac{\ln 2}{4.4\ h} = 0.158\ h^{-1}.$$

Substituting values into Eq. (13.13) gives:

$$t_b = \frac{-1}{0.158\ h^{-1}}\ \ln\left[1 - 0.158\ h^{-1}\left(\frac{8.9\ mM}{9\ mM\ h^{-1}}\ \ln\frac{12\ mM}{1.2\ mM} + \frac{(12 - 1.2)\ mM}{9\ mM\ h^{-1}}\right)\right]$$

$$= 5.0\ h.$$

With enzyme deactivation, the time required for 90% conversion increases from 3.5 h to 5.0 h.

For reactions with immobilised enzymes, Eq. (13.8) must be modified to account for mass-transfer effects:

$$\frac{ds}{dt} = -\eta_T\ \frac{v_{max}\,s}{K_m + s}$$

(13.14)

where η_T is the total effectiveness factor incorporating internal and external mass-transfer limitations (see Section 12.5), s is the bulk substrate concentration, and v_{max} and K_m are the intrinsic kinetic parameters. In principal, integration of

Eq. (13.14) allows evaluation of t_b; however, because η_T is a function of s, integration is not straightforward.

13.5.1.2 *Cell culture*

Similar analysis can be applied to fermentation processes to evaluate reaction times. Let us perform a mass balance on cells in a well-mixed batch fermenter using Eq. (6.5); a typical flow-sheet for this system is shown in Figure 13.20. $\hat{M}_i = \hat{M}_o = 0$ because cells do not flow into or out of the vessel. Mass of cells in the reactor, M, is equal to the cell concentration x multi-

Figure 13.20   Flowsheet for a stirred batch fermenter.

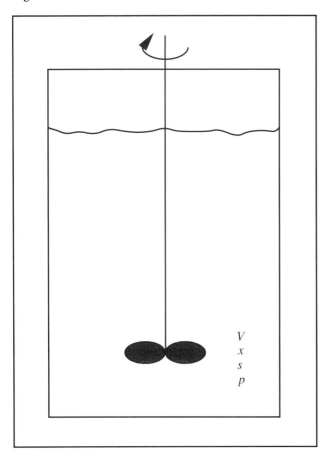

V
x
s
p

plied by the liquid volume V. The mass rate of cell growth R_G is equal to $r_X V$ where r_X is the volumetric rate of growth. From Eq. (11.52), $r_X = \mu x$ where μ is the specific growth rate. If cell death takes place in the reactor alongside growth, $R_C = r_d V$ where r_d is the volumetric rate of cell death. From Eq. (11.85), r_d can be expressed using the first-order equation: $r_d = k_d x$ where k_d is the specific death constant. Therefore, Eq. (6.5) for cells in a batch reactor is:

$$\frac{d(xV)}{dt} = \mu x V - k_d x V.$$

(13.15)

For V constant, Eq. (13.15) becomes:

$$\frac{dx}{dt} = (\mu - k_d)\, x.$$

(13.16)

Because μ in batch culture remains approximately constant and equal to μ_{max} for most of the growth period (see Section 11.7.3) if k_d likewise remains constant, we can integrate Eq (13.16) directly to find the relationship between batch time and cell concentration. Using the initial condition that $x = x_0$ at $t = 0$:

$$x = x_0 e^{(\mu_{max} - k_d)\, t}.$$

(13.17)

If x_f is the final biomass concentration after batch culture time t_b, rearrangement of Eq. (13.17) gives:

$$t_b = \frac{1}{\mu_{max} - k_d} \; \ln \frac{x_f}{x_0}.$$

(13.18)

If the rate of cell death is negligible compared with growth, $k_d \ll \mu_{max}$ and Eqs (13.17) and (13.18) reduce to:

$$x = x_0 e^{\mu_{max} t}$$

(13.19)

and:

$$t_b = \frac{1}{\mu_{max}} \; \ln \frac{x_f}{x_0}.$$

(13.20)

Therefore, we can calculate using Eq. (13.18) or Eq. (13.20) the time required to achieve cell density x_f starting from cell density x_0. Batch culture time can also be related to substrate conversion and product concentration using expressions for rates of substrate uptake and product formation derived in Chapter 11.

Let us apply Eq. (6.5) to the growth-limiting substrate in batch fermentation. $\hat{M}_i = \hat{M}_o = 0$ because substrate does not flow into or out of the reactor; the mass of substrate in the reactor, M, is equal to sV where s is substrate concentration and V liquid volume. Substrate is not generated; therefore $R_G = 0$. R_C is equal to $r_S V$ where r_S is the volumetric rate of substrate uptake. As discussed in Section 11.10, the expression for r_S depends on whether extracellular product is formed by the culture and the relationship between product synthesis and energy generation in the cell. If product is formed but not directly coupled with energy metabolism, r_S is given by Eq. (11.76) and:

$$R_C = r_S V = \left(\frac{\mu}{Y_{XS}} + \frac{q_P}{Y_{PS}} + m_S \right) xV$$

(13.21)

here μ is the specific growth rate, Y_{XS} is the true biomass yield from substrate, q_P is the specific rate of product formation, Y_{PS} is the true product yield from substrate and m_S is the maintenance coefficient. Therefore, from Eq. (6.5) the mass-balance equation is:

$$\frac{d(sV)}{dt} = -\left(\frac{\mu}{Y_{XS}} + \frac{q_P}{Y_{PS}} + m_S \right) xV.$$

(13.22)

For μ equal to μ_{max} and V constant, we can write Eq. (13.22) as:

$$\frac{ds}{dt} = -\left(\frac{\mu_{max}}{Y_{XS}} + \frac{q_P}{Y_{PS}} + m_S \right) x.$$

(13.23)

Because x is a function of time, we must substitute an expression for x into Eq. (13.23) before it can be integrated. Assuming cell death is negligible, x is given by Eq. (13.19). Therefore:

$$\frac{ds}{dt} = -\left(\frac{\mu_{max}}{Y_{XS}} + \frac{q_P}{Y_{PS}} + m_S \right) x_0 e^{\mu_{max}t}.$$

(13.24)

If all the bracketed terms are constant during culture, Eq. (13.24) can be integrated directly with initial condition $s = s_0$ at $t = 0$ to obtain the following equation:

$$t_b = \frac{1}{\mu_{max}} \ln \left[1 + \frac{s_0 - s_f}{\left(\dfrac{1}{Y_{XS}} + \dfrac{q_P}{\mu_{max} Y_{PS}} + \dfrac{m_S}{\mu_{max}} \right) x_0} \right]$$

(13.25)

where t_b is the batch culture time and s_f is the final substrate concentration.

For products indirectly coupled or not related at all to energy metabolism, evaluating q_P requires further analysis (see Sections 11.9.2 and 11.9.3). However, Eq. (13.25) can be simplified if no product is formed or if production is directly linked with energy metabolism; in these cases the expression for rate of substrate consumption does not contain a separate term for product synthesis (see Sections 11.10.1 and 11.10.2) and Eq. (13.25) reduces to:

$$t_b = \frac{1}{\mu_{max}} \ln \left[1 + \frac{s_0 - s_f}{\left(\dfrac{1}{Y_{XS}} + \dfrac{m_S}{\mu_{max}} \right) x_0} \right].$$

(13.26)

If, in addition, maintenance requirements can be neglected:

$$t_b = \frac{1}{\mu_{max}} \ln \left[1 + \frac{Y_{XS}}{x_0} (s_0 - s_f) \right].$$

(13.27)

To obtain an expression for batch culture time as a function of product concentration, we must apply Eq. (6.5) to the product. Again, $\hat{M}_i = \hat{M}_o = 0$; mass of product in the reactor, M, is equal to pV where p is product concentration and V is liquid volume. Assuming product is not consumed, $R_C = 0$. R_G is equal to $r_P V$ where r_P is the volumetric rate of product formation. According to Eq. (11.67), for all types of product $r_P = q_P x$ where q_P is the specific rate of product formation. Therefore:

$$R_G = r_P V = q_P x V$$

(13.28)

and, from Eq. (6.5), the mass-balance equation is:

$$\frac{d(pV)}{dt} = q_P x V.$$

(13.29)

If cell death is negligible x is given by Eq. (13.19). Therefore, for V constant, we can write Eq. (13.29) as:

$$\frac{dp}{dt} = q_P x_0 e^{\mu_{max}t}.$$

(13.30)

If q_P is constant, Eq. (13.30) can be integrated directly with initial condition $p = p_0$ at $t = 0$ to obtain the following equation for batch culture time as a function of the final product concentration p_f:

$$t_b = \frac{1}{\mu_{max}} \ln \left[1 + \frac{\mu_{max}}{x_0 q_P} (p_f - p_0) \right].$$

(13.31)

The batch culture time needed to achieve a certain biomass

density can be evaluated using Eq. (13.18) or (13.20). If cell death is negligible, the time required for a particular level of substrate conversion can be calculated using Eq. (13.25), (13.26) or (13.27) depending on the type of product and the importance of maintenance metabolism. For negligible cell death and constant q_P, the batch time required to achieve a particular product concentration can be found from Eq. (13.31). Application of these equations is illustrated in Example 13.3.

Example 13.3 Batch culture time

Zymomonas mobilis is used to convert glucose to ethanol in a batch fermenter under anaerobic conditions. The yield of biomass from substrate is 0.06 g g^{-1}; Y_{PX} is 7.7 g g^{-1}. The maintenance coefficient is 2.2 g g^{-1} h^{-1}; the specific rate of product formation due to maintenance is 1.1 h^{-1}. The maximum specific growth rate of *Z. mobilis* is approximately 0.3 h^{-1}. 5 g bacteria are inoculated into 50 litres of medium containing 12 g l^{-1} glucose. Determine batch culture times required to:

(a) produce 10 g biomass;
(b) achieve 90% substrate conversion; and
(c) produce 100 g ethanol.

Solution:
$Y_{XS} = 0.06$ g g^{-1}; $Y_{PX} = 7.7$ g g^{-1}; $\mu_{max} = 0.3$ h^{-1}; $m_S = 2.2$ g g^{-1} h^{-1}; $m_P = 1.1$ h^{-1}; $x_0 = {}^5 g/_{50 \, l} = 0.1$ g l^{-1}; $s_0 = 12$ g l^{-1}.

(a) If 10 g biomass are produced by reaction, the final amount of biomass present is $(10 + 5)$ g = 15 g. Therefore $x_f = {}^{15} g/_{50 \, l} = 0.3$ g l^{-1}. From Eq. (13.20):

$$
t_b = \frac{1}{0.3 \, h^{-1}} \, \ln \frac{0.3 \, g \, l^{-1}}{0.1 \, g \, l^{-1}} = 3.7 \, h.
$$

(b) If 90% of the substrate is converted, $s_f = (0.1 \, s_0) = 1.2$ g l^{-1}. Ethanol synthesis is directly coupled to energy metabolism in the cell; therefore, from Eq. (13.26):

$$
t_b = \frac{1}{0.3 \, h^{-1}} \, \ln \left[1 + \frac{(12 - 1.2) \, g \, l^{-1}}{\left(\dfrac{1}{0.06 \, g \, g^{-1}} + \dfrac{2.2 \, g \, g^{-1} \, h^{-1}}{0.3 \, h^{-1}} \right) 0.1 \, g \, l^{-1}} \right] = 5.7 \, h.
$$

(c) q_P is calculated using Eq. (11.70). In batch culture with $\mu = \mu_{max}$:

$$
q_P = 7.7 \, g \, g^{-1} \, (0.3 \, h^{-1}) + 1.1 \, h^{-1} = 3.4 \, h^{-1}.
$$

As no product is present initially, $p_0 = 0$. Production of 100 g ethanol corresponds to ${}^{100} g/_{50 \, l} = 2$ g l$^{-1} = p_f$. From Eq. (13.31):

$$
t_b = \frac{1}{0.3 \, h^{-1}} \, \ln \left[1 + \frac{0.3 \, h^{-1}}{(0.1 \, g \, l^{-1}) \, (3.4 \, h^{-1})} \, (2 \, g \, l^{-1}) \right] = 3.4 \, h.
$$

13.5.2 Total Time For Batch Reaction Cycle

In the above analysis, t_b represents the time required for batch cell or enzyme conversion. In practice, batch operations involve lengthy unproductive periods in addition to t_b. Following the fermentation or enzyme reaction, time t_{hv} is taken to harvest the contents of the reactor and time t_p is needed to clean, sterilise and otherwise prepare the reactor for the next batch. For cell culture, a lag time of duration t_l occurs after inoculation during which no growth or product forma-

Figure 13.21   Preparation, lag, reaction and harvest times in operation of a batch fermenter.

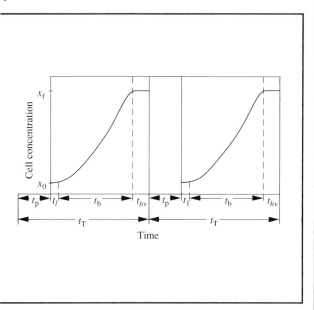

Figure 13.22   Flowsheet for a stirred fed-batch fermenter.

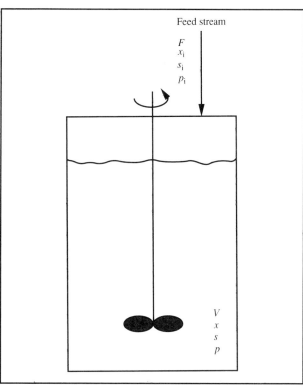

ion occurs. These time periods are illustrated for fermentation processes in Figure 13.21. Therefore, the total *downtime* t_{dn} associated with batch reactor operation is:

$$t_{dn} = t_{hv} + t_p + t_l \tag{13.32}$$

and the total batch reaction time t_T is:

$$t_T = t_b + t_{dn} . \tag{13.33}$$

13.5.3 Fed-Batch Operation of a Mixed Reactor

In fed-batch operation, intermittent or continuous feeding of nutrients is used to supplement the reactor contents and provide control over the substrate concentration. By starting with a relatively dilute solution of substrate and adding more nutrients as the conversion proceeds, high growth rates are avoided. This is important, for example, in cultures where the oxygen demand during fast growth is too high for the mass-transfer capabilities of the reactor. Alternatively, high substrate concentrations may be inhibitory or switch on undesirable metabolic pathways. Fed-batch culture is used extensively in production of bakers' yeast to overcome catabolite repression and control oxygen demand; it is also used routinely for peni-

cillin production. Space must be allowed in fed-batch reactors for addition of fresh medium; in some cases a portion of the broth is removed before injection of additional material. The flow rate and timing of the feed are often determined by monitoring parameters such as dissolved-oxygen level or exhaust gas composition. As enzyme reactions are rarely carried out as fed-batch operations, we will consider fed-batch reactors for fermentation only.

The flowsheet for a well-mixed fed-batch fermenter is shown in Figure 13.22. The volumetric flow rate of entering feed is F; the concentrations of biomass, growth-limiting substrate and product in this stream are x_i, s_i and p_i, respectively. We will assume F is constant. Owing to input of the feed, the liquid volume V is not constant. Equations for fed-batch culture are derived by carrying out unsteady-state mass balances.

The unsteady-state mass-balance equation for total mass in a flow reactor was derived in Chapter 6:

$$\frac{d(\rho V)}{dt} = F_i \rho_i - F_o \rho_o \tag{6.6}$$

where ρ is density of the reactor contents, V is liquid volume in the reactor, F_i and F_o are input and output mass flow rates, and ρ_i and ρ_o are densities of input and output streams, respectively. For the fed-batch reactor of Figure 13.22, $\dot{F}_o = 0$ and $F_i = F$. With dilute solutions such as those often used in bioprocessing, we can assume ρ is constant and that $\rho_i = \rho$; density can then be taken outside of the differential and cancelled through the equation. Therefore, Eq. (6.6) for fed-batch fermentation is:

$$\frac{dV}{dt} = F.$$

(13.34)

A similar mass balance based on Eq. (6.5) can be performed for cells. In fed-batch operation $\hat{M}_o = 0$; $\hat{M}_i$ is equal to the feed flow rate F multiplied by the cell concentration x_i in the feed. As in Section 13.5.1.2, mass of cells in the reactor, M, is equal to xV where x is cell concentration and V is liquid volume, rate of biomass generation R_G is equal to μxV where μ is the specific growth rate, and rate of cell death R_C is equal to $k_d xV$ where k_d is the specific death constant. Applying these terms in Eq. (6.5) gives:

$$\frac{d(xV)}{dt} = Fx_i + \mu xV - k_d xV.$$

(13.35)

Because V in fed-batch culture is a function of time, it cannot be cancelled through Eq. (13.35). Instead, we must expand the differential using the product rule of Eq. (D.22) in Appendix D. After grouping terms this gives:

$$x\frac{dV}{dt} + V\frac{dx}{dt} = Fx_i + (\mu - k_d)xV.$$

(13.36)

Applying Eq. (13.34) to Eq. (13.36):

$$xF + V\frac{dx}{dt} = Fx_i + (\mu - k_d)xV.$$

(13.37)

Dividing through by V and rearranging gives:

$$\frac{dx}{dt} = \frac{F}{V}x_i + x\left(\mu - k_d - \frac{F}{V}\right).$$

(13.38)

Let us define the *dilution rate D* with dimensions T^{-1}:

$$D = \frac{F}{V}.$$

(13.39)

In fed-batch systems V increases with time; therefore, if F is constant, D decreases as the reaction proceeds. Applying Eq. (13.39) to Eq. (13.38):

$$\frac{dx}{dt} = Dx_i + x(\mu - k_d - D).$$

(13.40)

Eq. (13.40) can be simplified for most applications. Usually the feed material is sterile so that $x_i = 0$. If, in addition, the rate of cell death is negligible compared with growth so that $k_d \ll \mu$, Eq. (13.40) becomes:

$$\frac{dx}{dt} = x(\mu - D).$$

(13.41)

Let us now apply Eq. (6.5) to the limiting substrate in our fed-batch reactor. In this case, $\hat{M}_o$ and R_G are zero; the mass flow rate of substrate entering the reactor $\hat{M}_i$ is equal to Fs_i. Mass of substrate in the reactor, M, is equal to concentration s multiplied by volume V. For fermentations producing product not directly coupled with energy metabolism, R_C is given by Eq. (13.21). Substituting these terms into Eq. (6.5) gives:

$$\frac{d(sV)}{dt} = Fs_i - \left(\frac{\mu}{Y_{XS}} + \frac{q_P}{Y_{PS}} + m_S\right)xV$$

(13.42)

where μ is the specific growth rate, Y_{XS} is the true biomass yield from substrate, q_P is the specific rate of product formation, Y_{PS} is the true product yield from substrate and m_S is the maintenance coefficient. Expanding the differential and applying Eqs (13.34) and (13.39) gives:

$$\frac{ds}{dt} = D(s_i - s) - \left(\frac{\mu}{Y_{XS}} + \frac{q_P}{Y_{PS}} + m_S\right)x.$$

(13.43)

Eqs (13.41) and (13.43) are differential equations for rates of change of cell and substrate concentrations in fed-batch reacors. Because D is a function of time, integration of these equations is more complicated than for batch reactors. However, we can derive analytical expressions for fed-batch culture if we simplify Eqs (13.41) and (13.43). Here, we will examine the situation where the reactor is operated first in batch until a high cell density is achieved and the substrate virtually exhausted. When this condition is reached, fed-batch operation is started with medium flow rate F. As a result, cell concentration x is maintained high and approximately constant so that $\frac{dx}{dt} \approx 0$. From Eq. (13.41), if $\frac{dx}{dt} \approx 0$, $\mu \approx D$. Therefore, substituting $\mu \approx D$ into the Monod expression of Eq. (11.60):

$$D \approx \frac{\mu_{max}s}{K_S + s}.$$

(13.44)

Rearrangement of Eq. (13.44) gives an expression for substrate concentration as a function of dilution rate:

$$s \approx \frac{DK_S}{\mu_{max} - D}.$$

(13.45)

Let us assume that the culture does not produce product, or, if there is product formation, that it is directly linked with energy generation. If maintenance requirements can also be neglected, Eq. (13.43) can be simplified to:

$$\frac{ds}{dt} = D(s_i - s) - \frac{\mu x}{Y_{XS}}.$$

(13.46)

At high cell density in the reactor, virtually all substrate entering the vessel is consumed immediately; therefore, $s \ll s_i$ and $\frac{ds}{dt} \approx 0$. Applying these relationships with $\mu \approx D$ to Eq. (13.46), we obtain:

$$x \approx Y_{XS} s_i.$$

(13.47)

For product synthesis directly coupled with energy metabolism, Eq. (13.47) allows us to derive an approximate expression for product concentration in fed-batch reactors. Assuming the feed does not contain product:

$$p \approx Y_{PS} s_i.$$

(13.48)

Even though cell concentration remains virtually unchanged with $\frac{dx}{dt} \approx 0$, because the liquid volume increases with time in fed-batch reactors, the total mass of cells also increases. Consider the rate of increase of total biomass in the reactor $\frac{dX}{dt}$, where X is equal to xV. Using the results of Eqs (13.34) and (13.47) with $\frac{dx}{dt} \approx 0$:

$$\frac{dX}{dt} = \frac{d(xV)}{dt} = x\frac{dV}{dt} + V\frac{dx}{dt} = Y_{XS} s_i F.$$

(13.49)

Eq. (13.49) can now be integrated with initial condition $X = X_0$ at the start of liquid flow to give:

$$X = X_0 + (Y_{XS} s_i F) t_{fb}$$

(13.50)

where t_{fb} is the fed-batch time after commencement of feeding. Eq. (13.50) indicates that, for Y_{XS}, s_i and F constant, total biomass in fed-batch fermenters increases as a linear function of time.

Under conditions of high biomass density and almost complete depletion of substrate, a *quasi-steady-state* condition prevails in fed-batch reactors where $\frac{dX}{dt} \approx 0$, $\frac{ds}{dt} \approx 0$ and $\frac{dp}{dt} \approx 0$. At quasi-steady state, Eqs (13.47) and (13.45) can be used to calculate biomass and substrate concentrations in reactors where cell death and maintenance requirements are negligible and product is either absent or directly coupled with energy metabolism. Eq. (13.48) allows calculation of product concentration for metabolites directly coupled with energy generation in the cell. At quasi-steady state, the specific growth rate μ and dilution rate F/V are approximately equal; therefore as V increases, the growth rate decreases. When fed-batch operation is used for production of biomass such as bakers' yeast, it is useful to be able to predict the total mass of cells in the reactor as a function of time. An expression for total biomass is given by Eq. (13.50). Note that under quasi-steady-state conditions, x, s and p are almost constant, but μ, V, D and X are changing. Further details of fed-batch operation are given by Pirt [37].

13.5.4 Continuous Operation of a Mixed Reactor

Bioreactors are operated continuously in a few bioprocess industries such as brewing, production of bakers' yeast and waste treatment; enzyme conversions can also be carried out using continuous systems. The flow sheet for a continuous

Figure 13.23 Flowsheet for a continuous stirred-tank fermenter.

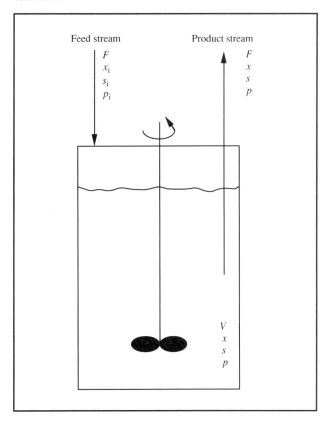

the dilution rate is therefore constant and steady state is achieved by concentrations in the chemostat adjusting themselves to the feed rate. In a *turbidostat*, the liquid volume is kept constant by setting the outlet flow rate equal to the inlet flow rate; however, the inlet flow rate is adjusted to keep the biomass concentration constant. Thus, in a turbidostat the dilution rate adjusts to its steady-state value corresponding to the set biomass concentration. Turbidostats require more complex monitoring and control systems than chemostats and are not used in large scale. Accordingly, we will concentrate here on chemostat operation.

Characteristic operating parameters for continuous reactors are the dilution rate D and the average *residence time* τ. These parameters are related as follows:

$$\tau = \frac{1}{D} = \frac{V}{F}$$

$$(13.51)$$

where D is defined by Eq. (13.39). In continuous reactor operation, the amount of material that can be processed over a given period of time is represented by the flow rate, F. Therefore, for a given throughput, the reactor size V and associated capital and operating costs are minimised when τ is made as small as possible. Continuous reactor theory allows us to determine relationships between τ (or D) and steady-state substrate, product and cell concentrations in the reactor. This theory is based on steady-state mass balances derived from Eq. (6.5).

reactor is shown in Figure 13.23. If the vessel is well mixed, the product stream has the same composition as the liquid in the reactor (see Section 6.4). Therefore, when continuous reactors are used with freely-suspended cells or enzymes, catalyst is continuously withdrawn from the vessel in the product stream. For enzymes this is a serious shortcoming as catalyst is not produced by the reaction; in cell culture, growth supplies additional cells to replace those removed. Continuous reactors are used with free enzymes only if the enzyme is inexpensive and can be added continuously to maintain the catalyst concentration. With more costly enzymes, continuous operation may be feasible if the enzyme is immobilised and retained inside the vessel. Well-mixed continuous reactors are often referred to using the abbreviation CSTR, meaning *continuous stirred-tank reactor*. The term CSTF is also sometimes used to denote *continuous stirred-tank fermenter*.

Different steady-state operating strategies are available for continuous fermenters. In a *chemostat* the liquid volume is kept constant by setting the inlet and outlet flow rates equal;

13.5.4.1 *Enzyme reaction*

Let us apply Eq. (6.5) to the limiting substrate in a continuous enzyme reactor operated at steady state. The mass flow rate of substrate entering the reactor $\hat{M}_i$ is equal to Fs_i; $\hat{M}_o$, the mass flow rate of substrate leaving, is Fs. As substrate is not generated by the reactor, $R_G = 0$. Rate of substrate consumption R_C is equal to the volumetric rate of reaction v multiplied by V; v is given by Eq. (11.31). The left-hand side of Eq. (6.5) is zero because the system is at steady state. Therefore, the steady-state substrate mass-balance equation for continuous enzyme reaction is:

$$Fs_i - Fs - \frac{v_{max}s}{K_m + s} V = 0$$

$$(13.52)$$

where v_{max} is the maximum rate of reaction and K_m is the Michaelis constant. For reactions with free enzyme, we assume that enzyme lost in the product stream is replaced continuously so that v_{max} remains constant and steady state is achieved. Dividing through by V and applying the definition of dilution rate from Eq. (13.39) gives:

$$D(s_i - s) = \frac{v_{\text{max}} s}{K_{\text{m}} + s}.$$

(13.53)

If v_{max}, K_{m} and s_i are known, Eq. (13.53) can be used directly to calculate the dilution rate required to achieve a particular level of substrate conversion. Steady-state product concentration can then be evaluated from stoichiometry.

For reactions with immobilised enzymes, Eq. (13.53) must be modified to account for mass-transfer effects:

$$D(s_i - s) = \frac{\eta_{\text{T}} v_{\text{max}} s}{K_{\text{m}} + s}$$

(13.54)

where η_{T} is the total effectiveness factor (see Section 12.5), s is the bulk substrate concentration, and v_{max} and K_{m} are intrinsic kinetic parameters. η_{T} can be calculated for constant s using the theory for heterogeneous reactions outlined in Chapter 12.

Example 13.4 Immobilised-enzyme reaction in a CSTR

Mushroom tyrosinase is immobilised in 2-mm spherical beads for conversion of tyrosine to DOPA in a continuous, well-mixed bubble column. The Michaelis constant for the immobilised enzyme is 2 gmol m^{-3}. A solution containing 15 gmol m^{-3} tyrosine is fed into the reactor; because of the high cost of the substrate, the desired conversion is 99%. The reactor is loaded with beads at a density of 0.25 m^3 m^{-3}; all enzyme is retained within the reactor. The intrinsic v_{max} for the immobilised enzyme is 1.5×10^{-2} gmol s^{-1} per m^3 beads. The effective diffusivity of tyrosine in the beads is 7×10^{-10} m^2 s^{-1}; external mass-transfer effects are negligible. Immobilisation stabilises the enzyme so that deactivation is minimal over the operating period. Determine the reactor volume needed to treat 18 m^3 tyrosine solution per day.

Solution:
$K_{\text{m}} = 2$ gmol m^{-3}; $v_{\text{max}} = 1.5 \times 10^{-2}$ gmol s^{-1} m^{-3}; $R = 10^{-3}$ m; $\mathcal{D}_{\text{Ae}} = 7 \times 10^{-10}$ m^2 s^{-1}; $s_i = 15$ gmol m^{-3}. Converting the feed flow rate to m^3 s^{-1}:

$$F = 18 \text{ m}^3 \text{ d}^{-1} \cdot \left| \frac{1 \text{ d}}{24 \text{ h}} \right| \cdot \left| \frac{1 \text{ h}}{3600 \text{ s}} \right| = 2.08 \times 10^{-4} \text{ m}^3 \text{ s}^{-1}.$$

For 99% conversion, the outlet and therefore the internal substrate concentration $s = (0.01 \, s_i) = 0.15$ gmol m^{-3}. As $s << K_{\text{m}}$, we can assume first-order kinetics (see Section 11.3.3) with $k_1 = v_{\text{max}}/K_{\text{m}} = 7.5 \times 10^{-3}$ s^{-1}. The first-order Thiele modulus for spherical catalysts is calculated from Table 12.2:

$$\phi_1 = \frac{R}{3} \sqrt{\frac{k_1}{\mathcal{D}_{\text{Ae}}}} = \frac{10^{-3} \text{ m}}{3} \sqrt{\frac{7.5 \times 10^{-3} \text{ s}^{-1}}{7 \times 10^{-10} \text{ m}^2 \text{ s}^{-1}}} = 1.09.$$

From Figure 12.8 or Table 12.3, $\eta_{i1} = 0.64$. As external mass-transfer resistance is negligible, $\eta_e = 1$ and, from Eq. (12.46), $\eta_{\text{T}} = 0.64$. Substituting values into Eq. (13.54) gives:

$$D(15 - 0.15) \text{ gmol m}^{-3} = \frac{0.64 \, (1.5 \times 10^{-2} \text{ gmol s}^{-1} \text{ m}^{-3})\,(0.15 \text{ gmol m}^{-3})}{2 \text{ gmol m}^{-3} + 0.15 \text{ gmol m}^{-3}}$$

$$D = 4.51 \times 10^{-5} \text{ s}^{-1}.$$

From Eq. (13.51):

$$V = \frac{F}{D} = \frac{2.08 \times 10^{-4} \text{ m}^3 \text{ s}^{-1}}{4.51 \times 10^{-5} \text{ s}^{-1}} = 4.6 \text{ m}^3.$$

13.5.4.2 Cell culture

Let us consider the reactor of Figure 13.23 operated as a continuous fermenter and apply Eq. (6.5) for steady-state mass balances on biomass, substrate and product.

For biomass, $\hat{M}_i$ in Eq. (6.5) is the mass flow rate of cells entering the reactor; $\hat{M}_i = Fx_i$. $\hat{M}_o$ is the mass flow rate of cells leaving: $\hat{M}_o = Fx$. The other terms in Eq. (6.5) are the same as in Section 13.5.1.2; $R_G = \mu\, xV$ where μ is the specific growth rate and V the reactor liquid volume; $R_C = k_d\, xV$ where k_d is the specific death constant. At steady state, the left-hand side of Eq. (6.5) is zero. Therefore, the steady-state mass-balance equation for biomass is:

$$Fx_i - Fx + \mu xV - k_d xV = 0.$$

$$(13.55)$$

Usually, the feed stream in continuous culture is sterile so that $x_i = 0$. If in addition the rate of cell death is negligible compared with growth, $k_d \ll \mu$ and Eq. (13.55) becomes:

$$\mu xV = Fx.$$

$$(13.56)$$

Cancelling x from both sides, dividing by V, and applying the definition of dilution rate from Eq. (13.39) gives:

$$\mu = D.$$

$$(13.57)$$

As in the derivation of Eq. (13.45), applying Eq. (13.57) to the Monod expression of Eq. (11.60) gives an equation for the steady-state concentration of limiting substrate in the reactor:

$$s = \frac{DK_S}{\mu_{max} - D}.$$

$$(13.58)$$

Let us now apply Eq. (6.5) at steady state to the limiting substrate. $\hat{M}_i = Fs_i$, $\hat{M}_o = Fs$, $R_G = 0$ and R_C is given by Eq. (13.21). Therefore:

$$Fs_i - Fs - \left(\frac{\mu}{Y_{XS}} + \frac{q_P}{Y_{PS}} + m_S \right) xV = 0$$

$$(13.59)$$

where μ is the specific growth rate, Y_{XS} is the true biomass yield from substrate, q_P is the specific rate of product formation not directly linked with energy metabolism, Y_{PS} is the

true product yield from substrate and m_S is the maintenance coefficient. In Eq. (13.59) we can divide through by V, substitute the definition of dilution rate from Eq. (13.39) and replace μ with D according to Eq. (13.57). Rearrangement then gives the following expression for the steady-state cell concentration x:

$$x = \frac{D(s_i - s)}{\dfrac{D}{Y_{XS}} + \dfrac{q_P}{Y_{PS}} + m_S}.$$

$$(13.60)$$

Eq. (13.60) can be simplified if there is no product synthesis or if production is directly linked with energy metabolism:

$$x = \frac{D(s_i - s)}{\dfrac{D}{Y_{XS}} + m_S}.$$

$$(13.61)$$

If, in addition, maintenance effects can be ignored, Eq. (13.61) becomes:

$$x = (s_i - s)\, Y_{XS}.$$

$$(13.62)$$

Substituting for s from Eq. (13.58) we obtain an expression for the steady-state cell concentration in a CSTR in terms of D and kinetic and yield parameters only:

$$x = \left(s_i - \frac{DK_S}{\mu_{max} - D} \right) Y_{XS}.$$

$$(13.63)$$

Eq. (13.63) is valid at steady state in the absence of maintenance requirements and when product synthesis is either absent or directly linked with energy metabolism.

We can also apply Eq. (6.5) for a steady-state mass balance on fermentation product. In this case, $\hat{M}_i = Fp_i$ and $\hat{M}_o = Fp$. R_G is given by Eq. (13.28); $R_C = 0$. Therefore, Eq. (6.5) becomes:

$$Fp_i - Fp + q_P xV = 0$$

$$(13.64)$$

where q_P is the specific rate of formation for all classes of product. Dividing through by V, substituting the definition of

Figure 13.24 Steady-state cell and substrate concentrations as a function of dilution rate in a chemostat. Curves correspond to $\mu_{max} = 0.5$ h^{-1}, $K_S = 0.2$ kg m^{-3}, $Y_{XS} = 0.5$ kg kg^{-1} and $s_i = 20$ kg m^{-3}.

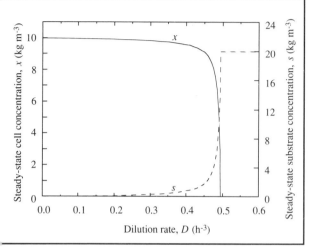

dilution rate from Eq. (13.39) and rearranging gives an expression for the steady-state product concentration as a function of biomass concentration x:

$$p = p_i + \frac{q_p x}{D}.$$

(13.65)

q_p in Eq. (13.65) can be evaluated from Eq. (13.60), (13.61) or (13.62). The nature of q_p depends on the type of product formed (see Section 11.9).

Eq. (13.58) is an explicit expression for steady-state substrate concentration in a chemostat. The steady-state biomass concentration x can be evaluated from Eq. (13.60), (13.61) or (13.62); the choice of expression for biomass depends on the relative significance of maintenance metabolism and the type of product, if any, produced. If q_p is known, product concentration in the reactor can be evaluated from Eq. (13.65). In the simplest case when products are either absent or directly linked to energy generation and maintenance effects can be neglected, the chemostat is represented by Eqs (13.58) and (13.63). The form of these equations is shown in Figure 13.24. At low feed rates, i.e. $D \to 0$, nearly all the substrate is consumed at steady state so that, from Eq. (13.62), $x \approx s_i Y_{XS}$. As D increases, s increases slowly at first and then more rapidly as D approaches μ_{max}; correspondingly, x decreases so that $x \to 0$ as $D \to \mu_{max}$. The condition at high dilution rate

whereby x reduces to zero is known as *washout*; washout of cells occurs when the rate of cell removal in the reactor outlet stream is greater than the rate of generation by growth. For systems with negligible maintenance requirements and either energy-associated or zero product formation, the critical dilution rate D_{crit} at which the steady-state biomass concentration just becomes zero can be estimated by substituting $x = 0$ into Eq. (13.63) and solving for D:

$$D_{crit} = \frac{\mu_{max} s_i}{K_S + s_i}.$$

(13.66)

For most cell cultures $K_S \ll s_i$; therefore $D_{crit} \approx \mu_{max}$. To avoid washout of cells from the chemostat, the operating dilution rate must always be less than D_{crit}. Near washout the system is very sensitive to small changes in D which cause relatively large shifts in x and s.

The rate of biomass production in a CSTR is equal to the rate at which cells leave the reactor: Fx. The volumetric productivity is therefore equal to Fx divided by V:

$$Q_X = \frac{Fx}{V} = Dx$$

(13.67)

where Q_X is the volumetric rate of biomass production. Similarly the volumetric rate of product formation Q_P is:

$$Q_P = \frac{Fp}{V} = Dp.$$

(13.68)

When maintenance requirements are negligible and product formation either absent or energy associated, we can substitute into Eq. (13.67) the expression for x from Eq. (13.63):

$$Q_X = D \left(s_i - \frac{K_S D}{\mu_{max} - D} \right) Y_{XS}.$$

(13.69)

This relationship between Q_X and D is shown in Figure 13.25. Rate of biomass output reaches a sharp maximum at the optimum dilution rate for biomass productivity D_{opt}; therefore, at D_{opt} the slope $dQ_X/dD = 0$. Differentiating Eq. (13.69) with respect to D and equating to zero provides an expression for D_{opt}:

$$D_{opt} = \mu_{max} \left(1 - \sqrt{\frac{K_S}{K_S + s_i}} \right).$$

(13.70)

Operation of a chemostat at D_{opt} gives the maximum rate of biomass production from the reactor. However, because D_{opt} is usually very close to D_{crit}, it may not be practical to operate at D_{opt}. Small variations of dilution rate in this region can cause large fluctuations in x and s and, unless the dilution rate is controlled very precisely, washout may occur.

Excellent agreement between chemostat theory and experimental results has been found for many culture systems. When deviations occur, they are due primarily to imperfect operation of the reactor. For example, if the vessel is not well mixed, some liquid will have higher residence time in the reactor than the rest and concentrations will not be uniform; under these conditions the equations derived in the section do not hold. Similarly, if cells adhere to glass or metal surfaces in the reactor and produce wall growth, biomass will be retained in the vessel and washout will not occur even at high dilution rates. Other deviations occur if inadequate time is allowed for the system to reach steady state.

Figure 13.25 Steady-state volumetric biomass productivity as a function of dilution rate in a chemostat. Curve corresponds to $\mu_{max} = 0.5 \text{ h}^{-1}$, $K_S = 0.2 \text{ kg m}^{-3}$, $Y_{XS} = 0.5 \text{ kg kg}^{-1}$ and $s_i = 20 \text{ kg m}^{-3}$.

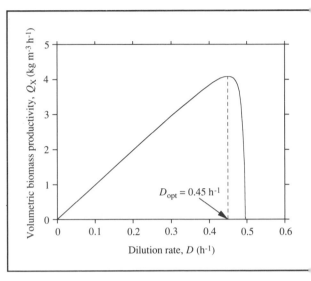

Example 13.5 Steady-state concentrations in a chemostat

The *Zymomonas mobilis* cells of Example 13.3 are used for chemostat culture in a 60 m³ fermenter. The feed contains 12 g l⁻ glucose; K_S for the organism is 0.2 g l^{-1}.

(a) What flow rate is required for a steady-state substrate concentration of 1.5 g l^{-1}?
(b) At the flow rate of (a), what is the cell density?
(c) At the flow rate of (a), what concentration of ethanol is produced?

Solution:
$Y_{XS} = 0.06 \text{ g g}^{-1}$; $Y_{PX} = 7.7 \text{ g g}^{-1}$; $\mu_{max} = 0.3 \text{ h}^{-1}$; $K_S = 0.2 \text{ g l}^{-1}$; $m_S = 2.2 \text{ g g}^{-1} \text{ h}^{-1}$; $s_i = 12 \text{ g l}^{-1}$; $V = 60 \text{ m}^3$. From Exampl 13.3, $q_P = 3.4 \text{ h}^{-1}$. From the general definition of yield in Section 11.6.1, $Y_{PS} = Y_{PX} Y_{XS} = 0.46 \text{ g g}^{-1}$.

(a) $s = 1.5 \text{ g l}^{-1}$. From Eq. (13.58):

$$D = \frac{\mu_{max} s}{K_S + s} = \frac{(0.3 \text{ h}^{-1}) (1.5 \text{ g l}^{-1})}{0.2 \text{ g l}^{-1} + 1.5 \text{ g l}^{-1}} = 0.26 \text{ h}^{-1}.$$

From the definition of dilution rate Eq. (13.39):

$$F = DV = (0.26 \text{ h}^{-1}) (60 \text{ m}^3) = 15.6 \text{ m}^3 \text{ h}^{-1}.$$

(b) When synthesis of product is coupled with energy metabolism as for ethanol, x is evaluated using Eq. (13.61). Therefore:

$$x = \frac{(0.26\,\text{h}^{-1})\,(12 - 1.5)\,\text{g l}^{-1}}{\left(\dfrac{0.26\,\text{h}^{-1}}{0.06\,\text{g g}^{-1}} + 2.2\,\text{g g}^{-1}\text{h}^{-1}\right)} = 0.42\,\text{g l}^{-1}.$$

) Assuming ethanol is not present in the feed, $p_i = 0$. Steady-state product concentration is given by Eq. (13.65):

$$p = \frac{(3.4\,\text{h}^{-1})\,(0.42\,\text{g l}^{-1})}{0.26\,\text{h}^{-1}} = 5.5\,\text{g l}^{-1}.$$

xample 13.6 Substrate conversion and biomass productivity in a chemostat

5 m³ fermenter is operated continuously with feed substrate concentration 20 kg m⁻³. The microorganism cultivated in the actor has the following characteristics: $\mu_{max} = 0.45\,\text{h}^{-1}$, $K_S = 0.8\,\text{kg m}^{-3}$, $Y_{XS} = 0.55\,\text{kg kg}^{-1}$.

) What feed flow rate is required to achieve 90% substrate conversion?
) How does the biomass productivity at 90% substrate conversion compare with the maximum possible?

lution:

) For 90% substrate conversion, $s = (0.1\,s_i) = 2\,\text{kg m}^{-3}$. From Eq. (13.58):

$$D = \frac{\mu_{max}\,s}{K_S + s} = \frac{(0.45\text{h}^{-1})\,(2\,\text{kg m}^{-3})}{0.8\,\text{kg m}^{-3} + 2\,\text{kg m}^{-3}} = 0.32\,\text{h}^{-1}.$$

om Eq. (13.39):

$$F = DV = 0.32\,\text{h}^{-1}\,(5\,\text{m}^3) = 1.6\,\text{m}^3\,\text{h}^{-1}.$$

) Assuming maintenance requirements and product formation are negligible, from Eq. (13.69):

$$Q_X = 0.32\,\text{h}^{-1}\left(20\,\text{kg m}^{-3} - \frac{(0.8\,\text{kg m}^{-3})\,(0.32\,\text{h}^{-1})}{(0.45\text{h}^{-1} - 0.32\,\text{h}^{-1})}\right) 0.55\,\text{kg kg}^{-1}$$

$$= 3.17\,\text{kg m}^{-3}\,\text{h}^{-1}.$$

aximum biomass productivity occurs at D_{opt} which can be evaluated using Eq. (13.70):

$$D_{opt} = 0.45\,\text{h}^{-1}\left(1 - \sqrt{\frac{0.8\,\text{kg m}^{-3}}{0.8\,\text{kg m}^{-3} + 20\,\text{kg m}^{-3}}}\right) = 0.36\,\text{h}^{-1}.$$

aximum biomass productivity is determined from Eq. (13.69) with $D = D_{opt}$:

$$Q_{X,max} = 0.36\,\text{h}^{-1}\left(20\,\text{kg m}^{-3} - \frac{(0.8\,\text{kg m}^{-3})\,(0.36\,\text{h}^{-1})}{(0.45\text{h}^{-1} - 0.36\,\text{h}^{-1})}\right) 0.55\,\text{kg kg}^{-1}$$

$$= 3.33\,\text{kg m}^{-3}\,\text{h}^{-1}.$$

herefore, biomass productivity at 90% substrate conversion is $^{3.17}/_{3.33} \times 100 = 95\%$ of the theoretical maximum.

Figure 13.26 Flowsheet for a continuous stirred-tank fermenter with immobilised cells.

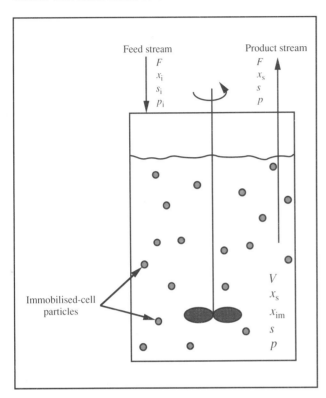

Feed stream
F
x_i
s_i
p_i

Product stream
F
x_s
s
p

V
x_s
x_{im}
s
p

Immobilised-cell particles

13.5.5 Chemostat With Immobilised Cells

Consider the continuous stirred-tank immobilised-cell fermenter shown in Figure 13.26. Spherical particles containing cells are kept suspended and well mixed by the stirrer. The concentration of immobilised cells per unit volume of liquid in the reactor is x_{im}. Let us assume that x_{im} is constant; this is achieved if all particles are retained in the vessel and all cells produced by immobilised-cell growth are released into the medium. The concentration of suspended cells is x_s. We will assume that the intrinsic specific growth rates of suspended and immobilised cells are the same and equal to μ. Suspended cells are removed from the reactor in the product stream; immobilised cells are retained inside the vessel. For simplicity, let us assume that cell death and maintenance requirements are negligible, the reactor feed is sterile, and any product synthesis is directly coupled with energy metabolism.

The system shown in Figure 13.26 reaches steady state. Relationships between operating variables and concentrations inside the reactor can be determined using mass balances.

Let us consider a mass balance on suspended cells. At steady state, the mass-balance equation is similar to Eq. (13.5 except that x_i is zero for sterile feed and cell death is assumed be negligible. In addition, as suspended cells are produced t growth of both suspended- and immobilised-cell population the equation must contain two generation terms instead one:

$$- Fx_s + \mu x_s V + \mu x_{im} V = 0.$$

(13.7

If diffusional limitations affect the growth rate of the immob ised cells, μx_{im} must be replaced by $\eta_T \mu x_{im}$ where η_T is t total effectiveness factor defined in Section 12.5. Dividir through by V and applying the definition of dilution rate fro Eq. (13.39) gives:

$$D x_s = \mu (x_s + \eta_T x_{im})$$

(13.7

or

$$D = \mu \left(1 + \frac{\eta_T x_{im}}{x_s} \right).$$

(13.7

For $x_{im} = 0$, Eq. (13.73) reduces to Eq. (13.57) for a chemost containing suspended cells only: $\mu = D$.

The steady-state mass-balance equation for limiting su strate can be derived from Eq. (6.5) with $\hat{M}_i = F s_i$, $\hat{M}_o = F s$ ar $R_G = 0$. Both cell populations consume substrate; in t absence of product- and maintenance-associated substra requirements, rate of substrate consumption R_C can be relat directly to the growth rates of immobilised and suspended ce using the biomass yield coefficient Y_{XS}. If we assume that t value of Y_{XS} is the same for all cells, by analogy with E (13.59) the mass-balance equation for limiting substrate is:

$$F s_i - F s - \frac{\mu x_s}{Y_{XS}} V - \frac{\eta_T \mu x_{im}}{Y_{XS}} V = 0.$$

(13.7

Dividing through by V, applying the definition of dilution ra from Eq. (13.39) and rearranging gives:

$$D(s_i - s) = \frac{\mu}{Y_{XS}} (x_s + \eta_T x_{im}).$$

(13.7

By manipulating Eqs (13.73) and (13.75) and substituting the Monod expression for μ from Eq. (11.60), the following relationship between steady-state substrate concentration, dilution rate and immobilised-cell concentration is obtained:

$$\frac{\mu_{max}\,s}{K_S + s} = \frac{D(s_i - s)\,Y_{XS}}{(s_i - s)\,Y_{XS} + \eta_T\,x_{im}}.$$

(13.76)

The form of Eq. (13.76) is shown in Figure 13.27. For a chemostat with suspended cells only, i.e. $x_{im} = 0$, at steady state $D = \mu$ and the maximum operating dilution rate D_{crit} is limited by the maximum specific growth rate of the cells. From Eq. (13.73), for any $x_{im} > 0$, D at steady state in the immobilised-cell reactor is greater than μ. Accordingly, dilution rate is no longer limited by the maximum growth rate and, as shown in Figure 13.27, immobilised-cell chemostats can be operated at D considerably greater than D_{crit} without washout. At a given dilution rate, presence of immobilised cells also improves substrate conversion and reduces the amount of substrate lost in the product stream. However, reaction rates with immobilised cells can be reduced significantly by the effects of mass transfer in and around the particles. As illustrated in Figure 13.27, at the same concentration of immobilised cells, substrate conversion at $\eta_T = 1.0$ is greater than at lower values of η_T when mass-transfer limitations are significant.

13.5.6 Chemostat Cascade

The joining together of two or more CSTRs in series produces a multi-stage process in which conditions such as pH, temperature and medium composition can be varied in each reactor. This is advantageous if reactor conditions required for growth are different from those for product synthesis, e.g. in production of recombinant proteins and many metabolites not directly linked with energy metabolism. One way of operating a two-stage chemostat cascade is shown in Figure 13.28; in this process the product stream from the first reactor feeds directly into the second. Substrate leaving the first reactor at concentration s_1 is converted in the second tank so that $s_2 < s_1$ and $p_2 > p_1$. In some applications, the second CSTR is supplemented with fresh medium containing nutrients, inducers or inhibitors for optimal product formation.

Design equations for cell and enzyme CSTR cascades can be derived as a simple extension of the theory developed in Section 13.5.4; the same mass-balance principles are applied

Figure 13.27 Steady-state substrate conversion as a function of dilution rate with and without immobilised cells. Curves were calculated with the following parameter values: $\mu_{max} = 0.1\ h^{-1}$, $K_S = 10^{-3}\ g\ l^{-1}$, $Y_{XS} = 0.5\ g\ g^{-1}$, $s_i = 8 \times 10^{-3}\ g\ l^{-1}$.

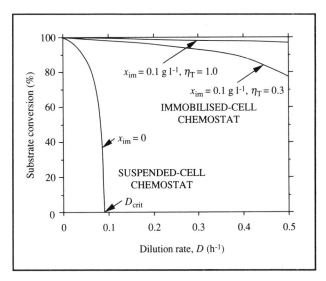

and steady state is assumed at each reactor stage. Details can be found in other references [5, 37, 38]. In fermenter cascades, cells entering the second and subsequent vessels may go through periods of unbalanced growth as they adapt to the new environmental conditions in each reactor. Therefore, use of simple unstructured metabolic models such as those outlined in this text does not always give accurate results. Nevertheless, it can be shown that the total reactor residence time required to achieve a given degree of substrate conversion is significantly smaller with two CSTRs in series than if only one CSTR were used. In other words, the total reactor volume required is reduced with two smaller tanks in series than with a single large tank. Usually however, only two to four reactors in series are justified as the benefits associated with adding successive stages diminish significantly [39].

13.5.7 Chemostat With Cell Recycle

Cell concentration in a single chemostat can be increased by recycling biomass from the product stream back to the reactor. With more catalyst present in the vessel, higher rates of substrate utilisation and product formation can be achieved. With cell recycle, the critical dilution rate for washout is also increased allowing greater operating flexibility.

There are several ways by which cells can be recycled in

Figure 13.28   Flowsheet for a cascade of two continuous stirred-tank fermenters.

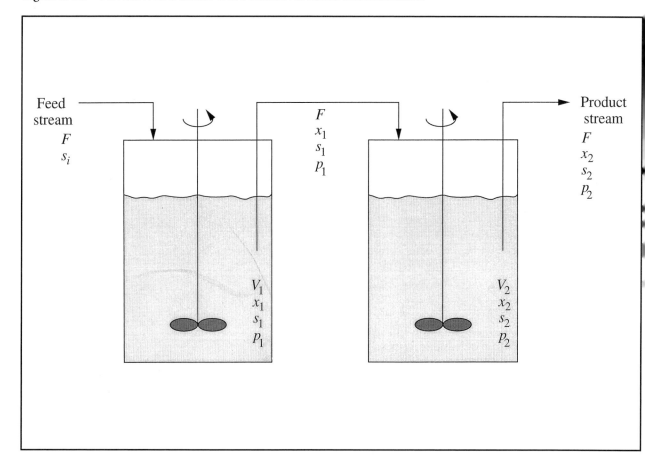

fermentation processes. *External biomass feedback* is illustrated in Figure 13.29; in this scheme, a cell separator such as a centrifuge or settling tank is used to concentrate biomass leaving the reactor. A portion of the concentrate is continuously recycled back to the CSTR with flow rate F_r and cell concentration x_r. Such systems can be operated under steady-state conditions and are used extensively in biological waste treatment. Another way of achieving biomass feedback is *perfusion culture* or *internal biomass feedback*. This operating scheme is often used for mammalian cell culture and is illustrated in Figure 13.30. Depletion of nutrients and accumulation of inhibitory products limit batch cell densities for many animal cell lines to about 10^6 cells ml^{-1}. Cell density and therefore the volumetric productivity of these cultures can be increased by retaining biomass in the reactor while fresh medium is continuously added and spent broth removed. As indicated in Figure 13.30, cells in a perfusion reactor are physically retained in the vessel by a mechanical device such as a filter.

Liquid throughput is thus achieved without continuous removal or dilution of the cells so that concentrations in excess of 10^7 cells ml^{-1} can be obtained. A common problem associated with perfusion systems is blocking or blinding of the filter.

With growing cells, if all the biomass is returned or retained in the reactor, the cell concentration will increase with time and steady state will not be achieved. Therefore, for steady-state operation, some proportion of the biomass must be removed from the system. Chemostat reactors with cell recycle can be analysed using the same mass-balance techniques applied in Section 13.5.4 [37, 38]. Typical results for biomass concentration and biomass productivity with and without cell recycle are shown in Figure 13.31. With cell feedback, because recycled cells are an additional source of biomass in the reactor, washout occurs at dilution rates greater than the maximum specific growth rate. If α is the recycle ratio:

Figure 13.29 Flowsheet for a continuous stirred-tank fermenter with external biomass feedback.

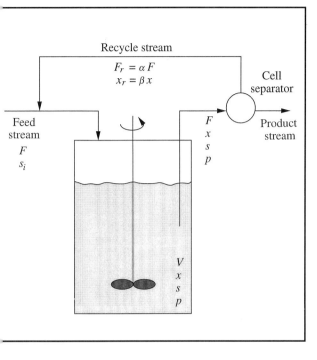

$$\alpha = \frac{F_r}{F}$$

(13.77)

and β is the biomass concentration factor:

$$\beta = \frac{x_r}{x}$$

(13.78)

the critical dilution rate with cell cycle is increased by a factor of $^1/_{(1 + \alpha - \alpha\beta)}$ relative to the simple chemostat. Figure 13.31 also shows that biomass productivity is greater in recycle systems by the same factor.

13.5.8 Continuous Operation of a Plug-Flow Reactor

Plug-flow operation is an alternative to mixed operation for continuous reactors. No mixing occurs in an ideal plug-flow reactor; liquid entering the reactor passes through as a discrete 'plug' and does not interact with neighbouring fluid elements. This is achieved at high flow rates which minimise backmixing and variations in liquid velocity. Plug flow is most readily achieved in column or tubular reactors such as that shown in Figure 13.32. Plug-flow reactors can be operated in upflow or

Figure 13.30 Flowsheet for a continuous perfusion reactor with internal biomass feedback.

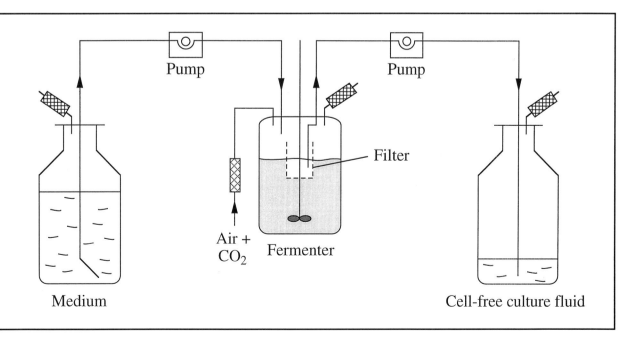

Figure 13.31 Steady-state biomass concentration x and volumetric biomass productivity Q_X for a chemostat with and without cell recycle. Curves were calculated with the following parameter values: $\mu_{max} = 0.5$ h^{-1}, $K_S = 0.01$ g l^{-1}, $Y_{XS} = 0.5$ g g^{-1}, $s_i = 2$ g l^{-1}, $\alpha = 0.5$, $\beta = 2.0$.

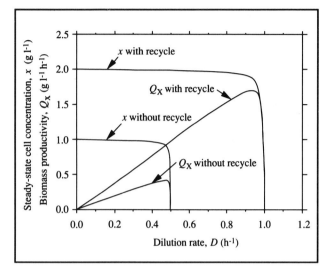

downflow mode or, in some cases, horizontally. Plug-flow tubular reactors are known by the abbreviation PFTR.

Liquid in a PFTR flows at constant velocity; thus all parts of the liquid have identical residence time in the reactor. As reaction in the vessel proceeds, concentration gradients of substrate and product develop in the direction of flow. At the feed-end of the PFTR the substrate concentration will be high and the product concentration low because the reaction mixture has just entered the vessel; at the end of the tube the substrate concentration will be low and the product level high.

Consider operation of the PFTR shown in Figure 13.32. The volumetric liquid flow rate through the vessel is F; the feed stream contains substrate at concentration s_i. At the reactor outlet, the substrate concentration is s_f. This exit concentration can be related to the inlet conditions and reactor residence time using mass-balance techniques. Let us first consider plug-flow operation for enzyme reaction.

13.5.8.1 Enzyme reaction

To develop equations for a plug-flow enzyme reactor, we must consider a small section of the reactor of length Δz as indicated in Figure 13.32. This section is located at distance z from the feed point. Let us perform a steady-state mass balance on substrate around the section using Eq. (6.5).

In Eq. (6.5), $\hat{M}_i$ is the mass flow rate of substrate entering the system; therefore $\hat{M}_i = Fs|_z$ where F is the volumetric flow rate through the reactor and $s|_z$ is the substrate concentration at z. Similarly, $\hat{M}_o$, the mass flow rate of substrate leaving the section, is $Fs|_{z+\Delta z}$. Substrate is not generated in the reaction; therefore $R_G = 0$. Rate of substrate consumption R_C is equal to the volumetric rate of reaction v multiplied by the volume of the section. v is given by Eq. (11.31); the section volume is equal to $A\Delta z$ where A is the cross-sectional area of the reactor. At steady state, the left-hand side of Eq. (6.5) is zero. Therefore, the mass-balance equation is:

$$Fs|_z - Fs|_{z+\Delta z} - \frac{v_{max}\, s}{K_m + s}\, A\Delta z = 0$$

(13.79)

where v_{max} is the maximum rate of enzyme reaction and K_m is the Michaelis constant. Dividing through by $A\Delta z$ and rearranging gives:

$$\frac{F(s|_{z+\Delta z} - s|_z)}{A\Delta z} = \frac{-v_{max}\, s}{K_m + s}.$$

(13.80)

The volumetric flow rate F divided by the reactor cross-sectional area A is equal to the superficial velocity through the column, u. Therefore:

$$\frac{u(s|_{z+\Delta z} - s|_z)}{\Delta z} = \frac{-v_{max}\, s}{K_m + s}.$$

(13.81)

For F and A constant, u is also constant. Eq. (13.81) is valid for any section in the reactor of thickness Δz. For it to be valid at any point in the reactor, we must take the limit as $\Delta z \to 0$:

$$u\left(\lim_{\Delta z \to 0} \frac{s|_{z+\Delta z} - s|_z}{\Delta z}\right) = \frac{-v_{max}\, s}{K_m + s}$$

(13.82)

and apply the definition of the differential from Eq. (D.13) in Appendix D:

$$u\frac{ds}{dz} = \frac{-v_{max}\, s}{K_m + s}.$$

(13.83)

Figure 13.32 Flowsheet for a continuous plug-flow tubular reactor.

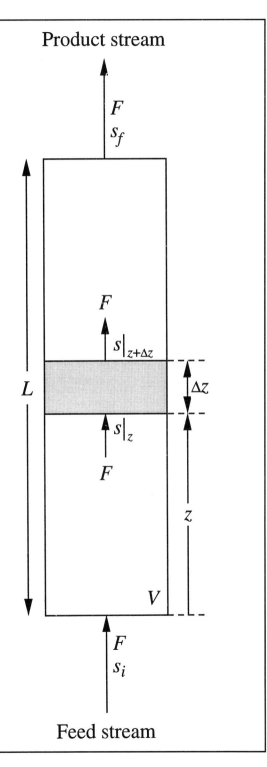

Product stream

F
s_f

F

$s\big|_{z+\Delta z}$

Δz

L

$s\big|_z$

F

z

V

F
s_i

Feed stream

Eq. (13.83) is a differential equation for the substrate concentration gradient through the length of a plug-flow reactor. Assuming u and the kinetic parameters are constant, Eq. (13.83) is ready for integration. Separating variables and integrating with boundary condition $s = s_i$ at $z = 0$ gives an expression for the reactor length L required to achieve an outlet concentration of s_f:

$$L = u \left[\frac{K_m}{v_{max}} \ln \frac{s_i}{s_f} + \frac{s_i - s_f}{v_{max}} \right].$$

(13.84)

Residence time τ for continuous reactors is defined in Eq. (13.51). If we divide V and F in Eq. (13.51) by A, we can express the residence time for plug-flow reactors in terms of parameters L and u:

$$\tau = \frac{V}{F} = \frac{\left(\dfrac{V}{A}\right)}{\left(\dfrac{F}{A}\right)} = \frac{L}{u}.$$

(13.85)

Therefore, Eq. (13.84) can be written as:

$$\tau = \frac{K_m}{v_{max}} \ln \frac{s_i}{s_f} + \frac{s_i - s_f}{v_{max}}.$$

(13.86)

Eqs (13.84) and (13.86) allow us to calculate the reactor length and residence time required to achieve conversion of substrate from concentration s_i to s_f at flow rate u. The form of Eq. (13.86) is identical to that of Eq. (13.10) for batch reactors. As in batch reactors where substrate and product concentrations vary continuously during the reaction period, concentrations in plug-flow reactors change continuously as material moves from inlet to outlet. Thus, plug-flow operation can be seen as a way of simulating batch culture in a continuous flow system.

Plug-flow operation is generally impractical for enzyme conversions unless the enzyme is immobilised and retained inside the vessel. For immobilised-enzyme reactions affected by diffusion, Eq. (13.83) must be modified to account for mass-transfer effects:

$$u \frac{ds}{dz} = -\eta_T \frac{-v_{max} s}{K_m + s}$$

(13.87)

where η_T is the total effectiveness factor representing internal and external mass-transfer limitations (see Section 12.5), s is the bulk substrate concentration, and v_{max} and K_m are intrinsic kinetic parameters. Because η_T is a function of s, we cannot integrate Eq. (13.87) directly as s varies with z in plug-flow reactors.

Plug-flow operation with immobilised enzyme is most likely to be approached in packed-bed reactors such as that shown in Figure 13.8. Packing in the column can cause substantial backmixing and axial dispersion of liquid, thus interfering with ideal plug flow. Nevertheless, application of the equations developed in this section can give satisfactory results for design of fixed-bed immobilised-enzyme reactors.

Example 13.7 Plug-flow reactor for immobilised enzymes

Immobilised lactase is used to hydrolyse lactose in dairy waste to glucose and galactose. Enzyme is immobilised in resin particles and packed into a 0.5 m³ column. The total effectiveness factor for the system is close to unity; K_m for the immobilised enzyme is 1.32 kg m⁻³; v_{max} is 45 kg m⁻³ h⁻¹. The lactose concentration in the feed stream is 9.5 kg m⁻³; a substrate conversion of 98% is required. The column is operated with plug flow for a total of 310 d per year.

(a) At what flow rate should the reactor be operated?
(b) How many tonnes of glucose are produced per year?

Solution:

(a) For 98% substrate conversion, $s_f = (0.02\, s_i) = 0.19$ kg m⁻³. Substituting into Eq. (13.86) gives:

$$\tau = \frac{1.32 \text{ kg m}^{-3}}{45 \text{ kg m}^{-3}\,\text{h}^{-1}} \ln\left(\frac{9.5 \text{ kg m}^{-3}}{0.19 \text{ kg m}^{-3}}\right) + \frac{(9.5 - 0.19) \text{ kg m}^{-3}}{45 \text{ kg m}^{-3}\,\text{h}^{-1}} = 0.32 \text{ h.}$$

From Eq. (13.51):

$$F = \frac{V}{\tau} = \frac{0.5 \text{ m}^3}{0.32 \text{ h}} = 1.56 \text{ m}^3\,\text{h}^{-1}.$$

(b) The rate of lactose conversion is equal to the difference between inlet and outlet mass flow rates of lactose:

$$F(s_i - s_f) = 1.56 \text{ m}^3\,\text{h}^{-1}\,(9.5 - 0.19) \text{ kg m}^{-3} = 14.5 \text{ kg h}^{-1}.$$

Converting this to an annual rate based on 310 d per year and a molecular weight for lactose of 342:

$$\text{Lactose converted} = 14.5 \text{ kg h}^{-1} \cdot \left|\frac{24 \text{ h}}{1 \text{ d}}\right| \cdot \left|\frac{310 \text{ d}}{1 \text{ yr}}\right| \cdot \left|\frac{1 \text{ kgmol}}{342 \text{ kg}}\right|$$

$$= 315 \text{ kgmol yr}^{-1}.$$

The enzyme reaction is:

lactose $+ H_2O \rightarrow$ glucose $+$ galactose.

Therefore, from reaction stoichiometry, 315 kgmol glucose are produced per year. The molecular weight of glucose is 180; therefore:

$$
\text{Glucose produced} = 315 \text{ kgmol yr}^{-1} . \left| \frac{180 \text{ kg}}{1 \text{ kgmol}} \right| . \left| \frac{1 \text{ tonne}}{1000 \text{ kg}} \right|
$$

$$
= 56.7 \text{ tonne yr}^{-1}.
$$

13.5.8.2 *Cell culture*

Analysis of plug-flow reactors for cell culture follows the same procedure as for enzyme reaction. If the cell specific growth rate is constant and equal to μ_{max} throughout the reactor and cell death can be neglected, the equations for reactor residence time are analogous to those derived in Section 13.5.1.2 for batch fermentation, e.g.:

$$
\tau = \frac{1}{\mu_{max}} \ln \frac{x_f}{x_i}
\tag{13.88}
$$

where τ is the reactor residence time defined in Eq. (13.51), x_i is the biomass concentration at the inlet and x_f is the biomass concentration at the outlet. The form of Eq. (13.88) is identical to that of Eq. (13.20) for batch reaction.

Plug-flow operation is not suitable for cultivation of suspended cells unless the biomass is recycled or there is continuous inoculation of the vessel. Plug-flow operation with cell recycle is used for large-scale wastewater treatment; however applications are limited. Plug-flow reactors are suitable for immobilised-cell reactions with catalyst packed into a fixed bed as shown in Figure 13.8. Even so, operating problems such as those mentioned in Section 13.2.5 mean that PFTRs are rarely employed for industrial fermentations.

13.5.9 Comparison Between Major Modes of Reactor Operation

The relative performance of batch, CSTR and PFTR reactors can be considered from a theoretical point of view in terms of the substrate conversion and product concentration obtained from vessels of the same size. Because the total reactor volume is not fully utilised at all times during fed-batch operation, it is difficult to include this mode of operation in a general comparison.

As indicated in Section 13.5.8, the kinetic characteristics of PFTRs are the same as batch reactors; the residence time required for conversion in a plug-flow reactor is therefore the same as in a mixed vessel operated in batch. It can also be shown theoretically that as the number of stages in a CSTR cascade increases, the conversion characteristics of the entire system approach those of an ideal plug-flow or mixed batch reactor. This is shown diagrammatically in Figure 13.33. The smooth dashed curve represents the progressive decrease in substrate concentration with time spent in a PFTR or batch reactor; concentration is reduced from s_i at the inlet to s_f at the outlet. In a single well-mixed CSTR operated with the same inlet and outlet concentrations, because conditions in the vessel are uniform there is a step change in substrate concentration as soon as the feed enters the reactor. In a cascade of CSTRs, the concentration is uniform in each reactor but there is a step-wise drop in concentration between each stage. As illustrated in Figure 13.33, the larger the number of units in a CSTR cascade, the closer the concentration profile approaches plug-flow or batch behaviour.

The benefits associated with particular reactor designs or modes of operation depend on the kinetic characteristics of the reaction. For zero-order reactions there is no difference between single batch, CSTR and PFTR reactors in terms of overall conversion rate. However, for most reactions including first-order and Michaelis–Menten conversions, rate of reaction decreases as the concentration of substrate decreases. Reaction rate is therefore high at the start of batch culture or at the entrance to a plug-flow reactor because the substrate level is greatest. Subsequently, the reaction velocity falls gradually as substrate is consumed. In contrast, substrate entering a CSTR is immediately diluted to the final or outlet steady-state concentration so that the rate of reaction is comparatively low for the entire reactor. Accordingly, for first-order and Michaelis–Menten reactions, CSTRs achieve lower substrate conversions and lower product concentrations than batch reactors or PFTRs of the same volume. In practice, batch processing is much preferred to PFTR systems because of the operating problems mentioned in Section 13.2.5. However, as discussed in Section 13.5.2, the total time for batch operation depends on the duration of the downtime between batches as well as on the actual conversion time. Because the length of downtime varies considerably from system to system, we cannot account for it here in a general way. Downtime between batches should be minimised as much as possible to maintain high overall production rates.

The comparison between reactors yields a different result if the reaction is autocatalytic. Catalyst is produced by the reaction in fermentation processes; therefore, the volumetric rate of reac-

Figure 13.33 Concentration changes in PFTR, single CSTR and multiple CSTR vessels.

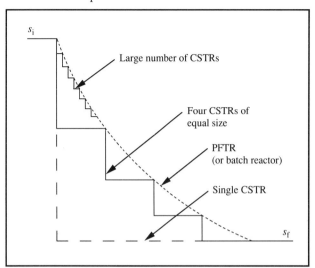

tion increases as the conversion proceeds because the amount of catalyst builds up. Volumetric reaction rate continues to increase until the substrate concentration becomes low, then it declines due to substrate depletion. At the beginning of batch culture, rate of substrate conversion is generally low because relatively few cells are present; it takes some time for cells to accumulate and the rate to pick up. However, in CSTR operation, substrate entering the vessel is immediately exposed to a relatively high biomass concentration so that the rate of conversion is also high. Rates of conversion in chemostats operated close to the optimum dilution rate for biomass productivity (see Section 13.5.4.2) are often 10–20 times greater than in PFTR or batch reactors. This rate advantage disappears if the steady-state substrate concentration in the CSTR is so small that, despite the higher biomass levels present, the conversion rate is lower than the average in a batch or PFTR device. For most fermentations however, CSTRs offer significant theoretical advantages over other modes of reactor operation.

Despite the productivity benefits associated with CSTRs, an overwhelming majority of commercial fermentations are conducted in batch. The reasons lie with the practical advantages associated with batch culture. Batch processes have a lower risk of contamination than continuous-flow reactors; equipment and control failures during long-term continuous operation are also potential problems. Continuous fermentation is feasible only when the cells are genetically stable; if developed strains revert to more rapidly-growing mutants the culture can become dominated over time by the revertant cells. In contrast, freshly-produced inocula are used in batch fermentations giving closer control over the genetic characteristics of the culture.

Continuous culture is not suitable for production of metabolites normally formed near stationary phase when the culture growth rate is low; as mentioned above, productivity in a batch reactor is likely to be greater than in a CSTR under these conditions. Continuous fermentations must be operated for lengthy periods to reap the full benefits of their high productivity. Production can be much more flexible with batch processing; for example, different products each with small market volumes can be made in different batches.

13.5.10 Evaluation of Kinetic and Yield Parameters in Chemostat Culture

In a steady-state chemostat with sterile feed and negligible cell death, the specific growth rate μ is equal to the dilution rate D. This relationship is useful for determining kinetic and yield parameters in cell culture. If growth can be modelling using Monod kinetics, for chemostat culture, Eq. (11.60) becomes:

$$D = \frac{\mu_{max}\, s}{K_S + s}$$

(13.89)

where μ_{max} is the maximum specific growth rate, K_S is the substrate constant and s is the steady-state substrate concentration in the reactor. Eq. (13.89) is analogous mathematically to the Michaelis–Menten expression for enzyme kinetics. If s is measured at various dilution rates, techniques described in Section 11.4 for determining v_{max} and K_m can be applied for evaluation of μ_{max} and K_S. Rearrangement of Eq. (13.89) gives the following linearised equations which can be used for Lineweaver–Burk, Eadie–Hofstee and Langmuir plots, respectively:

$$\frac{1}{D} = \frac{K_S}{\mu_{max}\, s} + \frac{1}{\mu_{max}}$$

(13.90)

$$\frac{D}{s} = \frac{\mu_{max}}{K_S} - \frac{D}{K_S}$$

(13.91)

and

$$\frac{s}{D} = \frac{K_S}{\mu_{max}} + \frac{s}{\mu_{max}}.$$

(13.92)

For example, according to Eq. (13.90), μ_{max} and K_S can be determined from the slope and intercept of a plot of $1/_D$ versus $1/_s$. The comments made in Sections 11.4.2–11.4.4 about distortion of experimental error apply also to Eqs (13.90)–(13.92).

Chemostat operation is also convenient for determining true yields and maintenance coefficients for cell cultures. An expression relating these parameters to the specific growth rate is given by Eq. (11.81). In chemostat culture with $\mu = D$, Eq. (11.81) becomes:

$$\frac{1}{Y'_{XS}} = \frac{1}{Y_{XS}} + \frac{m_S}{D}$$

(13.93)

where Y'_{XS} is the observed biomass yield from substrate, Y_{XS} is the true biomass yield from substrate and m_S is the maintenance coefficient. Therefore, as shown in Figure 13.34, a plot of $1/Y'_{XS}$ versus $1/D$ gives a straight line with slope m_S and intercept $1/Y_{XS}$. In a chemostat with sterile feed, the observed biomass yield from substrate Y'_{XS} is obtained as follows:

$$Y'_{XS} = \frac{x}{s_i - s}$$

(13.94)

where x and s are steady-state cell and substrate concentrations, respectively, and s_i is the inlet substrate concentration.

13.6 Sterilisation

Commercial fermentations typically require thousands of litres of liquid medium and millions of litres of air. For processes operated with axenic cultures, these raw materials must be provided free from contaminating organisms. Of all the methods available for sterilisation including chemical treatment, exposure to ultraviolet, gamma and X-ray radiation, sonication, filtration and heating, only the last two are used in large-scale operations. Aspects of fermenter design and construction for aseptic operation were considered in Sections 13.3.1 and 13.3.2. Here, we consider design of sterilisation systems for liquids and gases.

13.6.1 Batch Heat Sterilisation of Liquids

Liquid medium is most commonly sterilised in batch in the vessel where it will be used. The liquid is heated to sterilisation temperature by introducing steam into the coils or jacket of

Figure 13.34 Graphical determination of the maintenance coefficient m_S and true biomass yield Y_{XS} using data from chemostat culture.

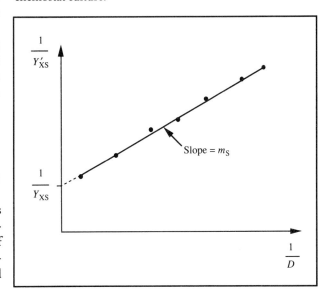

the vessel; alternatively, steam is bubbled directly into the medium, or the vessel is heated electrically. If direct steam injection is used, allowance must be made for dilution of the medium by condensate which typically adds 10–20% to the liquid volume; quality of the steam must also be sufficiently high to avoid contamination of the medium by metal ions or organics. A typical temperature–time profile for batch sterilisation is shown in Figure 13.35(a). Depending on the rate of heat transfer from the steam or electrical element, raising the temperature of the medium in large fermenters can take a significant period of time. Once the holding or sterilisation temperature is reached, the temperature is held constant for a period of time t_{hd}. Cooling water in the coils or jacket of the fermenter is then used to reduce the medium temperature to the required value.

For operation of batch sterilisation systems, we must be able to estimate the holding time required to achieve the desired level of cell destruction. As well as destroying contaminant organisms, heat sterilisation also destroys nutrients in the medium. To minimise this loss, holding times at the highest sterilisation temperature should be kept as short as possible. Cell death occurs at all times during batch sterilisation, including the heating-up and cooling-down periods. The holding time t_{hd} can be minimised by taking into account cell destruction during these periods.

Let us denote the number of contaminants present in the raw medium N_0. As indicated in Figure 13.35(b), during the

Figure 13.35 (a) Variation of temperature with time for batch sterilisation of liquid medium. (b) Reduction in number of viable cells during batch sterilisation.

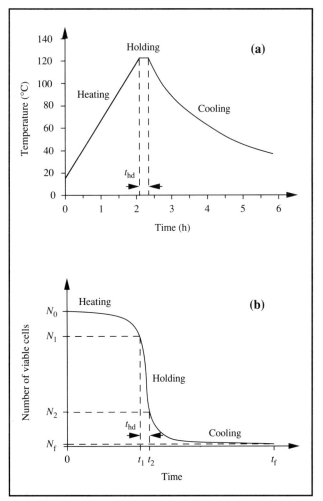

Rate of heat sterilisation is governed by the equations for thermal death outlined in Section 11.14. From Eq. (11.86) for first-order death kinetics, in a batch vessel where cell death is the only process affecting the number of viable cells:

$$\frac{dN}{dt} = -k_d N$$

(13.95)

where N is number of viable cells, t is time and k_d is the specific death constant. Eq. (13.95) applies to each stage of the batch sterilisation cycle: heating, holding and cooling. However, because k_d is a strong function of temperature, direct integration of Eq. (13.95) is valid only when the temperature is constant, i.e. during the holding period. The result is:

$$\ln \frac{N_1}{N_2} = k_d t_{hd}$$

(13.96)

or

$$t_{hd} = \frac{\ln \dfrac{N_1}{N_2}}{k_d}$$

(13.97)

where t_{hd} is the holding time, N_1 is the number of viable cells at the start of holding, and N_2 is the number of viable cells at the end of holding. k_d is evaluated as a function of temperature using the Arrhenius equation:

$$k_d = A\, e^{-E_d/RT}$$

(11.46)

where A is the Arrhenius constant or frequency factor, E_d is the activation energy for the thermal cell death, R is the ideal gas constant and T is absolute temperature.

To use Eq. (13.97) we must know N_1 and N_2. These numbers are determined by considering the extent of cell death during the heating and cooling periods when the temperature is not constant. Combining Eqs (13.95) and (11.46) gives:

$$\frac{dN}{dt} = -A\, e^{-E_d/RT} N.$$

(13.98)

Integration of Eq. (13.98) gives for the heating period:

heating period this number is reduced to N_1. At the end of the holding period, the cell number is N_2; the final number after cooling is N_f. Ideally, N_f is zero; at the end of the sterilisation cycle we want to have no contaminants present. However, because absolute sterility would require an infinitely-long sterilisation time, it is theoretically impossible to achieve. Normally, the target level of contamination is expressed as a fraction of a cell, which is related to the *probability* of contamination. For example, we could aim for an N_f value of 10^{-3}; this means we accept the risk that one batch in 1000 will not be sterile at the end of the process. If N_0 and N_f are known, we can determine the holding time required to reduce the number of cells from N_1 to N_2 by considering the kinetics of cell death.

Figure 13.36 Generalised temperature–time profiles for the heating and cooling stages of a batch sterilisation cycle. (From F.H. Deindoerfer and A.E. Humphrey, 1959, Analytical method for calculating heat sterilization times. *Appl. Microbiol.* 7, 256–264.)

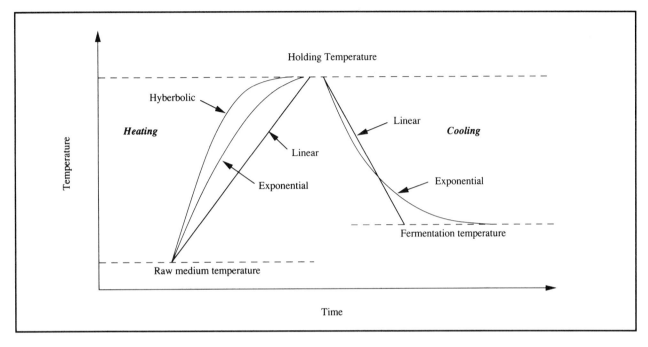

$$\ln \frac{N_0}{N_1} = \int_0^{t_1} A\, e^{-E_d/RT}\, dt$$

(13.99)

and, for the cooling period:

$$\ln \frac{N_2}{N_f} = \int_{t_2}^{t_f} A\, e^{-E_d/RT}\, dt$$

(13.100)

where t_1 is the time at the end of heating, t_2 is the time at the end of holding and t_f is the time at the end of cooling. We cannot complete integration of these equations until we know how the temperature varies with time during heating and cooling.

As outlined in Chapter 6, unsteady-state temperature profiles during heating and cooling can be determined from the heat transfer properties of the system. The general form of these equations is shown in Figure 13.36 and Table 13.3. Applying an appropriate expression for T in Eq. (13.99) from Table 13.3 allows us to evaluate the cell number N_1 at the start of the holding period. Similarly, substituting for T in Eq. (13.100) for cooling gives N_2 at the end of the holding period. Use of the resulting values for N_1 and N_2 in Eq. (13.97) completes the holding-time calculation.

Normally, cell death below about 100°C is minimal; however, when heating and cooling are relatively slow, temperatures remain close to the maximum for considerable periods of time and, as indicated in Figure 13.35(b), cell numbers can be reduced significantly outside of the holding period. Usually, holding periods are of the order of minutes whereas heating and cooling of large liquid volumes take hours. Sample design calculations for batch sterilisation are given by Aiba, Humphrey and Millis [40] and Richards [41].

The design procedures outlined in this section apply to batch sterilisation of medium when the temperature is uniform throughout the vessel. However, if the liquid contains contaminant particles in the form of flocs or pellets, temperature gradients may develop. Because heat transfer within solid particles is slower than in liquid, the temperature at the centre of the solid will be lower than that in the liquid for some proportion of the sterilising time. As a result, cell death inside the particles is not as effective as in the liquid. Longer holding times are required to treat solid-phase substrates and media containing particles.

When heat sterilisation is scaled up to larger volumes, longer treatment times are needed to achieve the same sterilisation results at the same holding temperature. For a given raw medium, the initial number of organisms N_0 is directly proportional to the liquid volume; therefore, to obtain the

Table 13.3 General equations for temperature as a function of time during the heating and cooling periods of batch sterilisation

(From F.H. Deindoerfer and A.E. Humphrey, 1959, Analytical method for calculating heat sterilization times. Appl. Microbiol. 7, 256–264)

Heat transfer method	Temperature–time profile

Heating

Direct sparging with steam

$$T = T_0 \left(1 + \frac{\dfrac{h \hat{M}_s t}{M_m C_p T_0}}{1 + \dfrac{\hat{M}_s}{M_m} t} \right)$$

(hyperbolic)

Electrical heating

$$T = T_0 \left(1 + \frac{\hat{Q} t}{M_m C_p T_0} \right)$$

(linear)

Heat transfer from isothermal steam

$$T = T_S \left[1 + \frac{T_0 - T_S}{T_S} \, e^{\left(\frac{-UAt}{M_m C_p} \right)} \right]$$

(exponential)

Cooling
Heat transfer to non-isothermal cooling water

$$T = T_{ci} \left\{ 1 + \frac{T_0 - T_{ci}}{T_{ci}} \, e^{\left[\left(\frac{-\hat{M}_w C_{pw} t}{M_m C_p} \right) \left(1 - e^{\left[\frac{-UA}{\hat{M}_w C_{pw}} \right]} \right) \right]} \right\}$$

(exponential)

A	=	surface area for heat transfer;
C_p	=	specific heat capacity of medium;
C_{pw}	=	specific heat capacity of cooling water;
h	=	specific enthalpy difference between steam and raw medium;
M_m	=	initial mass of medium;
$\hat{M}_s$	=	mass flow rate of steam;
$\hat{M}_w$	=	mass flow rate of cooling water;
$\hat{Q}$	=	rate of heat transfer;
T	=	temperature;
T_0	=	initial medium temperature;
T_{ci}	=	inlet temperature of cooling water;
T_S	=	steam temperature;
t	=	time; and
U	=	overall heat-transfer coefficient.

Figure 13.37 Continuous sterilising equipment: (a) continuous steam injection with flash cooling; (b) heat transfer using heat exchangers.

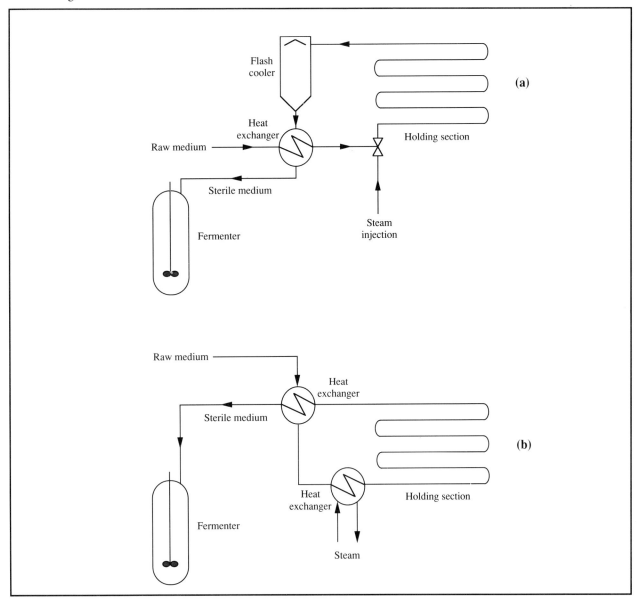

same final N_f, a greater number of cells must be destroyed. Scale-up also affects the temperature–time profiles for heating and cooling. Heat-transfer characteristics depend on the equipment used; heating and cooling of larger volumes usually take more time. Sustained elevated temperatures during heating and cooling are damaging to vitamins, proteins and sugars in nutrient solutions and reduce the quality of the medium [42]. Because it is necessary to hold large volumes of medium for longer periods of time, this problem is exacerbated with scale-up.

13.6.2 Continuous Heat Sterilisation of Liquids

Continuous sterilisation, particularly a high-temperature, short-exposure-time process, can significantly reduce damage

Figure 13.38 Variation of temperature with time in the continuous sterilisers of Figure 13.37.

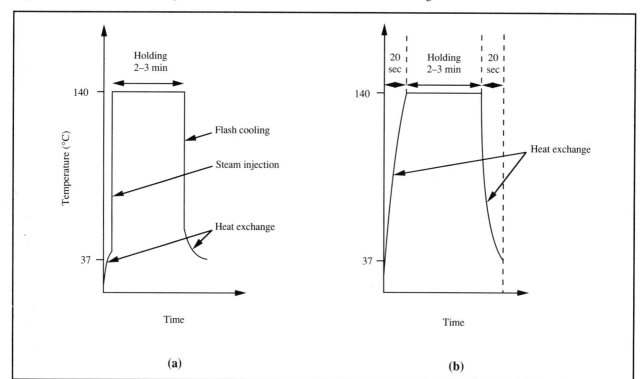

(a)   (b)

to medium ingredients while achieving high levels of cell destruction. Other advantages include improved steam economy and more reliable scale-up. The amount of steam needed for continuous sterilisation is 20–25% that used in batch processes; the time required is also significantly reduced because heating and cooling are virtually instantaneous.

Typical equipment configurations for continuous sterilisation are shown in Figure 13.37. In Figure 13.37(a), raw medium entering the system is first pre-heated by hot, sterile medium in a heat exchanger; this economises on steam requirements for heating and cools the sterile medium. Steam is then injected directly into the medium as it flows through a pipe; the liquid temperature rises almost instantaneously to the desired sterilisation temperature. The time of exposure to this temperature depends on the length of pipe in the holding section of the steriliser. After sterilisation, the medium is cooled instantly by passing it through an expansion valve into a vacuum chamber; further cooling takes place in the heat exchanger where residual heat is used to pre-heat incoming medium. Figure 13.37(b) shows an alternative sterilisation scheme based on heat exchange between steam and medium. Raw medium is pre-heated with hot, sterile medium in a heat exchanger then brought to the sterilisation temperature by further heat exchange with steam. The sterilisation temperature is maintained in the holding section before being reduced to the fermentation temperature by heat exchange with incoming medium. Heat-exchange systems are more expensive to construct than injection devices; fouling of the internal surfaces also reduces the efficiency of heat transfer between cleanings. On the other hand, a disadvantage associated with steam injection is dilution of the medium by condensate; foaming from direct steam injection can also cause problems with operation of the flash cooler. As indicated in Figure 13.38, rates of heating and cooling in continuous sterilisation are much more rapid than in batch; accordingly, in design of continuous sterilisers, contributions to cell death outside of the holding period are generally ignored.

An important variable affecting performance of continuous sterilisers is the nature of fluid flow in the system. Ideally, all fluid entering the equipment at a particular instant should spend the same time in the steriliser and exit the system at the same time; unless this occurs we cannot fully control the time spent in the steriliser by all fluid elements. No mixing should

Figure 13.39   Velocity distributions for flow in pipes. (a) In plug flow, fluid velocity is the same across the diameter of the pipe as indicated by the arrows of equal length. (b) In fully-developed turbulent flow, the velocity distribution approaches that of plug flow; however there is some reduction of flow speed at the walls. (c) In laminar flow, there is a continuous increase of velocity from the walls to the centre of the tube.

Figure 13.40   Correlation for determining the axial-dispersion coefficient in turbulent pipe flow. *Re* is Reynolds number, *D* is pipe diameter, *u* is average linear fluid velocity, ρ is fluid density, μ is fluid viscosity, and $\mathscr{D}_z$ is the axial-dispersion coefficient. Data were measured using single fluids in: (●) straight pipes; (■) pipes with bends; (□) artificially-roughened pipe; and (○) curved pipe. (From O. Levenspiel, 1958, Longitudinal mixing of fluids flowing in circular pipes. *Ind. Eng. Chem.* **50**, 343–346.)

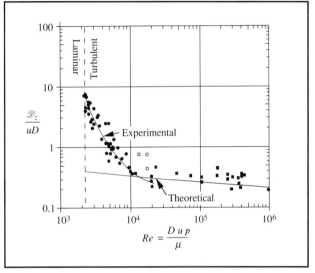

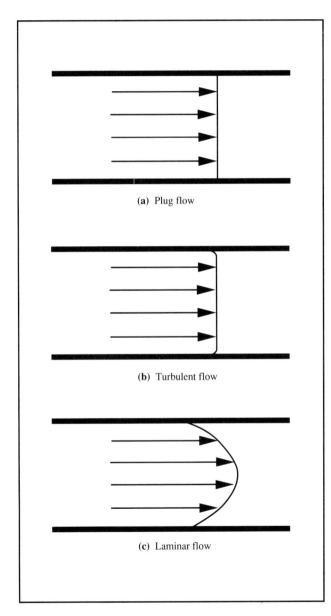

occur in the tubes; if fluid nearer the entrance of the pipe mixes with fluid ahead of it, there is a risk that contaminants will be transferred to the outlet of the steriliser. The type of flow in pipes where there is neither mixing nor variation in fluid velocity is called *plug flow* as already described in Section 13.5.8 for plug-flow reactors. Plug flow is an ideal flow pattern; in reality, fluid elements in pipes have a range of different velocities. As illustrated in Figure 13.39, flow tends to be faster through the centre of the tube than near the walls. However, plug flow is approached in pipes at turbulent Reynolds numbers (see Section 7.2.2) above about 2×10^4; operation at high Reynolds numbers minimises fluid mixing and velocity variation.

Deviation from plug-flow behaviour is characterised by the degree of *axial dispersion* in the system, i.e. the degree to which mixing occurs along the length or axis of the pipe. Axial dispersion is a critical factor affecting design of continuous sterilisers. The relative importance of axial dispersion and bulk flow in transfer of material through the pipe is represented by a dimensionless variable called the *Peclet number*:

$$Pe = \frac{uL}{\mathscr{D}_z}$$

(13.101)

where Pe is the Peclet number, u is the average linear fluid velocity, L is the pipe length and $\mathscr{D}_z$ is the *axial-dispersion coefficient*. For perfect plug flow, $\mathscr{D}_z$ is zero and Pe is infinitely large; in practice, Peclet numbers between 3 and 600 are typical. The value of $\mathscr{D}_z$ for a particular system depends on the Reynolds number and pipe geometry; a correlation from the engineering literature for evaluating $\mathscr{D}_z$ is shown in Figure 13.40. Once the Peclet number has been calculated from Eq. (13.101), the extent of cell destruction in the steriliser can be related to the specific death constant k_d using Figure 13.41. In Figure 13.41, N_1 is the number of viable cells entering the system, N_2 is the number of cells leaving, Pe is the Peclet number as defined by Eq. (13.101), and Da is another dimensionless number called the *Damköhler number*:

$$Da = \frac{k_d L}{u}$$

(13.102)

where k_d is the specific death constant, L is the length of the holding pipe and u is the average linear liquid velocity. The lower the value of N_2/N_1, the greater is the level of cell destruction. Figure 13.41 shows that, at any given sterilisation temperature defining the value of k_d and Da, performance of the steriliser declines significantly as the Peclet number decreases. Design calculations for a continuous steriliser are illustrated in Example 13.8.

Figure 13.41 Thermal destruction of contaminating organisms as a function of the Peclet number Pe and Damköhler number Da. N_1 is the number of viable cells entering the holding section of the steriliser; N_2 is the number of cells leaving. (From S. Aiba, A.E. Humphrey and N.F. Millis, 1965, *Biochemical Engineering*, Academic Press, New York.)

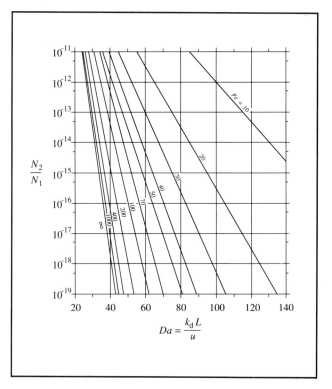

Example 13.8 Holding temperature in a continuous steriliser

Medium at a flow rate of 2 m³ h⁻¹ is to be sterilised by heat exchange with steam in a continuous steriliser. The liquid contains bacterial spores at a concentration of 5×10^{12} m⁻³; the activation energy and Arrhenius constant for thermal destruction of these contaminants are 283 kJ gmol⁻¹ and 5.7×10^{39} h⁻¹, respectively. A contamination risk of one organism surviving every 60 days' operation is considered acceptable. The steriliser pipe has an inner diameter of 0.1 m; the length of the holding section is 24 m. The density of the medium is 1000 kg m⁻³ and the viscosity is 3.6 kg m⁻¹ h⁻¹. What sterilising temperature is required?

Solution:

The desired level of cell destruction is evaluated using a basis of 60 days. Ignoring any cell death in the heating and cooling sections, the number of cells entering the holding section over 60 d is:

$$N_1 = 2 \text{ m}^3 \text{ h}^{-1} (5 \times 10^{12} \text{ m}^{-3}) . \left| \frac{24 \text{ h}}{1 \text{ d}} \right| . (60 \text{ d}) = 1.44 \times 10^{16}.$$

N_2, the acceptable number of cells leaving during this period is 1. Therefore:

$$\frac{N_2}{N_1} = \frac{1}{1.44 \times 10^{16}} = 6.9 \times 10^{-17}.$$

The linear velocity u in the steriliser is equal to the volumetric flow rate divided by the cross-sectional area of the pipe:

$$u = \frac{2 \text{ m}^3 \text{ h}^{-1}}{\pi \left(\frac{0.1 \text{ m}}{2}\right)^2} = 254.6 \text{ m h}^{-1}.$$

To calculate Pe we must first determine $\mathscr{D}_z$ using Figure 13.40:

$$Re = \frac{Du\rho}{\mu} = \frac{(0.1 \text{ m}) (254.6 \text{ m h}^{-1}) (1000 \text{ kg m}^{-3})}{3.6 \text{ kg m}^{-1} \text{ h}^{-1}} = 7.07 \times 10^3.$$

For $Re = 7.07 \times 10^3$ we can determine $\mathscr{D}_z$ from Figure 13.40 using either the experimental or theoretical curve. Let us choose the experimental curve as this gives a larger value of $\mathscr{D}_z$ and a smaller value of Pe; the steriliser design will thus be more conservative. Therefore, $\mathscr{D}_z/_{uD} = 0.65$:

$$\mathscr{D}_z = 0.65 \, (254.6 \text{ m h}^{-1}) \, (0.1 \text{ m}) = 16.6 \text{ m}^2 \text{ h}^{-1}.$$

From Eq. (13.101):

$$Pe = \frac{uL}{\mathscr{D}_z} = \frac{(254.6 \text{ m h}^{-1}) (24 \text{ m})}{16.6 \text{ m}^2 \text{ h}^{-1}} = 368.$$

Using Figure 13.41, we can determine the value of k_d for the desired level of cell destruction. Da corresponding to $N_2/_{N_1} = 6.9 \times 10^{-17}$ and $Pe = 368$ is about 42. Therefore:

$$k_d = \frac{u \, Da}{L} = \frac{(254.6 \text{ m h}^{-1}) (42)}{24 \text{ m}} = 445.6 \text{ h}^{-1}.$$

The sterilisation temperature can be evaluated after rearranging Eq. (11.46). Dividing both sides by A and taking natural logarithms gives:

$$\ln \frac{k_d}{A} = \frac{-E_d}{RT}.$$

Therefore:

$$T = \frac{\left(\frac{-E_d}{R}\right)}{\ln\left(\frac{k_d}{A}\right)}.$$

$E_d = 283 \text{ kJ gmol}^{-1} = 283 \times 10^3 \text{ J gmol}^{-1}$; $A = 5.7 \times 10^{39} \text{ h}^{-1}$; from Table 2.5, the ideal gas constant R is $8.3144 \text{ J K}^{-1} \text{ gmol}^{-1}$. Therefore:

$$T = \frac{\left(\frac{-283 \times 10^3 \text{ J gmol}^{-1}}{8.3144 \text{ J K}^{-1} \text{ gmol}^{-1}}\right)}{\ln\left(\frac{445.6 \text{ h}^{-1}}{5.7 \times 10^{39} \text{ h}^{-1}}\right)} = 398.4 \text{ K}.$$

Using the conversion between K and °C given in Eq. (2.24), $T = 125$°C.

Heating and cooling in continuous sterilisers are so rapid that in design calculations they are considered instantaneous. While reducing nutrient deterioration, this feature of the process can cause problems if there are solids present in the medium. During heating, the temperature at the core of solid particles remains lower than in the medium. Because of the extremely short contact times in continuous sterilisers compared with batch systems, there is a greater risk that particles will not be properly sterilised. It is important therefore that raw medium be clarified as much as possible before it enters a continuous steriliser.

13.6.3 Filter Sterilisation of Liquids

Sometimes, fermentation media or selected ingredients are sterilised by filtration rather than heat. For example, media containing heat-labile components such as enzymes and serum are easily destroyed by heat and must be sterilised by other means. Typically, membranes used for filter sterilisation of liquids are made of cellulose esters or other polymers and have pores between 0.2 and 0.45 μm in diameter. The membranes themselves must be sterilised before use, usually by steam. As medium is passed through the filter, bacteria and other particles with dimensions greater than the pore size are screened out and collect on the surface of the membrane. The small pore sizes used in liquid filtration mean that the membranes are readily blocked unless the medium is pre-filtered to remove any large particles. To achieve high flow rates, large surface areas are required.

Liquid filtration is generally not as effective or reliable as heat sterilisation. Viruses and mycoplasma are able to pass through membrane filters; care must also be taken to prevent holes or tears in the membrane. Usually, filter-sterilised medium is incubated for a period of time before use to test its sterility.

13.6.4 Sterilisation of Air

The number of microbial cells in air is of the order 10^3–10^4 m^{-3} [40]. Filtration is the most common method for sterilising air in large-scale bioprocesses; heat sterilisation of gases is economically impractical. *Depth filters* consisting of compacted beds or pads of fibrous material such as glass wool have been used widely in the fermentation industry. Distances between the fibres in depth filters are typically 2–10 μm, about 10 times greater than the dimensions of the bacteria and spores to be removed. Air-borne particles penetrate the bed to various depths before their passage through the filter is arrested; the depth of the filter medium required to produce air of sufficient

quality depends on the operating flow rate and the incoming level of contamination. Cells are collected in depth filters by a combination of impaction, interception, electrostatic effects, and, for particles smaller than about 1.0 μm, diffusion to the fibres. Depth filters do not perform well if there are large fluctuations in flow rate or if the air is wet; liquid condensing in the filter increases the pressure drop, causes channelling of the gas flow, and provides a pathway for organisms to grow through the bed.

Increasingly, depth filters are being replaced for industrial applications by membrane cartridge filters. These filters use steam-sterilisable polymeric membranes which act as *surface filters* trapping contaminants as on a sieve. Membrane filter cartridges typically contain a pleated, hydrophobic filter with small and uniformly-sized pores 0.45 μm or less in diameter. The hydrophobic nature of the surface minimises problems with filter wetting while the pleated configuration allows a high filtration area to be packed into a small cartridge volume. Pre-filters built into the cartridge or up-stream reduce fouling of the membrane by removing large particles, oil, water droplets and foam from the incoming gas.

Filters are also used to sterilise effluent gases leaving fermenters. In this application, the objective is to prevent release into the atmosphere of any microorganisms entrained in aerosols in the headspace of the reactor. The concentration of cells in fermenter off-gas is several times greater than in air. Containment is particularly important when organisms used in fermentation are potentially harmful to plant personnel or the environment; companies operating fermentations with pathogenic or recombinant strains are required by regulatory authorities to prevent escape of the cells.

13.7 Summary of Chapter 13

Chapter 13 contains a variety of qualitative and quantitative information about design and operation of bioreactors. After studying this chapter, you should:

(i) be able to assess in general terms the effect of reaction engineering on total production costs in bioprocessing;

(ii) be familiar with a range of bioreactor configurations in addition to the standard *stirred tank*, including *bubble-column*, *airlift*, *packed-bed*, *fluidised-bed* and *trickle-bed* designs;

(iii) understand the practical aspects of bioreactor construction, particularly those aimed at maintaining aseptic conditions;

(iv) be familiar with measurements used in *fermentation monitoring* and the problems associated with lack of on-line methods for important fermentation parameters;

v) be familiar with established and modern approaches to *fermentation control*;

vi) be able to predict *batch reaction times* for enzyme and cells reactions;

vii) be able to predict the performance of *fed-batch reactors* operated under quasi-steady-state conditions;

viii) be able to predict and compare the performance of *continuous stirred-tank reactors* and *continuous plug-flow reactors*;

ix) know how to use steady-state chemostat data to determine kinetic and yield parameters for cell culture; and

x) know how batch and continuous systems are designed for heat sterilisation of liquid medium and methods for filter sterilisation of fermentation gases.

Problems

3.1 Economics of batch enzyme conversion

An enzyme is used to convert substrate to a commercial product in a 1600-litre batch reactor. v_{max} for the enzyme is .9 g l^{-1} h^{-1}; K_m is 1.5 g l^{-1}. Substrate concentration at the start of the reaction is 3 g l^{-1}; according to the stoichiometry of the reaction, conversion of 1 g substrate produces 1.2 g product. The cost of operating the reactor including labour, maintenance, energy and other utilities is estimated at $4800 per day. The cost of recovering the product depends on the extent of substrate conversion and the resulting concentration of product in the final reaction mixture. For conversions between 70% and 100%, the cost of downstream processing can be approximated using the equation:

$$C = 155 - 0.33X$$

where C is cost in $ per kg product treated and X is the percentage substrate conversion. The market price for the product is $750 kg^{-1}. Currently, the enzyme reactor is operated with 75% substrate conversion; however it is proposed to increase this to 90%. Estimate the effect this will have on the economics of the process.

3.2 Batch production of aspartic acid using cell-bound enzyme

Aspartase enzyme is used industrially for manufacture of aspartic acid, a component of low-calorie sweetener. Fumaric acid and ammonia are converted to aspartic acid according to the equation:

$$C_4H_4O_4 + NH_3 \rightleftharpoons C_4H_7O_4N.$$
(fumaric acid)   (aspartic acid)

Under investigation is a process using aspartase in intact *Bacillus cadaveris* cells. In the substrate range of interest, the conversion can be described using Michaelis–Menten kinetics with K_m 4.0 g l^{-1}. The substrate solution contains 15% (w/v) ammonium fumarate; enzyme is added in the form of lyophilised cells and the reaction stopped when 85% of the substrate is converted. At 32°C, v_{max} for the enzyme is 5.9 g l^{-1} h^{-1} and its half-life is 10.5 d. At 37°C, v_{max} increases to 8.5 g l^{-1} h^{-1} but the half-life is reduced to 2.3 d.

(a) Which operating temperature would you recommend?
(b) The average downtime between batch reactions is 28 h. At the temperature chosen in (a), calculate the reactor volume required to produce 5000 tonnes of aspartic acid per year.

13.3 Prediction of batch culture time

A strain of *Escherichia coli* has been genetically engineered to produce human protein. A batch culture is started by inoculating 12 g cells into a 100-litre bubble-column fermenter containing 10 g l^{-1} glucose. The maximum specific growth rate of the culture is 0.9 h^{-1}; the biomass yield from glucose is 0.575 g g^{-1}.

(a) Estimate the time required to reach stationary phase.
(b) What will be the final cell density if the fermentation is stopped after only 70% of the substrate is consumed?

13.4 Fed-batch scheduling

Nicotiana tabacum cells are cultured to high density for production of polysaccharide gum. The reactor used is a stirred tank, containing initially 100 litres medium. The maximum specific growth rate of the culture is 0.18 d^{-1} and the yield of biomass from substrate is 0.5 g g^{-1}. The concentration of growth-limiting substrate in the medium is 3% (w/v). The reactor is inoculated with 1.5 g l^{-1} cells and operated in batch until the substrate is virtually exhausted; medium flow is then started at a rate of 4 l d^{-1}. Fed-batch operation occurs under quasi-steady-state conditions.

(a) Estimate the batch culture time and final biomass concentration.
(b) Fed-batch operation is carried out for 40 d. What is the final mass of cells in the reactor?
(c) The fermenter is available 275 d per year with a downtime between runs of 24 h. How much plant cell biomass is produced annually?

13.5 Fed-batch production of cheese starter culture

Lactobacillus casei is propagated under essentially anaerobic conditions to provide a starter culture for manufacture of Swiss cheese. The culture produces lactic acid as a by-product of energy metabolism. The system has the following characteristics:

$$
\begin{aligned}
Y_{XS} &= 0.23 \text{ kg kg}^{-1} \\
K_S &= 0.15 \text{ kg m}^{-3} \\
\mu_{max} &= 0.35 \text{ h}^{-1} \\
m_S &= 0.135 \text{ kg kg}^{-1} \text{ h}^{-1}
\end{aligned}
$$

A stirred fermenter is operated in fed-batch mode at quasi-steady state with a feed flow rate of 4 m^3 h^{-1} and feed substrate concentration of 80 kg m^{-3}. After 6 h, the liquid volume is 40 m^3.

(a) What was the initial culture volume?
(b) What is the concentration of substrate at quasi-steady state?
(c) What is the concentration of cells at quasi-steady state?
(d) What mass of cells is produced after 6 h fed-batch operation?

13.6 Continuous enzyme conversion in a fixed-bed reactor

A system is being developed to remove urea from the blood of patients with renal failure. A prototype fixed-bed reactor is set up with urease immobilised in 2-mm gel beads; buffered urea solution is recycled rapidly through the bed so that the system is well mixed. The urease reaction is:

$$(NH_2)_2CO + 3 H_2O \rightarrow 2 NH_4^+ + HCO_3^- + OH^-.$$

K_m for the immobilised urease is 0.54 g l^{-1}. The volume of beads in the reactor is 250 cm^3, the total amount of urease is 10^{-4} g, and the turnover number is 11 000 g NH$_4^+$ (g enzyme)$^{-1}$ s^{-1}. The effective diffusivity of urea in the gel is 7×10^{-6} cm^2 s^{-1}; external mass-transfer effects are negligible. The reactor is operated continuously with a liquid volume of 1 litre. The feed stream contains 0.42 g l^{-1} urea; the desired urea concentration is 0.02 g l^{-1}. Ignoring enzyme deactivation, what volume of urea solution can be treated in 30 min?

13.7 Batch and continuous biomass production

Pseudomonas methylotrophus is used to produce single-cell protein from methanol in a 1000 m^3 pressure-cycle airlift

fermenter. The biomass yield from substrate is 0.41 g g^{-1}, K_S is 0.7 mg l^{-1}, and the maximum specific growth rate is 0.44 h^{-1}. The medium contains 4% (w/v) methanol. A substrate conversion of 98% is desirable. The reactor may be operated either in batch or continuous mode. If operated in batch, an inoculum of 0.01% (w/v) is used and the downtime between batches is 20 h. If operated continuously, a downtime of 25 d is expected per year. Neglecting maintenance requirements, compare the annual biomass production from batch and continuous reactors.

13.8 Reactor design for immobilised enzymes

6-Aminopenicillanic acid used to produce semi-synthetic penicillins is prepared by enzymatic hydrolysis of fermentation-derived penicillin G. Penicillin acylase immobilised in alginate is being considered for the process; the immobilised-enzyme particles are small enough that mass-transfer effects can be ignored. The starting concentration of penicillin G is 10% (w/v); because of the high cost of the substrate, 99% conversion is required. Under these conditions, enzymatic conversion of penicillin G can be considered a first-order reaction. It has not been decided whether a batch, CSTR or plug-flow reactor would be most suitable. The downtime between batch reactions is expected to be 20 h. For the batch and CSTR reactors, the reaction rate constant is 0.8 $\times$ 10^{-4} s^{-1}; in the PFTR, the packing density of enzyme beads can be up to four times greater than in the other reactors. Determine the smallest reactor required to treat 400 tonnes penicillin G per year.

13.9 Two-stage chemostat for secondary metabolite production

A two-stage chemostat system is used for production of secondary metabolite. The volume of each reactor is 0.5 m^3; the flow rate of feed is 50 l h^{-1}. Mycelial growth occurs in the first reactor; the second reactor is used for product synthesis. The concentration of substrate in the feed is 10 g l^{-1}. Kinetic and yield parameters for the organism are:

$$
\begin{aligned}
Y_{XS} &= 0.5 \text{ kg kg}^{-1} \\
K_S &= 1.0 \text{ kg m}^{-3} \\
\mu_{max} &= 0.12 \text{ h}^{-1} \\
m_S &= 0.025 \text{ kg kg}^{-1} \text{ h}^{-1} \\
q_P &= 0.16 \text{ kg kg}^{-1} \text{ h}^{-1} \\
Y_{PS} &= 0.85 \text{ kg kg}^{-1}
\end{aligned}
$$

Assume that product synthesis is negligible in the first reactor and growth is negligible in the second reactor.

(a) Determine the cell and substrate concentrations entering the second reactor.

b) What is the overall substrate conversion?

c) What is the final concentration of product?

13.10 Kinetic analysis of bioremediating bacteria using a chemostat

A strain of *Ancylobacter* bacteria capable of growing on 1,2-dichloroethane is isolated from sediment in the river Rhine. The bacteria are to be used for on-site bioremediation of soil contaminated with chlorinated halogens. Kinetic parameters for the organism are determined using chemostat culture. A -litre fermenter is used with a 1,2-dichloroethane feed at a concentration of 100 μM. Steady-state substrate concentrations are measured as a function of flow rate.

Flow rate (ml h^{-1})	Substrate concentration (μM)
0	17.4
5	25.1
0	39.8
5	46.8
0	69.4
5	80.1
0	100

a) Determine μ_{max} and K_S for this organism.

b) Determine the maximum operating flow rate of the reactor.

13.11 Kinetic and yield parameters of an auxotrophic mutant

An *Enterobacter aerogenes* auxotroph capable of overproducing threonine has been isolated. The kinetic and yield parameters for this organism are investigated using a 2-litre chemostat with 10 g l^{-1} glucose in the feed. Steady-state cell and substrate concentrations are measured as a function of flow rate.

Flow rate (h^{-1})	Cell concentration (g l^{-1})	Substrate concentration (g l^{-1})
.0	3.15	0.010
.4	3.22	0.038
.6	3.27	0.071
.7	3.26	0.066
.8	3.21	0.095
.9	3.10	0.477

Determine the maximum specific growth rate, the substrate constant K_S, the maintenance coefficient, and the true biomass yield for this culture.

13.12 Continuous sterilisation

A 15-m^3 chemostat is operated with dilution rate 0.1 h^{-1}. A continuous steriliser with steam injection and flash cooling delivers sterilised medium to the fermenter. Medium in the holding section of the steriliser is maintained at 130°C. The concentration of contaminants in the raw medium is 10^5 ml^{-1}; an acceptable contamination risk is one organism every 3 months. The Arrhenius constant and activation energy for thermal death are estimated to be 7.5×10^{39} h^{-1} and 288.5 kJ gmol^{-1}, respectively. The steriliser pipe inner diameter is 12 cm. At 130°C the liquid density is 1000 kg m^{-3} and the viscosity is 4 kg m^{-1} h^{-1}.

(a) Assuming perfect plug flow, determine the length of the holding section.

(b) What length is required if axial-dispersion effects are taken into account?

(c) If the steriliser is constructed with length as determined in (a) and operated at 130°C as planned, what will be the rate of fermenter contamination?

References

1. Royce, P.N. (1993) A discussion of recent developments in fermentation monitoring and control from a practical perspective. *Crit. Rev. Biotechnol.* **13**, 117–149.

2. Curran, J.S., J. Smith and W. Holms (1989) Heat-and-power in industrial fermentation processes. *Appl. Energy* **34**, 9–20.

3. Maiorella, B.L., H.W. Blanch and C.R. Wilke (1984) Economic evaluation of alternative ethanol fermentation processes. *Biotechnol. Bioeng.* **26**, 1003–1025.

4. Heijnen, J.J. and K. van't Riet (1984) Mass transfer, mixing and heat transfer phenomena in low viscosity bubble column reactors. *Chem. Eng. J.* **28**, B21–B42.

5. van't Riet, K. and J. Tramper (1991) *Basic Bioreactor Design*, Marcel Dekker, New York.

6. Chisti, M.Y. (1989) *Airlift Bioreactors*, Elsevier Applied Science, Barking.

7. Verlaan, P., J. Tramper, K. van't Riet and K.Ch.A.M. Luyben (1986) A hydrodynamic model for an airlift-loop bioreactor with external loop. *Chem. Eng. J.* **33**, B43–B53.

8. Onken, U. and P. Weiland (1980) Hydrodynamics and

mass transfer in an airlift loop fermentor. *Eur. J. Appl. Microbiol. Biotechnol.* **10**, 31–40.

9. Verlaan, P., J.-C. Vos and K. van't Riet (1989) Hydrodynamics of the flow transition from a bubble column to an airlift-loop reactor. *J. Chem. Tech. Biotechnol.* **45**, 109–121.

10. Siegel, M.H., J.C. Merchuk and K. Schügerl (1986) Airlift reactor analysis: interrelationships between riser, downcomer, and gas–liquid separator behavior, including gas recirculation effects. *AIChE J.* **32**, 1585–1596.

11. Russell, A.B., C.R. Thomas and M.D. Lilly (1994) The influence of vessel height and top-section size on the hydrodynamic characteristics of airlift fermentors. *Biotechnol. Bioeng.* **43**, 69–76.

12. Atkinson, B. and F. Mavituna (1991) *Biochemical Engineering and Biotechnology Handbook*, 2nd edn, Macmillan, Basingstoke.

13. Morandi, M. and A. Valeri (1988) Industrial scale production of β-interferon. *Adv. Biochem. Eng./Biotechnol.* **37**, 57–72.

14. Cameron, B. (1987) Mechanical seals for bioreactors. *Chem. Engr.* No. 442 (November), 41–42.

15. Chisti, Y. (1992) Assure bioreactor sterility. *Chem. Eng. Prog.* **88** (September), 80–85.

16. Akiba, T. and T. Fukimbara (1973) Fermentation of volatile substrate in a tower-type fermenter with a gas entrainment process. *J. Ferment. Technol.* **51**, 134–141.

17. Stanbury, P.F. and A. Whitaker (1984) *Principles of Fermentation Technology*, Chapter 8, Pergamon Press, Oxford.

18. Kristiansen, B. (1987) Instrumentation. In: J. Bu'Lock and B. Kristiansen (Eds), *Basic Biotechnology*, pp. 253–281, Academic Press, London.

19. Carleysmith, S.W. (1987) Monitoring of bioprocessing. In: J.R. Leigh (Ed), *Modelling and Control of Fermentation Processes*, pp. 97–117, Peter Peregrinus, London.

20. Hall, E.A.H. (1991) *Biosensors*, Prentice-Hall, New Jersey.

21. Gronow, M. (1991) Biosensors — a marriage of biochemistry and microelectronics. In: V. Moses and R.E. Cape (Eds), *Biotechnology — the Science and the Business*, pp. 355–370, Harwood Academic, Chur.

22. Clarke, D.J., B.C. Blake-Coleman, R.J.G. Carr, M.R. Calder and T. Atkinson (1986) Monitoring reactor biomass. *Trends in Biotechnol.* **4**, 173–178.

23. Geisow, M.J. (1992) What's cooking? Optimizing bioprocess monitoring. *Trends in Biotechnol.* **10**, 230–232.

24. Royce, P.N. and N.F. Thornhill (1992) Analysis of noise and bias in fermentation oxygen uptake rate data. *Biotechnol. Bioeng.* **40**, 634–637.

25. Meiners, M. and W. Rapmundt (1983) Some practical aspects of computer applications in a fermentor hall. *Biotechnol. Bioeng.* **25**, 809–844.

26. van der Heijden, R.T.J.M., C. Hellinga, K.Ch.A.M. Luyben and G. Honderd (1989) State estimators (observers) for the on-line estimation of non-measurable process variables. *Trends in Biotechnol.* **7**, 205–209.

27. Montague, G.A., A.J. Morris and J.R. Bush (1988) Considerations in control scheme development for fermentation process control. *IEEE Contr. Sys. Mag.* **8** (April), 44–48.

28. Montague, G.A., A.J. Morris and M.T. Tham (1992) Enhancing bioprocess operability with generic software sensors. *J. Biotechnol.* **25**, 183–201.

29. Stephanopoulos, G. (1984) *Chemical Process Control: An Introduction to Theory and Practice*, Prentice-Hall, New Jersey.

30. Karim, M.N. and A. Halme (1989) Reconciliation of measurement data in fermentation using on-line expert system. In: N.M. Fish, R.I. Fox and N.F. Thornhill (Eds), *Computer Applications in Fermentation Technology: Modelling and Control of Biotechnological Processes*, pp. 37–46, Elsevier Applied Science, Barking.

31. Konstantinov, K.B. and T. Yoshida (1992) Knowledge-based control of fermentation processes. *Biotechnol. Bioeng.* **39**, 479–486.

32. Stephanopoulos, G. and G. Stephanopoulos (1986) Artificial intelligence in the development and design of biochemical processes. *Trends in Biotechnol.* **4**, 241–249.

33. Chen, Q., S.-Q. Wang and J.-C. Wang (1989) Application of expert system to the operation and control of industrial antibiotic fermentation process. In: N.M. Fish, R.I. Fox and N.F. Thornhill (Eds), *Computer Applications in Fermentation Technology: Modelling and Control of Biotechnological Processes*, pp. 253–261, Elsevier Applied Science, Barking.

34. Thibault, J., V. van Breusegem and A. Chéruy (1990) On-line prediction of fermentation variables using neural networks. *Biotechnol. Bioeng.* **36**, 1041–1048.

35. Willis, M.J., C. Di Massimo, G.A. Montague, M.T. Tham and A.J. Morris (1991) Artificial neural network in process engineering. *IEE Proc. D* **138**, 256–266.

36. Zhang, Q., J.F. Reid, J.B. Litchfield, J. Ren and S.-W. Chang (1994) A prototype neural network supervised control system for *Bacillus thuringiensis* fermentations. *Biotechnol. Bioeng.* **43**, 483–489.

37. Pirt, S.J. (1975) *Principles of Microbe and Cell Cultivation*, Blackwell Scientific, Oxford.

38. Shuler, M.L. and F. Kargi (1992) *Bioprocess Engineering: Basic Concepts*, Chapter 9, Prentice-Hall, New Jersey.

39. Reusser, F. (1961) Theoretical design of continuous antibiotic fermentation units. *Appl. Microbiol.* **9**, 361–366.

40. Aiba, S., A.E. Humphrey and N.F. Millis (1965) *Biochemical Engineering*, Academic Press, New York.

41. Richards, J.W. (1968) *Introduction to Industrial Sterilisation*, Academic Press, London.

42. Chopra, C.L., G.N. Qasi, S.K. Chaturvedi, C.N. Gaind and C.K. Atal (1981) Production of citric acid by submerged fermentation. Effect of medium sterilisation at pilot-plant level. *J. Chem. Technol. Biotechnol.* **31**, 122–126.

Suggestions for Further Reading

Reactor Configurations and Operating Characteristics (see also refs 5–12)

Chisti, M.Y. and M. Moo-Young (1987) Airlift reactors: characteristics, applications and design considerations. *Chem. Eng. Comm.* **60**, 195–242.

Cooney, C.L. (1983) Bioreactors: design and operation. *Science* **219**, 728–733.

Deckwer, W.-D. (1985) Bubble column reactors. In: H.-J. Rehm and G. Reed (Eds), *Biotechnology*, vol. 2, pp. 445–464, VCH, Weinheim.

Sittig, W. (1982) The present state of fermentation reactors. *J. Chem. Tech. Biotechnol.* **32**, 47–58.

Practical Considerations For Reactor Design (see also refs 14 and 15)

Chisti, Y. (1992) Build better industrial bioreactors. *Chem. Eng. Prog.* **88** (January), 55–58.

Fermentation Monitoring (see also refs 17–23)

Corcoran, C.A. and G.A. Rechnitz (1985) Cell-based biosensors. *Trends in Biotechnol.* **3**, 92–96.

Kennedy, M.J., M.S. Thakur, D.I.C. Wang and G.N. Stephanopoulos (1992) Estimating cell concentration in the presence of suspended solids: a light scatter technique. *Biotechnol. Bioeng.* **40**, 875–888.

Lübbert, A. (1992) Advanced methods for bioreactor characterization. *J. Biotechnol.* **25**, 145–182.

Mathewson, P.R. and J.W. Finley (Eds) (1992) *Biosensor Design and Application*, ACS Symposium Series 511, American Chemical Society, Washington DC.

North, J.R. (1985) Immunosensors: antibody-based biosensors. *Trends in Biotechnol.* **3**, 180–186.

Scheper, T., F. Plötz, C. Müller and B. Hitzmann (1994) Sensors as components of integrated analytical systems. *Trends in Biotechnol.* **12**, 42–46.

Parameter Estimation and Fermentation Control (see also refs 26–36)

Bastin, G. and D. Dochain (1990) *On-Line Estimation and Adaptive Control of Bioreactors*, Elsevier, Amsterdam.

Beck, M.B. (1986) Identification, estimation and control of biological waste-water treatment processes. *IEE Proc. D* **133**, 254–264.

Johnson, A. (1987) The control of fed-batch fermentation processes — a survey. *Automatica* **23**, 691–705.

Montague, G.A., A.J. Morris and A.C. Ward (1989) Fermentation monitoring and control: a perspective. *Biotechnol. Genet. Eng. Rev.* **7**, 147–188.

Munack, A. and M. Thoma (1986) Application of modern control for biotechnological processes. *IEE Proc. D* **133**, 194–198.

O'Connor, G.M., F. Sanchez-Riera and C.L. Cooney (1992) Design and evaluation of control strategies for high cell density fermentations. *Biotechnol. Bioeng.* **39**, 293–304.

Richards, J.R. (1987) Principles of control system design. In: J.R. Leigh (Ed), *Modelling and Control of Fermentation Processes*, pp. 189–214, Peter Peregrinus, London.

Shimizu, K. (1993) An overview on the control system design of bioreactors. *Adv. Biochem. Eng./Biotechnol.* **50**, 65–84.

Stephanopoulos, G. and K.-Y. San (1984) Studies on on-line bioreactor identification. Parts I and II. *Biotechnol. Bioeng.* **26**, 1176–1197.

Tarbuck, L.A., M.H. Ng, J. Tampion and J.R. Leigh (1986) Development of strategies for online estimation of biomass and secondary product formation in growth-limited batch fermentations. *IEE Proc. D* **133**, 235–239.

Sterilisation (see also refs 40 and 41)

Bader, F.G. (1986) Sterilization: prevention of contamination. In: A.L. Demain and N.A. Solomon (Eds), *Manual of Industrial Microbiology and Biotechnology*, pp. 345–362, American Society of Microbiology, Washington DC.

Conway, R.S (1985) Selection criteria for fermentation air filters. In: M. Moo-Young (Ed), *Comprehensive Biotechnology*, vol. 2, pp. 279–286, Pergamon Press, Oxford.

Cooney, C.L. (1985) Media sterilization. In: M. Moo-Young (Ed), *Comprehensive Biotechnology*, vol. 2, pp. 287–298, Pergamon Press, Oxford.

Deindoerfer, F.H. and A.E. Humphrey (1959) Analytical method for calculating heat sterilization times. *Appl. Microbiol.* **7**, 256–264.

Appendices

Conversion Factors

Note on Tables:
Entries in the same row are equivalent. For example, in Table A.1, 1 m = 3.281 ft, 1 mile = 1.609×10^3 m, etc. Exact numerical values are printed in bold type; others are given to four significant figures.

Table A.1 Length (L)

metre (m)	inch (in.)	foot (ft)	mile	micrometre (μm) (formerly micron)	angstrom (Å)
1	3.937×10^1	3.281	6.214×10^{-4}	10^6	10^{10}
2.54×10^{-2}	1	8.333×10^{-2}	1.578×10^{-5}	2.54×10^4	2.54×10^8
3.048×10^{-1}	1.2×10^1	1	1.894×10^{-4}	3.048×10^5	3.048×10^9
1.609×10^3	6.336×10^4	5.28×10^3	1	1.609×10^9	1.609×10^{13}
10^{-6}	3.937×10^{-5}	3.281×10^{-6}	6.214×10^{-10}	1	10^4
10^{-10}	3.937×10^{-9}	3.281×10^{-10}	6.214×10^{-14}	10^{-4}	1

Table A.2 Volume (L³)

cubic metre (m³)	litre (l or L) or cubic decimetre* (dm³)	cubic foot (ft³)	cubic inch (in.³)	imperial gallon (UKgal)	US gallon (USgal)
1	10^3	3.531×10^1	6.102×10^4	2.200×10^2	2.642×10^2
10^{-3}	1	3.531×10^{-2}	6.102×10^1	2.200×10^{-1}	2.642×10^{-1}
2.832×10^{-2}	2.832×10^1	1	1.728×10^3	6.229	7.481
1.639×10^{-5}	1.639×10^{-2}	5.787×10^{-4}	1	3.605×10^{-3}	4.329×10^{-3}
4.546×10^{-3}	4.546	1.605×10^{-1}	2.774×10^2	1	1.201
3.785×10^{-3}	3.785	1.337×10^{-1}	2.31×10^2	8.327×10^{-1}	1

* The litre was defined in 1964 as 1 dm³ exactly.

Table A.3 Mass (M)

kilogram (kg)	gram (g)	pound (lb)	ounce (oz)	tonne (t)	imperial ton (UK ton)	atomic mass unit* (u)
1	10^3	2.205	3.527×10^1	10^{-3}	9.842×10^{-4}	6.022×10^{26}
10^{-3}	1	2.205×10^{-3}	3.527×10^{-2}	10^{-6}	9.842×10^{-7}	6.022×10^{23}
4.536×10^{-1}	4.536×10^2	1	1.6×10^1	4.536×10^{-4}	4.464×10^{-4}	2.732×10^{26}
2.835×10^{-2}	2.835×10^1	6.25×10^{-2}	1	2.835×10^{-5}	2.790×10^{-5}	1.707×10^{25}
10^3	10^6	2.205×10^3	3.527×10^4	1	9.842×10^{-1}	6.022×10^{29}
1.016×10^3	1.016×10^6	2.240×10^3	3.584×10^4	1.016	1	6.119×10^{29}
1.661×10^{-27}	1.661×10^{-24}	3.661×10^{-27}	5.857×10^{-26}	1.661×10^{-30}	1.634×10^{-30}	1

* Atomic mass unit (unified); 1 u = $^1/_{12}$ of the rest mass of a neutral atom of the nuclide ^{12}C in the ground state.

Table A.4 Force (LMT^{-2})

newton (N, kg m s^{-2})	kilogram-force (kg$_f$)	pound-force (lb$_f$)	dyne (dyn, g cm s^{-2})	poundal (pdl, lb ft s^{-2})
1	1.020×10^{-1}	2.248×10^{-1}	**10^5**	7.233
9.807	1	2.205	9.807×10^5	7.093×10^1
4.448	4.536×10^{-1}	1	4.448×10^5	3.217×10^1
10^{-5}	1.020×10^{-6}	2.248×10^{-6}	1	7.233×10^{-5}
1.383×10^{-1}	1.410×10^{-2}	3.108×10^{-2}	1.383×10^4	1

Table A.5 Pressure and stress ($L^{-1}MT^{-2}$)

pascal (Pa, N m^{-2} J m^{-3} kg m^{-1} s^{-2})	pound force per inch^2 (psi, lb$_f$ in.$^{-2}$)	kilogram force per metre^2 (kg$_f$ m^{-2})	standard atmosphere (atm)	dyne per cm^2 (dyn cm^{-2})	torr (Torr, mmHg)*	inches of water** (inH$_2$O)	bar
1	1.450×10^{-4}	1.020×10^{-1}	9.869×10^{-6}	**10^1**	7.501×10^{-3}	4.015×10^{-3}	**10^{-5}**
6.895×10^3	1	7.031×10^2	6.805×10^{-2}	6.895×10^4	5.171×10^1	2.768×10^1	6.895×10^{-2}
9.807	1.422×10^{-3}	1	9.678×10^{-5}	9.807×10^1	7.356×10^{-2}	3.937×10^{-2}	9.807×10^{-5}
1.013×10^5	1.470×10^1	1.033×10^4	1	1.013×10^6	**7.6×10^2**	4.068×10^2	1.013
10^{-1}	1.450×10^{-5}	1.020×10^{-2}	9.869×10^{-7}	1	7.501×10^{-4}	4.015×10^{-4}	**10^{-6}**
1.333×10^2	1.934×10^{-2}	1.360×10^1	1.316×10^{-3}	1.333×10^3	1	5.352×10^{-1}	1.333×10^{-3}
2.491×10^2	3.613×10^{-2}	2.540×10^1	2.458×10^{-3}	2.491×10^3	1.868	1	2.491×10^{-3}
10^5	1.450×10^1	1.020×10^4	9.869×10^{-1}	**10^6**	7.501×10^2	4.015×10^2	1

*   mmHg refers to Hg at 0°C; 1 Torr = 1.00000 mmHg.

** in.H$_2$O refers to water at 4°C.

Table A.6 Surface tension (MT^{-2})

newton per metre (N m^{-1}, kg s^{-2}, J m^{-2})	dyne per centimetre (dyn cm^{-1}, g s^{-2}, erg cm^{-2})	kilogram-force per metre (kg$_f$ m^{-1})
1	**10^3**	1.020×10^{-1}
10^{-3}	1	1.020×10^{-4}
9.807	9.807×10^3	1

Table A.7 Energy, work and heat (L^2MT^{-2})

joule (J, N m, Pa m^3, W s, kg m^2 s^{-2})	kilocalorie* (kcal)	British thermal unit (Btu)	foot pound-force (ft lb$_f$)	litre atmosphere (l atm)	kilowatt hour (kW h)	erg (dyn cm)
1	2.388×10^{-4}	9.478×10^{-4}	7.376×10^{-1}	9.869×10^{-3}	2.778×10^{-7}	**10^7**
4.187×10^3	1	3.968	3.088×10^3	4.132×10^1	**1.163×10^{-3}**	4.187×10^{10}
1.055×10^3	2.520×10^{-1}	1	7.782×10^2	1.041×10^1	2.931×10^{-4}	1.055×10^{10}
1.356	3.238×10^{-4}	1.285×10^{-3}	1	1.338×10^{-2}	3.766×10^{-7}	1.356×10^7
1.013×10^2	2.420×10^{-2}	9.604×10^{-2}	7.473×10^1	1	2.815×10^{-5}	1.013×10^9
3.6×10^6	8.598×10^2	3.412×10^3	2.655×10^6	3.553×10^4	1	**3.6×10^{13}**
10^{-7}	2.388×10^{-11}	9.478×10^{-11}	7.376×10^{-8}	9.869×10^{-10}	2.778×10^{-14}	1

*   International Table kilocalorie (kcal$_{IT}$)

Table A.8 Power (L^2MT^{-3})

watt (W, J s^{-1}, kg m^2 s^{-3})	kilocalorie per min (kcal min^{-1})	foot pound-force per second (ft lb$_f$ s^{-1})	horsepower (British) (hp)	metric horsepower (–)	British thermal unit per minute (Btu min^{-1})	kilogram-force metre per second (kg$_f$ m s^{-1})
1	1.433×10^{-2}	7.376×10^{-1}	1.341×10^{-3}	1.360×10^{-3}	5.687×10^{-2}	1.020×10^{-1}
6.978×10^1	1	5.147×10^1	9.358×10^{-2}	9.487×10^{-2}	3.968	7.116
1.356	1.943×10^{-2}	1	1.818×10^{-3}	1.843×10^{-3}	7.710×10^{-2}	1.383×10^{-1}
7.457×10^2	1.069×10^1	5.5×10^2	1	1.014	4.241×10^1	7.604×10^1
7.355×10^2	1.054×10^1	5.425×10^2	9.863×10^{-1}	1	4.183×10^1	7.5×10^1
1.758×10^1	2.520×10^{-1}	1.297×10^1	2.358×10^{-2}	2.391×10^{-2}	1	1.793
9.807	1.405×10^{-1}	7.233	1.315×10^{-2}	1.333×10^{-2}	5.577×10^{-1}	1

Table A.9 Dynamic viscosity ($L^{-1}MT^{-1}$)

pascal second (Pa s, N s m^{-2}, kg m^{-1} s^{-1})	poise (g cm^{-1} s^{-1}, dyn s cm^{-2})	centipoise (cP)	kg m^{-1} h^{-1}	lb ft^{-1} h^{-1}
1	10^1	10^3	3.6×10^3	2.419×10^3
10^{-1}	1	10^2	3.6×10^2	2.419×10^2
10^{-3}	10^{-2}	1	3.6	2.419
2.778×10^{-4}	2.778×10^{-3}	2.778×10^{-1}	1	6.720×10^{-1}
4.134×10^{-4}	4.134×10^{-3}	4.134×10^{-1}	1.488	1

Table A.10 Plane angle (1)

radian (rad)	revolution (rev)	degree (°)	minute (′)	second (″)
1	1.592×10^{-1}	5.730×10^1	3.438×10^3	2.063×10^5
6.283	1	3.6×10^2	2.160×10^4	1.296×10^6
1.745×10^{-2}	2.778×10^{-3}	1	6×10^1	3.6×10^3
2.909×10^{-4}	4.630×10^{-5}	1.667×10^{-2}	1	6×10^1
4.848×10^{-6}	7.716×10^{-7}	2.778×10^{-4}	1.667×10^{-2}	1

Table A.11 Illuminance ($L^{-2}J$)

lux or lumen per metre^2 (lx or lm m^{-2})	foot-candle (fc, lm ft^{-2})
1	9.290×10^{-2}
1.076×10^1	1

B

Physical and Chemical Property Data

Table B.1 Atomic weights and numbers

Based on the atomic mass of ^{12}C. Values for atomic weights apply to elements as they exist in nature.

Name	Symbol	Relative atomic mass	Atomic number
Actinium	Ac	–	89
Aluminium	Al	26.9815	13
Americium	Am	–	95
Antimony	Sb	121.75	51
Argon	Ar	39.948	18
Arsenic	As	74.9216	33
Astatine	At	–	85
Barium	Ba	137.34	56
Berkelium	Bk	–	97
Beryllium	Be	9.0122	4
Bismuth	Bi	208.98	83
Boron	B	10.811	5
Bromine	Br	79.904	35
Cadmium	Cd	112.40	48
Caesium	Cs	132.905	55
Calcium	Ca	40.08	20
Californium	Cf	–	98
Carbon	C	12.011	6
Cerium	Ce	140.12	58
Chlorine	Cl	35.453	17
Chromium	Cr	51.996	24
Cobalt	Co	58.9332	27
Copper	Cu	63.546	29
Curium	Cm	–	96
Dysprosium	Dy	162.50	66
Einsteinium	Es	–	99
Erbium	Er	167.26	68
Europium	Eu	151.96	63
Fermium	Fm	–	100
Fluorine	F	18.9984	9
Francium	Fr	–	87
Gadolinium	Gd	157.25	64
Gallium	Ga	69.72	31
Germanium	Ge	72.59	32
Gold	Au	196.967	79
Hafnium	Hf	178.49	72
Helium	He	4.0026	2
Holmium	Ho	164.930	67
Hydrogen	H	1.00797	1
Indium	In	114.82	49
Iodine	I	126.9044	53
Iridium	Ir	192.2	77
Iron	Fe	55.847	26
Krypton	Kr	83.80	36
Lanthanum	La	138.91	57

Name	Symbol	Relative atomic mass	Atomic number
Lawrencium	Lr	–	103
Lead	Pb	207.19	82
Lithium	Li	6.939	3
Lutetium	Lu	174.97	71
Magnesium	Mg	24.312	12
Manganese	Mn	54.938	25
Mendelevium	Md	–	101
Mercury	Hg	200.59	80
Molybdenum	Mo	95.94	42
Neodymium	Nd	144.24	60
Neon	Ne	20.183	10
Neptunium	Np	–	93
Nickel	Ni	58.71	28
Niobium	Nb	92.906	41
Nitrogen	N	14.0067	7
Nobelium	No	–	102
Osmium	Os	190.2	76
Oxygen	O	15.9994	8
Palladium	Pd	106.4	46
Phosphorus	P	30.9738	15
Platinum	Pt	195.09	78
Plutonium	Pu	–	94
Polonium	Po	–	84
Potassium	K	39.102	19
Praseodymium	Pr	140.907	59
Promethium	Pm	–	61
Protactinium	Pa	–	91
Radium	Ra	–	88
Radon	Rn	–	86
Rhenium	Re	186.2	75
Rhodium	Rh	102.905	45
Rubidium	Rb	85.47	37
Ruthenium	Ru	101.07	44
Samarium	Sm	150.35	62
Scandium	Sc	44.956	21
Selenium	Se	78.96	34
Silicon	Si	28.086	14
Silver	Ag	107.868	47
Sodium	Na	22.9898	11
Strontium	Sr	87.62	38
Sulphur	S	32.064	16
Tantalum	Ta	180.948	73
Technetium	Tc	–	43
Tellurium	Te	127.60	52
Terbium	Tb	158.924	65
Thallium	Tl	204.37	81
Thorium	Th	232.038	90
Thulium	Tm	168.934	69
Tin	Sn	118.69	50
Titanium	Ti	47.90	22
Tungsten	W	183.85	74
Uranium	U	238.03	92
Vanadium	V	50.942	23
Wolfram (Tungsten)	W	183.85	74
Xenon	Xe	131.30	54
Ytterbium	Yb	173.04	70
Yttrium	Y	88.905	39
Zinc	Zn	65.37	30
Zirconium	Zr	91.22	40

Table B.2 Degree of reduction of biological materials

(Adapted from J.A. Roels, 1983, *Energetics and Kinetics in Biotechnology*, Elsevier Biomedical Press, Amsterdam)

Compound	Formula	Degree of reduction γ relative to NH_3	Degree of reduction γ relative to N_2
Acetaldehyde	C_2H_4O	5.00	5.00
Acetic acid	$C_2H_4O_2$	4.00	4.00
Acetone	C_3H_6O	5.33	5.33
Adenine	$C_5H_5N_5$	2.00	5.00
Alanine	$C_3H_7O_2N$	4.00	5.00
Ammonia	NH_3	0	3.00
Arginine	$C_6H_{14}O_2N_4$	3.67	5.67
Asparagine	$C_4H_8O_3N_2$	3.00	4.50
Aspartic acid	$C_4H_7O_4N$	3.00	3.75
n-Butanol	$C_4H_{10}O$	6.00	6.00
Butyraldehyde	C_4H_8O	5.50	5.50
Butyric acid	$C_4H_8O_2$	5.00	5.00
Carbon monoxide	CO	2.00	2.00
Citric acid	$C_6H_8O_7$	3.00	3.00
Cytosine	$C_4H_5ON_3$	2.50	4.75
Ethane	C_2H_6	7.00	7.00
Ethanol	C_2H_6O	6.00	6.00
Ethene	C_2H_4	6.00	6.00
Ethylene glycol	$C_2H_6O_2$	5.00	5.00
Ethyne	C_2H_2	5.00	5.00
Formaldehyde	CH_2O	4.00	4.00
Formic acid	CH_2O_2	2.00	2.00
Fumaric acid	$C_4H_4O_4$	3.00	3.00
Glucitol	$C_6H_{14}O_6$	4.33	4.33
Gluconic acid	$C_6H_{12}O_7$	3.67	3.67
Glucose	$C_6H_{12}O_6$	4.00	4.00
Glutamic acid	$C_5H_9O_4N$	3.60	4.20
Glutamine	$C_5H_{10}O_3N_2$	3.60	4.80
Glycerol	$C_3H_8O_3$	4.67	4.67
Glycine	$C_2H_5O_2N$	3.00	4.50
Graphite	C	4.00	4.00
Guanine	$C_5H_5ON_5$	1.60	4.60
Histidine	$C_6H_9O_2N_3$	3.33	4.83
Hydrogen	H_2	2.00	2.00
Isoleucine	$C_6H_{13}O_2N$	5.00	5.50
Lactic acid	$C_3H_6O_3$	4.00	4.00
Leucine	$C_6H_{13}O_2N$	5.00	5.50
Lysine	$C_6H_{14}O_2N_2$	4.67	5.67
Malic acid	$C_4H_6O_5$	3.00	3.00
Methane	CH_4	8.00	8.00
Methanol	CH_4O	6.00	6.00
Oxalic acid	$C_2H_2O_4$	1.00	1.00
Palmitic acid	$C_{16}H_{32}O_2$	5.75	5.75
Pentane	C_5H_{12}	6.40	6.40
Phenylalanine	$C_9H_{11}O_2N$	4.44	4.78
Proline	$C_5H_9O_2N$	4.40	5.00
Propane	C_3H_8	6.67	6.67
iso-Propanol	C_3H_8O	6.00	6.00
Propionic acid	$C_3H_6O_2$	4.67	4.67
Pyruvic acid	$C_3H_4O_3$	3.33	3.33
Serine	$C_3H_7O_3N$	3.33	4.33
Succinic acid	$C_4H_6O_4$	3.50	3.50

Compound	Formula	Degree of reduction γ relative to NH_3	Degree of reduction γ relative to N_2
Threonine	$C_4H_9O_3N$	4.00	4.75
Thymine	$C_5H_6O_2N_2$	3.20	4.40
Tryptophan	$C_{11}H_{12}O_2N_2$	4.18	4.73
Tyrosine	$C_9H_{11}O_3N$	4.22	4.56
Uracil	$C_4H_4O_2N_2$	2.50	4.00
Valeric acid	$C_5H_{10}O_2$	5.20	5.20
Valine	$C_5H_{11}O_2N$	4.80	5.40
Biomass	$CH_{1.8}O_{0.5}N_{0.2}$	4.20	4.80

Table B.3 Heat capacities

(Adapted from R.M. Felder and R.W. Rousseau, 1978, *Elementary Principles of Chemical Processes*, John Wiley and Sons, New York)

$$C_p\,(\mathrm{J\,gmol^{-1}\,°C^{-1}}) = a + bT + cT^2 + d\,T^3$$

Example. For acetone gas between 0°C and 1200°C:

$$C_p\,(\mathrm{J\,gmol^{-1}\,°C^{-1}}) = 71.96 + (20.10 \times 10^{-2})\,T - (12.78 \times 10^{-5})\,T^2 + (34.76 \times 10^{-9})\,T^3, \text{ where } T \text{ is in °C.}$$

Note that some equations require T in K, as indicated.

State: g = gas; l = liquid; c = crystal.

Compound	State	Temperature unit	a	$b.10^2$	$c.10^5$	$d.10^9$	Temperature range (units of T)
Acetone	g	°C	71.96	20.10	−12.78	34.76	0–1200
Air	g	°C	28.94	0.4147	0.3191	−1.965	0–1500
	g	K	28.09	0.1965	0.4799	−1.965	273–1800
Ammonia	g	°C	35.15	2.954	0.4421	−6.686	0–1200
Calcium hydroxide	c	K	89.5				276–373
Carbon dioxide	g	°C	36.11	4.233	−2.887	7.464	0–1500
Ethanol	l	°C	103.1				0
	l	°C	158.8				100
	g	°C	61.34	15.72	−8.749	19.83	0–1200
Formaldehyde	g	°C	34.28	4.268	0.000	−8.694	0–1200
Hydrogen	g	°C	28.84	0.00765	0.3288	−0.8698	0–1500
Hydrogen chloride	g	°C	29.13	−0.1341	0.9715	−4.335	0–1200
Hydrogen sulphide	g	°C	33.51	1.547	0.3012	−3.292	0–1500
Methane	g	°C	34.31	5.469	0.3661	−11.00	0–1200
	g	K	19.87	5.021	1.268	−11.00	273–1500
Methanol	l	°C	75.86				0
	l	°C	82.59				40
	g	°C	42.93	8.301	−1.87	−8.03	0–700
Nitric acid	l	°C	110.0				25
Nitrogen	g	°C	29.00	0.2199	0.5723	−2.871	0–1500
Oxygen	g	°C	29.10	1.158	−0.6076	1.311	0–1500
Sulphur (rhombic)	c	K	15.2	2.68			273–368
(monoclinic)	c	K	18.3	1.84			368–392
Sulphuric acid	l	°C	139.1	15.59			10–45
Sulphur dioxide	g	°C	38.91	3.904	−3.104	8.606	0–1500
Water	l	°C	75.4				0–100
	g	°C	33.46	0.6880	0.7604	-3.593	0–1500

Table B.4 Mean heat capacities of gases

(Adapted from D.M. Himmelblau, 1974, *Basic Principles and Calculations in Chemical Engineering*, 3rd edn, Prentice-Hall, New Jersey)

Reference state: $T_{ref} = 0°C$; $P_{ref} = 1$ atm.

$T(°C)$	C_{pm} (J gmol^{-1} °C^{-1})					
	Air	O_2	N_2	H_2	CO_2	H_2O
0	29.06	29.24	29.12	28.61	35.96	33.48
18	29.07	29.28	29.12	28.69	36.43	33.51
25	29.07	29.30	29.12	28.72	36.47	33.52
100	29.14	29.53	29.14	28.98	38.17	33.73
200	29.29	29.93	29.23	29.10	40.12	34.10
300	29.51	30.44	29.38	29.15	41.85	34.54
400	29.78	30.88	29.60	29.22	43.35	35.05
500	30.08	31.33	29.87	29.28	44.69	35.59

Table B.5 Specific heats of organic liquids

(From R.H. Perry, D.W. Green and J.O. Maloney, Eds, 1984, *Chemical Engineers' Handbook*, 6th edn, McGraw-Hill, New York)

Compound	Formula	Temperature (°C)	C_p (cal g^{-1} °C^{-1})
Acetic acid	$C_2H_4O_2$	26 to 95	0.522
Acetone	C_3H_6O	3 to 22.6	0.514
		0	0.506
		24.2 to 49.4	0.538
Acetonitrile	C_2H_3N	21 to 76	0.541
Benzaldehyde	C_7H_6O	22 to 172	0.428
Butyl alcohol (*n-*)	$C_4H_{10}O$	2.3	0.526
		19.2	0.563
		21 to 115	0.687
		30	0.582
Butyric acid (*n-*)	$C_4H_8O_2$	0	0.444
		40	0.501
		20 to 100	0.515
Carbon tetrachloride	CCl_4	0	0.198
		20	0.201
		30	0.200
Chloroform	$CHCl_3$	0	0.232
		15	0.226
		30	0.234
Cresol			
(*o-*)	C_7H_8O	0 to 20	0.497
(*m-*)		21 to 197	0.551
		0 to 20	0.477
Dichloroacetic acid	$C_2H_2Cl_2O_2$	21 to 106	0.349
		21 to 196	0.348
Diethylamine	$C_4H_{11}N$	22.5	0.516
Diethyl malonate	$C_7H_{12}O_4$	20	0.431
Diethyl oxalate	$C_6H_{10}O_4$	20	0.431
Diethyl succinate	$C_8H_{14}O_4$	20	0.450
Dipropyl malonate	$C_9H_{16}O_4$	20	0.431
Dipropyl oxalate (*n-*)	$C_8H_{14}O_4$	20	0.431
Dipropyl succinate	$C_{10}H_{18}O_4$	20	0.450

Compound	Formula	Temperature (°C)	C_p (cal g^{-1} °C^{-1})
Ethanol	C_2H_6O	0 to 98	0.680
Ether	$C_4H_{10}O$	−5	0.525
		0	0.521
		30	0.545
		80	0.687
		120	0.800
		140	0.819
		180	1.037
Ethyl acetate	$C_4H_8O_2$	20	0.457
		20	0.476
Ethylene glycol	$C_2H_6O_2$	−11.1	0.535
		0	0.542
		2.5	0.550
		5.1	0.554
		14.9	0.569
		19.9	0.573
Formic acid	CH_2O_2	0	0.436
		15.5	0.509
		20 to 100	0.524
Furfural	$C_5H_4O_2$	0	0.367
		20 to 100	0.416
Glycerol	$C_3H_8O_3$	15 to 50	0.576
Hexadecane (n-)	$C_{16}H_{34}$	0 to 50	0.496
Isobutyl acetate	$C_6H_{12}O_2$	20	0.459
Isobutyl alcohol	$C_4H_{10}O$	21 to 109	0.716
		30	0.603
Isobutyl succinate	$C_{12}H_{22}O_4$	0	0.442
Isobutyric acid	$C_4H_8O_2$	20	0.450
Lauric acid	$C_{12}H_{24}O_2$	40 to 100	0.572
		57	0.515
Methanol	CH_4O	5 to 10	0.590
		15 to 20	0.601
Methyl butyl ketone	$C_6H_{12}O$	21 to 127	0.553
Methyl ethyl ketone	C_4H_8O	20 to 78	0.549
Methyl formate	$C_2H_4O_2$	13 to 29	0.516
Methyl propionate	$C_4H_8O_2$	20	0.459
Palmitic acid	$C_{16}H_{32}O_2$	65 to 104	0.653
Propionic acid	$C_3H_6O_2$	0	0.444
		20 to 137	0.560
Propyl acetate (n-)	$C_5H_{10}O_2$	20	0.459
Propyl butyrate	$C_7H_{14}O_2$	20	0.459
Propyl formate (n-)	$C_4H_8O_2$	20	0.459
Pyridine	C_5H_5N	20	0.405
		21 to 108	0.431
		0 to 20	0.395
Quinoline	C_9H_7N	0 to 20	0.352
Salicylaldehyde	$C_7H_6O_2$	18	0.382
Stearic acid	$C_{18}H_{36}O_2$	75 to 137	0.550

Table B.6 Specific heats of organic solids

(From R.H. Perry, D.W. Green and J.O. Maloney, Eds, 1984, *Chemical Engineers' Handbook*, 6th edn, McGraw-Hill, New York)

Compound	Formula	Temperature (°C) T	C_p (cal g^{-1} °C^{-1})
Acetic acid	$C_2H_4O_2$	-200 to 25	$0.330 + 0.00080\,T$
Acetone	C_3H_6O	-210 to -80	$0.540 + 0.0156\,T$
Aniline	C_6H_7N		0.741
Capric acid	$C_{10}H_{20}O_2$	8	0.695
Chloroacetic acid	$C_2H_3ClO_2$	60	0.363
Crotonic acid	$C_4H_6O_2$	38 to 70	$0.520 + 0.00020\,T$
Dextrin	$(C_6H_{10}O_5)_x$	0 to 90	$0.291 + 0.00096\,T$
Diphenylamine	$C_{12}H_{11}N$	26	0.337
Erythritol	$C_4H_{10}O_4$	60	0.351
Ethylene glycol	$C_2H_6O_2$	-190 to -40	$0.366 + 0.00110\,T$
Formic acid	CH_2O_2	-22	0.387
		0	0.430
Glucose	$C_6H_{12}O_6$	0	0.277
		20	0.300
Glutaric acid	$C_5H_8O_4$	20	0.299
Glycerol	$C_3H_8O_3$	0	0.330
Hexadecane	$C_{16}H_{34}$		0.495
Lactose	$C_{12}H_{22}O_{11}$	20	0.287
	$C_{12}H_{22}O_{11}.H_2O$	20	0.299
Lauric acid	$C_{12}H_{24}O_2$	-30 to 40	$0.430 + 0.000027\,T$
Levoglucosane	$C_6H_{10}O_5$	40	0.607
Levulose	$C_6H_{12}O_6$	20	0.275
Malonic acid	$C_3H_4O_4$	20	0.275
Maltose	$C_{12}H_{22}O_{11}$	20	0.320
Mannitol	$C_6H_{14}O_6$	0 to 100	$0.313 + 0.00025\,T$
Oxalic acid	$C_2H_2O_4$	-200 to 50	$0.259 + 0.00076\,T$
	$C_2H_2O_4.2H_2O$	0	0.338
		50	0.385
		100	0.416
Palmitic acid	$C_{16}H_{32}O_2$	0	0.382
		20	0.430
Phenol	C_6H_6O	14 to 26	0.561
Succinic acid	$C_4H_6O_4$	0 to 160	$0.248 + 0.00153\,T$
Sucrose	$C_{12}H_{22}O_{11}$	20	0.299
Sugar (cane)	$C_{12}H_{22}O_{11}$	22 to 51	0.301
Tartaric acid	$C_4H_6O_6$	36	0.287
	$C_4H_6O_6.H_2O$	0	0.308
		50	0.366
Urea	CH_4N_2O	20	0.320

Table B.7 Normal melting points and boiling points, and standard heats of phase change

(From R.M. Felder and R.W. Rousseau, 1978, *Elementary Principles of Chemical Processes*, John Wiley, New York).

All thermodynamic data are at 1 atm.

Compound	Molecular weight	Melting temperature (°C)	Δh_f at melting point (kJ gmol^{-1})	Normal boiling point (°C)	Δh_v at boiling point (kJ gmol^{-1})
Acetaldehyde	44.05	−123.7		20.2	25.1
Acetic acid	60.05	16.6	12.09	118.2	24.39
Acetone	58.08	−95.0	5.69	56.0	30.2
Ammonia	17.03	−77.8	5.653	−33.43	23.351
Benzaldehyde	106.12	−26.0		179.0	38.40
Carbon dioxide	44.01	−56.6	8.33	(sublimates at −78°C)	
Chloroform	119.39	−63.7		61.0	
Ethanol	46.07	−114.6	5.021	78.5	38.58
Formaldehyde	30.03	−92		−19.3	24.48
Formic acid	46.03	8.30	12.68	100.5	22.25
Glycerol	92.09	18.20	18.30	290.0	
Hydrogen	2.016	−259.19	0.12	−252.76	0.904
Hydrogen chloride	36.47	−114.2	1.99	−85.0	16.1
Hydrogen sulphide	34.08	−85.5	2.38	−60.3	18.67
Methane	16.04	−182.5	0.94	−161.5	8.179
Methanol	32.04	−97.9	3.167	64.7	35.27
Nitric acid	63.02	−41.6	10.47	86	30.30
Nitrogen	28.02	−210.0	0.720	−195.8	5.577
Oxalic acid	90.04			(decomposes at 186°C)	
Oxygen	32.00	−218.75	0.444	−182.97	6.82
Phenol	94.11	42.5	11.43	181.4	
Phosphoric acid	98.00	42.3	10.54		
Sodium chloride	58.45	808	28.5	1465	170.7
Sodium hydroxide	40.00	319	8.34	1390	
Sulphur					
(rhombic)	256.53	113	10.04	444.6	83.7
(monoclinic)	256.53	119	14.17	444.6	83.7
Sulphur dioxide	64.07	−75.48	7.402	−10.02	24.91
Sulphuric acid	98.08	10.35	9.87	(decomposes at 340°C)	
Water	18.016	0.00	6.0095	100.00	40.656

Table B.8 Heats of combustion

(From *Handbook of Chemistry and Physics*, 1992, 73rd edn, CRC Press, Boca Raton; *Handbook of Chemistry and Physics*, 1976, 57th edn, CRC Press, Boca Raton; and R.M. Felder and R.W. Rousseau, 1978, *Elementary Principles of Chemical Processes*, John Wiley, New York)

Reference conditions: 1 atm and 25°C or 20°C; values marked with an asterisk refer to 20°C.
Products of combustion are taken to be CO_2 (gas), H_2O (liquid) and N_2 (gas); therefore, $\Delta h_c^\circ = 0$ for CO_2 (g), H_2O (l) and N_2 (g).
State: g = gas; l = liquid; c = crystal; s = solid.

Compound	Formula	Molecular weight	State	Heat of combustion Δh_c° (kJ gmol^{-1})
Acetaldehyde	C_2H_4O	44.053	l	−1166.9
			g	−1192.5
Acetic acid	$C_2H_4O_2$	60.053	l	−874.2
			g	−925.9
Acetone	C_3H_6O	58.080	l	−1789.9
			g	−1820.7

Compound	Formula	Molecular weight	State	Heat of combustion Δh_c° (kJ gmol^{-1})
Acetylene	C_2H_2	26.038	g	−1301.1
Adenine	$C_5H_5N_5$	135.128	c	−2778.1
			g	−2886.9
Alanine (D-)	$C_3H_7O_2N$	89.094	c	−1619.7
Alanine (L-)	$C_3H_7O_2N$	89.094	c	−1576.9
			g	−1715.0
Ammonia	NH_3	17.03	g	−382.6
Ammonium ion	NH_4^+			−383
Arginine (D-)	$C_6H_{14}O_2N_4$	174.203	c	−3738.4
Asparagine (L-)	$C_4H_8O_3N_2$	132.119	c	−1928.0
Aspartic acid (L-)	$C_4H_7O_4N$	133.104	c	−1601.1
Benzaldehyde	C_7H_6O	106.124	l	−3525.1
			g	−3575.4
Butanoic acid	$C_4H_8O_2$	88.106	l	−2183.6
			g	−2241.6
1-Butanol	$C_4H_{10}O$	74.123	l	−2675.9
			g	−2728.2
2-Butanol	$C_4H_{10}O$	74.123	l	−2660.6
			g	−2710.3
Butyric acid	$C_4H_8O_2$	88.106	l	−2183.6
			g	−2241.6
Caffeine	$C_8H_{10}O_2N_4$		s	−4246.5*
Carbon	C	12.011	c	−393.5
Carbon monoxide	CO	28.010	g	−283.0
Citric acid	$C_6H_8O_7$		s	−1962.0
Codeine	$C_{18}H_{21}O_3N.H_2O$		s	−9745.7*
Cytosine	$C_4H_5ON_3$	111.103	c	−2067.3
Ethane	C_2H_6	30.070	g	−1560.7
Ethanol	C_2H_6O	46.069	l	−1366.8
			g	−1409.4
Ethylene	C_2H_4	28.054	g	−1411.2
Ethylene glycol	$C_2H_6O_2$	62.068	l	−1189.2
			g	−1257.0
Formaldehyde	CH_2O	30.026	g	−570.7
Formic acid	CH_2O_2	46.026	l	−254.6
			g	−300.7
Fructose (D-)	$C_6H_{12}O_6$		s	−2813.7
Fumaric acid	$C_4H_4O_4$	116.073	c	−1334.0
Galactose (D-)	$C_6H_{12}O_6$		s	−2805.7
Glucose (D-)	$C_6H_{12}O_6$		s	−2805.0
Glutamic acid (L-)	$C_5H_9O_4N$	147.131	c	−2244.1
Glutamine (L-)	$C_5H_{10}O_3N_2$	146.146	c	−2570.3
Glutaric acid	$C_5H_8O_4$	132.116	c	−2150.9
Glycerol	$C_3H_8O_3$	92.095	l	−1655.4
			g	−1741.2
Glycine	$C_2H_5O_2N$	75.067	c	−973.1
Glycogen	$(C_6H_{10}O_5)_x$ per kg		s	−17530.1*
Guanine	$C_5H_5ON_5$	151.128	c	−2498.2
Hexadecane	$C_{16}H_{34}$	226.446	l	−10699.2
			g	−10780.5
Hexadecanoic acid	$C_{16}H_{32}O_2$	256.429	c	−9977.9
			l	−10031.3
			g	−10132.3
Histidine (L-)	$C_6H_9O_2N_3$	155.157	c	−3180.6
Hydrogen	H_2	2.016	g	−285.8
Hydrogen sulphide	H_2S	34.08		−562.6
Inositol	$C_6H_{12}O_6$		s	−2772.2*
Isoleucine (L-)	$C_6H_{13}O_2N$	131.175	c	−3581.1
Isoquinoline	C_9H_7N	129.161	l	−4686.5

Compound	Formula	Molecular weight	State	Heat of combustion Δh_c^o (kJ gmol^{-1})
Lactic acid (D,L-)	$C_3H_6O_3$		l	-1368.3
Lactose	$C_{12}H_{22}O_{11}$		s	-5652.5
Leucine (D-)	$C_6H_{13}O_2N$	131.175	c	-3581.7
Leucine (L-)	$C_6H_{13}O_2N$	131.175	c	-3581.6
Lysine	$C_6H_{14}O_2N_2$	146.189	c	-3683.2
Malic acid (L-)	$C_4H_6O_5$		s	-1328.8
Malonic acid	$C_3H_4O_4$		s	-861.8
Maltose	$C_{12}H_{22}O_{11}$		s	-5649.5
Mannitol (D-)	$C_6H_{14}O_6$		s	-3046.5^*
Methane	CH_4	16.043	g	-890.8
Methanol	CH_4O	32.042	l	-726.1
			g	-763.7
Morphine	$C_{17}H_{19}O_3N.H_2O$		s	-8986.6^*
Nicotine	$C_{10}H_{14}N_2$		l	-5977.8^*
Oleic acid	$C_{18}H_{34}O_2$		l	-11126.5
Oxalic acid	$C_2H_2O_4$	90.036	c	-251.1
Papaverine	$C_{20}H_{21}O_4N$		s	-10375.8^*
Pentane	C_5H_{12}	72.150	l	-3509.0
			g	-3535.6
Phenylalanine (L-)	$C_9H_{11}O_2N$	165.192	c	-4646.8
Phthalic acid	$C_8H_6O_4$	166.133	c	-3223.6
Proline (L-)	$C_5H_9O_2N$	115.132	c	-2741.6
Propane	C_3H_8	44.097	g	-2219.2
1-Propanol	C_3H_8O	60.096	l	-2021.3
			g	-2068.8
2-Propanol	C_3H_8O	60.096	l	-2005.8
			g	-2051.1
Propionic acid	$C_3H_6O_2$	74.079	l	-1527.3
			g	-1584.5
1,2-Propylene glycol	$C_3H_8O_2$	76.095	l	-1838.2
			g	-1902.6
1,3-Propylene glycol	$C_3H_8O_2$	76.095	l	-1859.0
			g	-1931.8
Pyridine	C_5H_5N	79.101	l	-2782.3
			g	-2822.5
Pyrimidine	$C_4H_4N_2$	80.089	l	-2291.6
			g	-2341.6
Salicylic acid	$C_7H_6O_3$	138.123	c	-3022.2
			g	-3117.3
Serine (L-)	$C_3H_7O_3N$	105.094	c	-1448.2
Starch	$(C_6H_{10}O_5)_x$ per kg		s	-17496.6^*
Succinic acid	$C_4H_6O_4$	118.089	c	-1491.0
Sucrose	$C_{12}H_{22}O_{11}$		s	-5644.9
Thebaine	$C_{19}H_{21}O_3N$		s	-10221.7^*
Threonine (L-)	$C_4H_9O_3N$	119.120	c	-2053.1
Thymine	$C_5H_6O_2N_2$	126.115	c	-2362.2
Tryptophan (L-)	$C_{11}H_{12}O_2N_2$	204.229	c	-5628.3
Tyrosine (L-)	$C_9H_{11}O_3N$	181.191	c	-4428.6
Uracil	$C_4H_4O_2N_2$	112.088	c	-1716.3
			g	-1842.8
Urea	CH_4ON_2	60.056	c	-631.6
			g	-719.4
Valine (L-)	$C_5H_{11}O_2N$	117.148	c	-2921.7
			g	-3084.5
Xanthine	$C_5H_4O_2N_4$	152.113	c	-2159.6
Xylose	$C_5H_{10}O_5$		s	-2340.5
Biomass	$CH_{1.8}O_{0.5}N_{0.2}$	25.9	s	-552

Steam Tables

(From R.W. Haywood, *Thermodynamic Tables in SI (Metric) Units*, 1972, 2nd edn, Cambridge University Press, Cambridge)

Table C.1 Enthalpy of saturated water and steam (Temperatures from 0.01°C to 100°C)

Reference state: Triple point of water: 0.01°C, 0.6112 kPa.

Temperature (°C)	Pressure (kPa)	Specific enthalpy (kJ kg^{-1})		
		Saturated liquid	Evaporation (Δh_v)	Saturated vapour
0.01 (Triple point)	0.611	+0.0	2501.6	2501.6
2	0.705	8.4	2496.8	2505.2
4	0.813	16.8	2492.1	2508.9
6	0.935	25.2	2487.4	2512.6
8	1.072	33.6	2482.6	2516.2
10	1.227	42.0	2477.9	2519.9
12	1.401	50.4	2473.2	2523.6
14	1.597	58.8	2468.5	2527.2
16	1.817	67.1	2463.8	2530.9
18	2.062	75.5	2459.0	2534.5
20	2.34	83.9	2454.3	2538.2
22	2.64	83.9	2454.3	2538.2
24	2.98	100.6	2444.9	2545.5
25	3.17	104.8	2442.5	2547.3
26	3.36	108.9	2440.2	2549.1
28	3.78	117.3	2435.4	2552.7
30	4.24	125.7	2430.7	2556.4
32	4.75	134.0	2425.9	2560.0
34	5.32	142.4	2421.2	2563.6
36	5.94	150.7	2416.4	2567.2
38	6.62	159.1	2411.7	2570.8
40	7.38	167.5	2406.9	2574.4
42	8.20	175.8	2402.1	2577.9
44	9.10	184.2	2397.3	2581.5
46	10.09	192.5	2392.5	2585.1
48	11.16	200.9	2387.7	2588.6
50	12.34	209.3	2382.9	2592.2
52	13.61	217.6	2378.1	2595.7
54	15.00	226.0	2373.2	2599.2
56	16.51	234.4	2368.4	2602.7
58	18.15	242.7	2363.5	2606.2
60	19.92	251.1	2358.6	2609.7
62	21.84	259.5	2353.7	2613.2
64	23.91	267.8	2348.8	2616.6
66	26.15	276.2	2343.9	2620.1
68	28.56	284.6	2338.9	2623.5
70	31.16	293.0	2334.0	2626.9

Temperature (°C)	Pressure (kPa)	Specific enthalpy (kJ kg^{-1})		
		Saturated liquid	Evaporation (Δh_v)	Saturated vapour
72	33.96	301.4	2329.0	2630.3
74	36.96	309.7	2324.0	2633.7
76	40.19	318.1	2318.9	2637.1
78	43.65	326.5	2313.9	2640.4
80	47.36	334.9	2308.8	2643.8
82	51.33	343.3	2303.8	2647.1
84	55.57	351.7	2298.6	2650.4
86	60.11	360.1	2293.5	2653.6
88	64.95	368.5	2288.4	2656.9
90	70.11	376.9	2283.2	2660.1
92	75.61	385.4	2278.0	2663.4
94	81.46	393.8	2272.8	2666.6
96	87.69	402.2	2267.5	2669.7
98	94.30	410.6	2262.2	2672.9
100 (Boiling point)	101.325	419.1	2256.9	2676.0

Table C.2 Enthalpy of saturated water and steam

(Pressures from 0.6112 kPa to 22 120 kPa)

Reference state: Triple point of water: 0.01°C, 0.6112 kPa.

Pressure (kPa)	Temperature (°C)	Specific enthalpy (kJ kg^{-1})		
		Saturated liquid	Evaporation (Δh_v)	Saturated vapour
0.6112 (Triple point)	0.01	+0.0	2501.6	2501.6
0.8	3.8	15.8	2492.6	2508.5
1.0	7.0	29.3	2485.0	2514.4
1.4	12.0	50.3	2473.2	2523.5
1.8	15.9	66.5	2464.1	2530.6
2.0	17.5	73.5	2460.2	2533.6
2.4	20.4	85.7	2453.3	2539.0
2.8	23.0	96.2	2447.3	2543.6
3.0	24.1	101.0	2444.6	2545.6
3.5	26.7	111.8	2438.5	2550.4
4.0	29.0	121.4	2433.1	2554.5
4.5	31.0	130.0	2428.2	2558.2
5.0	32.9	137.8	2423.8	2561.6
6	36.2	151.5	2416.0	2567.5
7	39.0	163.4	2409.2	2572.6
8	41.5	173.9	2403.2	2577.1
9	43.8	183.3	2397.9	2581.1
10	45.8	191.8	2392.9	2584.8
12	49.4	206.9	2384.3	2591.2
14	52.6	220.0	2376.7	2596.7
16	55.3	231.6	2370.0	2601.6
18	57.8	242.0	2363.9	2605.9
20	60.1	251.5	2358.4	2609.9

Pressure (kPa)	Temperature (°C)	Specific enthalpy (kJ kg^{-1})		
		Saturated liquid	Evaporation (Δh_v)	Saturated vapour
24	64.1	268.2	2348.6	2616.8
28	67.5	282.7	2340.0	2622.7
30	69.1	289.3	2336.1	2625.4
35	72.7	304.3	2327.2	2631.5
40	75.9	317.7	2319.2	2636.9
45	78.7	329.6	2312.0	2641.7
50	81.3	340.6	2305.4	2646.0
55	83.7	350.6	2299.3	2649.9
60	86.0	359.9	2293.6	2653.6
65	88.0	368.6	2288.3	2656.9
70	90.0	376.8	2283.3	2660.1
80	93.5	391.7	2274.1	2665.8
90	96.7	405.2	2265.6	2670.9
100	99.6	417.5	2257.9	2675.4
101.325 (Boiling point)	100.0	419.1	2256.9	2676.0
120	104.8	439.4	2244.1	2683.4
140	109.3	458.4	2231.9	2690.3
160	113.3	475.4	2220.9	2696.2
180	116.9	490.7	2210.8	2701.5
200	120.2	504.7	2201.6	2706.3
220	123.3	517.6	2193.0	2710.6
240	126.1	529.6	2184.9	2714.5
260	128.7	540.9	2177.3	2718.2
280	131.2	551.4	2170.1	2721.5
300	133.5	561.4	2163.2	2724.7
320	135.8	570.9	2156.7	2727.6
340	137.9	579.9	2150.4	2730.3
360	139.9	588.5	2144.4	2732.9
380	141.8	596.8	2138.6	2735.3
400	143.6	604.7	2133.0	2737.6
420	145.4	612.3	2127.5	2739.8
440	147.1	619.6	2122.3	2741.9
460	148.7	626.7	2117.2	2743.9
480	150.3	633.5	2112.2	2745.7
500	151.8	640.1	2107.4	2747.5
550	155.5	655.8	2095.9	2751.7
600	158.8	670.4	2085.0	2755.5
650	162.0	684.1	2074.7	2758.9
700	165.0	697.1	2064.9	2762.0
750	167.8	709.3	2055.5	2764.8
800	170.4	720.9	2046.5	2767.5
850	172.9	732.0	2037.9	2769.9
900	175.4	742.6	2029.5	2772.1
950	177.7	752.8	2021.4	2774.2
1000	179.9	762.6	2013.6	2776.2
1100	184.1	781.1	1998.5	2779.7
1200	188.0	798.4	1984.3	2782.7
1300	191.6	814.7	1970.7	2785.4
1400	195.0	830.1	1957.7	2787.8
1500	198.3	844.7	1945.2	2789.9
1600	201.4	858.6	1933.2	2791.7
1700	204.3	871.8	1921.5	2793.4
1800	207.1	884.6	1910.3	2794.8

Pressure (kPa)	Temperature (°C)	Specific enthalpy (kJ kg^{-1})		
		Saturated liquid	Evaporation (Δh_v)	Saturated vapour
1900	209.8	896.8	1899.3	2796.1
2000	212.4	908.6	1888.6	2797.2
2200	217.2	931.0	1868.1	2799.1
2400	221.8	951.9	1848.5	2800.4
2600	226.0	971.7	1829.6	2801.4
2800	230.0	990.5	1811.5	2802.0
3000	233.8	1008.4	1793.9	2802.3
3200	237.4	1025.4	1776.9	2802.3
3400	240.9	1041.8	1760.3	2802.1
3600	244.2	1057.6	1744.2	2801.7
3800	247.3	1072.7	1728.4	2801.1
4000	250.3	1087.4	1712.9	2800.3
4200	253.2	1101.6	1697.8	2799.4
4400	256.0	1115.4	1682.9	2798.3
4600	258.8	1128.8	1668.3	2797.1
4800	261.4	1141.8	1653.9	2795.7
5000	263.9	1154.5	1639.7	2794.2
5200	266.4	1166.8	1625.7	2792.6
5400	268.8	1178.9	1611.9	2790.8
5600	271.1	1190.8	1598.2	2789.0
5800	273.3	1202.3	1584.7	2787.0
6000	275.6	1213.7	1571.3	2785.0
6200	277.7	1224.8	1558.0	2782.9
6400	279.8	1235.7	1544.9	2780.6
6600	281.8	1246.5	1531.9	2778.3
6800	283.8	1257.0	1518.9	2775.9
7000	285.8	1267.4	1506.0	2773.5
7200	287.7	1277.6	1493.3	2770.9
7400	289.6	1287.7	1480.5	2768.3
7600	291.4	1297.6	1467.9	2765.5
7800	293.2	1307.4	1455.3	2762.8
8000	295.0	1317.1	1442.8	2759.9
8400	298.4	1336.1	1417.9	2754.0
8800	301.7	1354.6	1393.2	2747.8
9000	303.3	1363.7	1380.9	2744.6
10000	311.0	1408.0	1319.7	2727.7
11000	318.0	1450.6	1258.7	2709.3
12000	324.6	1491.8	1197.4	2689.2
13000	330.8	1532.0	1135.0	2667.0
14000	336.6	1571.6	1070.7	2642.4
15000	342.1	1611.0	1004.0	2615.0
16000	347.3	1650.5	934.3	2584.9
17000	352.3	1691.7	859.9	2551.6
18000	357.0	1734.8	779.1	2513.9
19000	361.4	1778.7	692.0	2470.6
20000	365.7	1826.5	591.9	2418.4
21000	369.8	1886.3	461.3	2347.6
22000	373.7	2011	185	2196
22120 (Critical point)	374.15	2108	0	2108

Table C.3 Enthalpy of superheated steam

Reference state: Triple point of water: 0.01°C, 0.6112 kPa.

Pressure (kPa)	10	50	100	500	1000	2000	4000	6000	8000	10000	15000	20000	22120*	30000	50000
Saturation temperature (°C)	45.8	81.3	99.6	151.8	179.9	212.4	250.3	275.6	295.0	311.0	342.1	365.7	374.15	–	–
State						Specific enthalpy at saturation (kJ kg^{-1})									
Water	191.8	340.6	417.5	640.1	762.6	908.6	1087.4	1213.7	1317.1	1408.0	1611.0	1826.5	2108	–	–
Steam	2584.8	2646.0	2675.4	2747.5	2776.2	2797.2	2800.3	2785.0	2759.9	2727.7	2615.0	2418.4	2108	–	–
Temperature (°C)						Specific enthalpy (kJ kg^{-1})									
0	0.0	0.0	0.1	0.5	1.0	2.0	4.0	6.1	8.1	10.1	15.1	20.1	22.2	30.0	49.3
25	104.8	104.8	104.9	105.2	105.7	106.6	108.5	110.3	112.1	114.0	118.6	123.1	125.1	132.2	150.2
50	2593	209.3	209.3	209.7	210.1	211.0	212.7	214.4	216.1	217.8	222.1	226.4	228.2	235.0	251.9
75	2640	313.9	314.0	314.3	314.7	315.5	317.1	318.7	320.3	322.0	326.0	330.0	331.7	338.1	354.2
100	2688	2683	2676	419.4	419.7	420.5	422.0	423.5	425.0	426.5	430.3	434.0	435.7	441.6	456.8
125	2735	2731	2726	525.5	525.5	526.2	527.6	529.0	530.4	531.8	535.3	538.8	540.2	545.8	560.1
150	2783	2780	2776	632.2	632.5	633.1	634.3	635.6	636.8	638.1	641.3	644.5	645.8	650.9	664.1
175	2831	2829	2826	2800	741.1	741.7	742.7	743.8	744.9	746.0	748.7	751.5	752.7	757.2	769.1
200	2880	2878	2875	2855	2827	852.6	853.4	854.2	855.1	855.9	858.1	860.4	861.4	865.2	875.4
225	2928	2927	2925	2909	2886	2834	967.2	967.7	968.2	968.8	970.3	971.8	972.5	975.3	983.4
250	2977	2976	2975	2961	2943	2902	1085.8	1085.8	1085.8	1085.8	1086.2	1086.7	1087.0	1088.4	1093.6
275	3027	3026	3024	3013	2998	2965	2886	1210.8	1210.0	1209.2	1207.7	1206.6	1206.3	1205.6	1206.7
300	3077	3076	3074	3065	3052	3025	2962	2885	2787	1343.4	1338.3	1334.3	1332.8	1328.7	1323.7
325	3127	3126	3125	3116	3106	3083	3031	2970	2899	2811	1486.0	1475.5	1471.8	1461.1	1446.0
350	3177	3177	3176	3168	3159	3139	3095	3046	2990	2926	2695	1647.1	1636.5	1609.9	1576.3
375	3228	3228	3227	3220	3211	3194	3156	3115	3069	3019	2862	2604	2319	1791	1716
400	3280	3279	3278	3272	3264	3249	3216	3180	3142	3100	2979	2820	2733	2162	1878
425	3331	3331	3330	3325	3317	3303	3274	3243	3209	3174	3075	2957	2899	2619	2068
450	3384	3383	3382	3377	3371	3358	3331	3303	3274	3244	3160	3064	3020	2826	2293
475	3436	3436	3435	3430	3424	3412	3388	3363	3337	3310	3237	3157	3120	2969	2522
500	3489	3489	3488	3484	3478	3467	3445	3422	3399	3375	3311	3241	3210	3085	2723
600	3706	3705	3705	3702	3697	3689	3673	3656	3640	3623	3580	3536	3516	3443	3248
700	3929	3929	3928	3926	3923	3916	3904	3892	3879	3867	3835	3804	3790	3740	3610
800	4159	4159	4158	4156	4154	4149	4140	4131	4121	4112	4089	4065	4055	4018	3925

* Critical isobar.

Mathematical Rules

In this Appendix, some simple rules for logarithms, differentiation and integration are presented. Further details of mathematical functions can be found in handbooks, e.g. [1–4].

D.1 Logarithms

The *natural logarithm* (ln or $\log_e$) is the inverse of the exponential function. Therefore, if:

$$y = \ln x \tag{D.1}$$

then

$$e^y = x \tag{D.2}$$

where the number e is approximately 2.71828. It also follows that:

$$\ln(e^y) = y \tag{D.3}$$

and

$$e^{\ln x} = x. \tag{D.4}$$

Natural logarithms are related to *common logarithms*, or logarithms to the base 10 (written as lg, log or $\log_{10}$), as follows:

$$\ln x = \ln 10 \, (\log_{10} x). \tag{D.5}$$

Since ln 10 is approximately 2.30259:

$$\ln x = 2.30259 \log_{10} x. \tag{D.6}$$

Zero and negative numbers do not have logarithms.

Rules for taking logarithms of products and powers are illustrated below. The logarithm of the product of two numbers is equal to the sum of the logarithms:

$$\ln(a x) = \ln a + \ln x. \tag{D.7}$$

When one term of the product involves an exponential function, application of Eqs (D.7) and (D.3) gives:

$$\ln(b e^{ax}) = \ln b + ax. \tag{D.8}$$

The logarithm of the quotient of two numbers is equal to the logarithm of the numerator minus the logarithm of the denominator:

$$\ln\left(\frac{a}{x}\right) = \ln a - \ln x.$$

(D.9)

As an example of this rule, because $\ln 1 = 0$:

$$\ln\left(\frac{1}{x}\right) = -\ln x.$$

(D.10)

The rule for taking the logarithm of a power function is as follows:

$$\ln(x^b) = b \ln x.$$

(D.11)

D.2 Differentiation

The derivative of y with respect to x, $\frac{dy}{dx}$, is defined as the limit of $\frac{\Delta y}{\Delta x}$ as Δx approaches zero, provided this limit exists:

$$\frac{dy}{dx} = \lim_{\Delta x \to 0} \frac{\Delta y}{\Delta x}.$$

(D.12)

That is:

$$\frac{dy}{dx} = \lim_{\Delta x \to 0} \frac{y|_{x+\Delta x} - y|_x}{\Delta x}$$

(D.13)

where $y|_x$ means the value of y evaluated at x, and $y|_{x+\Delta x}$ means the value of y evaluated at $x + \Delta x$. The operation of calculating the derivative is called differentiation.

There are simple rules for rapid evaluation of derivatives. Derivatives of various functions with respect to x are listed below; in all these equations A is a constant:

$$\frac{dA}{dx} = 0$$

(D.14)

$$\frac{dx}{dx} = 1$$

(D.15)

$$\frac{d}{dx}(e^x) = e^x$$

(D.16)

$$\frac{d}{dx}(e^{Ax}) = Ae^{Ax}$$

(D.17)

and

$$\frac{d}{dx}(\ln x) = \frac{1}{x}.$$

(D.18)

When a function is multiplied by a constant, the constant can be taken out of the differential. For example:

$$\frac{d}{dx}(Ax) = A\frac{dx}{dx} = A$$

(D.19)

and

$$\frac{d}{dx}(A\ln x) = A\frac{d}{dx}(\ln x) = \frac{A}{x}.$$

(D.20)

When a function consists of a sum of terms, the derivative of the sum is equal to the sum of the derivatives. Therefore, if $f(x)$ and $g(x)$ are functions of x:

$$\frac{d}{dx}[f(x) + g(x)] = \frac{df}{dx} + \frac{dg}{dx}.$$

(D.21)

To illustrate application of Eq. (D.21), for A and B constants:

$$\frac{d}{dx}[Ax + e^{Bx}] = \frac{d(Ax)}{dx} + \frac{d(e^{Bx})}{dx} = A + Be^{Bx}.$$

When a function consists of terms multiplied together, the *product rule* for derivatives is:

$$\frac{d}{dx}[f(x) \cdot g(x)] = f(x)\frac{dg}{dx} + g(x)\frac{df}{dx}$$

(D.22)

As an example of the product rule:

$$\frac{d}{dx}[(Ax) \cdot \ln x] = Ax \cdot \frac{d(\ln x)}{dx} + \ln x \cdot \frac{d(Ax)}{dx}$$

$$= Ax \cdot \frac{1}{x} + \ln x \cdot (A)$$

$$= A(1 + \ln x).$$

D.3 Integration

The integral of y with respect to x is indicated as $\int y\,dx$. The function to be integrated (y) is called the *integrand*; the symbol $\int$ is the *integral sign*. Integration is the opposite of differentiation; integration is the process of finding a function from its derivative.

From Eq. (D.14), if the derivative of a constant is zero, the integral of zero must be a constant:

$$\int 0\,dx = K$$

(D.23)

where K is a constant. K is called the *constant of integration*, and appears whenever a function is integrated. For example, the integral of constant A with respect to x is:

$$\int A\,dx = Ax + K.$$

(D.24)

We can check that Eq. (D.24) is correct by taking the derivative of the right-hand side and making sure it is equal to the integrand, A. Although the equation:

$$\int A\,dx = Ax$$

is also correct, addition of K in Eq. (D.24) makes solution of the integral complete. Addition of K accounts for the possibility that the answer we are looking for may have an added constant that disappears when the derivative is taken. Irrespective of the value of K, the derivative of the integral will always be the same because $\frac{dK}{dx} = 0$. Extra information is needed to evaluate the actual magnitude of K; this point is considered further in Chapter 6 where integration is used to solve unsteady-state mass and energy problems.

The integral of $\frac{dy}{dx}$ with respect to x is:

$$\int \frac{dy}{dx}\, dx = y + K.$$

(D.25)

When a function is multiplied by a constant, the constant can be taken out of the integral. For example, for $f(x)$ a function of x, and A a constant:

$$\int A f(x)\, dx = A \int f(x)\, dx.$$

(D.26)

Other rules of integration are:

$$\int \frac{dx}{x} = \int \frac{1}{x}\, dx = \ln x + K$$

(D.27)

and, for A and B constants:

$$\int \left(\frac{dx}{A + Bx} \right) = \int \left(\frac{1}{A + Bx} \right) dx = \frac{1}{B}\, \ln(A + Bx) + K.$$

(D.28)

The results of Eqs (D.27) and (D.28) can be confirmed by differentiating the right-hand sides of the equations with respect to x.

References

1. *CRC Standard Mathematical Tables*, CRC Press, Florida.
2. Cornish-Bowden, A. (1981) *Basic Mathematics for Biochemists*, Chapman and Hall, London.
3. Newby, J.C. (1980) *Mathematics for the Biological Sciences*, Oxford University Press, Oxford.
4. Arya, J.C. and R.W. Lardner (1979) *Mathematics for the Biological Sciences*, Prentice-Hall, New Jersey.

E

List of Symbols

Symbol	Definition	Dimensions	SI units
Roman symbols			
a	Area	L^2	m^2
a	Area per unit volume	L^{-1}	m^{-1}
a	Mass of dry gel	M	kg
A	Amplitude	L	m
A	Arrhenius constant	T^{-1}	s^{-1}
A	Area	L^2	m^2
A_c	Cross-sectional area	L^2	m^2
A_i	Inner surface area	L^2	m^2
A_o	Outer surface area	L^2	m^2
b	Length of bowl	L	m
b	Thickness	L	m
B	Thickness	L	m
c	Mass of cake solids deposited per volume of filtrate	$L^{-3}M$	$kg\,m^{-3}$
C	Geometry parameter in Eq. (8.44)	1	–
C	Concentration	$L^{-3}M$	$kg\,m^{-3}$
C_A	Concentration of component A	$L^{-3}M$	$kg\,m^{-3}$
C_{Ab}	Concentration of component A in the bulk fluid	$L^{-3}M$	$kg\,m^{-3}$
C_{AG}	Concentration of component A in gas	$L^{-3}M$	$kg\,m^{-3}$
C_{Al}	Concentration of component A in the lower phase	$L^{-3}M$	$kg\,m^{-3}$
C_{AL}	Concentration of component A in liquid	$L^{-3}M$	$kg\,m^{-3}$
$\bar{C}_{AL}$	Steady-state concentration of component A in liquid	$L^{-3}M$	$kg\,m^{-3}$
C^*_{AL0}	Solubility of component A in liquid at zero solute concentration, Eq. (9.45)	$L^{-3}M$	$kg\,m^{-3}$
$C_{A,min}$	Minimum concentration of component A	$L^{-3}M$	$kg\,m^{-3}$
C_{Ao}	Concentration of component A in the bulk fluid	$L^{-3}M$	$kg\,m^{-3}$
C_{As}	Surface concentration of component A	$L^{-3}M$	$kg\,m^{-3}$
C_{AS}	Average concentration of component A in the solid phase	$L^{-3}M$	$kg\,m^{-3}$
C_{ASm}	Maximum concentration of component A in the solid phase	$L^{-3}M$	$kg\,m^{-3}$
C_{Au}	Concentration of component A in the upper phase	$L^{-3}M$	$kg\,m^{-3}$
C_{crit}	Critical concentration	$L^{-3}M$	$kg\,m^{-3}$
C_e	Concentration at reaction equilibrium	$L^{-3}M$	$kg\,m^{-3}$

C_{iL}	Concentration of ionic component i in liquid	$L^{-3}M$	$kg\,m^{-3}$
C_{jL}	Concentration of non-ionic component j in liquid	$L^{-3}M$	$kg\,m^{-3}$
C_p	Specific heat capacity	$L^2T^{-2}\Theta^{-1}$	$J\,kg^{-1}\,K^{-1}$
C_{pc}	Specific heat capacity of cold fluid	$L^2T^{-2}\Theta^{-1}$	$J\,kg^{-1}\,K^{-1}$
C_{ph}	Specific heat capacity of hot fluid	$L^2T^{-2}\Theta^{-1}$	$J\,kg^{-1}\,K^{-1}$
C_{pm}	Mean heat capacity	$L^2T^{-2}\Theta^{-1}$	$J\,kg^{-1}\,K^{-1}$
C_{pw}	Specific heat capacity of cooling water	$L^2T^{-2}\Theta^{-1}$	$J\,kg^{-1}\,K^{-1}$
D	Dilution rate	T^{-1}	s^{-1}
D	Diameter	L	m
D_{b}	Bubble diameter	L	m
D_{crit}	Critical dilution rate for washout	T^{-1}	s^{-1}
D_{i}	Impeller diameter	L	m
D_{o}	Orifice diameter	L	m
D_{opt}	Dilution rate for optimum biomass productivity	T^{-1}	s^{-1}
D_{p}	Particle diameter	L	m
Da	Damköhler number defined in Eq. (13.102)	1	–
e	Base of natural logarithms	1	–
e	Enzyme concentration	$L^{-3}M$	$kg\,m^{-3}$
e_{a}	Concentration of active enzyme	$L^{-3}M$	$kg\,m^{-3}$
e_{k}	Kinetic energy per unit mass	L^2T^{-2}	$J\,kg^{-1}$
e_{p}	Potential energy per unit mass	L^2T^{-2}	$J\,kg^{-1}$
E	Quantity of enzyme (Section 11.1.3)		
E	Molar activation energy	$L^2MT^{-2}N^{-1}$	$J\,mol^{-1}$
E	Energy	L^2MT^{-2}	J
E_{d}	Molar activation energy for deactivation reaction	$L^2MT^{-2}N^{-1}$	$J\,mol^{-1}$
E_{k}	Kinetic energy	L^2MT^{-2}	J
E_{p}	Potential energy	L^2MT^{-2}	J
f	Function		
F	Shear force	LMT^{-2}	N
F	Fraction of cells carrying plasmid	1	–
F	Volumetric flow rate	L^3T^{-1}	$m^3\,s^{-1}$
F_{g}	Volumetric flow rate of gas	L^3T^{-1}	$m^3\,s^{-1}$
F_{L}	Volumetric flow rate of liquid	L^3T^{-1}	$m^3\,s^{-1}$
F_{r}	Volumetric flow rate of the recycle stream	L^3T^{-1}	$m^3\,s^{-1}$
g	Gravitational acceleration	LT^{-2}	$m\,s^{-2}$
g_{c}	Force unity bracket	1	–
G	Specific gravity	1	–
G	Free energy of formation	L^2MT^{-2}	J
G°	Standard free energy of formation	L^2MT^{-2}	J
$\Delta G^{\circ}_{\mathrm{rxn}}$	Change in molar free energy for reaction under standard conditions	$L^2MT^{-2}N^{-1}$	$J\,mol^{-1}$
Gr	Grashof number defined by Eq. (8.41) or (12.51)	1	–
h	Height	L	m
h	Specific enthalpy	L^2T^{-2}	$J\,kg^{-1}$
h	Heat-transfer coefficient	$MT^{-3}\Theta^{-1}$	$W\,m^{-2}\,K^{-1}$
h_{c}	Heat-transfer coefficient for cold fluid	$MT^{-3}\Theta^{-1}$	$W\,m^{-2}\,K^{-1}$
Δh_{c}	Molar heat of combustion	$L^2MT^{-2}N^{-1}$	$J\,mol^{-1}$
$\Delta h^{\circ}_{\mathrm{c}}$	Molar heat of combustion under standard conditions	$L^2MT^{-2}N^{-1}$	$J\,mol^{-1}$

Symbol	Description	Dimensions	Units
Δh_f	Specific latent heat of fusion	L^2T^{-2}	$J\,kg^{-1}$
h_{fc}	Fouling factor for cold fluid	$MT^{-3}\Theta^{-1}$	$W\,m^{-2}\,K^{-1}$
h_{fh}	Fouling factor for hot fluid	$MT^{-3}\Theta^{-1}$	$W\,m^{-2}\,K^{-1}$
h_h	Heat-transfer coefficient for hot fluid	$MT^{-3}\Theta^{-1}$	$W\,m^{-2}\,K^{-1}$
Δh_m	Molar integral heat of mixing	$L^2MT^{-2}N^{-1}$	$J\,mol^{-1}$
Δh_{rxn}	Specific heat of reaction	L^2T^{-2}	$J\,kg^{-1}$
Δh_s	Specific latent heat of sublimation	L^2T^{-2}	$J\,kg^{-1}$
Δh_v	Specific latent heat of vaporisation	L^2T^{-2}	$J\,kg^{-1}$
H	Henry's constant	L^2T^{-2}	$m^2\,s^{-2}$
H	Height	L	m
H	Enthalpy	L^2MT^{-2}	J
H_A^-	Enthalpy of component A	L^2MT^{-2}	J
H_B	Enthalpy of component B	L^2MT^{-2}	J
H_i	Parameter in Eq. (9.45)	L^3N^{-1}	$m^3\,mol^{-1}$
ΔH_m	Heat of mixing	L^2MT^{-2}	J
H_{ref}	Enthalpy of the reference state	L^2MT^{-2}	J
ΔH_{rxn}	Heat of reaction	L^2MT^{-2}	J
ΔH_{rxn}°	Heat of reaction under standard conditions	L^2MT^{-2}	J
$\Delta \dot{H}_{rxn}$	Rate of heat absorption or liberation by reaction	L^2MT^{-3}	W
J_A	Mass flux of component A	$L^{-2}MT^{-1}$	$kg\,m^{-2}\,s^{-1}$
k	Thermal conductivity	$LMT^{-3}\Theta^{-1}$	$W\,m^{-1}\,K^{-1}$
k	Geometry parameter in Eq. (7.11)	1	–
k	Mass-transfer coefficient	LT^{-1}	$m\,s^{-1}$
k	Capacity factor defined in Eq. (10.37)	1	–
k	Rate constant		
k_0	Zero-order rate constant	$L^{-3}MT^{-1}$	$kg\,m^{-3}\,s^{-1}$
k'_0	Specific zero-order rate constant for enzyme reaction	T^{-1}	s^{-1}
k''_0	Specific zero-order rate constant for cell reaction	T^{-1}	s^{-1}
k_1	First-order rate constant	T^{-1}	s^{-1}
k_1	Proportionality constant in Eq. (7.19)	1	–
k_{-1}	Reverse-reaction rate constant		
k_2	Turnover number	T^{-1}	s^{-1}
k_d	First-order deactivation rate constant	T^{-1}	s^{-1}
k_{fb}	Thermal conductivity of bulk fluid	$LMT^{-3}\Theta^{-1}$	$W\,m^{-1}\,K^{-1}$
k_G	Gas-phase mass-transfer coefficient	LT^{-1}	$m\,s^{-1}$
k_L	Liquid-phase mass-transfer coefficient	LT^{-1}	$m\,s^{-1}$
k_S	Liquid-phase mass-transfer coefficient for transfer to or from a solid	LT^{-1}	$m\,s^{-1}$
K	Constant of integration		
K	Consistency index for power-law fluids	$L^{-1}MT^{n-2}$	$Pa\,s^n$
K	Reaction equilibrium constant		
K	Partition coefficient	1	–
K_1	Parameter defined in Eq. (10.13)	$L^{-6}T$	$m^{-6}\,s$
K_2	Parameter defined in Eq. (10.14)	$L^{-3}T$	$m^{-3}\,s$
K_A	Constant in Eq. (10.30)	L^3M^{-1}	$m^3\,kg^{-1}$
K_F	Parameter in Eq. (10.31)		
K_G	Overall gas-phase mass-transfer coefficient	LT^{-1}	$m\,s^{-1}$
K_j	Parameter in Eq. (9.45)	L^3N^{-1}	$m^3\,mol^{-1}$
K_L	Overall liquid-phase mass-transfer coefficient	LT^{-1}	$m\,s^{-1}$
K_m	Michaelis constant	$L^{-3}M$	$kg\,m^{-3}$
K_p	Constant in Eqs (7.9) and (7.10)		
K_p	Partition coefficient	1	–
K_S	Substrate constant	$L^{-3}M$	$kg\,m^{-3}$

Symbol	Description	Dimensions	Units
K_v	Shape factor	1	–
L	Length	L	m
m	Distribution or partition coefficient	1	–
m_P	Specific rate of product formation due to maintenance activity	T^{-1}	s^{-1}
m_S	Maintenance coefficient	T^{-1}	s^{-1}
M	Torque	L^2MT^{-2}	N m
M	Mass	M	kg
$\hat{M}$	Mass flow rate	MT^{-1}	$kg\,s^{-1}$
M_A	Mass of component A	M	kg
M_c	Mass of cake solids	M	kg
$\hat{M}_c$	Mass flow rate of cold fluid	MT^{-1}	$kg\,s^{-1}$
$\hat{M}_h$	Mass flow rate of hot fluid	MT^{-1}	$kg\,s^{-1}$
M_m	Initial mass of medium	M	kg
$\hat{M}_s$	Mass flow rate of steam	MT^{-1}	$kg\,s^{-1}$
M_v	Mass of liquid evaporated	M	kg
$\hat{M}_v$	Mass flow rate of evaporated liquid	MT^{-1}	$kg\,s^{-1}$
$\hat{M}_w$	Mass flow rate of cooling water	MT^{-1}	$kg\,s^{-1}$
n	Mole	N	mol
n	Flow behaviour index for power-law fluids	1	–
n	Number of impellers	1	–
n	Number of generations	1	–
n	Adsorption parameter in Eq. (10.31)	1	–
N	Number of viable cells	1	–
N	Number of discs	1	–
N	Number of passes	1	–
N	Number of theoretical plates	1	–
N_1	Number of viable cells at the beginning of the holding period	1	–
N_2	Number of viable cells at the end of the holding period	1	–
N_A	Rate of mass transfer of component A	MT^{-1}	$kg\,s^{-1}$
N_A	Volumetric rate of mass transfer of component A	$L^{-3}MT^{-1}$	$kg\,m^{-3}\,s^{-1}$
N_i	Rotational speed	T^{-1}	s^{-1}
N_i^*	Minimum stirrer speed for suspension of solids	T^{-1}	s^{-1}
N_{iB}	Minimum stirrer speed for complete dispersion of gas	T^{-1}	s^{-1}
N_{iR}	Minimum stirrer speed for gross recirculation of gas	T^{-1}	s^{-1}
N_P	Power number defined by Eq. (7.17)	1	–
N_P'	Constant value of the power number in the turbulent regime	1	–
Nu	Nusselt number defined in Eq. (8.37)	1	–
p	Probability of plasmid loss	1	–
p	Pressure	$L^{-1}MT^{-2}$	Pa
p	Product concentration	$L^{-3}M$	$kg\,m^{-3}$
p_1	Product concentration in the first reactor	$L^{-3}M$	$kg\,m^{-3}$
p_2	Product concentration in the second reactor	$L^{-3}M$	$kg\,m^{-3}$
p_{AG}	Partial pressure of component A in gas	$L^{-1}MT^{-2}$	Pa
p_T	Total pressure	$L^{-1}MT^{-2}$	Pa
P	Power	L^2MT^{-3}	W
P	Mass of product	M	kg
P_0	Power consumption without sparging	L^2MT^{-3}	W
P_g	Power consumption with sparging	L^2MT^{-3}	W
Pe	Peclet number defined in Eq. (13.101)	1	–
Pr	Prandtl number defined in Eq. (8.40)	1	–

q	Heat evolved per mole of available electrons transferred to oxygen	$L^2MT^{-2}N^{-1}$	$J\,mol^{-1}$
$\hat{q}$	Heat flux	MT^{-3}	$W\,m^{-2}$
q_O	Specific oxygen-uptake rate	T^{-1}	s^{-1}
q_P	Specific rate of product formation	T^{-1}	s^{-1}
q_S	Specific rate of substrate consumption	T^{-1}	s^{-1}
Q	Volumetric flow rate	L^3T^{-1}	$m^3\,s^{-1}$
Q	Heat	L^2MT^{-2}	J
$\hat{Q}$	Rate of heat flow	L^2MT^{-3}	W
$\hat{Q}_c$	Rate of heat transfer to cold fluid	L^2MT^{-3}	W
$\hat{Q}_h$	Rate of heat transfer from hot fluid	L^2MT^{-3}	W
Q_O	Volumetric rate of oxygen uptake	$L^{-3}MT^{-1}$	$kg\,m^{-3}\,s^{-1}$
Q_P	Volumetric rate of product formation	$L^{-3}MT^{-1}$	$kg\,m^{-3}\,s^{-1}$
Q_X	Volumetric rate of biomass production	$L^{-3}MT^{-1}$	$kg\,m^{-3}\,s^{-1}$
$Q_{X,max}$	Maximum volumetric rate of biomass production	$L^{-3}MT^{-1}$	$kg\,m^{-3}\,s^{-1}$
r	Radius	L	m
r	Volumetric rate of reaction	$L^{-3}MT^{-1}$	$kg\,m^{-3}\,s^{-1}$
r_1	Inner radius of centrifuge disc	L	m
r_1	Radius of the liquid surface in a tubular centrifuge	L	m
r_2	Outer radius of centrifuge disc	L	m
r_2	Inner-wall radius of a tubular centrifuge	L	m
r_A	Volumetric rate of reaction with respect to component A	$L^{-3}MT^{-1}$	$kg\,m^{-3}\,s^{-1}$
$r_{Ab}^\star$	Volumetric rate of reaction with respect to component A at the bulk concentration	$L^{-3}MT^{-1}$	$kg\,m^{-3}\,s^{-1}$
$r_{A,obs}$	Observed volumetric rate of reaction with respect to component A	$L^{-3}MT^{-1}$	$kg\,m^{-3}\,s^{-1}$
$r_{As}^\star$	Volumetric rate of reaction with respect to component A at the surface concentration	$L^{-3}MT^{-1}$	$kg\,m^{-3}\,s^{-1}$
r_C	Volumetric rate of consumption by reaction	$L^{-3}MT^{-1}$	$kg\,m^{-3}\,s^{-1}$
r_d	Volumetric rate of deactivation	$L^{-3}MT^{-1}$	$kg\,m^{-3}\,s^{-1}$
r_m	Filter medium resistance	L^{-1}	m^{-1}
r_P	Volumetric rate of product formation	$L^{-3}MT^{-1}$	$kg\,m^{-3}\,s^{-1}$
r_S	Volumetric rate of substrate consumption	$L^{-3}MT^{-1}$	$kg\,m^{-3}\,s^{-1}$
r_X	Volumetric rate of cell growth	$L^{-3}MT^{-1}$	$kg\,m^{-3}\,s^{-1}$
r_{X^+}	Volumetric rate of growth of plasmid-carrying cells	$L^{-3}MT^{-1}$	$kg\,m^{-3}\,s^{-1}$
r_{X^-}	Volumetric rate of growth of plasmid-free cells	$L^{-3}MT^{-1}$	$kg\,m^{-3}\,s^{-1}$
r_Z	Volumetric rate of reaction with respect to component Z	$L^{-3}MT^{-1}$	$kg\,m^{-3}\,s^{-1}$
R	Ideal gas constant	$L^2MT^{-2}\Theta^{-1}N^{-1}$	$J\,mol^{-1}\,K^{-1}$
R	Radius	L	m
R	Thermal resistance	$L^{-2}M^{-1}T^3\Theta$	$K\,W^{-1}$
R	Amount of protein released in a homogeniser	M	kg
R	Total rate of reaction	MT^{-1}	$kg\,s^{-1}$
R_0	Radius at which substrate is depleted	L	m
R_A	Total rate of reaction with respect to component A	MT^{-1}	$kg\,s^{-1}$
R_c	Thermal resistance in cold fluid	$L^{-2}M^{-1}T^3\Theta$	$K\,W^{-1}$
R_C	Total rate of consumption by reaction	MT^{-1}	$kg\,s^{-1}$
R_G	Total rate of generation by reaction	MT^{-1}	$kg\,s^{-1}$
R_h	Thermal resistance in hot fluid	$L^{-2}M^{-1}T^3\Theta$	$K\,W^{-1}$
R_i	Inner radius	L	m

Symbol	Description	Dimensions	Units
R_m	Resistance to mass transfer	T	s
R_m	Maximum amount of protein available for release in a homogeniser	M	kg
R_{max}	Maximum particle radius	L	m
R_N	Resolution in chromatography	1	–
R_o	Outer radius	L	m
R_w	Thermal resistance of a wall	$L^{-2}M^{-1}T^3\Theta$	$K\,W^{-1}$
Re	Reynolds number defined in Eq. (7.1)	1	–
Re_i	Impeller Reynolds number defined in Eq. (7.2)	1	–
Re_{max}	Maximum Reynolds number	1	–
Re_p	Particle Reynolds number defined in Eq. (12.48)	1	–
RQ	Respiratory quotient	1	–
s	Cake compressibility	1	–
s	Substrate concentration	$L^{-3}M$	$kg\,m^{-3}$
s_1	Substrate concentration in the first reactor	$L^{-3}M$	$kg\,m^{-3}$
s_2	Substrate concentration in the second reactor	$L^{-3}M$	$kg\,m^{-3}$
s_b	Bulk substrate concentration	$L^{-3}M$	$kg\,m^{-3}$
S	Mass of substrate	M	kg
S	Molar entropy	$L^2MT^{-2}\Theta^{-1}N^{-1}$	$J\,K^{-1}\,mol^{-1}$
S_G	Mass of substrate consumed for growth	M	kg
ΔS°_{rxn}	Molar entropy change during reaction under standard conditions	$L^2MT^{-2}\Theta^{-1}N^{-1}$	$J\,K^{-1}\,mol^{-1}$
S_R	Mass of substrate consumed other than for growth	M	kg
S_T	Total mass of substrate consumed	M	kg
S_x	External surface area	L^2	m^2
Sc	Schmidt number defined by Eq. (12.49)	1	–
Sh	Sherwood number defined by Eq. (12.50)	1	–
t	Time	T	s
t_1	Time at the end of the heating period	T	s
t_2	Time at the end of the holding period	T	s
t_b	Batch reaction time	T	s
t_c	Circulation time	T	s
t_d	Doubling time	T	s
t_{dn}	Total downtime	T	s
t_{fb}	Fed-batch time	T	s
t_h	Half-life	T	s
t_{hd}	Holding time	T	s
t_{hv}	Time taken to harvest culture	T	s
t_l	Lag time	T	s
t_m	Mixing time	T	s
t_p	Reactor-preparation time	T	s
t_T	Total batch reaction time	T	s
T	Temperature	Θ	K
ΔT_A	Arithmetic-mean temperature difference	Θ	K
T_c	Cold-fluid temperature	Θ	K
T_{cw}	Cold-fluid temperature at the wall	Θ	K
T_F	Fermenter temperature	Θ	K
T_h	Hot-fluid temperature	Θ	K
T_{hw}	Hot fluid temperature at the wall	Θ	K
ΔT_L	Logarithmic-mean temperature difference	Θ	K
T_{ref}	Reference temperature	Θ	K
T_S	Steam temperature	Θ	K
u	Linear velocity	LT^{-1}	$m\,s^{-1}$
u	Interstitial velocity	LT^{-1}	$m\,s^{-1}$
u	Specific internal energy	L^2T^{-2}	$J\,kg^{-1}$
u_c	Sedimentation velocity in a centrifuge	LT^{-1}	$m\,s^{-1}$
u_g	Sedimentation velocity under gravity	LT^{-1}	$m\,s^{-1}$

u_G	Gas superficial velocity	LT^{-1}	$m\,s^{-1}$
u_L	Liquid linear velocity	LT^{-1}	$m\,s^{-1}$
u_L	Liquid superficial velocity	LT^{-1}	$m\,s^{-1}$
$u_L^\star$	Liquid velocity at which reaction rate becomes independent of liquid velocity	LT^{-1}	$m\,s^{-1}$
u_{pL}	Velocity of a particle relative to liquid	LT^{-1}	$m\,s^{-1}$
U	Internal energy	L^2MT^{-2}	J
U	Overall heat-transfer coefficient	$MT^{-3}\Theta^{-1}$	$W\,m^{-2}\,K^{-1}$
v	Specific volume	L^3M^{-1}	$m^3\,kg^{-1}$
v	Velocity	MT^{-1}	$m\,s^{-1}$
v	Volumetric rate of reaction	$L^{-3}MT^{-1}$	$kg\,m^{-3}\,s^{-1}$
v_{max}	Maximum volumetric rate of reaction	$L^{-3}MT^{-1}$	$kg\,m^{-3}\,s^{-1}$
V	Volume	L^3	m^3
V_1	Volume of the first reactor	L^3	m^3
V_2	Volume of the second reactor	L^3	m^3
V_e	Volume of eluant	L^3	m^3
V_f	Volume of filtrate	L^3	m^3
V_G	Volume of gas	L^3	m^3
V_i	Internal volume	L^3	m^3
V_l	Volume of lower phase	L^3	m^3
V_L	Volume of liquid	L^3	m^3
V_o	Void volume	L^3	m^3
V_p	Particle volume	L^3	m^3
V_s	Volume of solid	L^3	m^3
V_T	Total volume	L^3	m^3
V_u	Volume of upper phase	L^3	m^3
w	Baseline width	L	m
W_i	Impeller blade width	L	m
W_r	Water regain value	L^3M^{-1}	$m^3\,kg^{-1}$
W_s	Shaft work	L^2MT^{-2}	J
$\hat{W}_s$	Rate of shaft work	L^2MT^{-3}	W
W_f	Flow work	L^2MT^{-2}	J
x	Distance	L	m
x	Cell concentration	$L^{-3}M$	$kg\,m^{-3}$
x^+	Concentration of plasmid-carrying cells	$L^{-3}M$	$kg\,m^{-3}$
x^-	Concentration of plasmid-free cells	$L^{-3}M$	$kg\,m^{-3}$
$\bar{x}$	Mean value of x		
x_1	Cell concentration in the first reactor	$L^{-3}M$	$kg\,m^{-3}$
x_2	Cell concentration in the second reactor	$L^{-3}M$	$kg\,m^{-3}$
x_C	Number of carbon atoms in the molecular formula	1	–
x_{im}	Concentration of immobilised cells	$L^{-3}M$	$kg\,m^{-3}$
x_{max}	Maximum cell concentration	$L^{-3}M$	$kg\,m^{-3}$
x_r	Concentration of cells in the recycle stream	$L^{-3}M$	$kg\,m^{-3}$
x_s	Concentration of suspended cells	$L^{-3}M$	$kg\,m^{-3}$
X	Mass of cells	M	kg
y	Distance	L	m
y	Wave displacement	L	m
y_{AG}	Mole fraction of component A in gas	1	–
y_C	Weight fraction of cells	1	–
Y_l	Yield in the lower phase	1	–
Y_{PS}	True yield of product from substrate	1	–
Y'_{PS}	Observed yield of product from substrate	1	–
Y_{PX}	True yield of product from biomass	1	–
Y'_{PX}	Observed yield of product from biomass	1	–
Y_u	Yield in the upper phase	1	–
Y_{XO}	Yield of biomass from oxygen	1	–
Y_{XS}	True yield of biomass from substrate	1	–

Y'_{XS}	Observed yield of biomass from substrate	1	–
$Y_{XS,max}$	Maximum yield of biomass from substrate	1	–
z	Distance	L	m
z	Concentration of cellular constituent Z	$L^{-3}M$	$kg\,m^{-3}$
z_i	Valency of ionic component i	1	–
Z	g-number in centrifugation	1	–

Script symbols

$\mathscr{D}$	Diffusion coefficient	L^2T^{-1}	$m^2\,s^{-1}$
$\mathscr{D}_{AB}$	Binary diffusion coefficient of component A in component B	L^2T^{-1}	$m^2\,s^{-1}$
$\mathscr{D}_{Ae}$	Effective diffusivity of component A	L^2T^{-1}	$m^2\,s^{-1}$
$\mathscr{D}_{AL}$	Binary diffusion coefficient of component A in liquid	L^2T^{-1}	$m^2\,s^{-1}$
$\mathscr{D}_{Aw}$	Binary diffusion coefficient of component A in water	L^2T^{-1}	$m^2\,s^{-1}$
$\mathscr{D}_{Se}$	Effective diffusivity of substrate	L^2T^{-1}	$m^2\,s^{-1}$
$\mathscr{D}_z$	Axial-dispersion coefficient	L^2T^{-1}	$m^2\,s^{-1}$

Greek symbols

α	Specific cake resistance	LM^{-1}	$m\,kg^{-1}$
α	Exponent in Eq. (10.23)	1	–
α	Ratio defined in Eq. (11.65)	1	–
α	Recycle ratio defined in Eq. (13.77)	1	–
α'	Parameter defined in Eq. (10.2)		
β	Thermal coefficient of linear expansion	Θ^{-1}	K^{-1}
β	Dimensionless parameter equal to K_m/C_{As}	1	–
β	Biomass concentration factor defined in Eq. (13.78)	1	–
γ	Degree of reduction	1	–
$\dot{\gamma}$	Shear rate	T^{-1}	s^{-1}
$\dot{\gamma}_{av}$	Average shear rate	T^{-1}	s^{-1}
γ_B	Degree of reduction of biomass	1	–
γ_P	Degree of reduction of product	1	–
γ_S	Degree of reduction of substrate	1	–
δ	Selectivity in chromatography	1	–
δ_c	Concentration factor	1	–
Δ	Difference		
ε	Fractional gas hold-up	1	–
ε	Porosity	1	–
ε	Void fraction	1	–
ε	Rate of turbulent energy dissipation per mass of fluid	L^2T^{-3}	$W\,kg^{-1}$
ε	Time interval	T	s
ζ_B	Fraction of available electrons transferred to biomass	1	–
η_e	External effectiveness factor	1	–
η_{e0}	External effectiveness factor for zero-order reaction	1	–
η_{e1}	External effectiveness factor for first-order reaction	1	–
η_{em}	External effectiveness factor for Michaelis–Menten reaction	1	–
η_i	Internal effectiveness factor	1	–
η_{i0}	Internal effectiveness factor for zero-order reaction	1	–
η_{i1}	Internal effectiveness factor for first-order reaction	1	–
η_{im}	Internal effectiveness factor for Michaelis–Menten reaction	1	–
η_T	Total effectiveness factor	1	–

η_{T1}	Total effectiveness factor for first-order reaction	1	–
θ	Half-cone angle	1	rad
λ	Kolmogorov scale	L	m
μ	Specific growth rate	T^{-1}	s^{-1}
μ^+	Specific growth rate of plasmid-carrying cells	T^{-1}	s^{-1}
μ^-	Specific growth rate of plasmid-free cells	T^{-1}	s^{-1}
μ	Viscosity (dynamic)	$L^{-1}MT^{-1}$	Pa s
μ_a	Apparent viscosity	$L^{-1}MT^{-1}$	Pa s
μ_b	Bulk-fluid viscosity	$L^{-1}MT^{-1}$	Pa s
μ_f	Filtrate viscosity	$L^{-1}MT^{-1}$	Pa s
μ_L	Liquid viscosity	$L^{-1}MT^{-1}$	Pa s
μ_{max}	Maximum specific growth rate	T^{-1}	s^{-1}
μ_w	Fluid viscosity at the wall	$L^{-1}MT^{-1}$	Pa s
ν	Kinematic viscosity	L^2T^{-1}	$m^2\,s^{-1}$
ν_L	Liquid kinematic viscosity	L^2T^{-1}	$m^2\,s^{-1}$
π	3.14159	1	–
ρ	Density	$L^{-3}M$	$kg\,m^{-3}$
ρ_f	Fluid density	$L^{-3}M$	$kg\,m^{-3}$
ρ_g	Density of wet gel	$L^{-3}M$	$kg\,m^{-3}$
ρ_G	Gas density	$L^{-3}M$	$kg\,m^{-3}$
ρ_L	Liquid density	$L^{-3}M$	$kg\,m^{-3}$
ρ_p	Particle density	$L^{-3}M$	$kg\,m^{-3}$
ρ_w	Density of water	$L^{-3}M$	$kg\,m^{-3}$
σ	Surface tension	MT^{-2}	$N\,m^{-1}$
σ	Standard deviation		
Σ	Summation		
Σ	Centrifuge sigma factor defined in Eq. (10.18)	L^2	m^2
τ	Average residence time	T	s
τ	Shear stress	$L^{-1}MT^{-2}$	Pa
τ_0	Yield stress	$L^{-1}MT^{-2}$	Pa
ϕ	Angle	1	rad
ϕ	Thiele modulus	1	–
ϕ_0	Thiele modulus for zero-order reaction	1	–
ϕ_1	Thiele modulus for first-order reaction	1	–
ϕ_m	Thiele modulus for Michaelis–Menten reaction	1	–
Φ	Observable Thiele modulus	1	–
ψ	Volume fraction of solids	1	–
Ψ	Parameter defined in Table 12.3	1	–
ω	Angular velocity	T^{-1}	$rad\,s^{-1}$
Ω	Angular velocity	T^{-1}	$rad\,s^{-1}$
Ω	Observable modulus for external mass transfer	1	–

Subscripts

0	Initial
f	Final
i	Inlet
i	Interface
L	Logarithmic mean
o	Outlet

Superscripts

*	Equilibrium with prevailing value in the other phase

Index